Global Conference on Polymer and Composite Materials (PCM 2017)

IOP Conference Series: Materials Science and Engineering
Volume 213

Guangzhou, China
23 – 25 May 2017

ISBN: 978-1-5108-4415-5
ISSN: 1757-8981

Printed from e-media with permission by:

Curran Associates, Inc.
57 Morehouse Lane
Red Hook, NY 12571

Some format issues inherent in the e-media version may also appear in this print version.

Printed by Curran Associates, Inc. (2017)

For permission requests, please contact the Institute of Physics
at the address below.

Institute of Physics
Dirac House, Temple Back
Bristol BS1 6BE UK

Phone: 44 1 17 929 7481
Fax: 44 1 17 920 0979

techtracking@iop.org

Additional copies of this publication are available from:

Curran Associates, Inc.
57 Morehouse Lane
Red Hook, NY 12571 USA
Phone: 845-758-0400
Fax: 845-758-2633
Email: curran@proceedings.com
Web: www.proceedings.com

Table of contents

Volume 213

2017 Global Conference on Polymer and Composite Materials (PCM 2017)
23–25 May 2017, Guangzhou, China

Accepted papers received: 20 June 2017
Published online: 6 July 2017

Preface

Papers

IOP Conf. Series: Materials Science and Engineering **213** (2017) 011001 doi:10.1088/1757-899X/213/1/011001

PCM2017 Preface

The *2017 Global Conference on Polymer and Composite Materials (PCM 2017),* sponsored by Wuhan Advanced Materials Society, was held from May 23 – 25, 2017 in Guangzhou, China. The technical program consisted of 4 international keynote speakers, 62 invited speakers, 40 oral presentations, and a poster session with 42 works. The conference had parallel sessions with dedicated selected subjects related to Polymers and Composite Materials.

The present issue of IOP Conference Series: Materials Science and Engineering (MSE) records the proceedings of PCM 2017 and contains **50** specially selected manuscripts submitted to the PCM2017 conference. The electronic submission and handling of manuscripts via the conference website, including the selection of reviewers and evaluation of manuscripts, were identical to the procedures applied to manuscripts submitted as regular contributions for international peer-reviewed publications.

The organization of this conference and the preparation of proceedings volume would have been impossible without the tremendous efforts and dedication of many individuals, especially from Ms. Yajun Ye, who oversaw the organization of the conference and the program, and a large team of reviewers with their timely submission of quality reports. We express our sincere thanks to all authors and presenters for their contributions. We also thank very much our sponsors for their generous support.

The 2018 Global Conference on Polymer and Composite Materials (PCM2018) will be held during April 10-13, 2018, in Kitakyushu City, Japan. This is the best period to catch the Sakura Season in Japan. Located in the west of Japan, the city is situated conveniently close to many rapidly developing East Asian countries. With views over the Sea of Japan and the Seto Inland Sea, and with forty percent of the city's area comprised of rich forest, it is a city blessed with an abundance of nature. Being one of the four main industrial areas of Japan, Kitakyushu City has made great contributions towards the modernization of Japanese industry. Kitakyushu City's many notable attractions include Kokura Castle, the port town of Mojiko Retro, the Hiraodai Limestone Plateau (one of the three major karst plateaus of Japan), and many leisure facilities. Kitakyushu City also has a vast menu of appetizing dishes, such as Kokura beef, and Yaki-udon, as well as many other delicacies harvested from the mountains and sea.

All participants are welcome to the renewed and vibrant PCM2018! More information will be available at the conference website: www.cpcmconf.org.

Prof. Dr. Esteban Broitman

General Chairman PCM2017

Linköping, Sweden, June 2017

Peer review statement

All papers published in this volume of *IOP Conference Series: Materials Science and Engineering* have been peer reviewed through processes administered by the proceedings Editors. Reviews were conducted by expert referees to the professional and scientific standards expected of a proceedings journal published by IOP Publishing.

Effect of addition of Si on thermal and electrical properties of $Al\text{-}Si\text{-}Al_2O_3$ composites

R Cao[1,3], J X Jiang[1,3],C Wu[2] and X S Jiang[2]

[1]School of Mechanical Engineering, Chengdu Textile College, Chengdu 611731, P.R.China
[2]School of Materials Science and Engineering, Southwest Jiaotong University, Chengdu 610031, P.R.China

E-mail:crt2137@163.com; jiangjiaxin2004@126.com

Abstract. Al-5wt.%Si-Al_2O_3, Al-10wt.%Si-Al_2O_3, Al-20wt.%Si-Al_2O_3 composites were fabricated by powder metallurgy and in-situ reactive synthesis technology. The impact of the addition of Si on the thermal and electrical properties was tested and analysed for vary in silicon contentinAl-Si-Al_2O_3 composites. Results show that both thermal expansion coefficient and thermal conductivity decreased as silicon content increased because Si and Al_2O_3 dispersed in the Al matrix uniformly to suppress the high thermal expansion of Al to a large extent as well as the interfacial thermal resistance which led to the decline in thermal conductivity. Electrical resistivity increased when silicon content was increased because low thermal expansion coefficient particles of Si and Al_2O_3 severely damaged the continuity of the Al matrix which hindered movement of electron in the matrix.

Keywords: Al-Si-Al_2O_3 composites; Microstructure; Thermal properties; Electrical properties.

1. Introduction

Particle reinforced aluminum matrix composites have been recently developed due to the excellent performance of aluminum alloy combined with enhanced particles. With their high strength, wear resistance, thermal fatigue resistance and relatively inexpensive cost, Al-based composite materials are applied widely in the manufacture of optical and electronic parts, especially in the transport sector for price-sensitive automotive industry [1-3].Particle reinforced Al matrix composites are also favored in the aerospace and military sectors. Due to the requirement for satellite waveguide of small thermal expansion coefficient, Cr/Al composites have been manufactured[1, 4].Thus, particle reinforced aluminum matrix composites have a significant development potential and application prospects[5].One advantages of Al-based composite materials the ability to obtain excellent performance through proper control of the composition. Kasprzak W [6] has performed research on the properties of cast Al–Si based alloy, with the addition of Zr–V–Ti, which showed that this improved yield strength and cyclic yield strength at room temperature, while slightly improving hardness at elevated temperature. Mazahery Ali [7]fabricated aluminum matrix composites reinforced with boron carbide particles and found that the abrasion resistance of the composites increased with increasing volume fraction. Al-Si/SiCp (15 vol.%SiC particulate) composite was synthesized by WeiLi [8] and samples with high micro-hardness and low porosity volume fraction showed superior thermal fatigue resistance. At present, many countries have entered the stage of development and application of particle reinforced aluminum matrix composites. Also, many investigations have focused on the effect of the addition of Si on the properties of Al-based composites. Gao Ying [9] has discussed the mechanical properties of two kinds of Al-Si-Al_2O_3 composites with different Si content. The study by Wang Yongjin [10] indicated that the tensile strength of as-cast and as-extruded alloys could be improved with the increase in Si content. In this paper, the impact of the addition of Si on the thermal and electrical properties of Al-Si-Al_2O_3 composites were tested and discussed on the basis of analysis of the thermodynamic as well as the electrical properties of Si and microstructure of material, from the perspective of relation between the physical, chemical, and structural properties of composites. Al-Si-Al_2O_3 composites with different Si content exhibit much different properties. The effects of the addition of Si on the thermal and electrical properties of Al-Si-Al_2O_3 composites have been reported in very few studies. This investigation proposes to study the performance of Al-Si-Al_2O_3 composites as function of content of silicon.

2. Preparation of material

Al-based composites were fabricated by powder metallurgy and in-situ reactive synthesis technology, in which Al power with diameter of 5μm and nano-SiO_2 powder of purity99.99% were used.

Al-5wt.%Si-Al_2O_3, Al-10wt.%Si-Al_2O_3, Al-20wt.%Si-Al_2O_3composites were fabricated based on the reaction formula[11]:

$$(4+m)Al+3SiO_2=2Al_2O_3+3Si+mAl \quad (1)$$

The mixture ratios are shown in Table1.

Table 1. The mixture ratio of Al-Si-Al_2O_3 composites.

Si (wt. %)	Al : SiO_2 (g)
5	9.5:1
10	4.8 : 1
20	2.5 : 1

The process sequence for the composites, in this study, is as follows: Mixing→Freeze-drying→ Cold hydraulic pressing→Vacuum degassing→Hot hydraulic pressing. Hot hydraulic pressing process parameters are: 700 °C× 50Mpa × 1h.

The thermal expansion coefficient of the composite was tested by ZRPY-1400coefficient of thermal expansion tester; the thermal conductivity of the composite was tested by DRH thermal conductivity tester, and the electrical resistivity of the composite was tested by WSP-53 digital four probe tester.

3. Experimental results and analysis

3.1. Analysis of the microstructure of material

The microstructures of Al-Si-Al_2O_3 composite, with varying mass fraction of Si particles are shown in Figure 1.

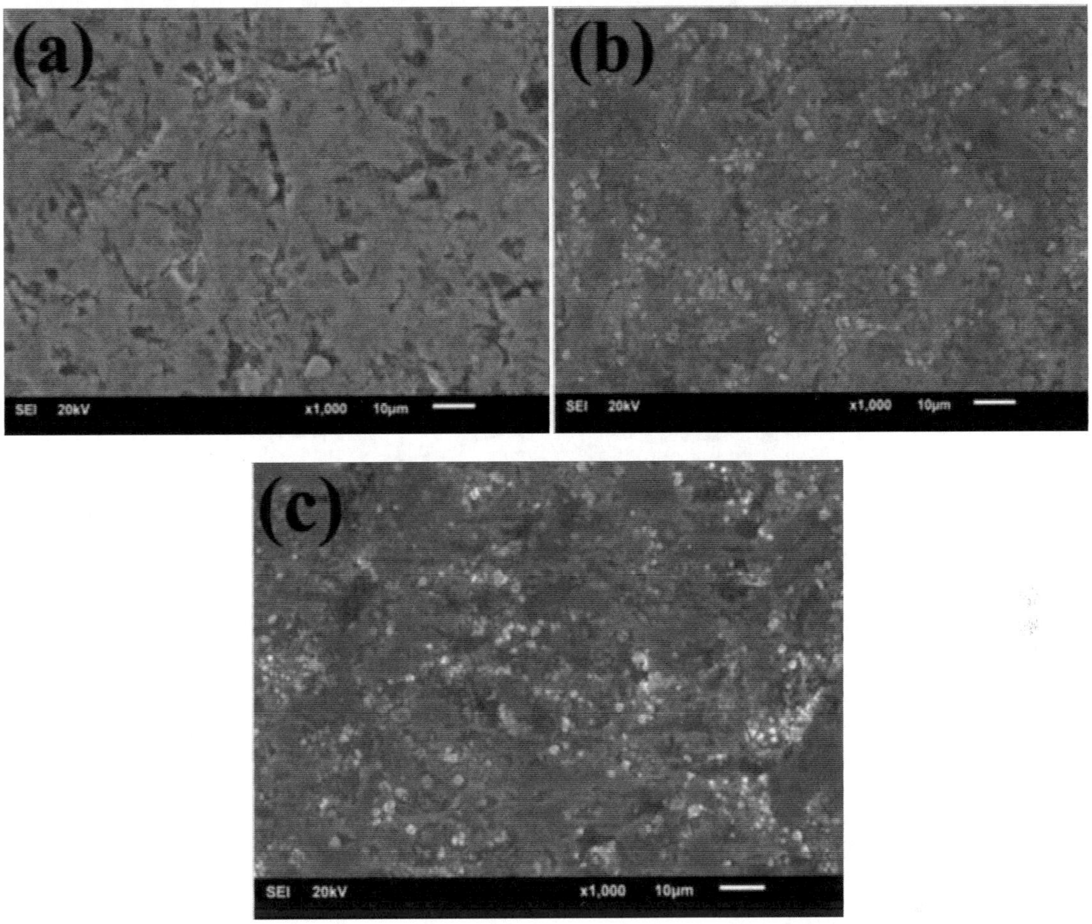

(a)Al-5wt.%Si-Al_2O_3; (b)Al-10wt.%Si-Al_2O_3; (c)Al-20wt.%Si-Al_2O_3

Figure 1. Secondary electron image ofAl-xwt.%Si-Al_2O_3.

The Al-5wt.%Si-Al_2O_3 composite (Figure 1a) is hypoeutectic.α-Al is firstly separated during condensation, and then the co-crystallized Al-Si alloy follows. As shown in Figure1a, the Si and Al_2O_3 reinforced phases have sheet-like appearance, and there are some pores located in the composite.

The Al-10wt.%Si-Al_2O_3 composite (Figure 1b) has more homogeneous distribution of Al_2O_3particles generated in situ in the Al matrix. Si, that is replaced, disperses uniformly in the mixture matrix of Al and Al_2O_3. It is apparent that the composite has high structural homogeneity and density without pores. In some areas, Al particles are agglomerated in large area as shown in Figure 1b, forming Al-Si eutectic microstructure in strip-like appearance. It may be caused by poor uniformity of the mixing or the existence of large aluminum particles.

Increasing the Si mass fraction to 20wt.% (Figure 1c)leads to some large pores created in the Al matrix. As shown in the micrograph, Si is of long strip and bulk shape, and the quantities and scale of primary Si increase, which is in accordance with the Al-Si binary phase diagram. The presence of eutectic Si around the Al_2O_3 particles shows that the nucleation and growth of eutectic Si are attached to alumina particles.

From the above SEM results, Al-Si-Al_2O_3composites which are fabricated by in-situ reactive synthesis technology show good microstructure uniformity and high density with the homogeneous distribution of the reinforced Al_2O_3 phase in the Al matrix. Si and AL, was fabricated by *in-situ* reactive synthesis technology, form Al-Si eutectic structure. From these microstructures, bulk primary silicon is observed by increasing the Si content from 5wt.% to 20wt.%, and the crystal morphology of particles is generally a polyhedron.

3.2. Effect of addition of Si on thermal expansion coefficient of composites

The linear expansion coefficient of the composites, with varying mass fraction of Si, is shown in Figure 2.

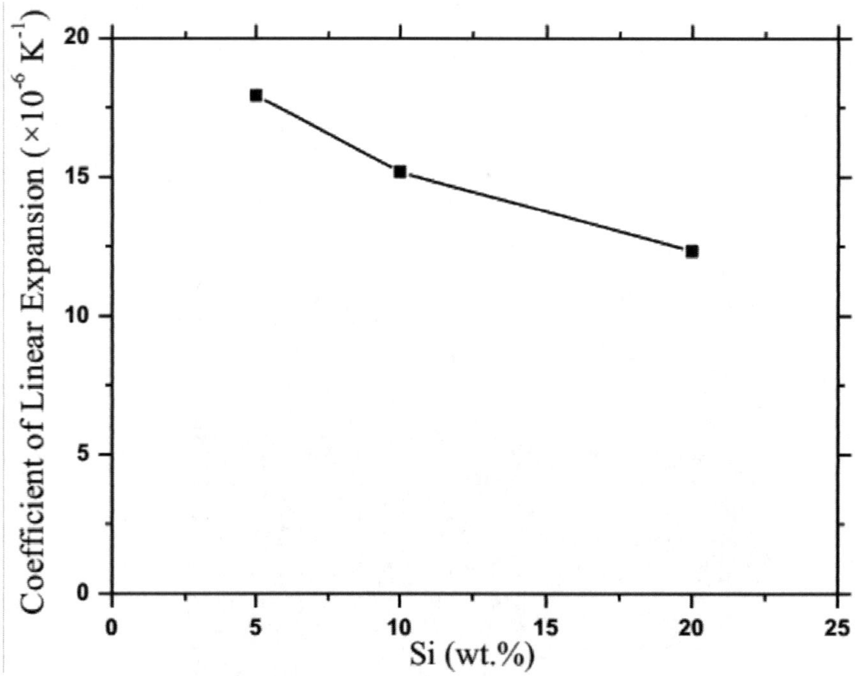

Figure 2. Coefficient of linear expansion versus Siwt.%.

As seen from Figure 2, with increasing Si content, the thermal expansion coefficient of composites decreases. This may be due to the following factors: (1) Physical properties of reinforcement and matrix: From Turner's model [12], the thermal expansion coefficient of composites mainly depends on the thermal expansion coefficient of the matrix and the degree of restraint on the matrix from reinforcement through the matrix/reinforcement interface. The high thermal expansion of Al is restrained to a large extent as silicon has low thermal expansion coefficient and Al_2O_3 particles disperse in Al matrix. The main factor influencing the thermal expansion coefficient is the volume content of reinforcement particles [13]. (2) Interfacial stress: Due to the mismatch in the expansion coefficients of the reinforcement and the matrix, within the constraints of the interface, this will result in residual stress in the matrix near the interface region which can lead to the occurrence of corresponding elastic strain in the matrix alloy. This is because when the composites solidify, a large number of dislocations occur at the Si-Al interface due to the stress, which causes diffusion along dislocation. The essence of thermal expansion is that the average distance in the lattice structure of particle increases with rising temperature [13], and the existence of the dislocation impedes expansion. (3) Binding force between atoms: The nature of thermal expansion in solids is that the average distance between atoms in the lattice structure increases with increasing temperature. In Al-Si-Al_2O_3 composites, there is a large area of Al accumulation in some areas resulting in a non-uniform composition distribution and particle size, so that the micro-structure is even worse and the binding energy among Al, Si and Al_2O_3 is reducing. The coefficient of thermal expansion increases.(4) Effect of grain size: Atoms restrain the expansion of the matrix mainly by the transfer of the interface zone. When the volume fraction is certain, the reinforcement/matrix interfacial area mainly depends on the particle size. With increasing Si from 5 wt.% to 20 wt.%, the restraints to matrix deformation reduces due to the increase in size of primary silicon. As a result, the variation in thermal expansion coefficient with change in composition [14] is not linear. (5) Impact of porosity: Pores could destroy the continuity of material microstructure so that the binding force between matrix alloys can be reduced. In this way, thermal vibration amplitude increases which can lead to poor thermal stability.

3.3. Effect of addition of Si on thermal conductivity of composites

The thermal conductivity value of the composites, with varying mass fraction of Si, is shown in Figure 3.

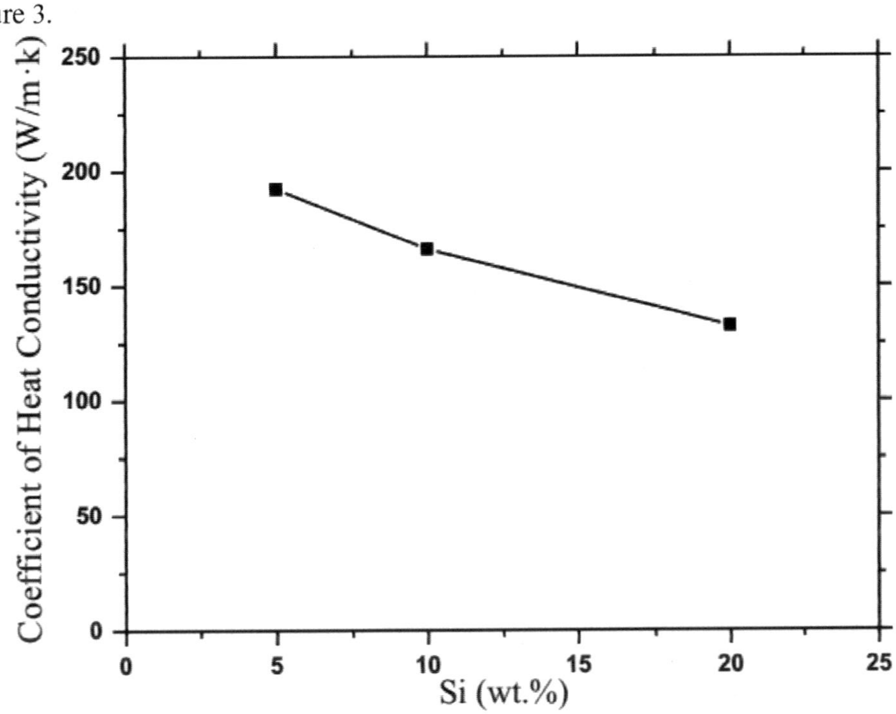

Figure 3. Thermal conductivity versus wt.% of Si.

From Figure 3, with increasing Si content, the thermal conductivity of the composites decreases. Heat in solid is conducted by lattice vibrations (phonon) and the movement of free electrons [13]. With the addition of Si and Al_2O_3 particles, the thermal conductivity of Al-Si-Al_2O_3 composites decreases. This is due to increase in heat conduction via phonons, decrease in heat conduction due to free electrons, and the influence of free electrons on thermal conductivity being higher than that of phonons. At the same time, the increase in volume content of the particles causes an increasing interfacial area per unit volume of the composites, and the interfacial thermal resistance increases, which leads to the decline in thermal conductivity [15]. In addition, the porosity of composites has a complex effect on thermal conductivity. The pores can be considered as dispersed phase and the increase in the porosity will decrease the thermal conductivity if the temperature and the porosity are not high, the pore size is very small and pores distribute uniformly. The presence of many pores in 20wt.%Si composite is one of the reasons for the decrease in thermal conductivity[13, 16].Otherwise, the dislocation in the Al matrix increases scattering, which decreases the thermal conductivity to some extent [17].

3.4. Effect of addition of Si on electrical resistivity of composites

The electrical resistivity of the composites with varying mass fraction of Si is shown in Figure 4.

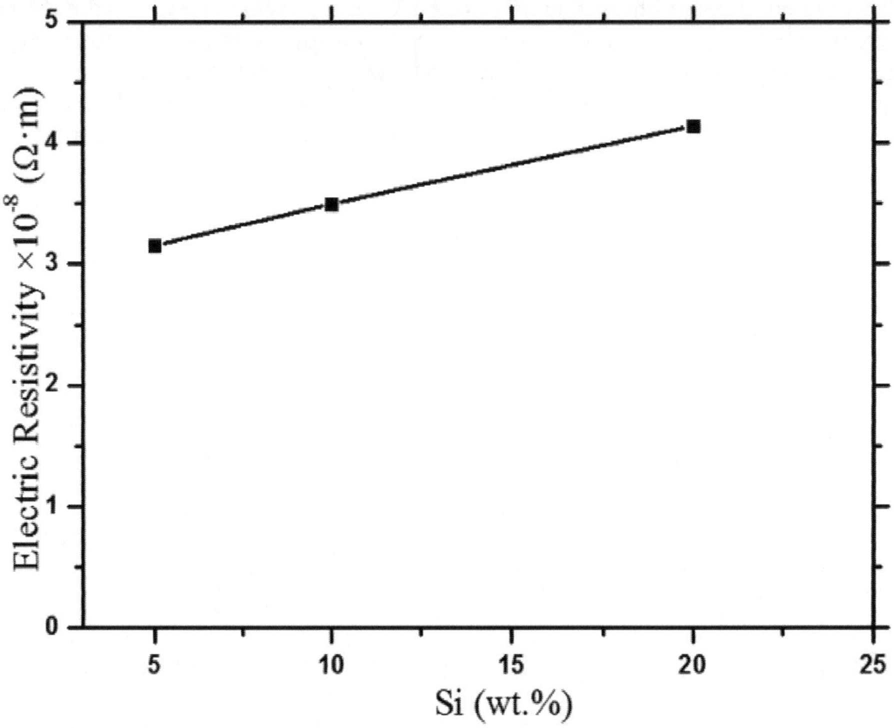

Figure 4. Variation of electrical resistivity with weight % of Si.

As can be seen from Figure 4, the electrical resistivity of the Al-Si-Al_2O_3 composites rises with increasing silicon content. The electrical resistance of a metal represents the integrity of the crystal lattice is due to the scattering of the electron wave [13]. In the conductive properties of Al-Si-Al_2O_3 composites, Al is in a highly conductive continuous phase which plays a leading role. Therefore, the continuous Al matrix directly determines the conductivity of the Al-Si-Al_2O_3 composite. With increase in the content of Si, Si and Al_2O_3 particles severely damage the continuity of the Al matrix, which destroys the periodic potential field of the matrix lattice, hinders the movement of electrons in the matrix and increases the scattering of the electrons. As a result, the electrical conductivity deteriorates and the electrical resistivity increases. The electric resistivity of the composite materials will increase with the addition of Si and Al_2O_3contentas Si andAl_2O_3 have high electrical resistivity, especially the electrical resistivity ofAl_2O_3 is up to 1013Ω•m. The electric resistivity is a micro structure-sensitive property; the degree of order and microstructure of particle has significant effect on the resistivity of materials and the change in resistivity often reflects the change in the inner microstructure of the materials [18]. The more uniform distribution of the microstructure of the composites, the smaller scattering of the electrons appear due to free electrons. This could lead to smaller resistivity. For the Al-20wt.%Si-Al_2O_3 composite, there are more porosities; pores reduce the average distance of the electrons and increase the electron scattering so that the resistance could be increased. Additionally, dislocations in the Al matrix cause lattice distortion and increase electron scattering. But it has a little influence on the increase in resistivity. In the reaction process between Al and SiO_2 the interface area will increase, and the interface can scatter electrons, to increase the electrical resistivity of the composites significantly [18].

4. Conclusions

(1)With increasing Si content, the coefficient of thermal expansion of Al-Si-Al_2O_3 composites decreases. This is because the coefficient of thermal expansion of the composite mainly depends on the coefficient of thermal expansion of the matrix and the degree of restraint on the matrix from reinforcement through the matrix/reinforcement interface. The high thermal expansion of Al is restrained to a large extent as Si has low thermal expansion coefficient and Al_2O_3 particles disperse in the Al matrix.

(2)With the addition of Si and Al_2O_3 particles, the thermal conductivity of Al-Si-Al_2O_3 composites decreases due to the increase in the heat conduction via phonons, the decrease in heat conduction by free electrons, and the effect of the free electrons on the thermal conductivity being far higher than that of phonons. At the same time, the increase in volume content of the particles causes an increase in the interfacial area per unit volume in the composites, and the interfacial thermal resistance increases, which leads to the decline in thermal conductivity.

(3)With increasing Si content, Si and Al_2O_3 particles severely damage the continuity of the Al matrix, which destroys the periodic potential field of the matrix lattice, hinders the movement of electrons in the matrix and increases the scattering of the electrons. As a result, the electrical conductivity deteriorates and the electrical resistivity increases.

References

[1] Hou L L, YinZ X and Fan X B 2008 Research status and development of aluminum matrix composite materials *Heat Treatment Technology* **4** 84-8

[2] Wu R J 1997 Status and prospect of metal matrix composites *Metal Journal* **33**(1)79-84

[3] Song W 2004 Development and application of metal matrix composites *Foundry Equipment Research* **5** 48-50

[4] Gong J G, TangB B and Han L 2014 Research status of Al matrix composites *Heat Processing* **43**(4) 23-6

[5] Nair S V, Tien J K and Bates R C 1985 SiCp reinforced aluminum metal matrix composites *Int. Met. Rev.* **30** 275-90

[6] Kasprzak W, Amirkhiz B S and Niewczas M 2014 Structure and properties of cast Al-Si based alloy with Zr-V-Ti additions and its evaluation of high temperature performance *J. Alloy compd.* **59** 567-79

[7] Mazahery A, Shabani M O, Rahimipour M R and Razavi M 2012 The numerical modeling of abrasion resistance in casting aluminum-silicon alloy matrix composites *J. Compos. Mater.* **46** 2647-58

[8] Li W, Chen Z H, Chen D, Fan C and Wang C R 2010 Thermal fatigue behavior of Al-Si/SiCp composite synthesized by spray deposition *J. Alloy compd.* **504** 522-6

[9] Gao Y, Zhu D G, Cheng L, Sun H L and Wang Q 2011 In situ fabrication of particles reinforced Al-Si-Al_2O_3 composites by hot isostatic pressing *Advanced Materials Research* **284-286** 38-42

[10] WangY J, Liao H C, Wu Y N andYang J 2014 Effect of Si content on microstructure and mechanical properties of Al-Si-Mg alloys *Mater. Des.* **53** 634-8

[11] Gao Y 2012 *Properties of Al-Si-Al_2O_3 Composites Fabricated by Powder Metallurgy* (Chengdu: Southwest Jiaotong University) P 13

[12] Makarian K, Santhanam S and Wing Z N 2016 Coefficient of thermal expansion of particulate composites with ceramic inclusions *Ceram. Inter.* **15** 17659-65

[13] Tian S 2004 *The Physical Properties of Materials* (Beijing: Beijing University of Aeronautics and Astronautics Press) P 11

[14] Zhu D Z, Li F Z, ChenG Q, Zhang Q and Wu G H 2005 Study on thermal expansion property of SiCp/Cu composite *Journal of Harbin University of Science and Technology* **10**(2) 125-8

[15] Chen C, Zhang G L and Yu H S 2012 The microstructure and thermal conductivity of the high volume fraction of SiCp/Al-Si Composites *Funct. Mater.* **19** 102-6

[16] Zhang Q, Wu G H, Sun D L and Han J C 2006 The microstructure and thermal conductivity of high volume fraction of SiCp/Al composite materialsscience and technology *Mater. Sci. Technol.* **14**(5) 474-7

[17] Zhang J Y, Li Y H, Cui X and Zhou X L 2010 Research on thermal conductivity of SiCp/Al composites *Heat Processing* **39**(16) 72-4

[18] Liu X C, Liu C W and Sun G 1990 Resistivity of SiC/Al composites *Journal of Astronautics* **2** 84-9

IOP Conf. Series: Materials Science and Engineering **213** (2017) 012002 doi:10.1088/1757-899X/213/1/012002

Effect of EVA on thermal stability, flammability, mechanical properties of HDPE/EVA/$Mg(OH)_2$ composites

R Cao[1,4], Z L Deng[2], Y H Ma[3] and X L Chen[3, 4]

[1]School of Mechanical Engineering, Chengdu Textile College, Chengdu 611731, P.R. China
[2]School of Economics and Management, Chengdu Textile College, Chengdu 611731, P.R. China
[3] Key Laboratory of Advanced Materials Technology Ministry of Education, School of Materials Science and Engineering, Southwest Jiaotong University, Chengdu 610031, P.R. China

E-mail: crt2137@163.com; chenxl612@sina.com

Abstract. In this work, ethylene vinyl acetate (EVA) is introduced to improve the properties of high-density polyethylene (HDPE)/magnesium hydroxide (MH) composites. The thermal stability, flame retardancy and mechanical properties of HDPE/EVA/MH composites are investigated and discussed. With increasing content of EVA, the limiting oxygen index (LOI) of the composites increases. The thermal stability analysis shows that the initial decomposition temperature begins at a low temperature; however, the residues of the composites at 600°C increase when HDPE is replaced by small amounts of EVA. The early degradation absorbs heat, dilute oxygen and residue. During this process, it protects the matrix inside. Compared with the HDPE/MH and EVA/MH composites, the ternary HDPE/EVA/MH composites exhibit better flame retardancy by increasing the LOI values, and reducing the heat release rate (HRR) and total heat release (THR). With increasing content of EVA, the mechanical properties can also be improved, which is attributed to the good affinity between EVA and MH particles.

1. Introduction

Polyethylene (PE) has numerous advantages such as balanced mechanical properties, chemical resistance, low cost and good process ability [1-2]. These properties have led to its wide use in many fields. PE is flammable and its limiting oxygen value is very low. Its flammability limits its applications to a certain extent. In the past decades, the flame retardant investigation of PE has attracted much attention.

In order to improve the flame retardancy of PE, an efficient method is to blend flame retardants with PE. Among the flame retardants, magnesium hydroxide (MH) has been the focus of many researchers due to its halogen-free, non-toxic and inhibiting smoke feature [3-4]. MH produces an endothermic reaction during combustion, which decreases the surface temperature of the polymer. In addition, free water can dilute the combustible gas, and the produced metal oxide (MgO) cuts off sources of heat between the matrix and the surface [5-6]. But the flame retardant efficient of MH is very low; more than 60% MH is required to reach combustion resistance [7]. Large amount of MH will lead to a sharp decrease in mechanical properties which results from poor compatibility.

In order to overcome the poor compatibility, elastomers, such as ethylene-propylene-diene monomer (EPDM), ethylene-octylene copolymer (POE), ethylene vinyl acetate (EVA), thermoplastic polyurethanes (TPU) etc., will be blended with the polymer [8-11]. Among the elastomers, EVA is suitable to blend with PE. This is because EVA can be well miscible with PE because of the presence of similar ethylene chain. In addition, it can also improve flexibility, toughness, environmental stress-cracking resistance due to the inner structure of EVA macromolecules [12]. At the same time, functionalized inorganic particles in the form of polymer/inorganic filler composites have exhibited remarkable performance [13-15]. Su, J. H.; et al. [16] have prepared graphene by the thermal reduction and the graphene is treated by stearic acid to improve its lipophilicity and dispersion in low density polyethylene. Actually, the inorganic fillers modified by the coating and surface treatment with various methods have become the research focus for inorganic particles/polymer composites with improved properties [17-18].

In this work, EVA is introduced to improve the properties of HDPE/MH composites, and the effect of the addition of EVA on the thermal stability, flammability, mechanical properties of HDPE/EVA/MH composites have been studied and discussed by utilizing thermal gravimetric analysis (TGA), limiting oxygen index (LOI), cone calorimeter tests (CCT), and tensile testing instruments.

2. Experiments

2.1. Materials

High-density polyethylene (HDPE, 5000S, MFR=0.923g/10min) was acquired from Lanzhou Petrochemical Company, China Petroleum. EVA (7240M, VA=15wt%, MFR=1.5g/10min) was supplied by TAISOX, Taiwan. Magnesium hydroxide ($Mg(OH)_2$, MH) (average particle size=1.5μm) was obtained from Yantai FR Chemical Technology Co., Ltd. The PE-g-MAH and silane coupling agent are commercial products.

2.2. Preparation of samples

Dried HDPE, EVA and $Mg(OH)_2$ powders are premixed according to the formulation (Table 1), and then compounded in a twin-screw extruder (ModelHFB-150/3300, made in Nanjing Ruiya Polymer Processing Equipment Co., Ltd., China). The temperatures from hopper to die are 140, 150, 160, 170 and 180°C, respectively, and the screw speed is 150rpm. All the samples are hot-pressed at 180°C under 8MPa into sheets of suitable thickness for tests and characterization. The samples used for mechanical properties are injected to model using EM80-V made by Zhende Company.

Table 1. The formulations of HDPE/EVA/MH flame retardant composites.

Samples	HDPE (phr)	EVA (phr)	MH (phr)
E0	100	0	100
E20	80	20	100
E40	60	40	100
E60	40	60	100
E80	20	80	100
E100	0	100	100

2.3. Measurements and characterization

Thermogravimetric analysis (TGA): the thermal properties of the composites are performed on a NETZSCH TG 209F1 thermogravimetric analyzer at a heating rate of 10 °C/min under nitrogen flow with a flow rate of 60 mL/min and the scan ranged from 30 to 700 °C. And the weight of each sample was about 10 mg.

Limited oxygen index (LOI)-LOI is tested by using a JF-4 oxygen index instrument (Jiangning Analysis Instrument Company, China) on the specimen of dimensions 120×6.5×3mm^3 according to the standard ASTM D2863-77.

Cone calorimeter test. The cone calorimeter (Stanton Redcroft, UK) tests are carried out on a sheet of (100×100×3 mm^3) according to ISO 5660 standard procedures. Each sample is wrapped in aluminium foil and exposed horizontally to an external heat flux of 35kW/m^2.

Mechanical property tests. Tensile properties are measured by a tensile testing machine AGS-J (Shimadzu Company, Suzhou) according to ASTM D412-1998 at room temperature. The test is performed at a crosshead speed of 50mm/min.

Phase morphology. The cryo-fractured samples are prepared under N_2 liquid for an hour, and then micrographs are taken under common high depth of field microscope.

3. Results and discussion

3.1. Thermal stability

The TGA and derivative thermogravimetric (DTG) curves of the HDPE/MH, EVA/MH, HDPE/EVA/MH composites are shown in Figure 1. The temperatures at 5wt% weight loss, 50wt% weight loss and the maximum decomposition rate are recorded as T_{onset}, $T_{50wt\%}$ and T_{peak}, respectively, and the detailed data are listed in Table 2. As shown in Figure 1, the HDPE/MH composites begin to decompose at 397.6°C while the EVA/MH composites begin to decompose at 335.2°C. The temperature of maximum mass loss also shifts to low temperature. These indicate that the thermal stability of EVA is lower than that of HDPE. The initial decomposition temperature of the ternary HDPE/EVA/MH composites is reduced by 37°C compared with HDPE/MH composites. This is attributed to the loss of acetic acid from EVA occurring at a low temperature [19].

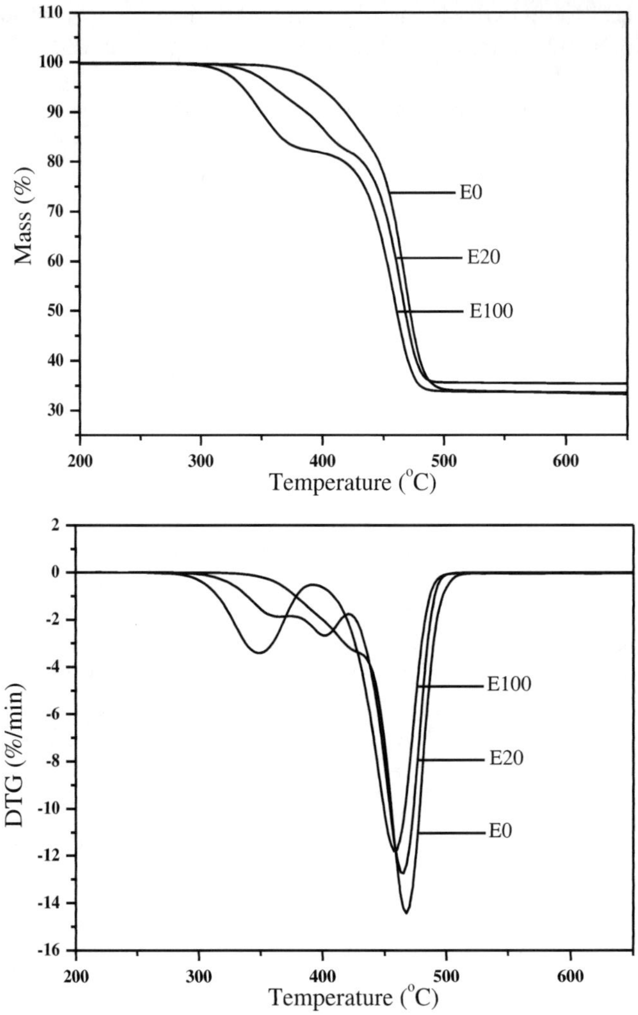

Figure 1. Thermogravimetric analysis (TGA) and derivative thermogravimetric (DTG) of HDPE/MH, HDPE/EVA/MH, and EVA/MH composites.

Table 2. Detailed data from TGA and DTG of HDPE/EVA/MH composites.

Samples	T_{onset}(°C)	$T_{50\%}$(°C)	T_{peak}(°C)	Residue at 600°C (wt%)
E0	397.6	471.8	467.4	33.2
E20	360.5	465.1	461.2	35.5
E100	335.2	461.2	458.2	33.4

However, HDPE/EVA/MH composites display two-step degradation. The first stage occurs at temperatures in the range of 300~400°C; the mass loss involves the deacylation of EVA phrase through radical and ionic β-elimination mechanism and degradation of MH[6, 20]. The second stage occurs at temperatures in the range of 400~550°C, which is attributed to the degradation of HDPE and the back bone of the EVA which forms in the first step and the formation of transvinylenes accompanied by main chain scission in the second step. Compared with the binary composites, the residue of the ternary composites at 600°C is increased by 2.3% and 2.1%. The above results indicate that the thermal stability of HDPE/MH composites decreases with the incorporation of EVA, but the residue of HDPE/EVA/MH composites increases after combustion. The residue may be in favor of good flame retardancy.

3.2. Flammability

Flammability of HDPE/EVA/MH composites is evaluated by LOI and Cone calorimeter test (CCT). The effect of EVA on the LOI of HDPE/EVA/MH composites is shown in Figure 2. When the EVA content increases from 0 to 20phr, a slight decrease in LOI is observed. The reason for this is not yet clear. However, more EVA can enhance LOI values of HDPE/EVA/MH composites. It can be observed from Figure 2 that the LOI values increase from 30.3% to 35.3% when the content of EVA increases from 20 to 100phr. This indicates that the addition of EVA can improve the flame retardancy of HDPE/EVA/MH composites. The positive effect of EVA on the flame retardancy is mainly attributed to the strong interaction between EVA and MH. The interaction of EVA and MH is shown in Figure 3. As is shown in Figure 3, the MH particles preferentially disperse in EVA. In addition, the strong interfacial bonding between particle and matrix improves the flame retardancy of the composites.

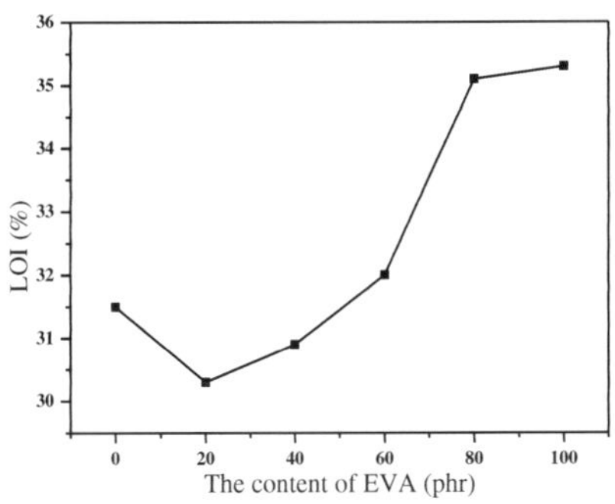

Figure 2. LOI values of HDPE/EVA/MH Hcomposites with various contents of EVA.

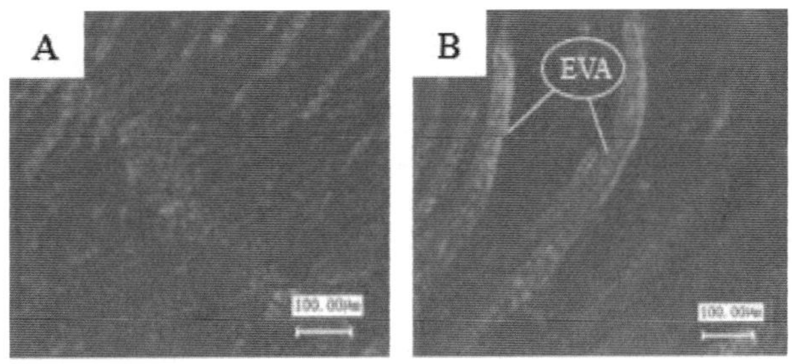

Figure 3. Micrographs of HDPE/MH and HDPE/EVA/MH composites with MH content of 5% and the ratio of HDPE/EVA is 4:1.

Besides, the cone calorimeter test has been widely used to evaluate the flammability of materials because it can predict the combustion behavior of materials during interaction with flames. Figure 4 presents the heat release rate (HRR) versus time curves of HDPE/EVA/MH composites. HDPE/MH composites are ignited for 157s, and have a flat heat release peak. Compared with HDPE/MH composites, EVA/MH composites have a short ignition time of 140s and the sharp peak heat release rate (PHRR) value reaches 183.7kW/m^2, and then decreases dramatically.

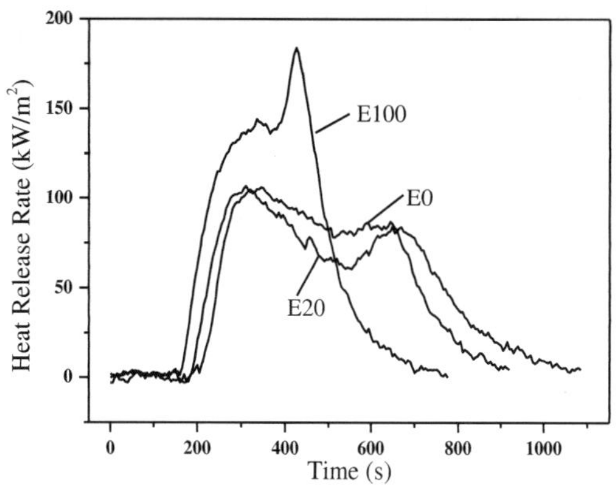

Figure 4. Heat release rate (HRR) curves of HDPE/MH, HDPE/EVA/MH, and EVA/MH composites.

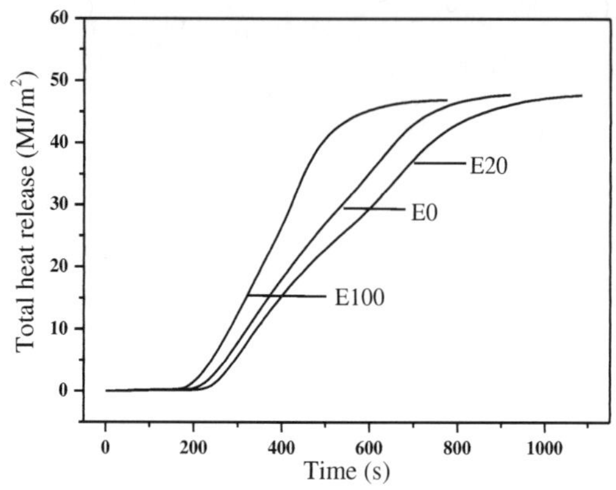

Figure 5. Total heat release (THR) curves of HDPE/MH, HDPE/EVA/MH, and EVA/MH composites.

However, when a small amount of HDPE is replaced by EVA, it is interesting to observe that HRR is suppressed. The HRR curve of HDPE/EVA/MH composites has two peaks. The ignition time is delayed by 183s and the PHRR value decreases to 104.99kW/m^2. The flammability of the composites is restrained, and the possible reason may be that MH disperses in the EVA which needs more oxygen to burn, and the phenomenon is also proved by LOI curves (Figure 2). This indicates that combination of HDPE, EVA and MH shows a better fire resistance.

The total heat release (THR) curves for the above sample are shown in Figure 5. The trend in THR is the same as HRR. The reduction in heat release can be explained by the amount of chars formed on the surface of the polymer matrix, and the formed char limits the combustion products in making the transition to the gas phase. Figure 6 shows the photographs of HDPE/MH, HDPE/EVA/MH, and EVA/MH composites after combustion. The thick and compact residues can protect the inside matrix and reduce heat release.

3.3. Tensile behaviors

The stress-strain curves of HDPE/EVA/MH composites, with various EVA content, are shown in Figure 7. The Young's modulus and tensile strength of HDPE/EVA/MH composites with various contents of EVA are shown in Figure 8. The previous study has shown that the incorporation of 50wt% MH to HDPE leads to a significant decrease in elongation at break. In order to solve this problem, elastomer EVA is added to HDPE/MH composites. It is clear that the tensile behaviors are strongly dependent on the EVA content. Elongation at break is directly ameliorated due to the addition of EVA and increases monotonically with increasing EVA concentration. However, EVA has negative effects on the Young's modulus and tensile strength. Compared with HDPE/MH composites, Young's modulus and tensile strength of HDPE/EVA/MH composites cause a more visible drop with the addition of EVA, and decrease with increasing EVA content. Actually, there is a close relationship between the locations of filler [21]. In the HDPE/EVA/MH composites, MH is usually situated in EVA phase. Behradfar et al. [22] reported a preferential dispersion of clay in the elastomer EVA phase. Thus the most plausible factor for the change in tensile properties is ascribed to the location of filler in the elastomer phase. Figure 3 is also a proof for preferential location in EVA.

IOP Conf. Series: Materials Science and Engineering **213** (2017) 012002 doi:10.1088/1757-899X/213/1/012002

Figure 6. Photographs of HDPE/MH, HDPE/EVA/MH, and EVA/MH composites after combustion.

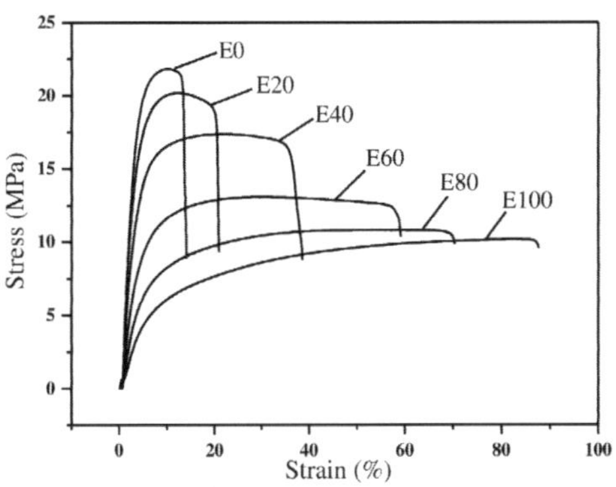

Figure 7. Stress-strain curves of HDPE/EVA/MH composites with various contents of EVA.

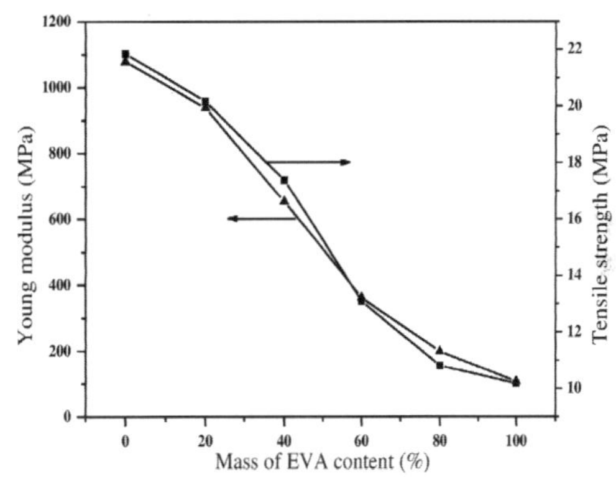

Figure 8. Young's modulus and tensile strength of HDPE/EVA/MH composites with various contents of EVA.

4. Conclusions

The HDPE/EVA/MH ternary composites were successfully prepared by using a twin-screw extruder. The thermal stability, flame retardancy and mechanical properties of HDPE/EVA/MH composites were studied and discussed. With increasing content of EVA, the LOI values of the composites increase. The thermal stability data show that the initial decomposition temperature of the composites begins at a low temperature while residue at 600°C increases when a small amount of HDPE is replaced by EVA. The early degradation absorbs heat; dilute oxygen and residue produced by early decomposition protects the matrix inside. Therefore, the HDPE/EVA/MH composites can show a better flame retardancy than the HDPE/MH composites as well as the EVA/MH composites. The tensile properties of the HDPE/EVA/MH ternary composites are strongly influenced by the incorporation of EVA elastomer. MH has a tendency for location in EVA. A good affinity between EVA and MH leads to good flame retardancy and mechanical properties of the composites.

Acknowledgements
The authors are grateful to the National Natural Science Foundation of China (51003088, 21266005) for financial support of this work.

References

[1] Faker M, Aghjeh M K R, Ghaffari M and Seyyedi S A 2008 *Eur. Polym. J.* **44** 1834
[2] Minkova L, Peneva Y, Tashev E, Filippi S, Pracella M and Magagnini P 2009 *Polym. Test.* **28** 528

[3] Yen Y Y, Wang H T and Guo W J 2012 *Polym. Degrad. Stabil.* **97** 863
[4] Witkowski A, Stec A A and Hull T R 2012 *Polym. Degrad. Stabil.* **97** 2231
[5] Oyama H T, Sekikawa M and Ikezawa Y 2011 *J. Macromol. Sci., Part B.* **50** 463
[6] Chen X L, Yu J, Guo S Y, Luo Z and He M 2009 *Polym. Compos.* **30** 941
[7] Lenza J, Merkel K and Rydarowski H 2012 *Polym. Degrad. Stabil.* **97** 2581
[8] Chen X L, Yu J, Luo Z, Guo S Y, He M and Zhou Z W 2011 *Polym. Advan. Technol.* **22** 657
[9] Xu Y, Chen M, Ning X, Chen X L, Sun Z D, Ma Y H, Yu J, Zhang Z B, Bo X J and Yang L 2014 *J. Therm. Anal. Calorim.* **115** 689
[10] Peeterbroeck S, Laoutid F, Taulemesse J M, Cuesta J B, Nagy J, Alexandre M and Dubbos P D 2007 *Adv. Funct. Mater.* **15** 2787
[11] Chen Y K, Xu C H, Cao L M, Wang Y W, Wan W P and Cao X D 2012 *Polym. Test.* **31** 728
[12] Shojaei A, Behradfar A and Sheikh N 2011 *Polym. Advan. Technol.* **22** 23-52
[13] Naoko S, Takuma A, Yoshiaki S and Hiroyoshi K 2012 *Polymer* **53** 2062
[14] Showkat A M, ZhangY P, Kim M S, Gopalan A I, Reddy K R and Lee K P 2007 *Bull. Korean. Chem. Soc.* **28** 1985
[15] Reddy K R, Sin B C, Ryu K S, Kim J C, Chung H and Lee Y 2009 *Synthetic. Met.* **159** 595
[16] Han S J, Lee H I, Jeong H M, Kim B K, Raghu A V and Reddy K R 2014 *J. Macromol. Sci., Part B: Phys.* **53** 1193
[17] Reddy K R, Jeong H M, Lee Y and Raghu A V 2010 *J. Polym. Sci., Part A: Polym. Chem.* **48** 1477
[18] Hassan M, Reddy K R, Haque E, Faisal S N, Ghasemi S, Minett A I and Gomes V G 2014 *Compos. Sci. Technol.* **98** 1
[19] Wu K, Shen M M and Hu Y 2010 *Polym-Plast. Technol. Eng.* **49** 1527
[20] Zhang J, Hereid J, Hagen M, Bakirtzis D, Delichatsios M A, Fina A, Castrovinci A, Camino G, Samyn F and Bourbigot S 2009 *Fire. Safety. J.* **44** 504
[21] Martins C G, Larocca N M, Paul D R and Pessan L A 2009 *Polymer* **50** 1743
[22] Behradfar A, Shojaei A and Sheikh N 2010 *Polym. Eng. Sci.* **50** 1315

IOP Conf. Series: Materials Science and Engineering **213** (2017) 012003 doi:10.1088/1757-899X/213/1/012003

Analysis of micro-failure behaviors in artificial muscles based on fishing line and sewing thread

J B Xu[1,3,6], K F Cheng[2], S L Tu[1], X M He[1], C Ma[1], Y Z Jin[1], X N Kang[4], T Sun[5,6] and Y Zhang[1,6]

[1]Department of Mechanical Engineering, Xi'an Jiaotong University, Xi'an, 710049, Shaanxi, China
[2]Department of Materials Science and Engineering, Xi'an Jiaotong University, Xi'an, 710049, Shaanxi, China
[3]Asia Pacific Research Center for Advanced Science and Technology, Jiabin Science Laboratory, Xi'an City, 710308, Shaanxi, China
[4]Department of Human Anatomy of Basic Medical College, Fuzhou Medical College of Nanchang University, Fuzhou City, 344000, Jiangxi, China
[5]Faculty of Materials Science and Engineering, Harbin Institute of Technology, Harbin City, 150006, Heilongjiang, China

E-mail: xujiabin114@yeah.net; clkexie2015@163.com; zy_xjtu@163.com

Abstract. The aim of the present study was to discuss a new and effective method for testing artificial muscles based on micro-failure behaviors analysis. Thermo-mechanical actuators based on fishing line and sewing thread, also, the capability of responding to ambient temperature variations producing a large amount of shrinkage ratio of a resulting variation in longitudinal length. The minimum micro-failure value is 0.02μm and the maximum value is 1.72μm with nylon twist pattern. The discovery of an innovative effective testing of artificial muscles based on polymeric fibers specimens on micro-failure, rupture, slippage, etc. This research finds out a micro-failure behavior analysis of thermo-mechanical actuators based on fishing line and sewing thread. The specimens show large deformations when heated together with warping performance in terms of shrinkage of energy and densities. With the purpose of providing useful analysis data for the further technology applications, we attempt micrometre-sized artificial muscles which were also tested was readily accessible and also can be applied to other polymeric fibers. Effective use of this technique achievement relies on rotate speed, temperature and tensile direction. The results of the tensile testing experiments were outstanding with respect to some important issues related to the response of micro-structure, twisted polymeric fibers and shrinkage ratio.

1. Introduction

Artificial muscle is a generic term used for advanced materials that can be expanded and rotated by external stimulus (such as voltage, pressure or ambient temperature) [1]. However, performance, stability and the cost have restricted their manufacture and application. When the sufficient thin metallic nano-wires has been made by specimen, we found the mechanical properties of their intrinsic bond strength. Seyed M. Mirvakili, *et al.* found that up to 49% upon heating for tensile tested thermal expansion [2-5]. Up to 29% linear actuation is reaching 840 kJ/m^{-3} (528 J/kg^{-1}) and 1.1 kW/kg^{-1} (operating at 0.1 Hz, 4% strain) [6]. In this work, we introduced a method for the specimens' micro-structuring twist spun with spun angle to explore micro-failure. These nylons average from 44.9 μ m to 45.3 μ m in diameter with an inserted twist of 300turns/m^{-1}. As with all fishing line and sewing thread materials, one outstanding method of approach is viable for their production: twisting. Nylon has been twisted and heated from fishing line and sewing thread to stretch with fixed angle, as described in the supplementary materials. Nylon is a general term for synthetic polymer, especially refers to aliphatic or semi-aromatic polyamides. They can be melt-processed into fibers, films or shapes [7]. The minimum micro-failure value is 1.06μm and the maximum value is 3.68μm with axial direction view of the nylon cross-section area (as shown in Figure 1a, Figure 1b).

Here we demonstrated large-stroke and high-work-capacity artificial muscle that provides millions of twist coils. These materials have demonstrated significant micro-failure when patterned in such a way as to convert into twist. In the present work we reported, for the first time, that artificial muscle micro-failure was analyzed using twist due to predict intrinsic bond strength. Also, the diameter of the material is measured by electron microscopy (as shown in Figure 1c). The twist-spun polymeric fibers confine this actuating in fishing line and sewing thread and provide the mechanical strength and enable large-stroke twist or torsional and tensile testing actuation. Dispersed carbon nano-tubes sheets have been applied for actuating materials to provide cantilever deflections [8-10].

The degree of nylon filament winding determines the range of overall volume changes. Twine exhibits a large volumetric shrinkage when heated (a Tm of 50%, between 25℃ and 80℃), and is very effective in micro-actuators since the thermal time constants are small [10]. We hope the artificial muscles can be applied by people who require abrasive resistance in prosthetic limbs, robotic exoskeletons and other medical devices. In addition, the application areas and opportunities for polymer muscles are vast when otherwise made to change in volume [12].

Fishing line and sewing thread share many properties with their micro-structure, such as the surface-area-to-volume ratio, and also provide cheap and easy machining property. As such, if once a particular phase has been formed, one of twisting of coiling twisting could generate micro-failure. The resultant polymer fibers twisting could be used as a heat template (Tm).

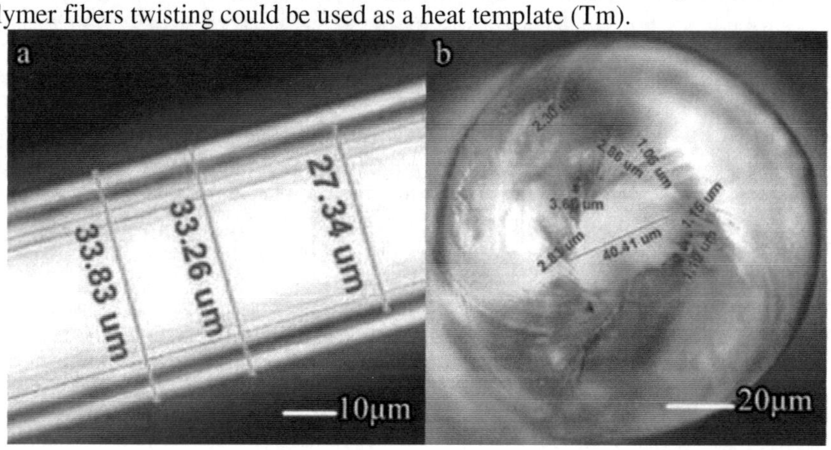

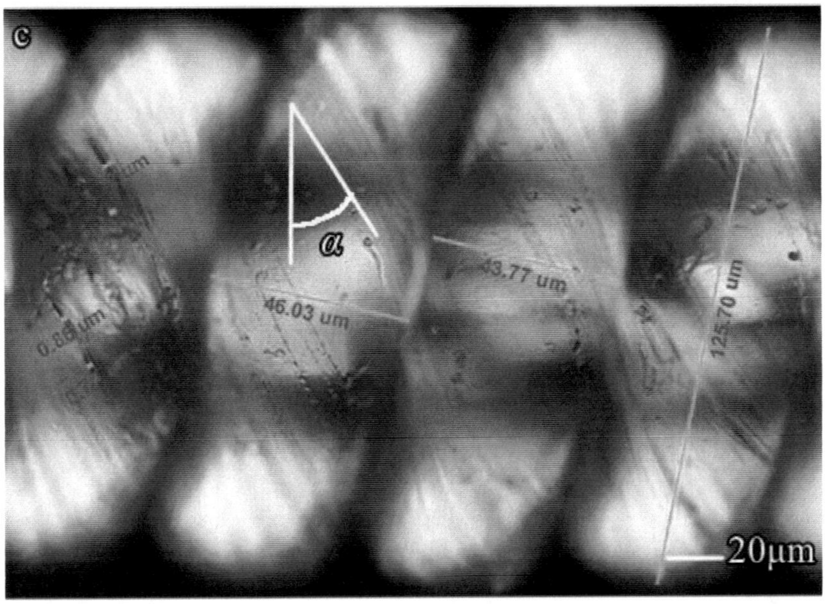

Figure 1. Micro-failure image of nylon 66. a) strands of nylon maintaining after ambient temperature heat start up; b) axial direction view of the nylon cross-section area; c) nylon twist-spun with spun angle spun angle α of 43°.

Based on the mono-filament winding method of winding around the center of the shaft, we measured the angle of winding and deflection (as shown in Figure 1c). Also, the rotation angle of nylon is 43°. As of 2014, the existing of most powerful artificial muscle fibers offered a hundredfold increase in power over equivalent lengths of natural muscle fibers [13]. Artificial muscles can be separated into several major groups based on their actuation mechanism, such as electric field actuation, ion-based actuation, pneumatic actuation and thermal actuation. For instance, the linear diameter of nylon mono-filament fishing line and sewing thread are 0.42mm, 0.54mm, 5.34mm and 5.63mm.

The polymer fibers which are finer than hair could give humanoid robots more life-like external expressions. Metal nano-wires are twisted to form yarns strong are from 0.4 to 1.1GPa, normalized to yarn length and torsional rotation was 12deg/mm^{-1} [14]. We attempted our study from low-cost and high-bond strength polymeric fibers, most often those used as fishing line and sewing thread (as shown in Table 1).

Table 1. Summary of the polymer fibers found to be most useful for coiled tensile testing. The tensile testing calculated results are given in Figure 2.

Material	Linear Diameter	K[a)] Value	Twist to coil
Nylon	0.42 mm	9.8 N/m	300 turns/m
mono-filament	0.54 mm	11.33 N/m	300 turns/m
fishing line and	5.34 mm	7.4 N/m	300 turns/m
sewing thread	5.63 mm	6.2 N/m	300 turns/m

[a)]Hooke's law: $F=K \cdot \Delta L$, where K is a constant factor characteristic of the spring, its stiffness.

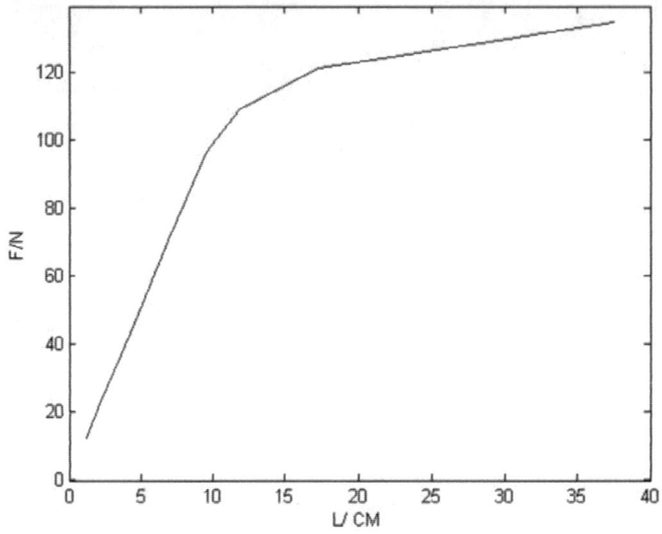

Figure 2. Thermomechanical analysis (TMA) of nylon mono-filament fishing line and sewing thread.

Figure 2 depicts that tensile tests direct correspondence between the variable force F from 12N to 140N and the variable length L from 2cm to 40cm. The tensile test measured is the degree of nylon fibers twist determines the range volume change that is converted to twine angle.

2. Experimental

Coiled and several-nylon actuators fibers were made by twisting nylon mono-filaments such as fishing line and sewing thread. The specimens were applied to the twisted nylon mono-filaments during the twisting process before coiling start up. After the specimens were coiled and heated (from 25℃ to 80℃), nylon 66 appeared to show micro-failure in diameter (as shown in Figure 3). Then by lowering the tensile testing, the nylon 66 twisted to coil is 300 turns/m. Depending on the diameter of the mono-filaments and structure, the nylon structure of K value is stretched under loads of around 6.2 to 11.33N/m (as shown in Table 1; main text).

The resulting coil wires are twisted to form a fishing line and sewing thread (as shown in Figure 1c). Longer samples can be obtained by passing longer fishing lines and nanowires during the experiment. The nylon weights are determined by a balance with a resolution of 1μg.

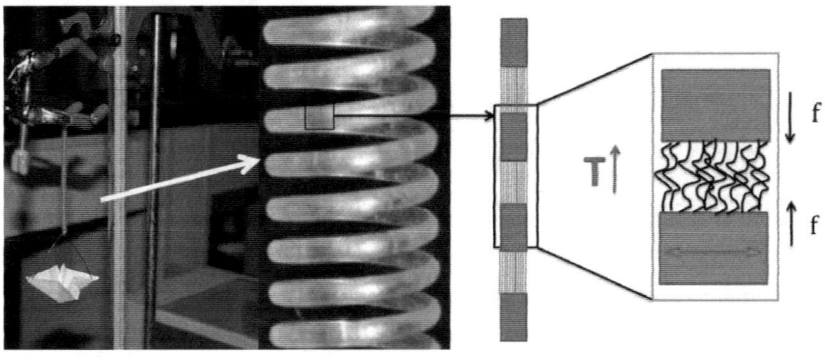

Figure 3. Micro-structure force and potential energy vary in entropy.

The nano-fibers' internal stress changes over time at constant temperature, and its deformation remains constant. (as shown in Figure 3). Stress relaxation proceeds slowly at ambient temperature but in case of heating, it will considerably shorten the time of the last stresses to 0. Stress reduction causes the fishing line and sewing thread to become a "spring-type" eternally. Also, due to the tensile and rotational effects of gravity, the fishing line and the sewing thread become spring.

Macro-molecule chains generate large increase in enthalpy and entropy. The fishing line and sewing thread internal energy will thus increase. Then internal energy releases heat and the polymeric fibers twisting and coiling attain the lowest potential energy. Heat-Shrinkage versus temperature trace on a graph is shown in Figure 4.

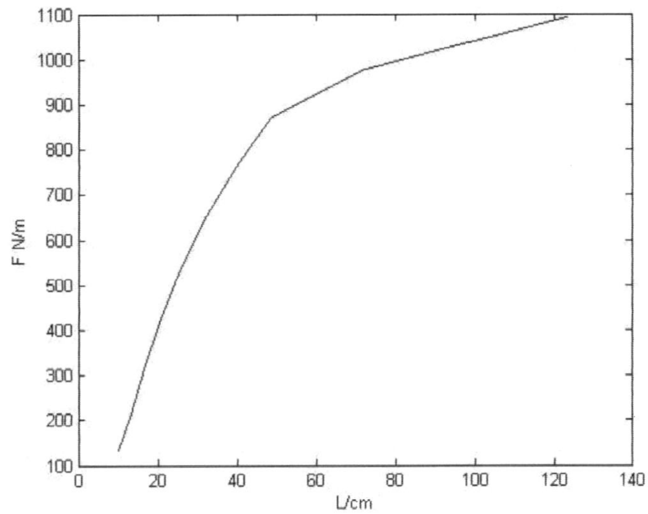

Figure 4. Heat-Sh6rinkage versus temperature trace on a graph.

The structure and micro-structure of the polymer fibers were characterized by the test parameters in the experiment.

3. Results and discussion

As known to all, the artificial muscle technologies have comprehensively applied in bio-mimetic machines, medicine, industrial actuators and powered exoskeletons, etc. Haines et al. suggested that the twist-aligned polymer nanofibers in a helical experimental model should be improved in terms of heating and torque direction and by adding micro-fixtures in the fixed nylon fibers [15]. In the horizontal direction, the material is contracted and stretched by stretching and heating to change the material's performance. When the nylon is coiled, the specimens have changed coil length from 25℃ (ambient temperature) to 80℃ (as shown in Figure 5b).

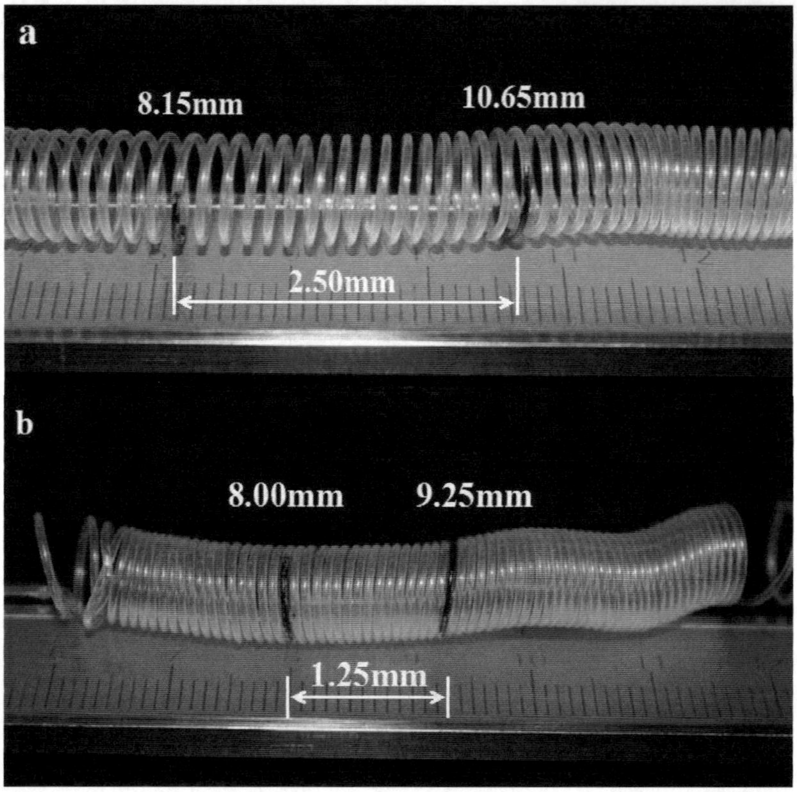

Figure 5. Tensile testing of nylon 66. a) Ambient temperature 25℃; b) heat temperature from 25℃ to 80℃.

The artificial muscles are powered by ambient temperature changes; Figure 3 depicts tensile tests from 25℃ (ambient temperature) to 80℃. Nylon 66 diameter is 5.46mm, and the shrinkage ratio around a Tm of 50% of deformation in length, which resulted from a heat temperature increases twist model volume caused by external micro-force in diameter, a change in length and twisting.

Artificial muscles increase in diameter occurs due to the polymer fibers interior macro-molecule chains shrinkage and coiling. During the heating process, the artificial muscle materials were contracted and its micro-structure changed. The lowest law of energy principle that increases in internal energy is equal to the total heat added. Artificial muscles lessen internal energy achieve steady state. Under the condition of external heating, the materials would yield shrinkage performance and the surface energy would change too (as shown in Figure 4). Artificial muscles with heat contractility and electrical contractility are sensitive to change in the external environment. It is found that the effect of rotation and temperature changes the performance of flexure and deformation of mono-filament materials during the experiment.

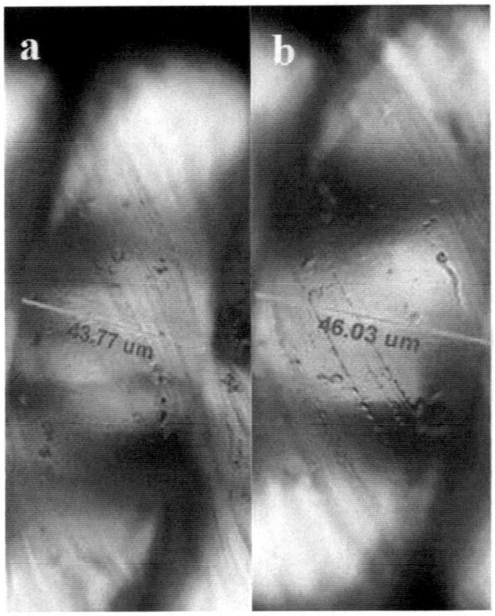

Figure 6. Artificial muscles increase in diameter. a) Before heating; b) After heating (The temperature error is ±2℃).

4. Conclusions

This article investigates a micro-failure behaviors analysis of thermo-mechanical actuators based on fishing line and sewing thread. These micro-failure behaviors have the capability of responding to ambient temperature variations producing a large amount of shrinkage ratio of a resulting variation in longitudinal length. The minimum micro-failure value is 0.02μm and the maximum value is 1.72μm with nylon twist pattern (as shown in Figure 1c).

(1) Heat setting is the process of internal material stress release, making the artificial muscle of K factor (as shown in Table 1; main text), thermal shrinkage and other performance more superior.

(2) Artificial muscles increase in diameter because of interior macro-molecule chains' shrinkage and coiling.

(3) In artificial muscles, the greater the diameter, the smaller the coiling, and also better water absorbing.

(4) Artificial muscles with heat conductivity, electrical conductivity are sensitive to change in the external environment.

Future work should investigate some possible micro-critical changes such as: twist failure, surface stress, heat treatment temperature (Tm), or effect of twist strain on the performance and on the energy density, dynamic behaviors, etc. Future work should also address other unexplored areas such as mechanical-medicine cross research, fatigue lifetime, or effect tensile testing of hygroscope on the specimen performance.

5. Acknowledgements

This research is funded and instrument support by Faculty of Materials Science and Engineering, Harbin University of Science and Technology, China. The innovation project grant from Jiabin Science Laboratory, China. The authors wish to express Harbin University of Science &Technology and Jiabin Science Laboratory researcher gratitude for the generous support.

IOP Conf. Series: Materials Science and Engineering **213** (2017) 012003 doi:10.1088/1757-899X/213/1/012003

References

[1] Mirvakili S M, Pazukha A, Sikkema W, Sinclair C W, Spinks G M, Baughman R H and Madden J D W 2013 Niobium nanowire yarns and their application as artificial muscle *Adv. Funct. Mater* **23** 4311-6 DOI: 10.1002/adfm.201203808

[2] Mehan R and Herzog J 1970 *Whisker Technology* ed A Levitt (New York: John Wiley & Sons) pp157-96

[3] Webb W W and Forgeng W D 1958 *Acta. Metal.* **6** 462-9

[4] Salkind M, Lemkey F and George F *Whisker Technology* ed A Levitt (New York: John Wiley & Sons) Chapter 10

[5] Yoshida K, Goto Y, Yamamoto M and Phys J 1966 *Soc. Jpn.***21** 825-6

[6] Seyed M M *et al,* 2014 Simple and strong: Twisted silver painted nylon artificial muscle actuated by Joule heating electroactive polymer actuators and devices (EAPAD) *Proc. SPIE* **9056** 90560I-1

[7] Kohan M 1995 *Nylon Plastics Handbook* (Munich: Carl Hanser Verlag) p 2

[8] Sellinger A T, Wang D H, Tan L S and Vaia R A 2010 *Adv. Mater.* **22** 3430

[9] Chen L *et al* 2011 *ACS Nano* **5** 1588

[10] Hu Y, Chen W, Lu L H, Liu J H and Chang C R 2010 *ACS Nano* **4** (6) 3498-502

[11] Klintberg L, Karlsson M, Stenmark L, Schweitz J Å and Thornell G 2002 *Sens. Actuators. A* **96** 189-92

[12] Lima M D *et al* 2012 *Science* **338** 928-32

[13] "Scientists just created some of the most powerful muscles in existence". 20 February 2014. MATERIALS SCIENCE

[14] Seyed M M, Alexey P, William S, Chad W S, Geoffrey M S, Ray H B and John D W M 2013 Niobium Nanowire Yarns and their Application as Artificial Muscles *Adv. Funct. Mater.* DOI: 10.1002/adfm.201203808

[15] Haines C S *et al* 2014 Artificial Muscles from Fishing Line and Sewing Thread *Science* **343**(6173) 868-72

IOP Conf. Series: Materials Science and Engineering **213** (2017) 012004 doi:10.1088/1757-899X/213/1/012004

Developing analytical solutions for transverse, matrix strain magnification and fibre strain reduction in uniaxially aligned continuous fibre reinforced composites, based on the principle of conservation of strain energy and the Reuss rule

M Maringa[1,3] and L M Masu[2]

[1]Central University of Technology, Free State, Private Bag, X20539, Bloemfontein, 9300, South Africa
[2]Vaal University of Technology, Andries Potgieter Blvd, Private Bag X021, Vanderbijlpark, 1900, Gauteng, South Africa

E-mail: maina_maringa@yahoo.com

Abstract. Analytical solutions for transverse matrix strain magnification and transverse fibre strain reduction in uniaxially aligned continuous fibre reinforced composites developed here for a Representative Volume Element (RVE), based on the Reuss rule and the principle of constant strain energy. The results obtained show respective matrix and fibre transverse strain magnification and reduction that increase with increasing volume fraction of the reinforcing fibre for both square and hexagonal fibre packing geometries. The square arrays register higher values of matrix and fibre transverse strain magnification and reduction, respectively, than hexagonal arrays. The values of transverse stress ratio between the fibre and matrix in the RVE central sub-region, and between the two and the RVE central sub-region composite are seen to vary with the volume fraction of reinforcing fibre. The results obtained here denote a dependency of respective matrix and fibre transverse strain magnification and reduction as well a stress ratio on the reinforcing fibre packing geometry and volume fraction.

1. Background

Existing literature on the phenomenon of matrix transverse strain magnification is very scanty and is mainly based on the simple rule developed by Kies J. L in 1962 [1-3]. Moreover, and to the best knowledge of the author, very little work has been done on the effect of transverse matrix strain magnification on the transverse strain reduction in the reinforcing fibre. This creates a need to undertake rigorous studies of these two phenomena.

The objective of this work is to use the strain energy method based on an RVE to develop analytical solutions for use in studying the variation of matrix transverse strain magnification and transverse fibre strain reduction with volume fraction of reinforcing fibre, in uniaxially aligned continuous fibre reinforced composites. The developed solutions are also expected to help in studying the dependence of the two phenomena on the reinforcing fibre packing geometry, while providing insight into the applicability of the Reuss rule in studying the phenomena. The term, matrix strain magnification, as used in this paper refers to the ratio between the transverse strain in the matrix adjacent to the reinforcing fibres and the transverse strain in the bulk matrix. The term, fibre strain reduction, as used in this paper refers to the ratio between the strains in the fibre, taking into account the effect of matrix strain magnification, to the strain in the bulk fibre in the absence of this consideration.

The properties of composite materials, both natural and manufactured, are dependent on the relative proportions and properties of their constituents, as well as the quality of the interface between constituents [1-4]. The various micromechanical rules that have used to try to predict the mechanical properties of composite materials can group into the four categories of phenomenological, elasticity, semi-empirical and homogenisation [5]. The Rule of Mixtures (ROM) also referred to as the iso-strain or Voigt rule and the Rule of Hybrid Mixtures for (RoHM) are known to give good predictions for the longitudinal composite mechanical properties of strength, stiffness and of the Poisson's ratios (ν_{12} & ν_{23}) for two phase materials and hybrid composites, respectively [5]. The Inverse Rule of Mixtures (IRoM) also referred to as iso-stress or Reuss rule for both transverse and shear properties however challenged in accuracy, and is only good as a first estimate [2-5]. Recourse is therefore sought in the Modified Rule of Mixtures (MRoM) corrected to increase the accuracy of predicting the values of transverse and shear moduli [5], in the semi-empirical rules of Halpin –Tsai and separately Chamis C. C. for dual phase materials [1-5] and also in the modified Halpin-Tsai semi-empirical rules for hybrid composite systems [6]. The properties of composite materials can also be determined through experimental testing, using for instance the very useful off-axis test that generates values of longitudinal, transverse, and shear stiffness and strength, as well as Poisson's ratios of composites [3, 7].

The results arising from the aforementioned theoretical and semi-empirical models, as well as experimental testing represent composites well at a macro scale but are unable to account for micro-variation of properties that may exist, which is particularly important for the interfacial bond as well as transverse stresses and strains.

The Kies model showed transverse strain magnification in the matrix in between the fibres to be a function of the reinforcing fibre volume fraction, matrix/fibre stiffness ratio, fibre diameter and geometric arrangement of the reinforcing fibres [2, 4].

The loss of stiffness, decrease in fatigue strength and initiation of weepage degradation in composite pipes arising from transverse failure of composites, because of low transverse properties of fibre-reinforced composites noted to severely limit use of composite materials [8]. A transition of glass fibre/epoxy composite failure mode from adhesive rupture at the fibre/matrix interface to cohesive rupture in the matrix, at a global strain of 1.15%, has also observed in the same work. This low failure strain is likely to be a result of matrix transverse strain magnification.

The transverse tensile strength of unidirectional composites observes to be much lower than the tensile strength of the pure matrix material because of transverse strain magnification related stress concentration arising from the introduction of reinforcing fibres in the matrix [9]. Additionally, it was further observed that failure of composites under compression did not occur along a line orthogonal to the direction of application of load, as was the case in tension, but along a line inclined to the direction of application load.

It was observed that despite the Kies transverse strain magnification model predicting lower failure strains for higher reinforcing fibre volume fraction, failure in polyester glass composites with a fibre volume fraction between 37 - 43% occurred at strains of 0.13 to 0.6% [10]. It was further observed in the same work that failure of epoxy glass fibre composites with higher values of fibre volume fraction between 55-60%, occurred at considerably much higher strains of 0.55 to 0.60. It is noted here that the elastic moduli for the two matrices ranges between 2 – 4.5 and 3 – 6 GPa, respectively. This from Kies transverse strain magnification equation would tend to lower the transverse strain magnification effect for the polyester resin, which has a lower stiffness range thus raising the strain at failure. Clearly, there must have been other factors contributing to failure in this reported study other than that of transverse strain magnification. Whether this phenomenon is peculiar to the reported work or is universally applicable to composite materials is not clear.

The importance of the matrix transverse strain magnification phenomenon is evident from the foregoing review and advices the attempts made here to develop analytical expressions that can be used to determine respective values of matrix and fibre transverse strain/stress magnification and reduction in continuous fibre reinforced composites. The analytical models developed here are also intended for use in studying their variation of the phenomenon with the volume fraction of reinforcing fibre. Due to the noted shortcomings of the Resuss rule and the lack of other alternative theoretical models for determining the transverse elastic properties of composite materials, recourse is sought in the RVE used by of Gibson [3] in order to develop expressions for the transverse elastic modulus of longitudinally aligned continuous fibre reinforced composites.

The solutions developed here assume isotropy and homogeneity of the composite and its constituents. It is understood however, that variations from this idealisation do exist such as the known anisotropy of carbon and aramid fibres [2, 3, 5] as well as polyethylene fibres [5]. This is clear from the values shown in Table 1.

Table 1. Elastic properties of some composite constituents.

Fibres	$\mathbf{E_{11}}$ **(GPa)**	$\mathbf{E_{22}}$ **(GPa)**	$\mathbf{G_{12}}$ **(GPa)**	ν_{12}	ν_{23}
E-Glass	73.1	73.1	29.95	0.22	0.22
Carbon	232	15	24	0.279	0.49
Polyethylene	60.4	4.68	1.65	0.38	0.55
Matrix	E_m (GPa)		G**(GPa)**	ν_m	
Epoxy	3.45	3.45	1.28	0.35	0.35
Epoxy	5.35	5.35	1.97	0.354	0.354
Epoxy	5.5	5.5	1.28	0.37	0.37

It is evident from Table 1 [5] that elastic modulus and Poisson's ratio of some types of epoxy are higher and lower than the transverse elastic modulus of polyethylene fibres, which raises interesting possibilities with respect to the subject of this study. In the foregoing table, the subscripts "11" and "22", refer to the longitudinal and orthogonal directions of the reinforcing fibres, respectively, while the subscripts "12" and "23" refer to the 12 and 13 planes, respectively. The symbols E, G and in the table refer to the elastic modulus, shear modulus and Poisson's ratio, respectively.

The theory of RVEs and their definition are covered well in literature and are discussed here in brief. RVEs are selected based on use, statistical representativeness of the whole (thus the same elastic constants, stresses and strains, and fibre volume fraction for the case of composites) and periodic repetitiveness. They must also be sufficiently larger than a typical structural macroscopic dimension, while being sufficiently smaller than the microscopic size. This makes them useful in simplifying analysis by focusing on a small item with known geometric and material properties including grains, flaws, inclusions, voids and fibres [1, 11-15]. Based on the statistical nature of the microstructure, statistical properties of the RVE, and based on homogenisation of material properties without reference to their statistical fluctuations over finite domains, the definitions of RVEs can classify into three broad areas [14]. An RVE can define as a volume of heterogeneous material that is large enough to include an effective sampling of all microstructural heterogeneities in the material and is therefore statistically representative of the whole, and which is small enough to be considered as an element of a continuum mechanics model [13]. An RVE can also been defined as, "the smallest material volume element of the composite for which the usual spatially constant (overall modulus) macroscopic constitutive representation is a sufficiently accurate model to represent mean constitutive response" [16]. A good overarching definition of an RVE is that of an element that can be used to determine the properties of a homogenised macroscopic model, with size between the macroscopic body and the microstructural size, and which has enough information about the microstructure and is a good representation of the continuum of a material [11].

Based on the principle of conservation of strain energy, it was demonstrated using E-glass fibre/epoxy resin that the assumption of constant transverse stress through the matrix and reinforcing fibre depth, which is used in arriving at the Reuss rule was wrong and that the stresses in the two constituent components of a matrix were different [3]. He then adapted the Hopkins and Chamis model of the RVE applied to a square array of reinforcing fibres to develop an alternative expression for the transverse elastic modulus of a fibre reinforced composite. However, it is noted that the iso-stress or Reuss rule was still used to determine the value of transverse stiffness in sub-region (B) of the RVE. This raises concerns of the same constraints of the Reuss rule pointed out before, though at the much smaller scale of a single fibre/surrounding matrix cell.

2. Developing analytical solutions

In this section, the principle of constant strain energy is applied first, followed by the iso-stress rule, Hooke's law and finally the assumptions of isotropy and homogeneity.

2.1. Applying the principle of constant strain energy to an RVE

Consider the RVE shown in the left end diagram of Figure 1, and its derivatives going to the right of the figure. The thick arrow in Figure 1 shows the direction of transverse deformation. The RVE is considered here to be the smallest volume element whose properties are representative of the macroscopic properties of the material. Figure 1b shows the reinforcing fibre of circular cross-section in Figure 1a, represented by a reinforcing fibre of square cross-section with the same effective cross-sectional area. Figure 1c consists of two sub-regions; the central one with the reinforcing fibre that has a square cross-sectional, and the two regions on either side of it.

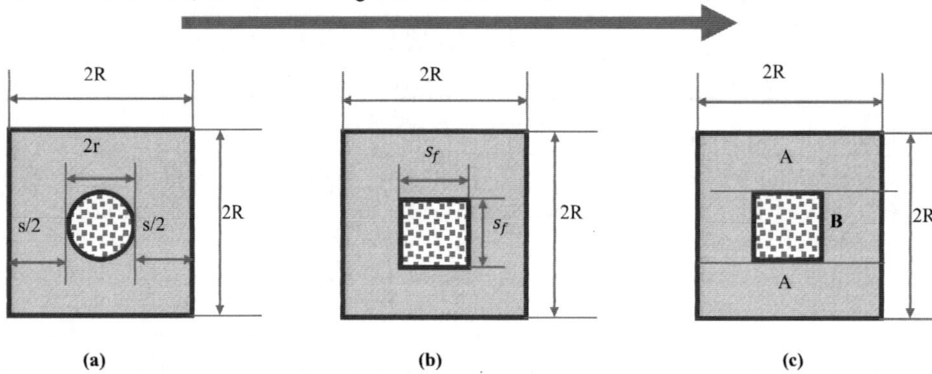

Figure 1. Representative Volume Element.

In Figure 1, the composite longitudinal and transverse axes are in the out of plane and in the plane directions, respectively.

In order for the area of the square central reinforcing fibre in Figure 1b to be the same as that of the round central reinforcing fibre in Figure 1a according to Equation 1, the following relationship must hold.

$$s_f^2 = \pi r^2 \Rightarrow s_f = r\sqrt{\pi} \tag{1}$$

The reinforcing fibre volume fractions ($v_f|_{sq}$ and $v_f|_{hex}$) of square and hexagonal arrays of reinforcing fibre, respectively, with fibres of radius (r) and fibre center to centre distance (R), are given by the expressions [2,3].

$$v_f|_{sq} = \frac{\pi}{4}\left(\frac{r}{R}\right)^2 \tag{2a}$$

$$v_f|_{hex} = \frac{\pi}{2\sqrt{3}}\left(\frac{r}{R}\right)^2 \tag{2b}$$

υ_f') of the reinforcing fibre in sub-region (B) of the right end diagram is The volume fraction (given by the expression:

$$\upsilon_f' = \frac{s_f \times s_f}{s_f \times 2R} = \frac{s_f}{2R} \tag{3}$$

Substituting Equation 1 into Equation 3 changes it to:

$$\upsilon_f' = \frac{\sqrt{\pi}}{2}\left(\frac{r}{R}\right) = \sqrt{\upsilon_f} \tag{4}$$

Substituting Equation 4 into each one of the two expressions in Equation 2 generates the following expressions for the reinforcing fibre volume fraction in sub-region (B) for the two fibre geometries:

$$\upsilon_f'\big|_{sq} = \sqrt{\upsilon_f}_{\,sq} \tag{5a}$$

$$\upsilon_f'\big|_{hex} = \sqrt[4]{3}\left(\sqrt{0.5\upsilon_f\big|_{hex}}\right) \tag{5b}$$

For the central sub-region (*B*) in Figure 1c, geometric compatibility of deformation in the transverse direction implies that:

$$\varepsilon_{c2}' = \varepsilon_f'\upsilon_f' + \varepsilon_m'\upsilon_m' \tag{6}$$

Introducing parameters *a* and *b* to relate the strains in the central sub-region composite, to the strains in the fibre and matrix produces the expression:

$$\varepsilon_f' = a\varepsilon_{c2}' \tag{7a}$$

$$\varepsilon_m' = b\varepsilon_{c2}' \tag{7b}$$

Substituting Equations 7a and 7b into Equation 6 gives rise to:

$$\varepsilon_{c2}' = a\varepsilon_{c2}'\upsilon_f' + b\varepsilon_{c2}'\upsilon_m' \tag{8}$$

Equation 8 can be rewritten as:

$$1 = a\upsilon_f' + b\upsilon_m' \tag{9}$$

It is evident from Equation 9 that:

$$a = \frac{1 - b\upsilon_m'}{\upsilon_f'} \tag{10a}$$

$$b = \frac{1 - a\upsilon_f'}{\upsilon_m'} \tag{10b}$$

Assuming that the strain energy (U) in the composite is equal to the sum of the strain energies in the reinforcing fibre and matrix constituents of the composite [3], thus:

$$\mathrm{U}_{c2}' = \mathrm{U}_f' + \mathrm{U}_m' \tag{11}$$

Equation 11 can be rewritten in terms of stresses and strains as:

$$0.5\sigma_{c2}'\varepsilon_{c2}'V_{c2}' = 0.5\sigma_f'a^2\varepsilon_f'V_f' + 0.5\sigma_m'a^2\varepsilon_m'V_m' \tag{12}$$

Dividing out the entire expression by 0.5 and substituting in the known relationship between stress and strain for linear elastic deformation transforms the foregoing equation to:

$$E_{c2}'\varepsilon_{c2}'^2V_{c2}' = E_f'\varepsilon_f'^2V_f' + E_m'\varepsilon_m'^2V_m' \tag{13}$$

Substituting Equations 7a and 7b into Equation 13 gives rise to:

$$E'_{c2}\varepsilon'^2_{c2}V'_{c2} = E'_f a^2 \varepsilon'^2_{c2} V'_f + E'_m b^2 \varepsilon'^2_{c2} V'_m \tag{14}$$

Dividing out Equation 14 with the term $(\varepsilon'^2_{c2}V'_{c2})$ leads to:

$$E'_{c2} = E'_f a^2 \upsilon'_f + E'_m b^2 \upsilon'_m \tag{15}$$

Substituting Equations 10a and 10b into Equation 15 converts it to:

$$E'_{c2} = E'_f a^2 \upsilon'_f + E'_m \left(\frac{1-a\upsilon'_f}{\upsilon'_m}\right)^2 \upsilon'_m \tag{16}$$

Equation 16 can be rewritten in the form of a quadratic equation in terms of the parameter a as follows:

$$a^2\left(E'_f\upsilon'_f + \frac{E'_m\upsilon'^2_f}{\upsilon'_m}\right) - \frac{2aE'_m\upsilon'_f}{\upsilon'_m} + \left(\frac{E'_m}{\upsilon'_m} - E'_{c2}\right) = 0 \tag{17}$$

The solutions for parameter a in Equation 17 are obtained from the general solution for a quadratic equation thus:

$$a = \frac{\frac{2E'_m\upsilon'_f}{\upsilon'_m} \pm \sqrt{\left(\frac{2E'_m\upsilon'_f}{\upsilon'_m}\right)^2 - 4\left(E'_f\upsilon'_f + \frac{E'_m\upsilon'^2_f}{\upsilon'_m}\right)\left(\frac{E'_m}{\upsilon'_m} - E'_{c2}\right)}}{2\left(E'_f\upsilon'_f + \frac{E'_m\upsilon'^2_f}{\upsilon'_m}\right)} \tag{18}$$

Following a similar procedure, the solutions for parameter b are determined by the equation:

$$\mathrm{b} = \frac{\frac{2E'_f\upsilon'_m}{\upsilon'_f} \pm \sqrt{\left(\frac{2E'_f\upsilon'_m}{\upsilon'_f}\right)^2 - 4\left(E'_m\upsilon'_m + \frac{E'_f\upsilon'^2_m}{\upsilon'_f}\right)\left(\frac{E'_f}{\upsilon'_f} - E'_{c2}\right)}}{2\left(E'_m\upsilon'_m + \frac{E'_f\upsilon'^2_m}{\upsilon'_f}\right)} \tag{19}$$

An examination of Equations 18 and 19 shows them to contain the parameter (E'_{c2}) with unknown values. Therefore, this parameter is rewritten in terms of the known parameters by introducing the Hookean stress/strain relationship into Equation 6 for geometric compatibility, thus:

$$\frac{\sigma'_{c2}}{E'_{c2}} = \frac{\sigma'_f}{E'_f}\upsilon'_f + \frac{\sigma'_m}{E'_m}\upsilon'_m \tag{20}$$

Equation 20 may be rewritten as:

$$\frac{1}{E'_{c2}} = \frac{(\sigma'_f/\sigma'_{c2})}{E'_f}\upsilon'_f + \frac{(\sigma'_m/\sigma'_{c2})}{E'_m}\upsilon'_m \tag{21}$$

Making parameter (E'_{c2}) the subject converts Equation 21 into:

$$E'_{c2} = \frac{E'_m E'_f}{E'_m(\sigma'_f/\sigma'_{c2})\upsilon'_f + E'_f(\sigma'_m/\sigma'_{c2})\upsilon'_m} \tag{22}$$

Equation 22 can be rewritten as:

$$E'_{c2} = \frac{E'_m}{(1-\upsilon'_f)\left(\frac{\sigma'_m}{\sigma'_{c2}}\right) + \upsilon'_f\left(\frac{\sigma'_f}{\sigma'_{c2}}\right)\frac{E'_m}{E'_f}} \tag{23}$$

Equation 23 can be rewritten as:

IOP Conf. Series: Materials Science and Engineering **213** (2017) 012004 doi:10.1088/1757-899X/213/1/012004

$$E'_{c2} = \frac{E'_m}{\left(\frac{\sigma'_m}{\sigma'_{c2}}\right) - \upsilon'_f\left(\frac{\sigma'_m}{\sigma'_{c2}} - \frac{\sigma'_f}{\sigma'_{c2}} \times \frac{E'_m}{E'_f}\right)} \tag{24}$$

Substituting Equations 5a and 5b into Equation 24 gives rise to the following two equations:

$$E'_{c2}|_{sq} = \frac{E'_m}{\left(\frac{\sigma'_m}{\sigma'_{c2}}\right) - \sqrt{\upsilon_f}\left(\frac{\sigma'_m}{\sigma'_{c2}} - \frac{\sigma'_f}{\sigma'_{c2}} \times \frac{E'_m}{E'_f}\right)} \tag{25a}$$

$$E'_{c2}|_{hex} = \frac{E'_m}{\left(\frac{\sigma'_m}{\sigma'_{c2}}\right) - \sqrt[4]{3}\left(\sqrt{0.5\upsilon_f}\right)\left(\frac{\sigma'_m}{\sigma'_{c2}} - \frac{\sigma'_f}{\sigma'_{c2}} \times \frac{E'_m}{E'_f}\right)} \tag{25b}$$

Equations 25a and 25b can be rewritten as:

$$E'_{c2}|_{sq} = \frac{E'_m\left(\frac{\sigma'_{c2}}{\sigma'_m}\right)}{1 - \sqrt{\upsilon_f}\left(1 - \frac{\sigma'_f}{\sigma'_m} \times \frac{E'_m}{E'_f}\right)} \tag{26a}$$

$$E'_{c2}|_{hex} = \frac{E'_m\left(\frac{\sigma'_{c2}}{\sigma'_m}\right)}{1 - \sqrt[4]{3}\left(\sqrt{0.5\upsilon_f}\right)\left(1 - \frac{\sigma'_f}{\sigma'_m} \times \frac{E'_m}{E'_f}\right)} \tag{26b}$$

2.1.1. Introducing the Reuss rule. Introducing the iso-stress rule for the central sub-region, which is represented by the relationship, $\boldsymbol{\sigma'_{c2} = \sigma'_f = \sigma'_m}$, into the two foregoing equations converts them to:

$$\boldsymbol{E'_{c2}|_{sq} = \frac{E'_m}{1 - \sqrt{\upsilon_f}\left(1 - \frac{E'_m}{E'_f}\right)}} \tag{27a}$$

$$\boldsymbol{E'_{c2}|_{hex} = \frac{E'_m}{1 - \sqrt[4]{3}\left(\sqrt{0.5\upsilon_f}\right)\left(1 - \frac{E'_m}{E'_f}\right)}} \tag{27b}$$

Equation 27a is familiar as the equation that developed by Gibson [2] for the central sub-region of the right hand diagram in Figure 1, adapted for an isotropic reinforcing fibre. It is a special case of the Chamis equation for predicting the transverse stiffness of a composite [5] for isotropic reinforcing fibres.

Substituting the Equation 27a and 27b into Equations 18 and 19 converts them to a form solely dependent on known values of the parameters of the composite constituent components.

Substituting the known relationship for the volume fractions of the composite constituents, $\boldsymbol{\upsilon'_m = 1 - \upsilon'_f}$, into the expressions arising from this substitution produces the following expressions:

$$a_{sq} = \frac{\frac{2E'_m\sqrt{\upsilon_f}}{1-\sqrt{\upsilon_f}} \pm \sqrt{\left(\frac{2E'_m\sqrt{\upsilon_f}}{1-\sqrt{\upsilon_f}}\right)^2 - 4\left(E'_f\sqrt{\upsilon_f} + \frac{E'_m\upsilon_f}{1-\sqrt{\upsilon_f}}\right)\left(\frac{E'_m\left(\left(1-\sqrt{\upsilon_f}\left(1-\frac{E'_m}{E'_f}\right)\right)-\left(1-\sqrt{\upsilon_f}\right)\right)}{\left(1-\sqrt{\upsilon_f}\right)\left(1-\sqrt{\upsilon_f}\left(1-\frac{E'_m}{E'_f}\right)\right)}\right)}}{2\left(E'_f\sqrt{\upsilon_f} + \frac{E'_m\upsilon_f}{1-\sqrt{\upsilon_f}}\right)} \tag{28}$$

$$b_{sq}=\frac{\frac{2E'_f\left(1-\sqrt{\upsilon_f}\right)}{\sqrt{\upsilon_f}}\pm\sqrt{\left(\frac{2E'_f\left(1-\sqrt{\upsilon_f}\right)}{\sqrt{\upsilon_f}}\right)^2-4\left(E'_m\left(1-\sqrt{\upsilon_f}\right)+\frac{E'_f\left(1-\sqrt{\upsilon_f}\right)^2}{\sqrt{\upsilon_f}}\right)\left(\frac{E'_f\left(\left(1-\sqrt{\upsilon_f}\left(1-\frac{E'_m}{E'_f}\right)\right)-\frac{E'_m}{E'_f}\sqrt{\upsilon_f}\right)}{\sqrt{\upsilon_f}\left(1-\sqrt{\upsilon_f}\left(1-\frac{E'_m}{E'_f}\right)\right)}\right)}}{2\left(E'_m\left(1-\sqrt{\upsilon_f}\right)+\frac{E'_f\left(1-\sqrt{\upsilon_f}\right)^2}{\sqrt{\upsilon_f}}\right)} \quad (29)$$

$$a_{hex}=\frac{\frac{2E'_m\sqrt[4]{3}\left(\sqrt{0.5\upsilon_f}\right)}{1-\sqrt[4]{3}\left(\sqrt{0.5\upsilon_f}\right)}\pm\sqrt{\left(\frac{2E'_m\sqrt[4]{3}\left(\sqrt{0.5\upsilon_f}\right)}{1-\sqrt[4]{3}\left(\sqrt{0.5\upsilon_f}\right)}\right)^2-4\left(E'_f\sqrt[4]{3}\left(\sqrt{0.5\upsilon_f}\right)+\frac{E'_m 0.5\sqrt{3}\upsilon_f}{1-\sqrt[4]{3}\left(\sqrt{0.5\upsilon_f}\right)}\right)\left(\frac{E'_m\left(\left(1-\sqrt[4]{3}\left(\sqrt{0.5\upsilon_f}\right)\left(1-\frac{E'_m}{E'_f}\right)\right)-\left(1-\sqrt[4]{3}\left(\sqrt{0.5\upsilon_f}\right)\right)\right)}{\left(1-\sqrt[4]{3}\left(\sqrt{0.5\upsilon_f}\right)\right)\left(1-\sqrt[4]{3}\left(\sqrt{0.5\upsilon_f}\right)\left(1-\frac{E'_m}{E'_f}\right)\right)}\right)}}{2\left(E'_f\sqrt[4]{3}\left(\sqrt{0.5\upsilon_f}\right)+\frac{E'_m 0.5\sqrt{3}\upsilon_f}{1-\sqrt[4]{3}\left(\sqrt{0.5\upsilon_f}\right)}\right)} \quad (30)$$

$$b_{hex}=\frac{\frac{2E'_f\left(1-\sqrt[4]{3}\left(\sqrt{0.5\upsilon_f}\right)\right)}{\sqrt[4]{3}\left(\sqrt{0.5\upsilon_f}\right)}\pm\sqrt{\left(\frac{2E'_f\left(1-\sqrt[4]{3}\left(\sqrt{0.5\upsilon_f}\right)\right)}{\sqrt[4]{3}\left(\sqrt{0.5\upsilon_f}\right)}\right)^2-4\left(E'_m\left(1-\sqrt[4]{3}\left(\sqrt{0.5\upsilon_f}\right)\right)+\frac{E'_f\left(1-\sqrt[4]{3}\left(\sqrt{0.5\upsilon_f}\right)\right)^2}{\sqrt[4]{3}\left(\sqrt{0.5\upsilon_f}\right)}\right)\left(\frac{E'_f\left(\left(1-\sqrt[4]{3}\left(\sqrt{0.5\upsilon_f}\right)\left(1-\frac{E'_m}{E'_f}\right)\right)-\frac{E'_m}{E'_f}\sqrt[4]{3}\left(\sqrt{0.5\upsilon_f}\right)\right)}{\sqrt[4]{3}\left(\sqrt{0.5\upsilon_f}\right)\left(1-\sqrt[4]{3}\left(\sqrt{0.5\upsilon_f}\right)\left(1-\frac{E'_m}{E'_f}\right)\right)}\right)}}{2\left(E'_m\left(1-\sqrt[4]{3}\left(\sqrt{0.5\upsilon_f}\right)\right)+\frac{E'_f\left(1-\sqrt[4]{3}\left(\sqrt{0.5\upsilon_f}\right)\right)^2}{\sqrt[4]{3}\left(\sqrt{0.5\upsilon_f}\right)}\right)} \quad (31)$$

2.1.2. Introducing isotropy and homogeneity of the *composite constituents.* Assuming isotropy and homogeneity of the composites constituents converts Equations 28 and 31 to:

$$\boldsymbol{a_{sq}}=\frac{\frac{2E_m\sqrt{\upsilon_f}}{1-\sqrt{\upsilon_f}}\pm\sqrt{\left(\frac{2E_m\sqrt{\upsilon_f}}{1-\sqrt{\upsilon_f}}\right)^2-4\left(E_f\sqrt{\upsilon_f}+\frac{E_m\upsilon_f}{1-\sqrt{\upsilon_f}}\right)\left(\frac{E_m\left(\left(1-\sqrt{\upsilon_f}\left(1-\frac{E_m}{E_f}\right)\right)-\left(1-\sqrt{\upsilon_f}\right)\right)}{\left(1-\sqrt{\upsilon_f}\right)\left(1-\sqrt{\upsilon_f}\left(1-\frac{E_m}{E_f}\right)\right)}\right)}}{2\left(E_f\sqrt{\upsilon_f}+\frac{E_m\upsilon_f}{1-\sqrt{\upsilon_f}}\right)} \quad (32)$$

$$\boldsymbol{b_{sq}}=\frac{\frac{2E_f\left(1-\sqrt{\upsilon_f}\right)}{\sqrt{\upsilon_f}}\pm\sqrt{\left(\frac{2E_f\left(1-\sqrt{\upsilon_f}\right)}{\sqrt{\upsilon_f}}\right)^2-4\left(E_m\left(1-\sqrt{\upsilon_f}\right)+\frac{E_f\left(1-\sqrt{\upsilon_f}\right)^2}{\sqrt{\upsilon_f}}\right)\left(\frac{E_f\left(\left(1-\sqrt{\upsilon_f}\left(1-\frac{E_m}{E_f}\right)\right)-\frac{E_m}{E_f}\sqrt{\upsilon_f}\right)}{\sqrt{\upsilon_f}\left(1-\sqrt{\upsilon_f}\left(1-\frac{E_m}{E_f}\right)\right)}\right)}}{2\left(E_m\left(1-\sqrt{\upsilon_f}\right)+\frac{E_f\left(1-\sqrt{\upsilon_f}\right)^2}{\sqrt{\upsilon_f}}\right)} \quad (33)$$

$$a_{hex}=\frac{\frac{2E_m\sqrt[4]{3}\left(\sqrt{0.5v_f}\right)}{1-\sqrt[4]{3}\left(\sqrt{0.5v_f}\right)}\pm\sqrt{\left(\frac{2E_m\sqrt[4]{3}\left(\sqrt{0.5v_f}\right)}{1-\sqrt[4]{3}\left(\sqrt{0.5v_f}\right)}\right)^2-4\left(E_f\sqrt[4]{3}\left(\sqrt{0.5v_f}\right)+\frac{E_m0.5\sqrt{3}v_f}{1-\sqrt[4]{3}\left(\sqrt{0.5v_f}\right)}\right)\left(\frac{E_m\left(\left(1-\sqrt[4]{3}\left(\sqrt{0.5v_f}\right)\left(1-\frac{E_m}{E_f}\right)\right)-\left(1-\sqrt[4]{3}\left(\sqrt{0.5v_f}\right)\right)\right)}{\left(1-\sqrt[4]{3}\left(\sqrt{0.5v_f}\right)\right)\left(1-\sqrt[4]{3}\left(\sqrt{0.5v_f}\right)\left(1-\frac{E_m}{E_f}\right)\right)}\right)}}{2\left(E_f\sqrt[4]{3}\left(\sqrt{0.5v_f}\right)+\frac{E_m0.5\sqrt{3}v_f}{1-\sqrt[4]{3}\left(\sqrt{0.5v_f}\right)}\right)} \quad \mathbf{(34)}$$

$$b_{hex}=\frac{\frac{2E_f\left(1-\sqrt[4]{3}\left(\sqrt{0.5v_f}\right)\right)}{\sqrt[4]{3}\left(\sqrt{0.5v_f}\right)}\pm\sqrt{\left(\frac{2E_f\left(1-\sqrt[4]{3}\left(\sqrt{0.5v_f}\right)\right)}{\sqrt[4]{3}\left(\sqrt{0.5v_f}\right)}\right)^2-4\left(E_m\left(1-\sqrt[4]{3}\left(\sqrt{0.5v_f}\right)\right)+\frac{E_f\left(1-\sqrt[4]{3}\left(\sqrt{0.5v_f}\right)\right)^2}{\sqrt[4]{3}\left(\sqrt{0.5v_f}\right)}\right)\left(\frac{E_f\left(\left(1-\sqrt[4]{3}\left(\sqrt{0.5v_f}\right)\left(1-\frac{E_m}{E_f}\right)\right)-\frac{E_m}{E_f}\sqrt[4]{3}\left(\sqrt{0.5v_f}\right)\right)}{\sqrt[4]{3}\left(\sqrt{0.5v_f}\right)\left(1-\sqrt[4]{3}\left(\sqrt{0.5v_f}\right)\left(1-\frac{E_m}{E_f}\right)\right)}\right)}}{2\left(E_m\left(1-\sqrt[4]{3}\left(\sqrt{0.5v_f}\right)\right)+\frac{E_f\left(1-\sqrt[4]{3}\left(\sqrt{0.5v_f}\right)\right)^2}{\sqrt[4]{3}\left(\sqrt{0.5v_f}\right)}\right)} \quad (35)$$

Equations 32 – 35 can be simplified in this order to:

$$a_{sq}=\frac{\frac{2E_m\sqrt{v_f}}{1-\sqrt{v_f}}\pm\sqrt{\left(\frac{2E_m\sqrt{v_f}}{1-\sqrt{v_f}}\right)^2-4\left(E_f\sqrt{v_f}+\frac{E_mv_f}{1-\sqrt{v_f}}\right)\left(\frac{\frac{E_m^2}{E_f}\sqrt{v_f}}{\left(1-\sqrt{v_f}\right)\left(1-\sqrt{v_f}\left(1-\frac{E_m}{E_f}\right)\right)}\right)}}{2\left(E_f\sqrt{v_f}+\frac{E_mv_f}{1-\sqrt{v_f}}\right)} \quad (36)$$

$$b_{sq}=\frac{\frac{2E_f\left(1-\sqrt{v_f}\right)}{\sqrt{v_f}}\pm\sqrt{\left(\frac{2E_f\left(1-\sqrt{v_f}\right)}{\sqrt{v_f}}\right)^2-4\left(E_m\left(1-\sqrt{v_f}\right)+\frac{E_f\left(1-\sqrt{v_f}\right)^2}{\sqrt{v_f}}\right)\left(\frac{E_f\left(1-\sqrt{v_f}\right)}{\sqrt{v_f}\left(1-\sqrt{v_f}\left(1-\frac{E_m}{E_f}\right)\right)}\right)}}{2\left(E_m\left(1-\sqrt{v_f}\right)+\frac{E_f\left(1-\sqrt{v_f}\right)^2}{\sqrt{v_f}}\right)} \quad (37)$$

$$a_{hex}=\frac{\frac{2E_m\sqrt[4]{3}\left(\sqrt{0.5v_f}\right)}{1-\sqrt[4]{3}\left(\sqrt{0.5v_f}\right)}\pm\sqrt{\left(\frac{2E_m\sqrt[4]{3}\left(\sqrt{0.5v_f}\right)}{1-\sqrt[4]{3}\left(\sqrt{0.5v_f}\right)}\right)^2-4\left(E_f\sqrt[4]{3}\left(\sqrt{0.5v_f}\right)+\frac{E_m0.5\sqrt{3}v_f}{1-\sqrt[4]{3}\left(\sqrt{0.5v_f}\right)}\right)\left(\frac{\frac{E_m^2}{E_f}\sqrt[4]{3}\left(\sqrt{0.5v_f}\right)}{\left(1-\sqrt[4]{3}\left(\sqrt{0.5v_f}\right)\right)\left(1-\sqrt[4]{3}\left(\sqrt{0.5v_f}\right)\left(1-\frac{E_m}{E_f}\right)\right)}\right)}}{2\left(E_f\sqrt[4]{3}\left(\sqrt{0.5v_f}\right)+\frac{E_m0.5\sqrt{3}v_f}{1-\sqrt[4]{3}\left(\sqrt{0.5v_f}\right)}\right)} \quad (38)$$

$$\mathbf{b_{hex}} = \frac{\frac{2E_f\left(1-\sqrt[4]{3}\left(\sqrt{0.5\upsilon_f}\right)\right)}{\sqrt[4]{3}\left(\sqrt{0.5\upsilon_f}\right)} \pm \sqrt{\left(\frac{2E_f\left(1-\sqrt[4]{3}\left(\sqrt{0.5\upsilon_f}\right)\right)}{\sqrt[4]{3}\left(\sqrt{0.5\upsilon_f}\right)}\right)^2 - 4\left(E_m\left(1-\sqrt[4]{3}\left(\sqrt{0.5\upsilon_f}\right)\right)+\frac{E_f\left(1-\sqrt[4]{3}\left(\sqrt{0.5\upsilon_f}\right)\right)^2}{\sqrt[4]{3}\left(\sqrt{0.5\upsilon_f}\right)}\right)\left(\frac{E_f\left(1-\sqrt[4]{3}\left(\sqrt{0.5\upsilon_f}\right)\right)}{\sqrt[4]{3}\left(\sqrt{0.5\upsilon_f}\right)\left(1-\sqrt[4]{3}\left(\sqrt{0.5\upsilon_f}\right)\left(1-\frac{E_m}{E_f}\right)\right)}\right)}}{2\left(E_m\left(1-\sqrt[4]{3}\left(\sqrt{0.5\upsilon_f}\right)\right)+\frac{E_f\left(1-\sqrt[4]{3}\left(\sqrt{0.5\upsilon_f}\right)\right)^2}{\sqrt[4]{3}\left(\sqrt{0.5\upsilon_f}\right)}\right)} \quad (39)$$

Equations 36 and 37, as well as Equations 38 and 39 facilitate the variation of parameters a and b with volume fraction of the reinforcing fibre to be investigated for square and hexagonal arrays, respectively. Examination of Equations 7a and 7b show the parameters to represent the strain reduction and magnification of the central sub-region reinforcing fibre and matrix, respectively, with reference to the central sub-region composite. Looking at the overall deformation of the RVE consisting of the central sub-region as one material and the two side matrix regions as another, reduces the problem to one of the deformation of a matrix/central sub-region composite in the RVE's longitudinal direction, which is the overall composite transverse direction. Since the iso-strain or Voigt rule applies in the longitudinal direction of a fibre reinforced composite, the strain in the central sub-region is equal to the strain in the RVE in its longitudinal direction, which is the transverse direction of the composite. The absolute strain reduction and magnification of the reinforcing fibre $(\varepsilon_f'/\varepsilon_m)$ and matrix $(\varepsilon_m'/\varepsilon_m)$, respectively, are therefore equal to the strain ratios $(\varepsilon_f'/\varepsilon_{c2}')$ and $(\varepsilon_f'/\varepsilon_{c2}')$, represented by the parameters a and b, respectively. In the foregoing strain ratios the strain (ε_m) stands for the strain in the matrix that does not experience transverse strain magnification. These matrix regions are adjacent to the central sub-region in Figure 1.
Introducing the relationship between stress and strain for linear elastic deformation into Equations 7a and 7b converts them to:

$$\varepsilon_f' = a\varepsilon_{c2}' \Rightarrow \frac{\sigma_f'}{E_f'} = a\frac{\sigma_{c2}'}{E_{c2}'} \quad (40a)$$

$$\varepsilon_m' = b\varepsilon_{c2}' \Rightarrow \frac{\sigma_m'}{E_m'} = b\frac{\sigma_{c2}'}{E_{c2}'} \quad (40b)$$

Equations 40a and 40b can be rewritten as:

$$\frac{\sigma_f'}{\sigma_{c2}'} = a\frac{E_f'}{E_{c2}'} \quad (41a)$$

$$\frac{\sigma_m'}{\sigma_{c2}'} = b\frac{E_m'}{E_{c2}'} \quad (41b)$$

For isotropic and homogeneous reinforcing fibres and matrix, their respective elastic moduli are constant, which allows the superscript to be taken of the terms for elastic modulus.
Substituting Equations 27a and 27b into the Equations 40a and 40b, in this order, assuming isotropy and homogeneity for both fibre and matrix, changes them to:

$$\left.\frac{\sigma_f'}{\sigma_{c2}'}\right|_{sq} = a\frac{E_f}{E_m}\left(1-\sqrt{\upsilon_f}\left(1-\frac{E_m}{E_f}\right)\right) \quad (42a)$$

$$\left.\frac{\sigma_m'}{\sigma_{c2}'}\right|_{sq} = b\left(1-\sqrt{\upsilon_f}\left(1-\frac{E_m}{E_f}\right)\right) \quad (42b)$$

$$\left.\frac{\sigma'_f}{\sigma'_{c2}}\right|_{hex} = a\frac{E_f}{E_m}\left(1 - \sqrt[4]{3}\left(\sqrt{0.5v_f}\right)\left(1 - \frac{E_m}{E_f}\right)\right) \tag{43a}$$

$$\left.\frac{\sigma'_m}{\sigma'_{c2}}\right|_{hex} = b\left(1 - \sqrt[4]{3}\left(\sqrt{0.5v_f}\right)\left(1 - \frac{E_m}{E_f}\right)\right) \tag{43b}$$

Equations 42a and 43a, and 42b and 43b are statements of the stress magnification in the central sub-region reinforcing fibre and matrix, respectively, with respect to the central sub-region composite stress. The ratio of stress between the reinforcing fibre and matrix of the central sub-region for both square and hexagonal arrays of reinforcing fibre is obtained by equating Equations 42a to 42b, and Equations 43a to 43b, respectively, thus:

$$\left.\frac{\sigma'_f}{\sigma'_m}\right|_{sq} = \frac{a}{b}\frac{E_f}{E_m} \tag{44}$$

It is important to note here that the values of the parameters a and b are different for the two reinforcing fibre geometries, as is evident from an examination of their defining equations 36 – 39.

3. Results, analysis and discussion

Results of strain and stress ratios are now presented, analysed and discussed, first for square arrays of reinforcing fibre and finally for hexagonal reinforcing fibre arrays.

3.1. *Square arrays of reinforcing fibre*

Plots of fibre strain reduction and matrix strain magnification in the central sub-region, with respect to the central sub-region composite represented by the respective parameters a and b. And plots of the ratio (b/a) for stress magnification represented by the ratios shown in Equations 42 (a & b) and 44, for various values of reinforcing fibre volume, are shown in Figures 2-9 for square arrays of reinforcing.

Figures 2-3 show the variation of the transverse strain in the matrix in between the reinforcing fibres with volume fraction of the reinforcing fibres. Two roots (the term roots are used throughout the paper to denote solutions to a polynomial) were obtained with different magnitudes and trends and are plotted separately in Figures 2 and 3. The roots are seen to have magnitudes that are significantly different, with ratios between the two that increase at an increasing rate with increasing volume fraction of the reinforcing fibre ($\boldsymbol{v_f}$), and which vary between 10.31 at $\boldsymbol{v_f}$ = 1% and 1329.89 at $\boldsymbol{v_f}$ = 99%. The different curve profiles in the two figures denote different trends of the phenomena of matrix strain magnification with increasing volume fraction of the reinforcing fibre. The two values of root ratios given above for two different volume fractions of reinforcing fibre emphasises the change in the scale of the difference of the roots with increasing volume fraction of the reinforcing fibre.

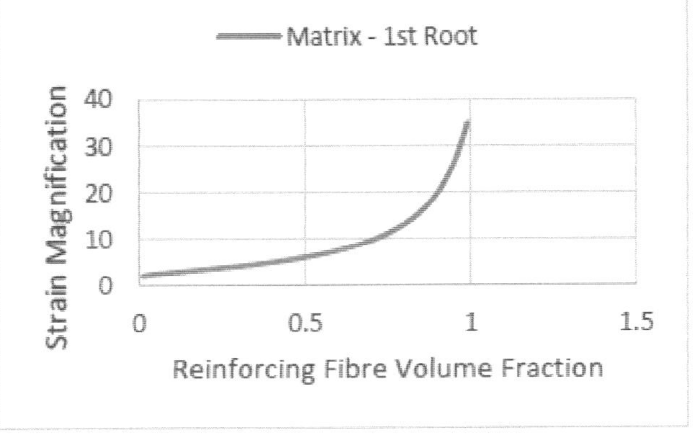

Figure 2. Variation of the magnification of transverse strain in the matrix with reinforcing fibre volume fraction of a square array– 1st root (Equation 37).

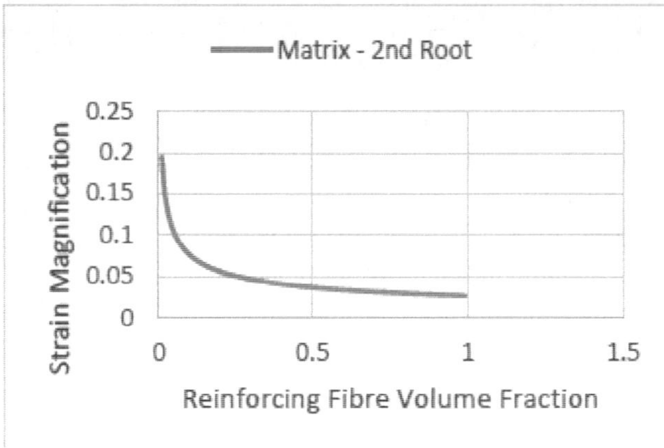

Figure 3. Variation of the reduction of transverse strain in the matrix with reinforcing fibre volume fraction of a square array – 2nd root (Equation 37).

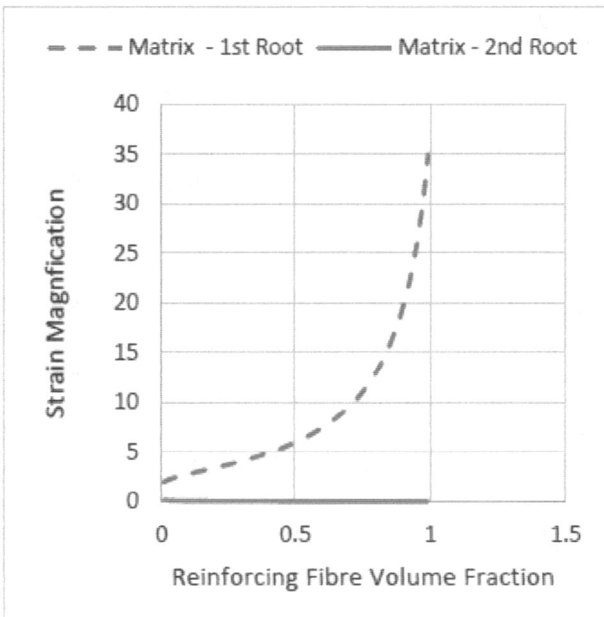

Figure 4. Variation of the magnification / reduction of transverse strain in the in the matrix with reinforcing fibre volume fraction for a square array – 1st and 2nd roots (Equation 37).

The curve showed in Figure 2 displays a magnification of transverse strain in the matrix that increases at an increasing rate with increasing volume fraction of the reinforcing fibre, with sample values of 2.0142 at υ_f =1%, 12.2361 at υ_f =78%, to 20.8064 at υ_f =91%. The physical meaning of the 1st root is plotted in Figures 2 and 4, which easily explained from the fact that the matrix with its elastic modulus is smaller than that of the reinforcing fibre deforms more than the reinforcing fibre under the same applied load. Therefore, the resulting strains are expected to be greater for lower volume fractions of the matrix.

The curves showed in Figures 3 and 4 manifests a reduction of transverse strain in the matrix whose magnitude of reduction increases at a decreasing rate with increasing volume fraction of the reinforcing fibre, with sample values of 0.1952 at υ_f = 1%, 0.0295 at υ_f = 78%, and 0.0274 at

$\boldsymbol{v_f}$ =91%. The values and trend of the 2nd root shown in Figure 3 do not have any physical meaning that is evident to the authors. Be it in terms of the magnitudes (which wrongly impute a higher elastic modulus for the matrix than the reinforcing fibre) and trend (which implies an increasing reduction of strain with increasing fibre volume fraction which is converse to the expectation) and can therefore be ignored.

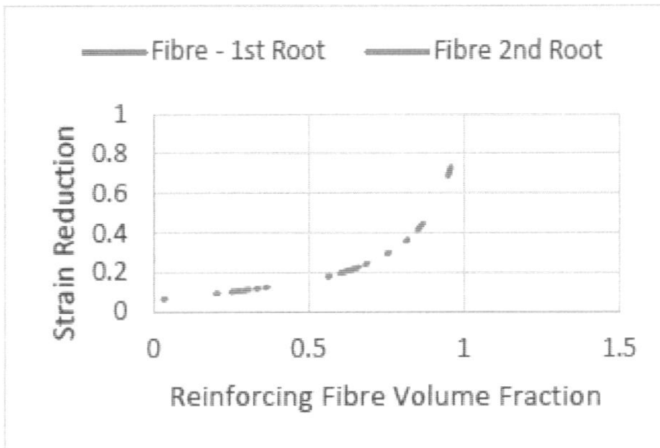

Figure 5. Variation of the reduction of transverse strain in the fibre with reinforcing fibre volume fraction of a square array – 1st and 2nd roots (Equation 36).

Figure 5 shows two superimposed curves that exhibit a strain reduction from 0.0578 at $\boldsymbol{v_f} = 1\%$, 0.3210 at $\boldsymbol{v_f} = 78\%$, and 0.5453 at $\boldsymbol{v_f} = 91\%$ that decreases at an increasing rate with increasing volume of the reinforcing fibre. This is consistent with the fact of reduced matrix content with its attendant higher strains because of its lower stiffness, leading to a transfer of load to the reinforcing fibres with their higher stiffness. The effect of the increase in volume fraction of reinforcing fibres is an increase of strain in the fibres, at a reducing rate. A high strain reduction of 0.321 at $\boldsymbol{v_f} = 91\%$ is seen to prevail near the maximum theoretically possible fibre reinforcing volume fraction of 78.5% for square arrays. This is consistent with the fact that as the matrix content reduces, the response of the composite tends to that of the reinforcing fibres, which because of the higher stiffness of fibres translates to lower values of strain when compared to that of the unreinforced matrix.

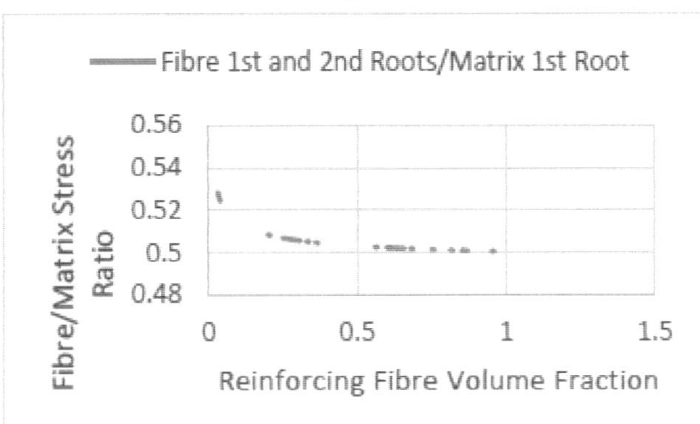

Figure 6. Variation of the fibre 1st and 2nd roots / matrix 1st root transverse stress ratio with reinforcing fibre volume fraction for a square array (Equation 44).

The fibre 1st and 2nd roots/matrix 2nd root transverse stress ratio are seen in Figure 6 to decrease at a decreasing rate with increasing volume fraction of the reinforcing fibre from 0.5484 at ($\boldsymbol{v_f} = 1\%$),

0.5012 at ($\boldsymbol{v_f}$ = 78%) and 0.5007 at ($\boldsymbol{v_f}$ = 91%). The fact that the curve in Figures 6 is not horizontal and equal to unity implies that the stresses in the matrix and reinforcing fibre are different for all volume fractions of reinforcing fibre, which contradicts the assumption of iso-stress.

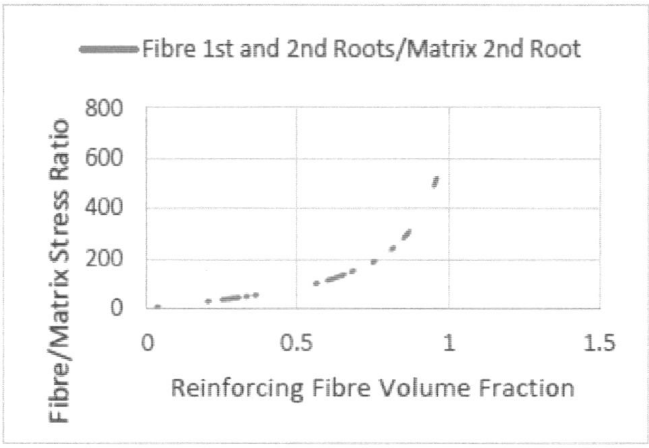

Figure 7. Variation of the fibre 1st and 2nd roots/matrix 2nd root transverse stress ratio with the volume fraction of the reinforcing fibre for a square array (Equation 44).

The curve in Figure 7 should be ignored as it is based on values of the 2nd root of the matrix strain, which was recommended earlier to be ignored because of a lack of physical meaning to justify the values of the root and trend.

Analysis of the curves in Figures 6 & 7 may also be approached from considerations of the linear elastic relationship between stress and strain, the curves in Figures 6 and 7 represent the relationship, $(\sigma_f/\sigma_m) = (\varepsilon_f E_f/\varepsilon_m E_m)$, which may be rewritten as, $(\sigma_f/\sigma_m) = (\varepsilon_f/\varepsilon_m)(E_f/E_m)$. Noting that the values of stiffness of the fibre and matrix are constants, their ratio may be represented by a constant ($\boldsymbol{E_R}$), which then converts the foregoing relationship to, $(\sigma_f/\sigma_m) = \boldsymbol{E_R}(\varepsilon_f/\varepsilon_m)$. It is evident from this relationship that the curve in Figure 6 implies that the values of fibre strain are less than those in the matrix for all volume fractions of reinforcing fibre, and further that the rate of increase of strain in the fibre with increasing volume fraction of the reinforcing is greater than in the matrix. The curve in Figure 7 indicates values of strain and a rate of increase of strain in the fibre that is greater than that in the matrix at all volume fractions of the reinforcing fibre, which is inconsistent with the expected behaviour of the composite given the relative values of stiffness of the matrix and fibre.

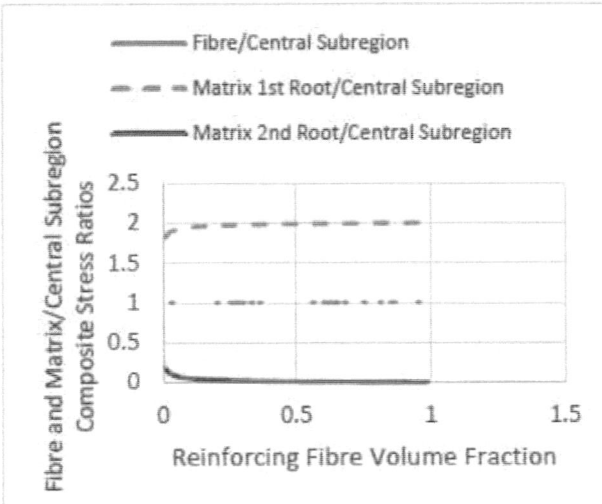

Figure 8. Variation of the fibre and matrix / central sub-region composite transverse stress ratios with the volume fraction of the reinforcing fibre for a square array (Equations 42a & 42b).

That stresses in the matrix and fibre within the central sub-region are different is clear from the three curves in Figure 8. It is evident from the curves in this figure that the ratio of the central sub-region matrix 1st root stress to the central sub-region composite stress is greater than 1.82 for all volume fractions of reinforcing fibre. The ratio is seen in the figure to increase at a decreasing rate with increasing volume fraction of the reinforcing fibre, tending to a ratio of two. The curve for the 2nd root for the matrix may be ignored for reasons noted previously in this work. Moreover, the trend of the curve towards zero with increasing volume fraction of reinforcing fibre would indicate stresses in the matrix and therefore strains that tend to zero as well, which is inconsistent with the prevailing physical situation. Furthermore, the stress ratios that are less than unity, seen in the curve for the 2nd root for the matrix, denote a reduction in stress and therefore strain in the matrix. This is contrary to the known fact of the existence of transverse strain magnification in the matrix. The reinforcing fibre / central sub-region composite transverse stress ratio is seen to be equal to unity for all volume fractions of reinforcing fibre. It is evident from Figure 8 that the stresses in the reinforcing fibre and the matrix within the composite central sub-region are different at all volume fractions of the reinforcing fibre. This is contrary to the assumption of iso-stress made earlier in the paper.

3.1.1. Hexagonal reinforcing fibre arrays. Plots of fibre strain reduction and matrix strain magnification in the central sub-region fibre, with respect to the central sub-region composite represented by the respective parameters a and b. And plots of the ratio (b/a) for stress magnification represented by the ratios shown in Equations 43 (a & b), and 44 for various values of reinforcing fibre volume, are shown in Figures 10-17 for hexagonal reinforcing arrays.

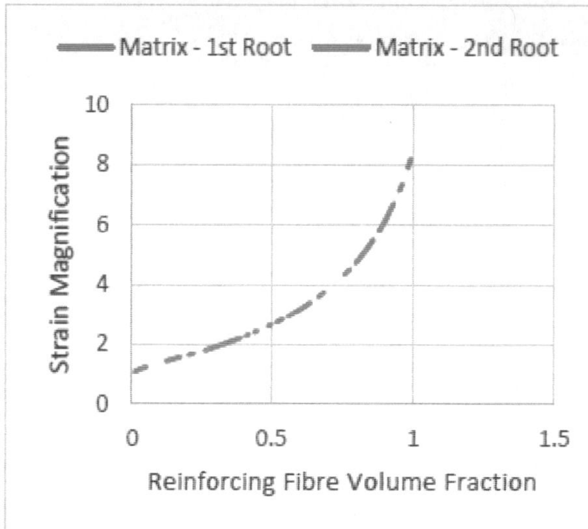

Figure 9. Variation of the magnification of transverse strain in the matrix with reinforcing fibre volume fraction of a hexagonal array – 1st and 2nd roots (Equation 39).

The two roots for Equation 44 in this case are equal therefore the overlaid curves shown in Figure 9. The curves exhibit a magnification of strain in the matrix that increases with increasing volume fraction of the reinforcing fibre from 1.0967 at $\boldsymbol{v_f}$ = 1%, 4.52219 at $\boldsymbol{v_f}$ = 78%, to 6.2999 at $\boldsymbol{v_f}$ = 91%. This increase in magnification is explainable based on "diminishing" matrix as the volume fraction of reinforcing fibre increases, coupled with the higher ductility of the matrix compared to that of the reinforcing fibre.

The curves in Figure 10 and Figure 11 both manifest strain reductions in the reinforcing fibre that decrease at increasing rates with increasing volume fractions of the reinforcing fibre.

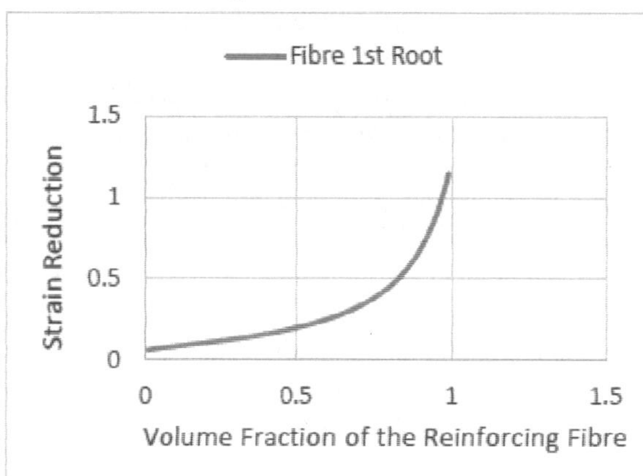

Figure 10. Variation of the reduction of transverse strain in the fibre with reinforcing fibre volume fraction of a hexagonal array – 1st root (Equation 38).

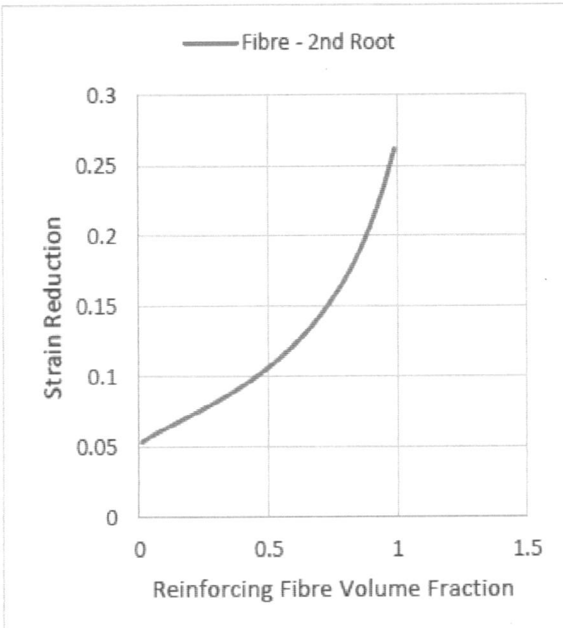

Figure 11. Variation of the reduction transverse strain in the fibre with reinforcing fibre volume fraction of a hexagonal array – 2nd root (Equation 38).

The values of reinforcing fibre strain reduction shown in the curve plotted in Figure 10 for the 1st root vary from 0.0619 at $\boldsymbol{v_f}$ = 1%, 0.4235 at $\boldsymbol{v_f}$ = 78%, and 0.7186 at $\boldsymbol{v_f}$ = 91%. The values of reinforcing fibre strain reduction for the 2nd root that are shown in Figure 11 vary from 0.0535 at $\boldsymbol{v_f}$ = 1%, 0.1643 at $\boldsymbol{v_f}$ = 78%, and 0.2140 at $\boldsymbol{v_f}$ = 91%. The variations in the two curves are consistent with expectations arising from the known relative values of stiffness of the composite constituents and the increasing volume fraction of the reinforcing fibre. The 2nd root predicts higher strain reductions than the 1st root.

The basis for choice between the curves for the 1st and 2nd root shown in Figures 10 – 12 is not evident from the information contained in the figures. It is notable however, that the strain reduction exhibited by the 2nd root of 0.2140 at $\boldsymbol{v_f}$ = 91%, which is near the maximum theoretically possible volume fraction of reinforcing fibre of 90.7% for hexagonal arrays, is considerably much higher than the value of 0.7186 at $\boldsymbol{v_f}$ = 91% for the 1st root. This is significant in light of the fact that the strain in the composite is expected to converge to the value for the reinforcing fibre as the content of matrix diminishes, which is expected to be lower than that of the unreinforced matrix because of the higher stiffness for the reinforcing fibre. This argument tips the balance of choice to the 2nd root.

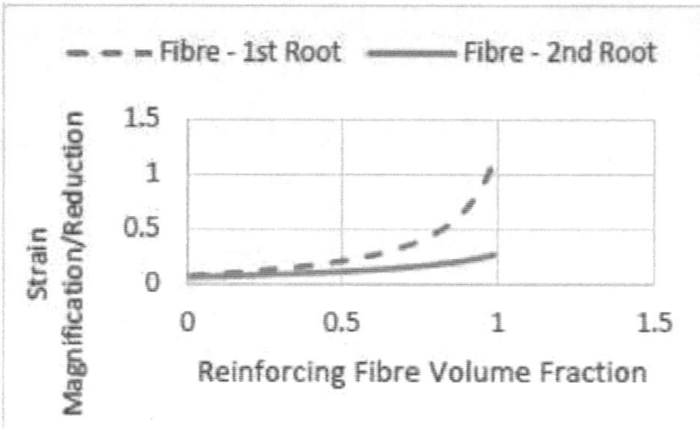

Figure 12. Variation of the magnification / reduction of transverse strain in the fibre with reinforcing fibre volume fraction of a hexagonal array – 1st and 2nd roots (Equation 38).

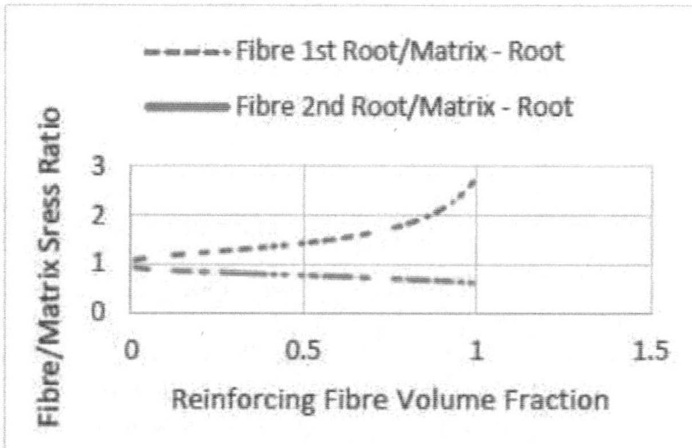

Figure 13. Variation of the fibre 1st and 2nd roots / matrix root transverse stress ratio of a hexagonal array with the volume fraction of the reinforcing fibre (Equation 44).

The curves shown in Figure 13 are both different in magnitudes and trends. The curve for the 1st root show values of fibre/matrix stress ratio that are greater than unity for all volume fractions of reinforcing fibre, while those for the 2nd root show values that are less than unity over the same range. While the curve for the 1st root exhibits a fibre/matrix stress ratio that increases at an increasing rate with increasing volume fraction of reinforcing fibre, the curve for the 2nd root shows a ratio that decreases at a more or less constant rate. From the previously noted relationship, $(\sigma_f/\sigma_m) = \boldsymbol{E_R}(\varepsilon_f/\varepsilon_m)$, it is evident that a higher fibre/matrix stress ratio is indicative of higher fibre/matrix strain ratio, which is contrary to the prevailing physical state in the transverse direction of a fibre reinforced composite. Moreover, an increase in the fibre/matrix stress ratio with increasing volume fraction of the reinforcing fibre is contrary to the expected trend as the fibre content increase, given they are stiffer than the matrix. These considerations move the balance of choice between the two curves for fibre strain reduction to the curve for the 2nd root, a choice that ties in with the one made purely on considerations of transverse strain magnification. The fact that none of the two curves is equal to unity at any point implies that the iso-stress state does not exist at any volume fraction of the reinforcing fibre.

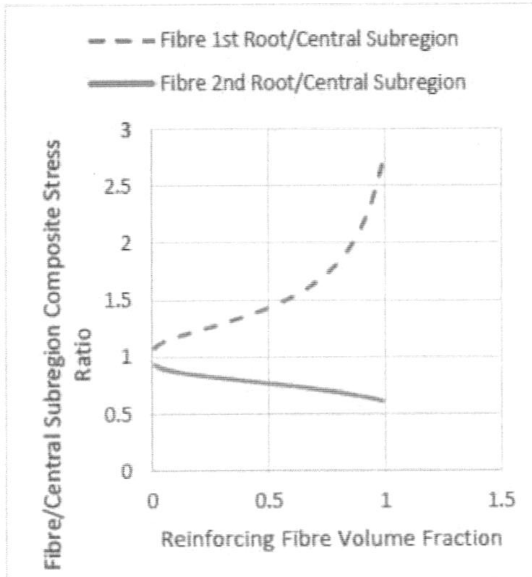

Figure 14. Variation of the fibre / central sub-region composite transverse stress ratio of a hexagonal array with the volume fraction of the reinforcing fibre (Equation 43a).

For the same reasons given in the analysis of the curves shown in Figure 13, the choice of curves in Figure 14 is for the 2nd root. The stresses in the reinforcing fibre are seen in Figure 14 to be less than those in the central sub-region composite and to reduce with increasing volume fraction of the reinforcing fibre. The fact that none of the two curves in this figure is equal to unity implies that the iso-stress state does not exist at any volume fraction of the reinforcing fibres.

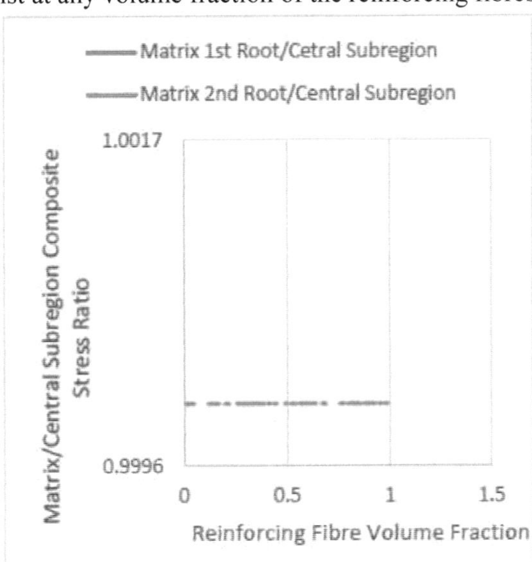

Figure 15. Variation of the matrix / central sub-region composite transverse stress ratio of a hexagonal array with the volume fraction of the reinforcing fibre (Equation 43b).

The curve shown in Figure 15 depicts equality of the stress in the central sub-region composite and matrix at all volume fractions of the reinforcing fibre. This implies a direct, constant relationship between the product of the transverse strain and stiffness of the central sub-region composite and the transverse strain and stiffness of the matrix, since for, ($\boldsymbol{\sigma_{c2} = k\sigma_m}$), then($\boldsymbol{\varepsilon_{c2}E_{c2} = k\varepsilon_m E_m}$).

It is noted hear that whilst the stiffness of the matrix remains constant, composite theory shows that the stiffness of the central sub-region composite increases with increasing volume fraction of reinforcing fibre. The foregoing equality therefore implies that the transverse strain of the central sub-region composite increases at a lower rate than that of the surrounding matrix.

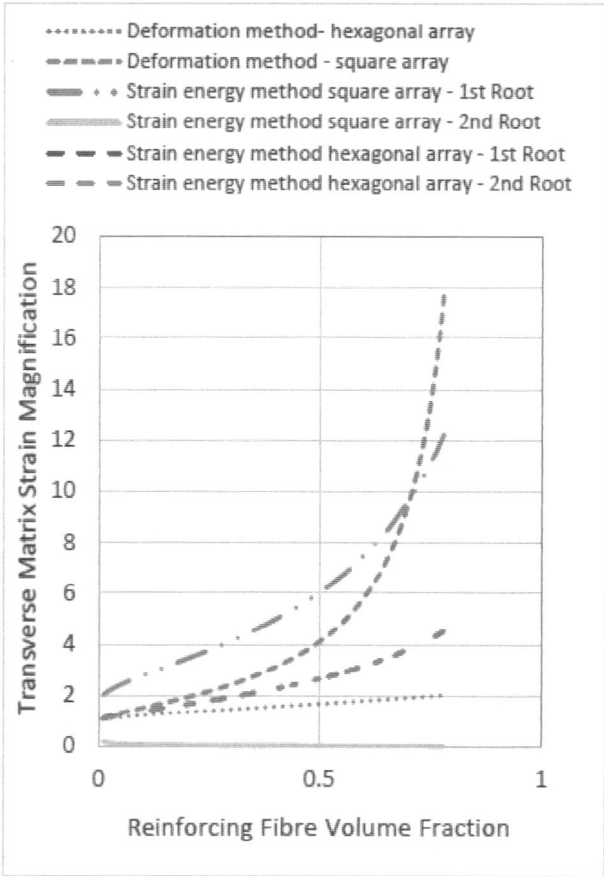

Figure 16. Comparison of the variation of matrix transverse strain magnification based on the displacement and strain energy methods for both square and hexagonal arrays of reinforcing fibre.

The values of matrix transverse strain magnification for the 1st and 2nd roots of the strain energy method for the hexagonal arrays are the same and their curves in Figure 16 are therefore coincident. With an exception of the curve for the 2nd root of the square array which shows a decreasing trend with increasing volume fraction of the reinforcing fibre (whose values were recommended earlier in this paper to be ignored), all other curves in Figure 16 exhibit increasing values of matrix transverse strain magnification with increasing volume fraction of the reinforcing fibre. With the same exception, all the curves tend to converge and diverge with decreasing and increasing volume fraction of the reinforcing fibre, respectively. The curves for the deformation method [17] applied to square and hexagonal arrays and the curves for the two roots of the strain energy method applied to hexagonal arrays are closer to one another in magnitude when compared with the other two curves, up to a volume fraction of reinforcing fibre of about 40%. The lowest values of matrix transverse strain magnification for all volume fractions of the reinforcing fibre, with the exception of the curve noted here, occur for the deformation method [17] applied to the hexagonal array. The highest values of matrix transverse strain magnification on the other hand occur for the 1st root of the strain energy method applied to the square array up until a volume fraction of reinforcing fibre just above 70%, after which the deformation method [17] applied to the square array shows the highest values. For each one of the two types of fibre geometries the values of matrix transverse strain magnification are lower for the deformation method [15] than for the strain energy method, over most of the reinforcing the fibre volume fraction.

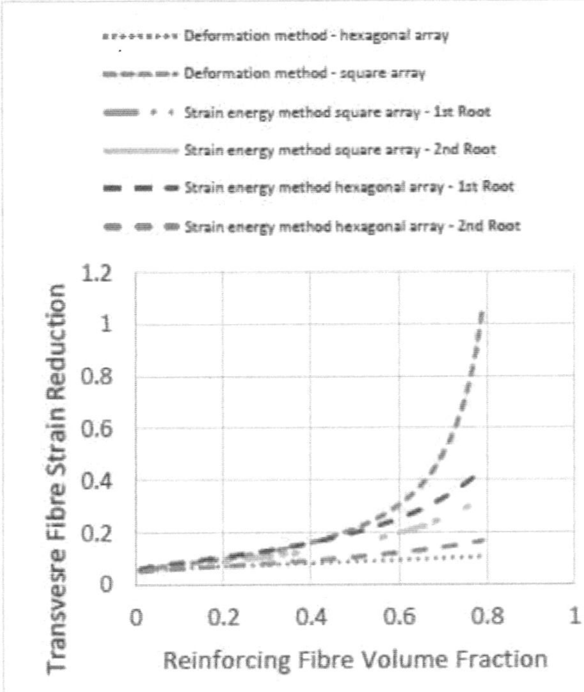

Figure 17. Comparison of the variation of transverse fibre strain reduction based on the displacement and strain energy methods for both square and hexagonal arrays of reinforcing fibre.

The values transverse fibre strain reduction for the 1st and 2nd roots of the strain energy method for the square array are the same and their curves in Figure 17 are therefore coincident. The curves all exhibit strain reductions that decrease with increasing volume fraction of the reinforcing fibre. It is noted from the figure that, the curve for the 1st root of the strain energy method applied to hexagonal arrays coincides with that of the deformation method [17], applied to square arrays, up to a volume fraction of the reinforcing fibre of about 42%. It is recalled that the balance of choice between the 1st and 2nd root of the strain energy method applied to hexagonal arrays was argued in the second paragraph below Figure 11 to tip towards the 2nd root. The curve for the 2nd root of the strain energy method applied to hexagonal arrays coincides with that of the deformation method [17] applied to the hexagonal array up to a volume fraction of the reinforcing fibre of about 20%. The curves all tend to converge and diverge with reducing and increasing volume fraction of the reinforcing fibre, respectively. The curves for the square arrays are lower for the strain energy method while those for the hexagonal arrays are lower for the deformation method [17].

4. Conclusions

It can be concluded from the foregoing work that in the transverse direction of uniaxially aligned fibre reinforced composites:

- The matrix lying in between the reinforcing fibres experiences transverse strain magnification whose magnitude increases non-linearly (at an increasing rate) with the increasing volume fraction of reinforcing fibres.

- The matrix transverse strain magnification for the square arrays is greater than that for the hexagonal arrays based on the strain energy method, as was the case in the method of direct deformation.
- The reinforcing fibres experience strain reduction whose magnitude decreases non-linearly (at an increasing rate) with the increasing volume fraction of reinforcing fibres.
- The fibre/matrix stress ratio is less than unity and decreases with increasing volume fraction of the reinforcing fibres for both reinforcing fibre geometries, being higher for hexagonal than for square arrays of reinforcing fibre.
- The curves for the 2nd root of matrix transverse strain magnification for the square arrays of reinforcing fibre and the 1st root of fibre strain reduction for the hexagonal arrays of reinforcing fibre should be investigated further, in order to determine the reasons for their inconsistency with expected trends and limits, respectively.
- The approach of direct deformation generates respective solutions of matrix and fibre transverse strain magnification and reduction that are very similar in magnitude to those arising from the more rigorous strain energy methods referred to the RVE.
- In the transverse direction, the respective strain reduction and magnification of the reinforcing fibre and matrix with respect to the central sub-region are equal to the respective strain reduction and magnification of the reinforcing fibre and matrix with respect to the matrix in between the reinforcing fibres in the longitudinal direction.
- This strain reduction is in turn is equal to the strain in the overall composite that does not experience strain magnification; the matrix sub-regions (A) in Figure 1c.
- The assumption of iso-stress (the Reuss rule) is shown not to apply for the fibre and matrix, fibre and central sub-region composite or matrix and central sub-region composite.

5. Recommendations

- Solutions should be sought that are independent of the iso-stress rule in order to quantify the magnitude of errors arising from use of the Reuss rule in the present work based on the strain energy method and that of the deformation method.
- Expressions should be developed for the solutions presented here with reference to the (s/r) ratio in order to facilitate comparison of these solutions with others in literature that have the same reference.

References

[1] Wacker G, Blezdki A K and Chate A 1998 Effect of Interface on the Transverse Young's Modulus of Glass/Epoxy Composites *Composites Part A* **29A** 619-26
[2] Hull D 1990 *An Introduction to Composite Materials* (Cambridge: Cambridge University Press)
[3] Gibson R F 1994 *Principles of Composite Material Mechanics* (New York: McGraw-Hill Inc)
[4] Harris B 1999 *Engineering Composite Materials* (London: The Institute of Materials)
[5] Younes R, Hallal A, Fardoun F and Chehade F H 2012 *Composites and Their Properties* ed N Hu (Vienna: InTech) Chapter Composites and their Properties – Comparative Review Study on Elastic Properties Modeling for Unidirectional Composite Materials
[6] Banerjee S and Sankar G V 2012 Mechanical Properties of Hybrid Composites Using Finite Element Method Based Micro Mechanics *27th Annual Technical Conference of the American Society for Composites(Arlington*) p 153
[7] Aghdam M M 2001 Stress Interactions in Unidirectional Fibre Reinforced Composites Subjected to Off-axis Loading *Proc. 13th Int. Conf. on Composite Materials (Beijing)* Paper code 1573
[8] Benzarti K, Cangemi L nd Dal M F 2001 Transverse properties of unidirectional glass/epoxy composites: influence of fibre surface treatments *Composites: Part A* **32** 197-206
[9] Huang Z M and Xin L M 2014 Stress Concentration Factors of Matrix in a Composite Subjected to Transverse Loads *ICCM2014 (Cambridge)*

[10] Phifer S P 1998 Quasi-Static and Fatigue Evaluation of Pultruded Vinyl Ester/E-Glass Composites *MSc.-Eng. Sci. Thesis* (USA: Virginia Polytechnic Institute and State University)
[11] Gitman I M, Askes H, Sluys L J and Valls O L 2004 The concept of Representative Volume for elastic, hardening and softening materials *Proc. XXXII Int. Summer School–Conference Advanced Problems in Mechanics* pp 180-4
[12] Gitman I M, Askes H and Sluys L J 2007 Representative volume: Existence and size determination *Engineering Fracture Mechanics* **74** 2518-34,
[13] Kanit T, Forest S, Galliet I, Mounoury V and Jeulin D 2003 Determination of the size of the representative volume element for random composites: statistical and numerical approach *Int. J. Solids and Structures* **40** 3647-79
[14] Sun and Vaidya R S 1996 Prediction of Composite Properties from a Representative Volume Element Compos. Sci. Technol. 56 171-9
[15] Pelissou C, Baccou J, Monerie Y and Perales F 2009 Determination of the size of the representative volume element for random quasi-brittle composites *Int. J. Solids and Structures*, **46** 2842-55
[16] Drugan W F and Wills J R 1996 A Micromechanics-Based Nonlocal Constitutive Equation and Estimates of Representative Volume Element Size for Elasti composites *J. Mech. Phys. Solids* **44** 497-524
[17] Maringa M and Masu L M 2016 The effects of different fibre packing geometries on the transverse matrix strain magnification and fibre strain reduction in uniaxially aligned continuous fibre-reinforced composites *J. Compos. Mater.* DOI: 10.1177/0021998316631701

IOP Conf. Series: Materials Science and Engineering **213** (2017) 012005 doi:10.1088/1757-899X/213/1/012005

Research of spin-orbit interaction in organic conjugated polymers

H Li[1,2], M Y Zhou[1], S Y Wu[1] and X R Liang[1]

[1]Naval Aeronautical and Astronautical University Department of Basic Sciences, Yantai, Shandong 264001, P.R.China

E-mail: 346521161@qq.com

Abstract. The effect of spin-orbit interaction on the one-dimensional organic polymer was investigated theoretically. Spin-orbital interaction led to the spatial separation of energy band but did not eliminate spin degeneration, which was different from energy level splitting in the Zeeman Effect. Spin-orbit interaction had little effect on the energy band structure, charge density, and lattice position, etc.; Spin precession was obtained when a polaron was transported along the polymer chain, which theoretically proved that it was feasible to control the spin precession of polaron in organic polymers by the use of external electric field.

1. Introduction

In 1990, Datta and Das proposed the idea of FET field-effect transistor, which controlled the electron spin precession effectively through the Rashba spin-orbit interaction [1]. Since then, the spin-orbit interaction had become the hot research topic in the semiconductor spin electronics [2-11]. The strength of the effect of Rashba has been realized by using an applied electric field in the experiment. The energy band structure, electronic wave function, spin angular velocity and the linear velocity, electric conductivity of the quasi one-dimensional electron gas system can all be influenced by the Rashba spin-orbit interaction.

At present, the research on organic spin electronics has just started [12-16]. In theory, some new phenomena related to spin are predicted, and the experimental results also confirmed that the spin polarized current could exist in the organic body. In organic semiconductors there are few previous studies on the spin-orbit interaction, because it is generally believed that the spin-orbit interaction in organic semiconductors is slight, but its impact is inevitable. This paper will discuss the change of one-dimensional organic polymers under spin-orbit interaction.

2. Spin-orbit coupling effect on the electronic structure of organic conjugated polymers

The system which we are studying is the one-dimensional organic polymers with non-degenerate ground state. Assuming that one-dimensional organic polymer chain points to the x axis, we add an electric field (Rashba electric field) in the z axis, Rashba electric field is used to regulate the spin-orbit interaction, and the Hamiltonian of spin-orbit coupling effect is expressed as:

$$H_{SO} = i\gamma\left(\sigma_x \frac{\partial}{\partial y} - \sigma_y \frac{\partial}{\partial x}\right) \quad (1)$$ [17-18].

IOP Conf. Series: Materials Science and Engineering **213** (2017) 012005 doi:10.1088/1757-899X/213/1/012005

In the equation (1) we only retain the second term in the brackets, because the one-dimensional organic polymer chains points to the x direction. In which $\gamma = |e|\hbar^2 E_{SO} / 4m_e^2 c^2$ [19], is a constant representing spin-orbit interaction. c, e, m_e are the speed of light, the electronic quantity and the electron quality respectively. E_{SO} describes the quantities of Rashba electric field, and σ_y is the Pauli matrix.

We can get the one-dimensional tight binding model through the second quantization:

$$H_{SO} = -t_{so}\sum_n \left[C^+_{n+1,\uparrow}C_{n,\downarrow} - C^+_{n+1,\downarrow}C_{n,\uparrow} + C^+_{n,\downarrow}C_{n+1,\uparrow} - C^+_{n,\uparrow}C_{n+1,\downarrow} \right] \tag{2}$$

in which $t_{SO} = \gamma / 2a$, t_{SO} is the quantities of the Rashba spin-orbit coupling, which describes the transition integral of electrons in adjacent grid points, and a is the lattice constant. $C^+_{n,s}\left(C_{n,s}\right)$ is the generation (disappearance) operator of an electron at a lattice point n.

The Hamiltonian of electronic partion includes the polymer that is represented by the extended one-dimensional tight binding model and the spin-orbit coupling portion represented by one-dimensional tight binding model.

$$H_e = H_{SSH} + H_{so} \tag{3}$$

$$H_{SSH} = -\sum_{n,s}\left[t_0 - \alpha(u_{n+1} - u_n) - (-1)^n t_e\right]\left(C^+_{n+1,s}C_{n,s} + C^+_{n,s}C_{n+1,s}\right) \tag{4}$$

$$H_{SO} = -t_{so}\sum_n \left[C^+_{n+1,\uparrow}C_{n,\downarrow} - C^+_{n+1,\downarrow}C_{n,\uparrow} + C^+_{n,\downarrow}C_{n+1,\uparrow} - C^+_{n,\uparrow}C_{n+1,\downarrow} \right] \tag{5}$$

in which α is the electron-phonon coupling constant, t_e is degenerate parameter, t_0 is the transition integral of electrons in adjacent grid points n and $n+1$, u_n is the position deviation of the n lattice point(CH group).

The electronic energy spectrum can be obtained by Fourier transform:

$$E(k) = \pm 2t_{SO}\sin(ka) + E_0(k) \tag{6}$$

$$E_0(k) = \left[(-2t_0\cos(ka))^2 + (4\alpha u \sin(ka))^2\right]^{\frac{1}{2}} \tag{7}$$

Rashba spin-orbit interaction leads the spatial separation between "y^+" (along the $\bar{y}$ axis) and "y^-" (along the $-\bar{y}$ axis) spinor components (One-dimensional organic polymer points to the x direction, and Rashba electric field is along the z direct).The Zeeman effect can also lead to the spatial separation, but the spatial separation which is induced by Rashba spin-orbit coupling is different from the energy splitting caused by the Zeeman effect, because the spin merger is not eliminated. Spin-orbit interaction induces wider energy band of organic conjugated polymer and narrower energy gap, but the change is not very obvious as a result of the weak spin-obit coupling in the organic polymer. The lattice configuration and the charge density of the carriers in organic polymer such as polarons and bipolarons, do not change significantly.

3. Spin-orbit coupling effect on the polaron transport in organic polymer

Spin-orbit interaction will produce when the electrons are moving in the influence of asymmetric potential, which is closely connected to the charge motion. The physical image of the transport dynamics of organic polymers is needed to be studied so as to understand the effect of spin-orbit coupling deeply. In this paper, we study the effect of spin-orbit interaction on polaron (carrying the single charge and 1/2 spin) transport in one dimensional organic polymer.

The electric field is added in the direction parallel to the one dimensional organic polymer chain, which drives the movement of polaron along the chain direction. In order to study long time movement of the polaron, we adopt the periodic boundary condition.

The Hamiltonian of organic semiconductors includes two parts lattice part and electronic part.

$$H = H_e + H_{lattice} \tag{8}$$

The electronic part of the Hamiltonian is expressed as the following form because of the existence of the driven electric field:

$$H_e = H_{SSH} + H_{so} \tag{9}$$

$$H_{SSH} = -\sum_{n,s}\left[t_0 - \alpha\left(u_{n+1} - u_n\right) - (-1)^n t_e\right]\left(e^{-i\gamma A}C^+_{n+1,s}C_{n,s} + e^{i\gamma A}C^+_{n,s}C_{n+1,s}\right) \tag{10}$$

$$H_{SO} = -t_{so}\sum_n\left[e^{-i\gamma A}C^+_{n+1,\uparrow}C_{n,\downarrow} - e^{-i\gamma A}C^+_{n+1,\downarrow}C_{n,\uparrow} + e^{i\gamma A}C^+_{n,\downarrow}C_{n+1,\uparrow} - e^{i\gamma A}C^+_{n,\uparrow}C_{n+1,\downarrow}\right] \tag{11}$$

Driving electric field is introduced in the form of the index terms of transition integrals, in which $\gamma = \frac{ea}{\hbar c}$, a, e, c are the lattice constant, the electcronic quantity and the speed of light respectively. A is vector-potential, which is a scalar in the one dimensional system, and $A = A(t)$. The driven electric field is $E(t) = -\partial_t A(t)/c$.

The Hamiltonian of the lattice is expressed in a classical way:

$$H_{lattice} = \frac{K}{2}\sum_n\left(u_{n+1} - u_n\right)^2 + \frac{M}{2}\sum_n \dot{u}_n^2 \tag{12}$$

in which M is the effective quality of the lattice point , and K represents the elastic coefficient of crystal lattice.

Electron wave function $\left|\Phi_\mu(t)\right\rangle = \begin{pmatrix}\Phi_{\mu,\uparrow}(t)\\ \Phi_{\mu,\downarrow}(t)\end{pmatrix} = \begin{pmatrix}\sum_n \varphi_{\mu,\uparrow}(n,t)\left|n\right\rangle\\ \sum_n \varphi_{\mu,\downarrow}(n,t)\left|n\right\rangle\end{pmatrix}$ satisfies the time-dependent Schrodinger equation:

$$i\hbar\frac{\partial\Phi_\mu(t)}{\partial t} = H_e\Phi_\mu(t) = \left(H_{SO} + H_{SSH}\right)\Phi_\mu(t) \tag{13}$$

We can get:

$$\begin{aligned}i\hbar\dot{\varphi}_{\mu,\uparrow}(n,t) = &-t_{n,n+1}e^{i\gamma A}\varphi_{\mu,\uparrow}(n+1,t) - t_{n-1,n}e^{-i\gamma A}\varphi_{\mu,\uparrow}(n-1,t)\\ &+ t_{SO}\left(e^{-i\gamma A}\varphi_{\mu,\downarrow}(n-1,t) - e^{i\gamma A}\varphi_{\mu,\downarrow}(n+1,t)\right)\end{aligned} \tag{14a}$$

$$i\hbar\dot{\varphi}_{\mu,\downarrow}(n,t)=-t_{n,n+1}e^{i\gamma A}\varphi_{\mu,\downarrow}(n+1,t)-t_{n-1,n}e^{-i\gamma A}\varphi_{\mu,\downarrow}(n-1,t) \\ -t_{SO}(e^{-i\gamma A}\varphi_{\mu,\uparrow}(n-1,t)-e^{i\gamma A}\varphi_{\mu,\uparrow}(n+1,t)) \quad (14b)$$

The evolution of the lattice shape of the polymer $\{u_n\}$ is determined by the classical equation of motion:

$$M\ddot{u}_n(t)=-K[2u_n(t)-u_{n+1}(t)-u_{n-1}(t)]+\alpha[e^{i\gamma A}(\rho_{n,n+1}(t)-\rho_{n-1,n}(t))+c.c], \quad (15)$$

in which density matrix elements in the Equation (15) are determined by the following two equation:

$$\rho^{s}_{n,n'}(t)=\sum_{\mu}\psi^{*}_{\mu,s}(n,t)f_{\mu}\psi_{\mu,s}(n',t) \quad (16)$$

$$\rho_{n,n'}(t)=\rho^{s}_{n,n'}(t)+\rho^{-s}_{n,n'}(t) \quad (17)$$

in which f_μ is the time independent distribution function(Fermi distribution function), which is determined by the occupied state (0 or 1) of the initial moment.

The electron wave function of initial moment $|\Phi_\mu(t=0)\rangle=\begin{pmatrix}\Phi_{\mu,\uparrow}(t=0)\\ \Phi_{\mu,\downarrow}(t=0)\end{pmatrix}=\begin{pmatrix}\sum_n\varphi_{\mu,\uparrow}(n,t=0)|n\rangle\\ \sum_n\varphi_{\mu,\downarrow}(n,t=0)|n\rangle\end{pmatrix}$ is obtained by solving the Schrodinger equation in the moment t=0:

$$H_e(t=0)\Phi_\mu(t=0)=\varepsilon_\mu(t=0)\Phi_\mu(t=0) \quad (18)$$

We can get:

$$\varepsilon_\mu\varphi_{\mu,\uparrow}(t=0,n)=-t_{n-1,n}\varphi_{\mu,\uparrow}(t=0,n-1)-t_{n,n+1}\varphi_{\mu,\uparrow}(t=0,n+1) \\ +t_{SO}(\varphi_{\mu,\downarrow}(t=0,n-1)-\varphi_{\mu,\downarrow}(t=0,n+1)) \quad (19a)$$

$$\varepsilon_\mu\varphi_{\mu,\downarrow}(t=0,n)=-t_{n-1,n}\varphi_{\mu,\downarrow}(t=0,n-1)-t_{n,n+1}\varphi_{\mu,\downarrow}(t=0,n+1) \\ -t_{SO}(\varphi_{\mu,\uparrow}(t=0,n-1)-\varphi_{\mu,\uparrow}(t=0,n+1)) \quad (19b)$$

Before the addition of driving electric field in the one dimensional organic polymer, the most stable position of the system (the static configuration of a polaron) is obtained by the energy optimization:

$$u_n=\frac{1}{2}\left[u_{n-1}+u_{n+1}+\frac{2\alpha}{K}\sum_{\mu,s}{}'(\psi_{\mu,s}(n+1,t=0)\psi_{\mu,s}(n,t=0)-\psi_{\mu,s}(n,t=0)\psi_{\mu,s}(n-1,t=0))\right] \quad (20)$$

The static configuration of a polaron is the initial configuration of the evolution of lattice dynamics, in which $\sum_{\mu,s}{}'$ describes the sum of the occupied states of the electron. Self-consistent Equation (14) and (15) are obtained through 8 order Runge-Kutta method [20].

One dimensional organic polymer chain is composed of 100 CH groups (N=100). Preferences [21]: $a=1.22\mathring{A}$, $\alpha=4.1\text{eV}/\mathring{A}$, $M=1349.14\,\text{eV}\cdot\text{fs}^2/\mathring{A}^2$, $t_0=2.5\text{eV}$, $K=21\text{eV}/\mathring{A}^2$, $t_e=0.05\text{eV}$, $t_{so}=0.02\text{eV}$ (This value is a small amount compared to the other interactions).

According to the linear distribution in the calculation the driven electric field is applied, i.e. when time $0<t<t_c, E(t)=(E_0/t_c)\cdot t$; when $t\geq t_c$, $E(t)=E_0$, in which t_C is the starting period of electric field, $t_c=30\text{fs}$. Suppose that the driven electric field points to the $-\vec{x}$ direction, then polaron will move along the $\vec{x}$ direction. The driven electric field $E_0=5\times10^3$ V/cm , which is far less than polaron dissociation field. The driven electric field will reach the maximum value when $t=t_C=30\,fs$.

If spin-orbit interaction is not considered, i.e. the Rashba electric field is not added, when $t_{SO}=0$, the polaron spin will remain constant, which is moving along the x direction under the effect of the driven electric field. Spin evolution $S_n^Z(t)=\frac{1}{2}\left(\rho_{n,n}^{\uparrow}(t)-\rho_{n,n}^{\downarrow}(t)\right)$ is shown in the Figure 1.

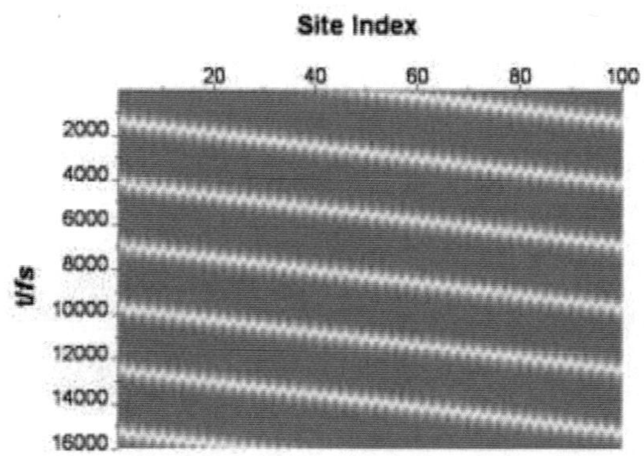

Figure 1. The evolution of spin polaron without spin-orbit interaction.

When the Rashba electric field is added ($t_{SO}=0.02$), we find spin evolution has changed with the polaron motion, as shown in the Figure 2. The black part is the projection of the spin down (along $-\vec{z}$ axis direction) and the white part is the projection of the spin up (along $\vec{z}$ axis direction) in the Figure 2. Obviously, the spin of polaron is precessing. As time goes on, the spin projection in the $\vec{z}$ axis direction gradually becomes dark. At the time $t\approx5000\,fs$, the spin projection has been completely black, which indicates that the spin projection of the $\vec{z}$ axis is downward, soon after the spin projection slowly turns to light. When the driven electric field remains unchanged, the movement velocity of the polaron remains unchanged, which is irrelevant to the Rashba electric field (spin-orbit coupling).

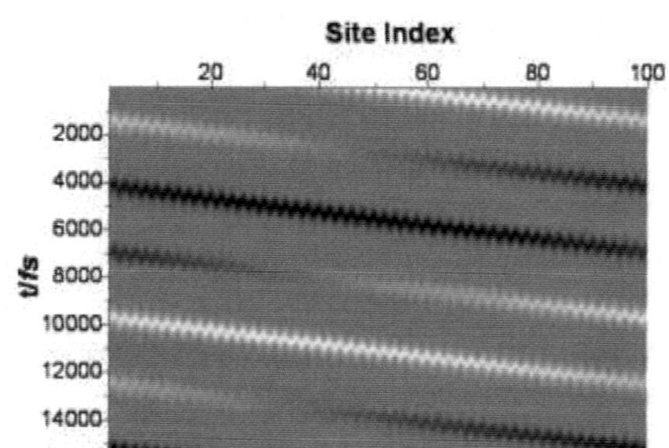

Figure 2. The evolution of spin polaron with spin-orbit interaction $(t_{SO}=0.02\text{eV})$.

4. Conclusion

The effect of spin-orbit coupling on one dimensional organic polymer is calculated theoretically. Spin-orbit coupling makes the energy splitting of organic polymers split, but does not erase the spin merger. Due to the weak spin-orbit coupling in organic matter, the band structure of the whole system does not change significantly. When polaron is moving along the one dimensional organic polymer chain under the impact of driven electric field, the spin-orbit coupling interaction can lead to the spin precession. When the driven electric field remains unchanged, the movement velocity of the polaron remains unchanged, and it has nothing to do with the Rashba electric field (t_{SO}). This paper theoretically proves that it is feasible to control the spin precession of polaron in organic polymers by the use of external electric field.

References

[1] Datta S and Das B 1990 *App. Phys. Lett.* **56** 665
[2] Jatiyanon K and Soodchomshom B 2016 *Physica. E.* **10** 1016
[3] Faizabadi E and Molavi M 2016 *Current Applied Physics* **11** 6
[4] Khordad R 2014 *Journal of Computational Electronics* **2** 13
[5] Karaaslan Y, Gisi B, Sakiroglu S, Kasapoglu E, Sari H and Sokmen I 2016 *Superlattices and Microstructures* **2** 43
[6] Eslami L, Faizabadi E and Ahmadi S 2016 *Physics. Lett. A* **9** 22
[7] Heremans J J, Kallaher R L, Rudolph M and Santos M B 2015 *Integrated Ferroelectrics* **166** 10
[8] Derakhshan V and Ketabi S A 2016 *Physica E* **9** 3
[9] Boda A, Boyacioglu B , Erkaslan U and Chatterjee A 2016 *Physica B* **6** 12
[10] Bai L, Zhang R and Duan C L 2012 *Nanoscale Research Letters* **7** 1
[11] Sakr M R 2016 *Physics. Lett.* A **7** 29
[12] González D M, Körstgens V , Yao Y, Song L, Santoro G, Roth S V and Buschbaum P M 2015 *Adv. Energy. Mater.* **5** 8
[13] Zhang X M, Shigemi M, Takahide K, Ma Q L, Mikihiko O, Hiroshi N, Yasuo A and Terunobu M 2013 *Nature Communication* **4** 1392
[14] Yoo J W, Chen C Y, Jang H W, Bark C W, Prigodin V N, Eom C B and Epstein A J 2010 *Nature Materials* **9** 638
[15] Chen G P 2010 *Electronics Newsweekly* **11** 56
[16] Stefano S 2008 *Nature Nanotechnology* **2** 204
[17] Rashba E I 1960 *Sov. Phys. Solid State* **2** 1109
[18] Bychkov Y A and Rashba E I 1984 *JETP. Lett.* **39** 78
[19] Lee C H 2005 *Phys. Rev.* B **72** 245-338
[20] Heeger A J, Kivelson S, Schrieffer J R and Su W P 1988 *Rev. Mod. Phys* **60** 781
[21] Brankin R W, Gladwell I and Shampine L F 2001 *RKSUITE: Software for ODE IVPS* http://www.netlib.org

IOP Conf. Series: Materials Science and Engineering **213** (2017) 012006 doi:10.1088/1757-899X/213/1/012006

Blasting methods for heterogeneous rocks in hillside open-pit mines with high and steep slopes

Y J Chen[1], Z G Chang[1,3], X H Chao[1] and J F Zhao[2]

[1]Department of Mining, Xinjiang Institute of Engineering, Urumqi, 830000, China
[2]Xinjiang Hong Shun Mining Investment Co. Ltd. Urumqi, 830054, China

E-mail:changzg@cumt.edu.cn

Abstract. In the arid desert areas in Xinjiang, most limestone quarries are hillside open-pit mines (OPMs) where the limestone is hard, heterogeneous, and fractured, and can be easily broken into large blocks by blasting. This study tried to find effective technical methods for blasting heterogeneous rocks in such quarries based on an investigation into existing problems encountered in actual mining at Hongshun Limestone Quarry in Xinjiang. This study provided blasting schemes for hillside OPMs with different heights and slopes. These schemes involve the use of vertical deep holes, oblique shallow holes, and downslope hole-by-hole sublevel or simultaneous detonation techniques. In each bench, the detonations of holes in a detonation unit occur at intervals of 25-50 milliseconds. The research findings can offer technical guidance on how to blast heterogeneous rocks in hillside limestone quarries.

1. Background

Hongshun Limestone Quarry is a hillside quarry in Dabancheng District, Urumqi City. This hill has an average height of 58 m and a slope of approximately 45°. This quarry is situated in a windy, rainless desert region in inland China, where the highest temperature in summer can reach above 40 ℃. Due to the influence of structural planes, the surface 6-9 m limestone in this hill is heavily weathered and eroded, contains a lot of joints and fissures, and can be easily broken into large blocks by blasting. It is feasible to characterize the structure of the heterogeneous rock mass by studying the rock surfaces exposed by blasting and the current status of this quarry.

2. Analysis of factors affecting the effectiveness of blasting

The hillside limestone quarries in Dabancheng District are mostly small quarries. Field survey and analysis of these quarries show that limestone is locally exposed at the surface of hills and the surface limestone can be easily broken into large blocks by blasting, which are a potential safety hazard.

2.1. A subsection

Small-scale millisecond delay blasting with blast holes arranged in lines is commonly used in hillside OPMs. The effect of blasting between rows is apparently directional. When the fracture orientations are parallel to the directions of blasting between rows, the blast wave and gases generated by blasting can cause further dislocation and expansion of primary fractures along the free surfaces of bench [1]. Consequently, the rock mass is split into several large blocks.

2.2. Effect of inclined bedding planes

Previous experiences in blasting demonstrate that during blasting of rock masses with non-structural fractures, the fractures created by the dynamic stress resulting from blasting vary in density, orientations, and length. The stratiform structural planes in this quarry are dominated by inclined bedding planes [2]. The non-structural fractures mostly extend along the dip direction of the hill. The presence of these structural planes reduces the rock's fracture strength. The superhigh pressure gases produced by blasting can enter the joints and fissures in rock masses and pull fracture surfaces apart. The surrounding rock masses, squeezed by the fracture surfaces, tend to break into large blocks and be thrown towards the inclined free surface [3]. The inclined bedding planes contribute to the widely scattered distribution of the rock blocks all over the inclined free surface.

3. Blasting methods

The study suggests that blasting methods should be determined based on hill slope and the mode and parameters of blasting operation. In order to ensure safe blasting, downslope bench mining from hill top to bottom is recommended.

3.1. A blasting scheme for a hillside OPM with a height greater than 50 m and a dip larger than 45°

For high and steeply sloping hillside OPMs, millisecond delay blasting or chamber blasting with concentrated charge is adoptable. The fracture and deformation of rock masses are controlled by the strength and orientations of weak structural planes. As rock blocks are cast towards the free surfaces of the hill, the blasting can hardly produce desirable results [4]. To prevent cast of rock blocks towards the free surfaces, a hole-by-hole millisecond-delay detonation network was designed. It consists of nonels arranged in V-shaped oblique lines and deep blast holes in a triangular layout. In this blasting scheme, the deep holes are loaded with uncoupled cylindrical deck charge. After initiation, the detonation propagates along the V-shaped oblique lines from the outside in. The blasting process can create three lateral free surfaces inside the rock mass blasted. The vertical holes are drilled to proper depths based on bench heights and the three rows of holes are detonated by one-shot initiation at the hole in the middle of the first row. The explosives in each row are detonated hole by hole at intervals of 25 ms and the detonations along each oblique line occur at intervals of 100 ms. The rock is quarried by downslope sublevel mining. Figure 1 shows the schematic of this blasting scheme.

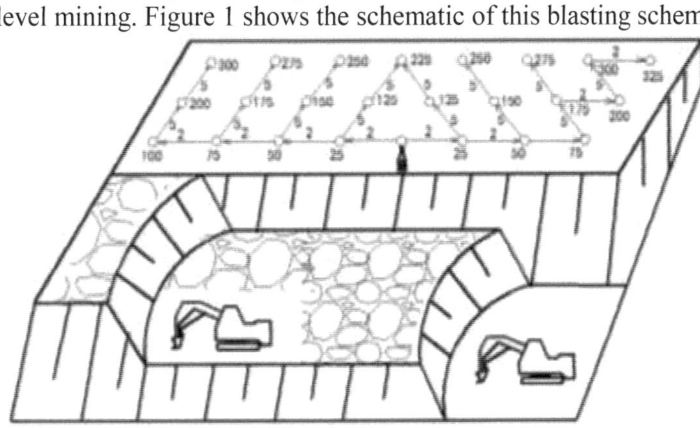

Figure 1. A blasting scheme for a hillside OPM with a height greater than 50m and a dip larger than 45°.

3.2. A blasting scheme for a hillside OPM with a height of 35-50 m and a dip of 30-45°

For an OPM with a relatively gentle hill slope, rock blocks are thrown towards the free surfaces on a smaller scale. Using vertical deep holes and one-shot initiation will increase the bench bottom burden, weaken the explosive effect at hole bottoms, and cause hole bottom undercut and formation of large rock blocks. This can reduce the effectiveness of blasting. A multistage millisecond-delay simultaneous detonation network was designed for such OPMs. It consists of nonels arranged in V-shaped oblique lines and blast holes in a triangular layout [5]. A group of shallow vertical holes are drilled in two adjacent benches and loaded with concentrated and coupled column charge. Along the strike of the working face, the benches are divided into sections by the V-shaped oblique lines. Three lateral free surfaces can be formed in a blasted rock mass. The initiation points are located at the rightmost holes in the second rows in the two benches. Detonations occur at intervals of 50 ms along each row (blasting line) and at intervals of 25 ms along each V-shaped oblique line. Figure 2 shows the schematic of this design.

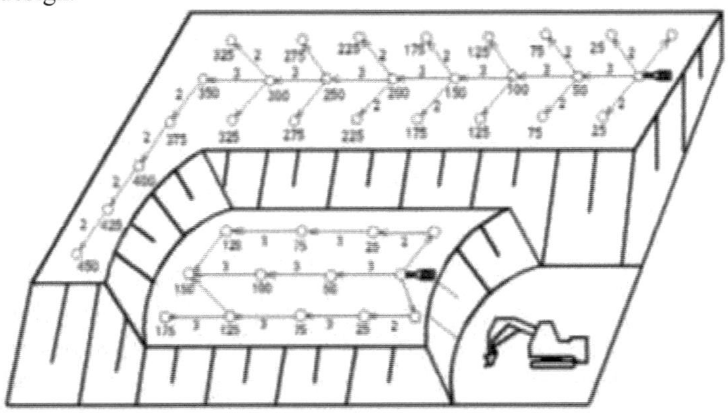

Figure 2. A blasting scheme for a hillside OPM with a height of 35-50 m and a dip of 30-45°.

3.3. A blasting scheme for a hillside OPM with a height less than 35 m and a dip smaller than 30°

Small hillside OPMs are normally lower than 35 m and slope at angles smaller than 30°. Given the relatively gentle hill slopes, the blasting scheme designed for such OPMs uses a group of shallow oblique holes drilled in two adjacent benches. These holes are arranged in a triangular layout and loaded with concentrated and coupled column charge [6]. Six adjacent holes in three rows constitute a detonation unit. Adjacent detonation units are connected by dominant detonating cords through one interval hole. These dominant detonating cords are connected perpendicular to the nonels in detonation units. Sublevel detonation can be achieved using nonel detonators, and 2 or 4 lateral free surfaces can be formed at the interval holes [7]. The initiation point is located at the leftmost hole of the first row in the lower bench, and the holes in the upper and lower benches are detonated simultaneously. In each detonation unit, detonations occur at intervals of 25 ms. This blasting scheme is illustrated in Figure 3. The blasting parameters are presented in Table 1.

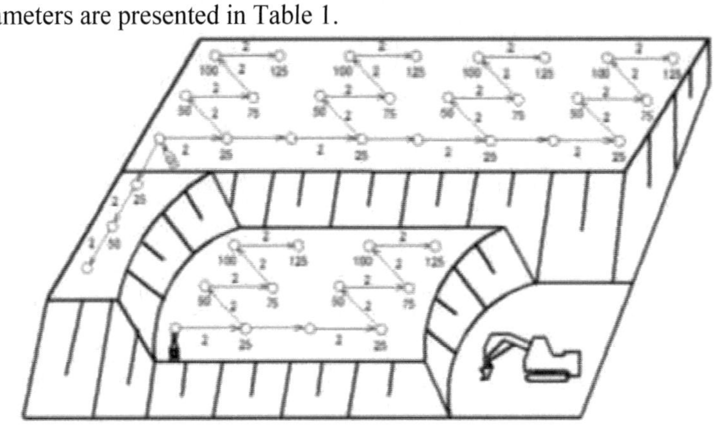

Figure 3. A blasting scheme for a hillside OPM with a height less than 35 m and a slope smaller than 30°.

IOP Conf. Series: Materials Science and Engineering **213** (2017) 012006 doi:10.1088/1757-899X/213/1/012006

The blasting parameters are presented in Table 1.

Table 1. Design parameters for these blasting schemes.

Drilling and blasting parameters	Hill height: > 50 m Hill dip: > 45°	Hill height: 35-50 m Hill dip: 30-45°	Hill height: < 35 m Hill dip: < 30°
Bench height (m)	8	4	4
Hole spacing (m)	4	2.5	2.5
Row spacing (m)	3	2	2
Unit explosive	0.38	0.42	0.4
consumption (km/m^3)	36	8.4	8.0
Explosive charge (kg)	90	90	90

4. Analysis of results

Higher hillside OPMs are more easily weathered and eroded and thus more likely to break into large rock blocks during blasting. In heterogenous rock masses, non-structural fractures are not the results of extension of pre-existing fractures, but are induced by the radial tensile stress exerted by the tensile waves reflected from the free surfaces created by blasting. An analysis of the rock surfaces blasted reveals that the blast holes along blasting lines can produce stronger explosive effect and break the rock more effectively due to the existence of the free surfaces. Therefore, hole-by-hole, millisecond-delay detonation networks that allow early detonations to create free surfaces for subsequent holes were designed for hills with different heights and slopes. The results of field experiments and statistical analysis of the experimental data suggest that an increase of one free surface in a rock mass can lead to a 4-6% decrease in block rate, as shown in Table 2.

Table 2. Analysis of block rate measurements.

Blasting scheme	Percentages of blocks of different sizes in rock masses (%)								
	Rock mass with two lateral free surfaces			Rock mass with three lateral free surfaces			Rock mass with four lateral free surfaces		
	>30cm	>70cm	>110cm	>30cm	>70cm	>110cm	>30cm	>70cm	>110cm
Hill height: >50m				48	23	11			
Hill height: 35-50m				29	17	7			
Hill height: <35m	26	17	11				15	8	3

5. Conclusions

Hillside OPMs in the arid desert areas of Xinjiang are characterized by heterogeneous rock masses with non-structural fractures and poor structural integrity, due to the presence of structural planes. These rock masses can be easily broken into large blocks by blasting. It is necessary to design proper blasting schemes for such OPMs based on the distribution of structural planes, development of joints and fissures, and the morphology of hill. This study provided different blasting schemes for different hill conditions and the effectiveness of blasting experiments using these schemes were assessed in terms of block rate (percentage of large blocks in rock mass). The study found that using hole-by-hole, millisecond-delay detonation networks can effectively reduce block rate.

For hillside OPMs that are higher than 50 m and have a slope larger than 45°, a hole-by-hole millisecond-delay detonation network consisting of nonels arranged in V-shaped oblique lines and vertical deep holes was designed. This scheme uses one-shot initiation and downslope small-bench excavation.

For hillside OPMs with a height of 35-50m and a slope of 30-45°, a multistage and millisecond-delay simultaneous detonation network was designed. This design consisting of nonels arranged in V-shaped oblique lines and vertical shallow holes distributed in two benches.

In the blasting scheme designed for hillside OPMs with a height less than 35 m and a slope smaller than 30°, a group of shallow oblique holes are drilled in two adjacent benches. The detonations in each detonation unit occurred at intervals of 25 ms.

Acknowledgments

This study was supported by the National Natural Science Foundation of China(51464043) and the Research foundation of Xinjiang Institute of Engineering(2015xgy331712). The authors are grateful for the supports.

References

[1] Liu D, Xu Q J and Wen Z L 2016 Analysis on the effect of joints and fractures of rock mass on blasting fragmentation distribution *West-China Exploration Engineering* **3** 188-91

[2] Ren Y L, Cai Q X, Shu J S, Zhou W and Han L 2014 Influence of blasting vibration and structural plane progressive failure on slope stability *J. Min. Safe. Eng.* **3** 435-40

[3] Valko P and Economdes M J 1994 Propagation of hydraulically induced fractures acontinuum damage mechnics approach *Int. J. Rock. Mech. Min. Sci* **31** 221-9

[4] Chen Y J, Chang Z G and Mao J F 2015 Blasting effect analysis of hole-by-hole millisecond minute difference initiation network along V-Shaped oblique line *Electron. J. Geotech. Eng.* **20** 2533-8

[5] Yang R S, Ding C X and Wang Y B 2016 Action-effect study of medium under loading of explosion stress wave and explosion gas *Chinese J. Rock Mech. Eng.* **2** 3501-6

[6] Yu J H and Song Z L 2015 Research on the application of deep hole blasting in open-pit coal mine with hole-by-hole blasting technique *Journal of Liaoning Technical University Natural Science* **34** 438-41

[7] Liu G X and Mu Y Q 2012 Application of hole-by-hole blasting technology in Sandaozhuang open-pit mine *China Mining Magazine* **21** 495-7

Electrospinning chitosan blends for nonwovens with morphologies between nanofiber mat and membrane

N Grimmelsmann[1], S V Homburg[1] and A Ehrmann[1,2]

[1]Bielefeld University of Applied Sciences, Faculty of Engineering and Mathematics, Bielefeld, Germany

E-mail: andrea.ehrmann@fh-bielefeld.de

Abstract. Chitosan belongs to the biopolymers possessing a broad spectrum of intrinsic physical and chemical properties which make it useful for diverse applications, such as filters or tissue engineering. For both areas it is necessary to control not only the chemical composition of the polymer, but also the shape of the surface which is in contact with the filtered medium or the growing cells, respectively. Depending on the desired form, chitosan and other biopolymers can be sprayed, coated, or spun, with few possibilities to vary their morphology between droplets, thin layers, and fibres. One possibility to mix thin films and fibres consists in using an electrospinning process which normally produces fine fibres, but depending on spinning solution and process, also membranes can be created. The article gives a short overview of the possibilities to vary a chitosan nonwoven between nanofiber mat and membrane, resulting in significantly different surface shapes.

1. Introduction

Chitosan is used in a broad field of different applications. Due to its antifouling properties under visible light, chitosan was used for the production of filters which showed high methylene blue degradation efficiency. To optimize mechanical properties and the porous structure, the chitosan was mixed with Cr-doped TiO_2 [1].

In air filtration, chitosan nanofibers blended with TiO_2 or Ag nanoparticles showed a significant increase in filtration of nanoparticle aerosols, compared with commercially available filters, combined with antibacterial activity [2]. For water filtration, composites of a PHBV (poly (hydroxybutyrate-cohydroxyvalerate)) nanospun mat and a chitosan casted barrier layer were found to reach up to 99 % filter efficiency for different disperse dyes [3]. As a coating on a cotton fabric, chitosan was used as a biological water filter for Gram-positive and Gram-negative bacteria [4-5]. Blended with PCL (polycaprolactone), chitosan nanofiber membranes were used for antibacterial water filtration [6].

More special filter demands can be found, e.g., in filtering silver from the waste of X-ray film processing. Here, chitosan membranes were found to have filtration coefficients of up to 99.9 % [7]. In solid phase extraction cartridges, chitosan can be used for perchlorate removal [8]. Dissolved pharmaceuticals could be removed from water by chitosan better than by cellulose or sodium alginate [9].

On the other hand, chitosan belongs to the biopolymers which are typically used for production of scaffolds in biotechnological or medical applications. Human mesenchymal stem cells, e.g., were grown on chitosan thin films after enzymatic modification [10]. Genipin-crosslinked chitosan is used as a hydrogel due to its antibacterial activity for different medical purposes, such as scaffolds for human mesenchymal stem cell growth, but also for cartilage engineering and similar applications [11].

In several investigations, chitosan was used for electrospinning nanofiber mats. For chitosan/ poly (ethylene oxide) (PEO) blends, some research groups found nanofiber membranes with varying morphologies, depending on the chitosan:PEO ratio [12]. Another group, working with needle-electrospinning, produced fibres and/or beads, depending on the chitosan:PEO ratio [13].

Most of the aforementioned investigations, however, were performed for special morphologies of the chitosan, such as nanofibers, membranes, or hydrogels, without the possibility to examine the influence of the surface structure in detail. In an earlier investigation, however, we found a first indication that nanofibers as well as thin film membranes could be produced by needleless electrospinning, depending on the chitosan:PEO ratio [14].

This is why in the recent project, we studied the influence of different spinning solution and process parameters on the created chitosan nonwovens, resulting in structures between nanofiber mats and thin film membranes.

2. Experimental

The needleless nanospinning machine "Nanospider Lab" (Elmarco, Czech Republic) was used to produce electrospun nanofiber mats. The spinning parameters were as follows: voltage 80 kV, carriage speed 250 mm/s, nozzle diameter 0.9 mm, electrode-substrate distance 160 mm, relative humidity in the chamber 38 %, temperature 22.5 °C. The substrate speed was varied between 0 and 30 mm/min to create mats of different densities.

For the spinning solutions, a chitosan solution of 2 wt% medium molecular weight chitosan (190-310 kDa; 200-800 mPa·s; Sigma Aldrich) with 1 % acetic acid was prepared. The chitosan was solved in aqueous solution with sodium bicarbonate ($NaHCO_3$), using a concentration of 20.16 g/l. PEO (poly (ethylene glycol)) with a molecular weight of 600,000 daltons (concentration 8 %) purchased from S3 Chemicals was added as a spinning agent. Water was added to modify the viscosity of the polymer solution; the respective amounts of H_2O are given in the descriptions of the results. Viscosities were not measured but varied in a range which was suitable for electrospinning with the chosen nozzle.

The base parameters were chosen as follows: 1 ml chitosan solution (2 wt%), 1 ml $NaHCO_3$, 3 ml H_2O; 0.5 mL PEO solution (8 wt%). The amounts of water and of PEO solution were varied in the experiments.

The resulting samples had dimensions of approx. 30 cm x 50 cm of which the middle area was chosen for further investigation. Images were taken using a confocal laser scanning microscope VK-9000 (Keyence), using a nominal magnification of 2000 x.

3. Results and discussion

In a first test series, chitosan/PEO nonwovens were produced using different amounts of H_2O with 1 ml chitosan solution, 1 ml $NaHCO_3$ solution, and 0.5 ml PEO solution. In Figure 1, the results are depicted. For the smallest amount of water (1 ml H_2O), a relatively closed mat with some pores is produced, partly showing the rainbow colours typical for diffraction in thin films, while other parts seem to be more solid.

A middle quantity of water (3 ml H_2O) results in evenly thin films, showing rainbow diffraction colours in several areas, with apparently more holes in which only very fine fibres are formed. The highest amount of water (5 ml H_2O) leads to an open-pore structure with several thicker and thinner fibres and a few membrane areas. Apparently, the middle amount of water results in creation of thin membranes, while less water produces more fibrous structures.

As visible by the scales in the images, the pores in the membranes are of the order of magnitude of 10-50 μm, while the fibre diameters vary between approx. 100 nm and 1 μm.

In the next test series, the quantity of PEO solution was modified, while the other parameters (1 ml chitosan solution, 1 ml $NaHCO_3$ solution, 3 ml H_2O) were kept constant. The results are presented in Figure 2. For the lowest amount of PEO solution (0.2 ml), only a web of fine fibres is shown, covered with small droplets. In some parts of the web, however, thin films are visible. By increasing the fraction of PEO solution to 0.5 ml, the already known membrane structure with some holes is achieved. By increasing the quantity of PEO solution further (0.8 ml), more areas are covered with fibre structures and fewer thin film areas. Finally, adding 1.1 ml PEO solution, most areas are covered with fine fibres, while the areas covered with thin films are again reduced. Apparently it is possible to tailor the ratio of nanofiber and membrane areas by modifying the amount of PEO in the spinning solution.

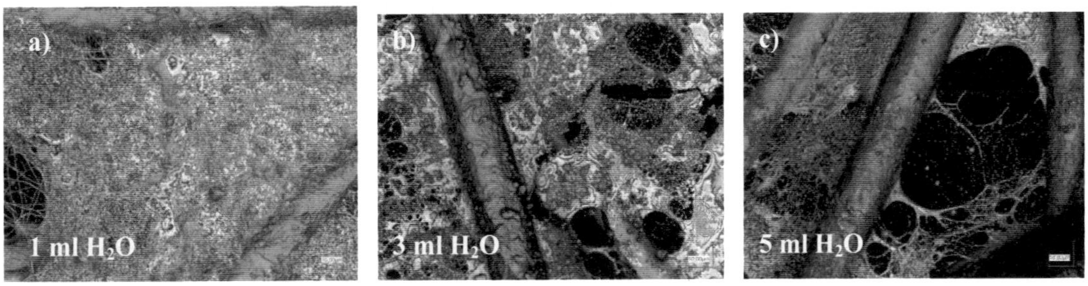

Figure 1. Chitosan nanofiber mats, produced from 1 ml chitosan solution, 1 ml $NaHCO_3$ solution, 0.5 ml PEO solution, and different amounts of H_2O.

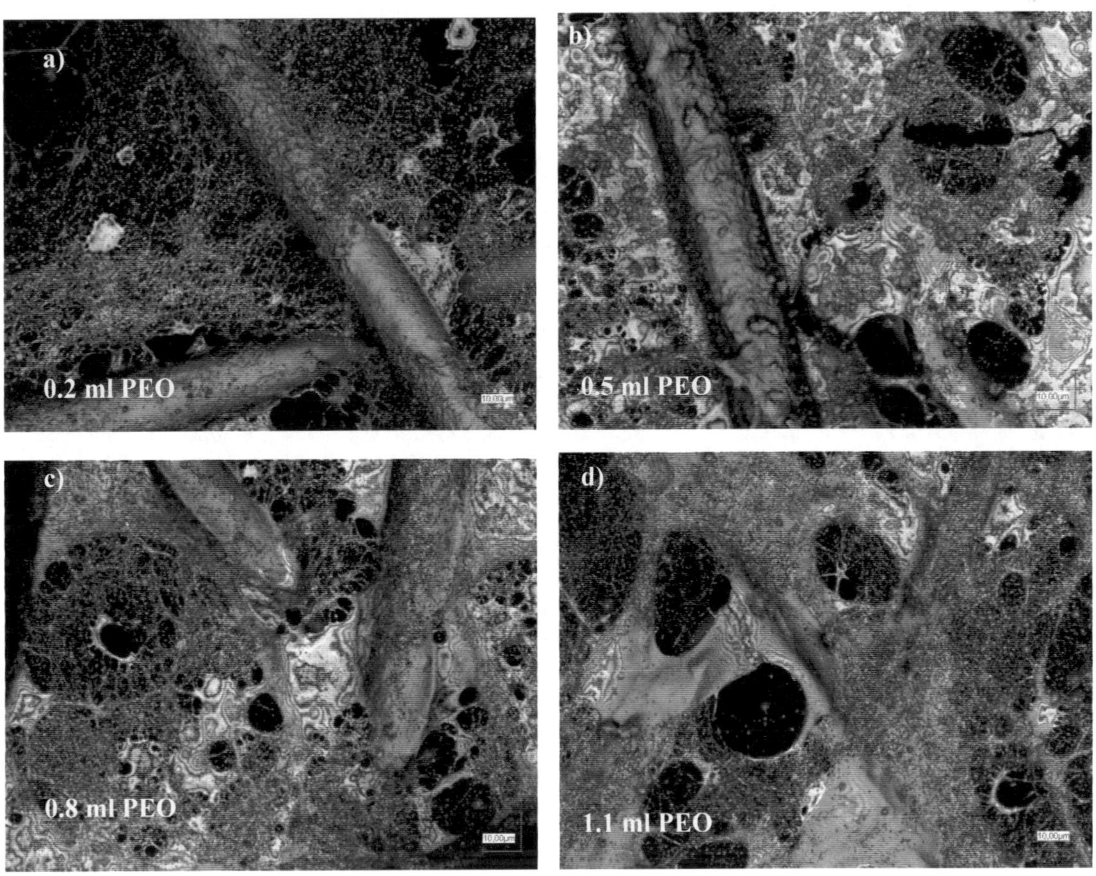

Figure 2. Chitosan nanofiber mats, produced from 1 ml chitosan solution, 1 ml $NaHCO_3$ solution, 3 ml H_2O, and different amounts of PEO solution.

Interestingly, the formation of droplets seems to be independent from the amount of PEO in the polymer solution. This is correlated with the observation found in literature that chitosan tends to forming droplets or beads along the nanospun fibres [13]. Apparently this behaviour cannot be changed by adding more PEO as a spinning agent. Future tests will show whether modifications in the preparation of the spinning solution may reduce the number of these droplets.

Another interesting effect, influencing the ratio between thin film and nanofiber parts in the produced nonwoven, is depicted in Figure 3. Nonwovens of different density were produced from 1 ml chitosan solution, 1 ml $NaHCO_3$ solution, 3 ml H_2O, and 0.5 ml PEO solution, by varying the substrate speed between 0 and 30 mm/min. In this way, very fine as well as very dense chitosan mats were created.

For the finest mat shown in Figure 3a, only fibres are visible with droplets adhered on them, as recognized before. By increasing the density slightly by enhandinc the spinning time, first membrane areas become visible (Figure 3b). Further increase of the density leads to larger areas with membranes, still showing several completely open parts (Figure 3c). Finally, for the densest mat, the nonwoven does not show open areas anymore, but completely consists of a mixture of fibres and thin films (Figure 3d).

As depicted here, modifications of the mat density significantly change the structure of the created nonwoven. Future experiments will reveal which of these structures are ideally suited for applications as filters or in tissue engineering.

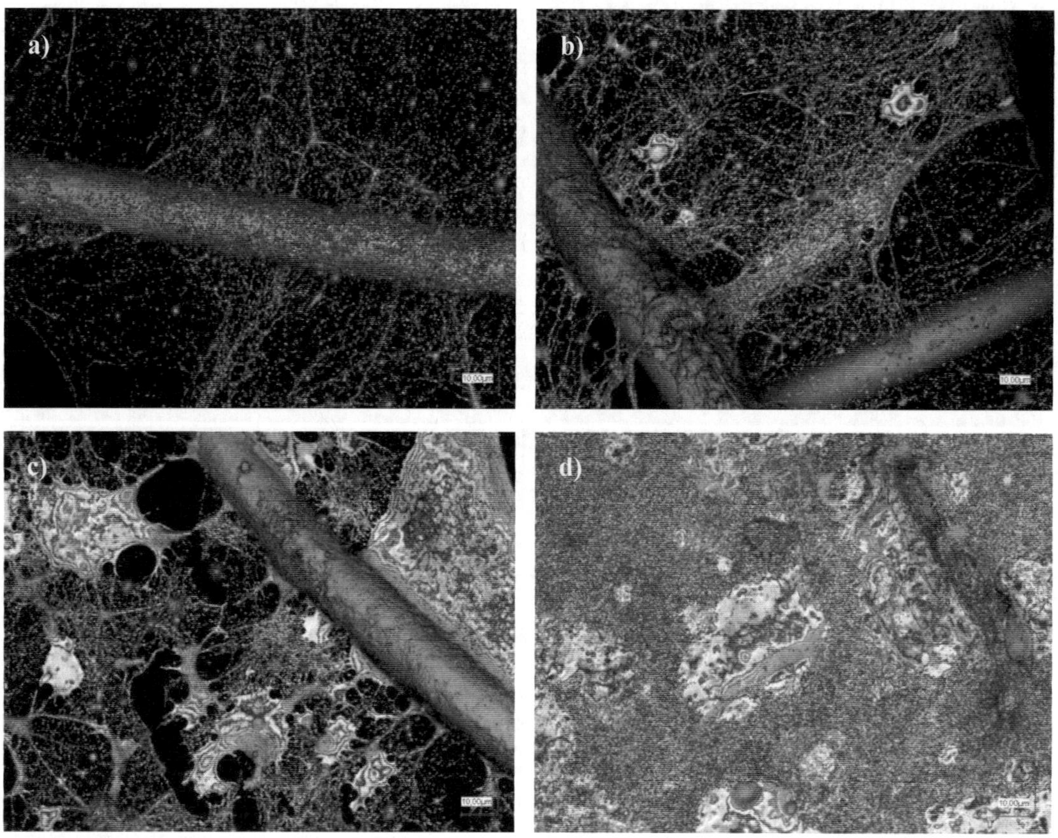

Figure 3. Correlation between mat density and membrane formation. Chitosan nanofiber mats of different densities, produced from 1 ml chitosan solution, 1 ml $NaHCO_3$ solution, 3 ml H_2O, and 0.5 ml PEO solution.

4. Conclusion

Needleless electrospinning was used to create chitosan/PEO blended nonwovens. Depending on the amount of water, the chitosan: PEO ratio and even the mat density, different surface morphologies between nanofiber mats and thin film membranes were created.

This possibility to tailor the structure of a nonwoven enables new applications in filter technology or for tissue engineering, allowing the creation of multi-layer filters with desired flow lines or choosing the ideal morphology for cells to grow. The "ideal" morphology thus depends on the desired application.

Future research will concentrate on measuring the mechanical and chemical properties of these chitosan/PEO nonwovens before and after stabilization of the chitosan for application in aqueous environments and testing the materials as filters and biotechnological scaffolds, depending on the created morphology.

References

[1] Lim J K 2016 Simple fabrication of a photocatalyst hybridized porous chitosan-based antifouling active filter under visible light *J. Porous Mater.* **23** 1163-8

[2] Wang L, Zhang C, Gao F and Pan G 2016 Needleless electrospinning for scaled-up production of ultrafine chitosan hybrid nanofibers used for air filtration *RSC Advances* **6** 105988-95

[3] Zhao X, Liu Y, Wang C and Liu Q S 2015 Structure and filtration performance of fibrous composite membranes containing environmentally friendly materials for water purification *Fibers and Polymers* **16** 2586-2592

[4] Ferrero F, Periolatto M, Vineis C and Varesano A 2014 Chitosan coated cotton gauze for antibacterial water filtration *Carbohydr. Polym.* **103** 207-12

[5] Periolatto M, Ferrero F, Vineis C and Varesano A 2014 Antibacterial Water Filtration by Cationized or Chitosan Coated Cotton Gauze *Chem. Eng. Trans.* **38** 235-40

[6] Cooper A, Oldinski R, Ma H Y, Bryers J D and Zhang M Q 2013 Chitosan-based nanofibrous membranes for antibacterial filter applications *Carbohydr. Polym.* **92** 254-9

[7] Rupiasih N N, Purnomo R R and Sumadiyasa M 2016 Preparation and application of chitosan membranes to filter silver from X-ray film processing wastes *J. Phys. Conf. Ser.* **710** 012009

[8] Sayed S and Jardine A 2015 Chitosan derivatives utilized in anion exchange chromatography for perchlorate removal *Int. J. Mol.Sci* **16** 9064-77

[9] Ahmed A E I, Moustafa H Y, El-Masry A and Hassan S A 2014 Natural and synthetic polymers for water treatment against dissolved pharmaceuticals *J. Appl. Polym. Sci.* **131** 40458

[10] Aljawish A, Muniglia L and Chevalot I 2016 Growth of human mesenchymal stem cells (MSCs) on films of enzymatically modified chitosan *Biotechnol. Progr.* **32** 491-500

[11] Muzzarelli R A A, Mehtedi M, Bottegoni C, Aquili A and Gigante A 2015 Genipin-crosslinked chitosan gels and scaffolds for tissue engineering and regeneration of cartilage and bone *Marine Drugs* **13** 7314-38

[12] Wei H Q, Zhang F H, Zhang D W, Liu Y J and Leng J S 2015 Shape-memory behaviors of electrospun chitosan/poly(ethylene oxide) composite nanofibrous membranes *J Appl. Polymer Sci.* **132** 42532

[13] Lemma S M, Bossard F and Rinaudo M 2016 Preparation of pure and stable chitosan nanofibers by electrospinning in the presence of Poly(ethylene oxide) *Int. J. Mol. Sci.* **17** 1790

[14] Grimmelsmann N, Homburg S V and Ehrmann A 2016 Needleless electrospinning of pure and blended chitosan *Proc. Int. Conf. Advanced Material Technologies (Visakhapatnam / India), accepted*

IOP Conf. Series: Materials Science and Engineering **213** (2017) 012008 doi:10.1088/1757-899X/213/1/012008

Effect of chain extenders on mechanical and thermal properties of recycled poly (ethylene terephthalate) and polycarbonate blends

Y Srithep[1,3], D Pholharn[2], A Dassakorn[1] and J Morris[1]

[1]Manufactuing and Materials Research Unit, Faculty of Engineering, Mahasarakham University, Mahasarakham 44150, Thailand
[2]Department of Chemistry, Faculty of Science and Technology, Rajabhat Mahasarakham University, Mahasarakham 44000, Thailand

E-mail: yottha.s@msu.ac.th

Abstract. Poly(ethylene terephthalate) (PET) based thermoplastics are common, particularly in packaging. Due to both thermal and hydrolytic degradation, recycled PET (rPET) shows poor mechanical properties. The effect of adding 30% polycarbonate (PC) and chain extender (CE) on mechanical, thermal and morphological properties was investigated. rPET with PC and CE was melt blended in a single screw extruder, which was then tensile test specimens were formed by injection molding rPET and PC compatibility was improved via chain extending reactions which in turn improved the rPET and PC blend mechanical properties. With the addition of 30%PC the tensile modulus of rPET increased by 18% and it increased by 223% when 2%CE was added to rPET and 30%PC blends. Also, the CE reaction affected crystallization, monitored by differential scanning calorimetry: the melting enthalpy ofblends was lowered by increased CE content.

1. Introduction

Poly (ethylene terephyhalate) (PET) has become common in soft drink bottles because it is transparent, shows high chemical resistance and low diffusibility. Beverage bottles are generally thrown away and degrade the environment. Thus, to reduce environmental hazards and burden created by landfill dumping of PET, large amounts of PET is recycled and reused in various ways[1]. However, PET recycling is a complicated process because of chemical and mechanical degradation during reprocessing [2].

Polycarbonate (PC) has high thermal stability and impact resistance which lead to wide use as an engineering plastic. To improve the properties and therefore application of PET, PET/PC blends have been found to show active interphase interactions, which causes them to be described as partly compatible [3, 4]. Nevertheless, simple melt blended PET and PC mixtures in a mixer generally lead to a phase-separated PET/PC mixture; therefore there is considerable interest in improving the compatibility of PET and PC. Further, at compounding and processing of PET/PC molten blends, it is desirable to maintain the molecular weight and intrinsic viscosity which are both reduced by hydrolytic and thermal degradation that, in turn, degrade mechanical properties of recycled material [2, 5].

Many papers have proposed chemical strategies to enhance the properties of virgin, reprocessed or recycled polycondensates [6]. Commonly, bi- or multi-functional chemical compounds which can lengthen chains and increase the molecular weight and physical properties of degraded thermoplastics by bridging functional chain end groups and thus lengthening or ''chain extending'' them. Chain extenders are usually low molecular weight compounds which react rapidly with the polymer end groups linking them to increase chain length [1]. Chain extension mechanisms in reactive extrusion have been surveyed [1, 7, 8].

This study investigated the effects of chain extender (CE) on mechanical and thermal properties of recycled poly (ethylene terephthalate) (rPET) and PC blends. As the multifunctional epoxide group of the CE can react with hydroxyl and carboxyl end group of rPET and PC, it also may improve the compatibility of this blending system.

2. Experiments

2.1. Materials

For rPET, we used PET bottles collected from urban waste. The measured melting temperature (T_m) was 245 °C and melt flow index (MFI) (with 2.16 kg weight) was 30 g/10 min at 260 °C and 60 g/10 min at 270 °C. Mitsubishi Lupilon S3000R (MFI 16.5 g/10min) PC was purchased from Mitsubishi Engineering Plastics Corporation. The chain extender (BASF Corporation Joncryl ADR4370) used was a styrene acrylic copolymer (oligomeric coupling agent) that has epoxy functional groups that react with condensation polymers.

2.2. Processing

rPET/PC (70/30 w/w), with the CE contents from 0 to 2wt%, were melt compounded using a single screw extruder (HAAKE Polylab OS system) with screw diameter, D = 20 mm, and L/D = 40, where L is the screw length. Screw speed was set at 100 rpm. The barrel temperature was at 270°C. Before processing, PET and PC were dried in an oven for 12 h at 80°C to remove water to prevent hydrolysis in the molten state. After blending, the extruded strands were cooled in a water bath, palletized and dried in a vacuum at 80 °C for 10 h. Injection molding used an injection molding machine (ING-58T, Chareon Tut Co., Ltd) to obtain tensile bars (ASTM D638 Type I). The neat rPET (reprocessed PET without PC addition) and PC were processed and used as reference materials.

2.3. Material characterization

2.3.1. Differential Scanning Calorimetry (DSC). Thermal properties of the blended materials were measured with a differential scanning calorimeter (DSC 4000 PerkinElmer). Specimens of 4 to 5 mg were placed in aluminium sample pans and heated from -10°C to 280°C at a 10 degC/min rate and held for 3 min at 280°C to erase any prior thermal history before cooling at 10 degC/min to -10°C. The specimens were then reheated to 280 °C using at 10 degC/min. The *cold crystallization temperature* (T_{cc}) *and its enthalpy* (ΔH_{cc}), melting temperature (T_m), and apparent melting enthalpy (ΔH_m), crystallization temperature (T_c), were determined from DSC curves. T_m was taken from the peak temperature and ΔH_m from the area of the melting endotherm.

The absolute degree of crystallinity (χ_c) of the rPET phase was calculated from

$$\chi_c\,(\%) = \frac{\Delta H_m(rPET)}{\Delta H^\circ(rPET)} \times \frac{100}{w} \quad (1)$$

where ΔH°(rPET) is the melting enthalpy (J per gram) of 100% crystalline material (i.e. perfect crystal) (120 J/g) and *w* is the weight fraction of rPET in the blend [9].

To determine the original crystallinity of the injection molded sample that was subjected to rapid cooling during the molding process. We noted that, during heating, some additional crystallites form from the amorphous material and we must subtract their heat released from the total endothermic heat flow generated from all the crystallites (i.e., initial+formed by heating) to calculate the correct % crystallinity, χ_c. The equation for the original crystallinity of the injection molded sample became [1]:

IOP Conf. Series: Materials Science and Engineering **213** (2017) 012008 doi:10.1088/1757-899X/213/1/012008

$$\chi_c(\%) = \frac{\Delta H_m(rPET) - \Delta H_{cc}(rPET)}{\Delta H°(rPET)} \times \frac{100}{w} \tag{2}$$

2.3.2. Mechanical testing. Tensile tests on injection molded samples followed ASTM D638. We measured static tensile strength and strain-at-break at 25 °C and relative humidity of 50 ± 5% on an NRI-TS501 universal testing instrument. For all specimens, an initial load of 0.5 N and a constant crosshead speed of 50 mm/min were used. Averages were calculated from five specimens taken from each sample group.

2.3.3. Thermogravimetric Analysis (TGA). Thermogravimetric analysis (TGA) samples were conditioned at 50% humidity at 25ºC in the testing room and used a TGA 4000 (Perkin-Elmer) from 25 to 600°C at 10degC/min. Each test used ~10 mg of rPET, PC and rPET/PC, with varying chain extender. Weight loss was noted and normalized versus the initial weight.

2.3.4. Scanning Electron Microscopy (SEM). SEM images of the fracture surfaces generated by tensile tests were acquired using a Hitachi SU-1500 machine at 3 kV. The samples were frozen in liquid nitrogen and then quickly impact-fractured. A thin layer of gold (~20nm) was sputter-coated onto specimens prior to imaging.

3. Results and discussions

3.1. Thermal properties

To measure thermal properties of rPET/PC blends, including crystallization and melting behaviors. The temperature plots from the first and second heating cycles appear in Figures 1 and 2. Cooling scans are shown in Figures 3. Derived thermal data is in Table 1. Values obtained from the first heating cycle were affected by the prior sample thermal history, whereas the second heating cycle allowed direct comparison of the crystallization behavior of different materials after the first heating scan thermal had been erased [1].

3.1.1. First heating cycle. DSC measurements in Figure 1 show that the glass transition temperature (T_g) around 60 to 80 ºC associated with the rPET phase in all the blends was slightly higher than for neat rPET and T_g was slightly higher as the amount of CE increased. With 2%CE content, T_g was near to 74ºC, while T_g of neat rPET was ~72 ºC and for neat PC was ~147 ºC (see Figure 1 and Table 1). A similar effect of PC and chain extender on T_g of rPET have been reported previously [9]. The first exothermic peak for all rPET blend specimens (in the temperature range 110 to 130 ºC) is labelled as the cold crystallization peak. The enthalpy of cold crystallization (ΔH_{cc}) and of melting enthalpy (ΔH_m) was lower as the amount of CE increased (ΔH_{cc} varied from 23.0 J/g to 11.9 J/g and ΔH_m varied from 17.50 to 12.32 J/g). In general, when CE was added in the blending system, the chain-extending reaction for rPET increased its molecular weight, whereas the rPET-PC reaction restrained rPET chain movements. Decreased rPET chain mobility constrained rPET crystallization compared to rPET without CE, and its cold crystallization temperature increased consistent with the decreased degree of crystallinity. Similar results have been reported for PLA/PC blends with chain extender [10].

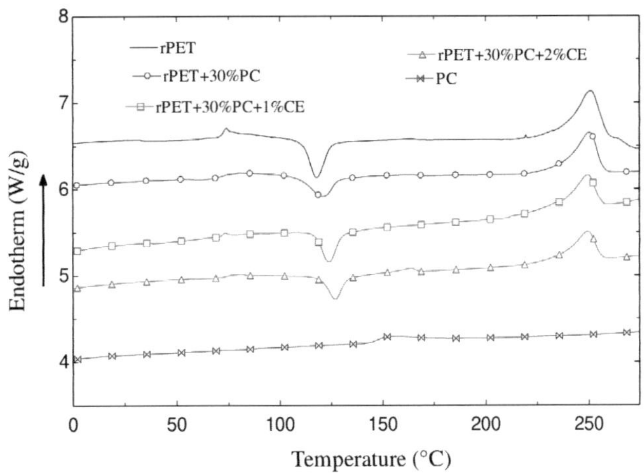

Figure 1. First heating cycle DSC curves for rPET blends.

Table 1. DSC data for rPET/PC (70/30) blends with varying CE content.

Samples	Glass Transition	Cold Crystallization		Melting		Degree of crystallinity
	T_g(°C)	T_{cc} (°C)	ΔH_{cc} (J/g)	T_m (°C)	ΔH_m (J/g)	χ_c (%)
First heating						
rPET	72.2	117.8	22.1	250.6	53.4	26.0
rPET+30%PC	72.8	121.0	18.0	250.1	39.5	25.5
rPET+30%PC+1%CE	73.2	123.8	15.9	249.2	32.2	19.6
rPET+30%PC+2%CE	73.9	126.6	11.9	249.1	27.7	19.3
PC	147.7	-				
Second heating						
rPET	81.5	-	-	245.9	41.2	34.3
rPET+30%PC	90.5	-	-	230.6	28.3	33.6
rPET+30%PC+1%CE	88.7	-	-	222.6	17.8	21.4
rPET+30%PC+2%CE	88.9	-	-	224.1	17.4	21.3
PC	143.6	-	-	-	-	

3.1.2. Cooling cycle. Figure 2 and Table 1 show the thermograms and data derived from the cooling cycle of rPET, PC and rPET/PC blends with 1% and 2%CE. It was found that only rPET had showed a crystallization temperature at about 215°C,indicating that blending rPET with 30%PC decelerated the crystallization rate owing to its interactions with PC [4].

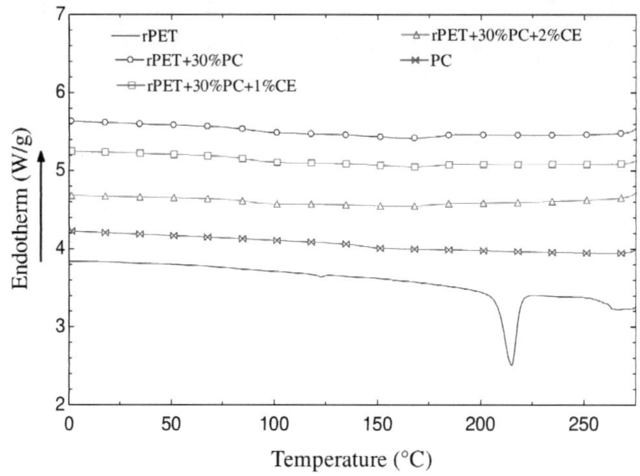

Figure 2. Cooling cycle DSC curves of the rPET blends.

3.1.3. Second heating cycle. Figure 3 shows thermograms from the second heating cycle of rPET, PC and rPET/PC blends and Table 1 also shows data derived from the second heating cycle. Unlike the first heating cycle, due to the slower cooling rate during the DSC test than injection molding process, cold crystallization disappeared and the degree of crystallinity for all samples was higher than that observed during the first cycle [1]. Moreover, melting enthalpy ΔHm decreased significantly with added chain extender (ΔHm was 28.3 J/g without CE to 17.4 J/g with 2%CE). The interfacial reaction between the epoxy group and the carboxyl end group of rPET at the interface [11] discourages rPET chain rearrangement (motion) and consequently cold crystallization [9].

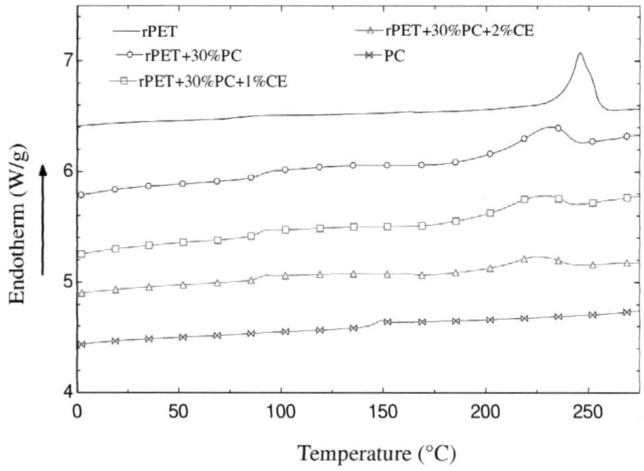

Figure 3. Second heating cycle DSC curves of the rPET blends.

3.2. Morphology
As rPET and PC are partly miscible, it is expected that simple blending would produce a phase-separated structure, leading to low mechanical properties. The CE, being a third component with active groups, can react with both end groups of rPET and PC and thus lead to better compatibility between rPET and PC. Representative SEM images of rPET blends are in Figure 4. They reveal information on the microstructure and the fracture behavior of the specimens. The rPET/PC blend clearly showed distinct phases (Figure 4(b)). With CE fed, compatibility between rPET and PC improved and the dispersed phase size reduced (Figure (c) - 1% CE and (d) -2%CE).

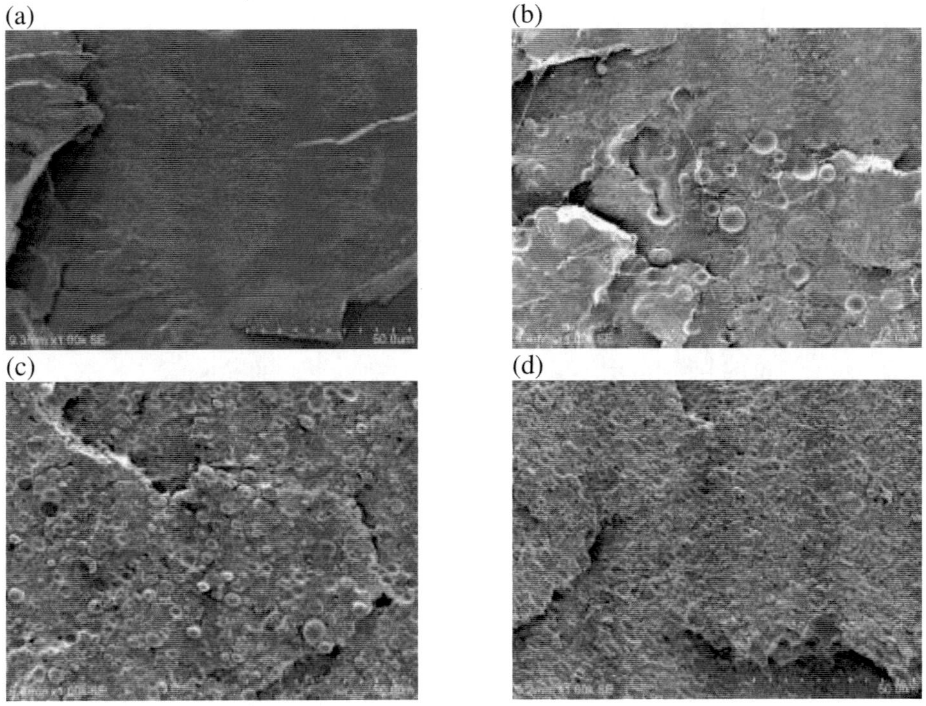

Figure 4. Scanning electron micrographs (a) rPET (b) rPET+30%PC (c) rPET+30%PC+1%CE and (d) rPET+30%PC+2%CE. All images were taken at the same magnification (scale bar: 50μm).

3.3. Tensile Properties

Tensile tests followed ASTM-D-638-02 on the injection molded specimens of the rPET blends. Clearly, the size of the dispersed phase of rPET/PC (70/30) blends decreased with higher CE content. This is consistent with the mechanical properties of the rPET/PC (70/30) blends, i.e. that higher mechanical strength and strain-at-break was observed with larger amounts of CE added to the blends. As shown in Figure 5, rPET had lowest tensile strength (~ 10 MPa) suggesting a reduced molecular weight as a consequence of degradation and it increased by addition of 30%PC to 17 MPa. With the addition of 1% and 2%CE, the tensile strength of rPET/PC increased to 39 and 55 MPa, respectively. As with tensile strength, a similar trend was observed for the elongation at break (Figure 6). Therefore, the blend particle size decrease suggests increased compatibility between the PC and rPET components [12].

Figure 7 shows the tensile modulus of rPET, PC and rPET/PC blends. With the addition of 30%PC the tensile modulus of rPET increased by 18%. Moreover, the addition of CE increased the tensile modulus further. The tensile modulus of rPET+30%PC was found to be 1760 MPa and with added 2%CE the tensile modulus increased to 2300 MPa.

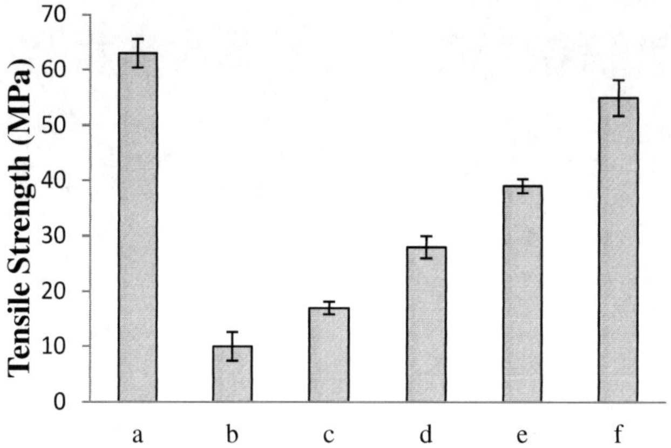

Figure 5. Tensile strength of the rPET/PC blends (a) PC (b) rPET (c) rPET+30%PC (d) rPET+30%PC+0.5%CE (e) rPET+30%PC+1%CE (f) rPET+30%PC+2%CE.

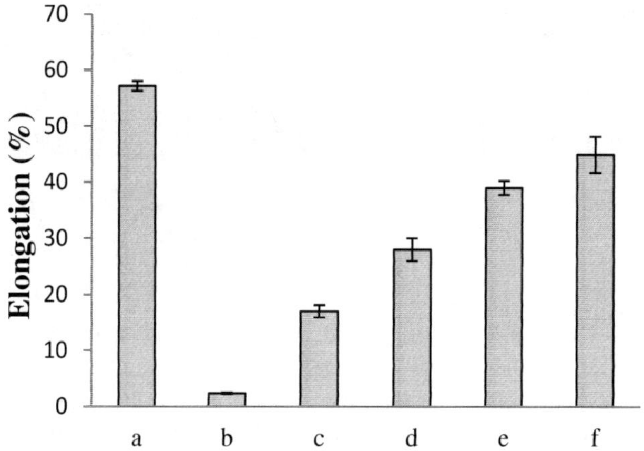

Figure 6. Elongation at break of the rPET/PC blends (a) PC (b) rPET (c) rPET+30%PC (d) rPET+30%PC+0.5%CE (e) rPET+30%PC+1%CE (f) rPET+30%PC+2%CE.

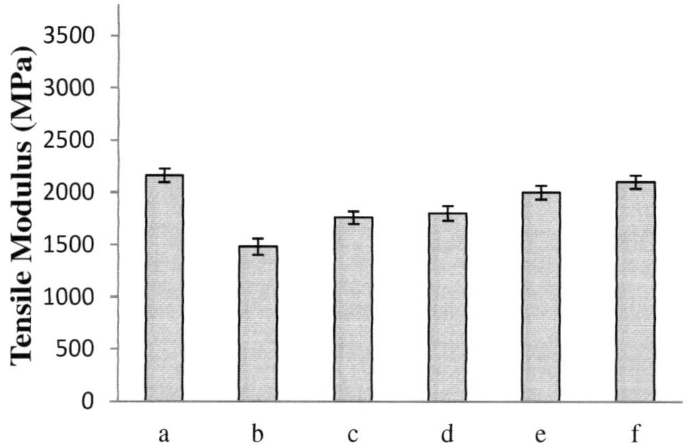

Figure 7. Tensile modulus of the rPET/PC blends (a) PC (b) rPET (c) rPET+30%PC (d) rPET+30%PC+0.5%CE (e) rPET+30%PC+1%CE (f) rPET+30%PC+2%CE.

3.4. Thermal stability

TGA was used to assess the thermal stability of rPET, PC, and rPET/PC blends at a heating rate of 10 degC/min with results shown in Figure 8. PC had much higher thermal stability than rPET. The degradation of rPET/PC blend lay between rPET and PC and any difference in the temperature loss of rPET/PC and rPET/PC with added chain extender is not obvious. However, a weight loss step at the temperature ~318°C were observed for rPET/PC blends. We attribute this to interaction between rPET and PC after blending, which leads to small amounts of short molecular weight segments, which are lost more readily [4].

Table 2 shows the temperature at 10% and 20% weight loss ($T_{10\%}$ and $T_{20\%}$) of rPET, PC, and rPET/PC blends. T_{10}% of rPET/PC blends was lower than that of neat PC and neat rPET due to the small amount of low molecular weight polymer blends produced from the interaction between PC and rPET. Moreover, $T_{20\%}$ of the rPET/PC blends lay between neat PC and rPET. Therefore, it could be concluded that thermal stability of rPET improved on addition of PC. Figure 9 shows a schematic mechanism illustrating the interfacial reaction between the epoxy group of CE, the carboxyl end group of rPET and hydroxyl end group of PC.

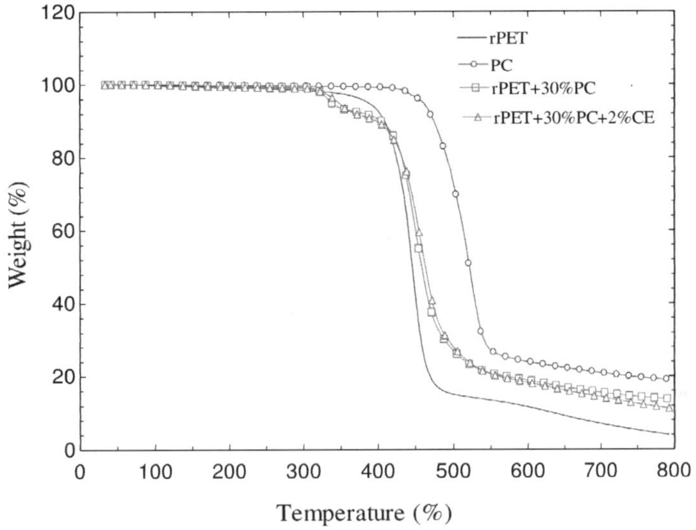

Figure 8. TGA curves for rPET blends.

Table 2. Temperature at 10% and 20% weight loss of rPET, PC, and rPET/PC blends.

Samples	$T_{10\%}$(°C)	$T_{20\%}$(°C)
rPET	409.1	425.2
rPET+30%PC	397.4	432.7
rPET+30%PC+2%CE	404.1	432.9
PC	474.1	491.2

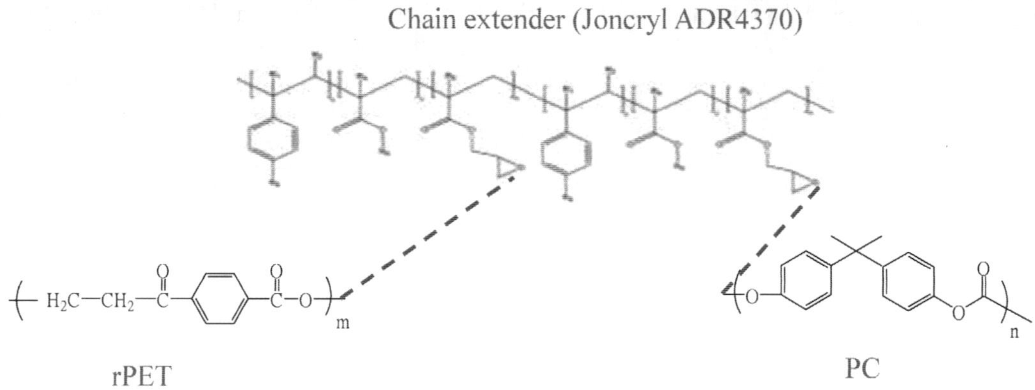

Figure 9. The interfacial reaction between CE, rPET and PC.

4. Conclusions

rPET/PC alloys were prepared by reactive extrusion with addition of chain-extender (CE). By adding CE, compatibility between rPET and PC was improved leading to enhanced mechanical properties. In addition, CE affected crystallization of rPET/PC blends as it reduced the degree of crystallinity. The interfacial reaction between the epoxy group of CE, the carboxyl end group of rPET and the hydroxyl end group of PC discourages rPET/PC chain rearrangement. So we concluded that the chain extender had a dual function in these alloys as both chain extender and compatibility enhancer.

Acknowledgement

Financial support from the Faculty of Engineering, Mahasarakham University, and the Mahasarakham University Development Fund is gratefully acknowledged.

References

[1] Srithep Y, Javadi A, Pilla S, Turng L S, Gong S, Clemons C and Peng J 2011 *Polym. Eng. Sci.* **51** 1023-32
[2] Fraisse F, Verney V, Commereuc S and Obadal M 2005 *Polym. Degrad. Stab.* **90** 250-5
[3] Soroudi A and Jakubowicz I 2013 *Eur. Polym. J.* **49** 2839-58
[4] Nampoothiri K M, Nair N R, and John R P 2010 *Bioresour. Technol.* **101** 8493-501
[5] Zhang N, Wang Q, Ren J and Wang L 2009 *J. Mater. Sci.* **44** 250-6
[6] Villalobos M, Awojulu A, Greeley T, Turco G and Deeter G 2006 *Energy.* **31** 3227-34
[7] Bikiaris D N and Karayannidis G P 1996 *J. Appl. Polym. Sci.* **60** 55-61
[8] Pilla S, Kramschuster A, Yang L, Lee J, Gong S and Turng L S 2009 *Mater. Sci. Eng. C Mater. Biol. Appl.* **29** 1258-65
[9] Tang X, Guo W, Yin G, Li B and Wu C 2007 *J. Appl. Polym. Sci.* **104** 2602-7
[10] Srithep Y, Rungseesantivanon W, Hararak B and Suchiva K 2014 *J. Polym. Eng.* **34** 665-72
[11] Pesetskii S, Jurkowski B, Filimonov O, Koval V and Golubovich V 2011 *J. Appl. Polym. Sci.* **119** 225-34
[12] Lee J B, Lee Y K, Choi G D, Na S W, Park T S and Kim W N 2011 *Polym. Degrad. Stabil.* **96** 553-60

IOP Conf. Series: Materials Science and Engineering **213** (2017) 012009 doi:10.1088/1757-899X/213/1/012009

Synthesis and photoluminescence properties of ZnO nanohelices

M Y Zhou[1,3], L S Qu[1] and H Gao[2]

[1]Naval Aeronautical and Astronautical University Department of Basic Sciences, Yantai, Shandong 26400, P.R. China
[2]Department of Physics, Harbin Normal University, Harbin, Heilongjiang 150080, P.R. China

E-mail: zhoumingyu_2004@163.com

Abstract. Sb-doped zinc oxide nanohelices were synthesized on silicon substrates by chemical vapor deposition. The morphologies and structures of the samples have been investigated by XRD，SEM，and HR-TEM. The ZnO nanohelices have a single crystallinewurtzite structure. Its length is up to tens of micrometres as a whole. It grows in the direction of (0001). There are six sticks in each period of the helices, which grow al-ong six directions of the equivalent $\langle 0\bar{1}11 \rangle$. In addition, its growth mechanism is discussed. Optical properties of the nanohelices were demonstrated by Photo-luminescence (PL) spectra.

1. Introduction

ZnO with a high band gap (3.37eV) and exciton binding energy (60 meV) has been researche-d widely because of its broad range of applications. In particular, the application of low voltage and short wave optoelectronic devices has attracted more and more attention. ZnO has the structure of wurtzite, which is simply depicted schematically as lots of alternating planes consisted of tetrahedrally coordinated O^{2-} and Zn^{2+} ions, Selective stacking growth along the c-axis [1]. Because of their originality properties and application possible in electronics, optoelectr-onics, electromechanics[2], one-dimensional ZnO nanostructures has created wide comment. Forone-dimensional nanostructures, their morphology determines their possible applications.

Different morphologies were synthesized successfully, such as nanorings [3], nanocastles [4] and nanocombs [5]. Among these morphologies, ZnO nanohelices have been reported also. It is likely that different growth mechanisms are dominant for different materials but the addition of some foreign materials into the sources can help to form nanohelices[6-8]. In our experiment, Zinc Oxide source was added to Sb and the nanohelical structures were produced. The nanohelice is a part of a family of ZnO ribbon-like nanobelts. But unlike the elastic nanosprings it has been reported, the nanohelice is rigid. These uniform nanohelices may have the potential utilization of nano-size gas sensor, piezoelectricity and wave absorption. Its spread foreground in application was prospected.

2. Experimental

Put the ZnO and Sb precursors in alumina plane, which was located at the center of a horizontal tube furnace. The furnace chamber was heated up to 13000C, and had been mingled withmolar ratio 70:1 before, The alumina tube was first evacuated to vapor pressure of 6.0 Pa to remove the residual oxygen under an N2 carrier gas of 75 sccm. In order to collect gray–black ZnO precursors, Covered with a 3nm thick Au layer the Silicon substrates were placed in the downstream area of carrier gas. The pressure was maintained at 3000 Pa and N2 was blown at a speed of about 100 sccm for ten minutes. When the furnace temperature dropped to room temperature, it can be find the white growth on the Au coated Si substrates. The as-prepared products were characterized and analyzed by X-ray diffraction(XRD), the scanning electron microscopy(SEM), Transmission electron microscopy(TEM) and high-resolution transmission nelectron microscopy(HRTEM). The Photoluminescence measurement was accomplished under the condition of room temperature by He-Cd laser.

3. Results and discussions

XRD spectra of ZnO nanohelices (Figure 1). The main diffraction peaks corresponds to the faces of ZnO crystal. In the samples, the diffraction peaks were not found from Zn or Sb.

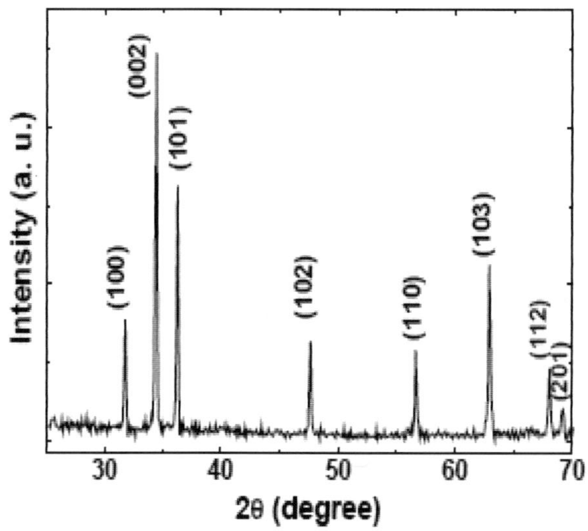

Figure 1. XRD pattern of the as-formated ZnO nanostructures.

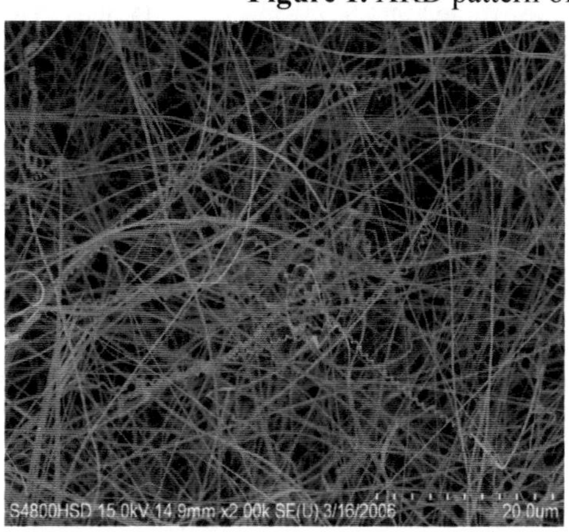

(a)

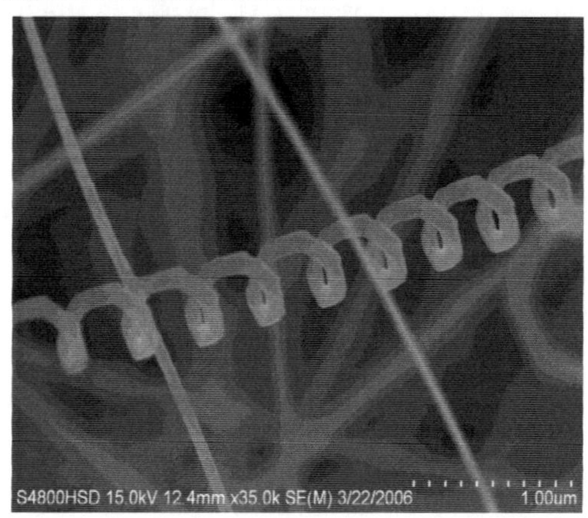

(b)

Figure 2. (a) SEM image of a ZnO nanohelice; (b) A uniform ZnO nanohelix with right hand roller.

Figure 2(a) shows the general morphology of the as-formatted sample from the low magnification SEM image. 1D ZnO nanowires were synthetized, yield up to 8%. The nanostructures are helices with lengths of several tens of micrometers. It shows distinctly that the nanohelices are concordant by High magnification SEM image from Figure 2(b). The pitch distance L, mean diameter D and the width of the nanowires of one ZnO nanohelice are uniform. They are range between 400-1000nm, 200-300nm, 100-300nm, respectively. The nanohelices are almost hexagonal in cross-sections indicating that the axis of the nanohelix maybe grows along <0001> directions of Zinc Oxide. A period of six equal length pieces forming a helix. It may be pointed out that nanohelices can convert from right to left, or contrary. These helical structures gradually become loose, and even becomes curved nanowires.

Detailed structural of the nanohelices was further carried out using TEM. The transmission electron microscope images (Figure 3(a)) show that the angle between the direction of growth and the direction of axial is about 47^0.

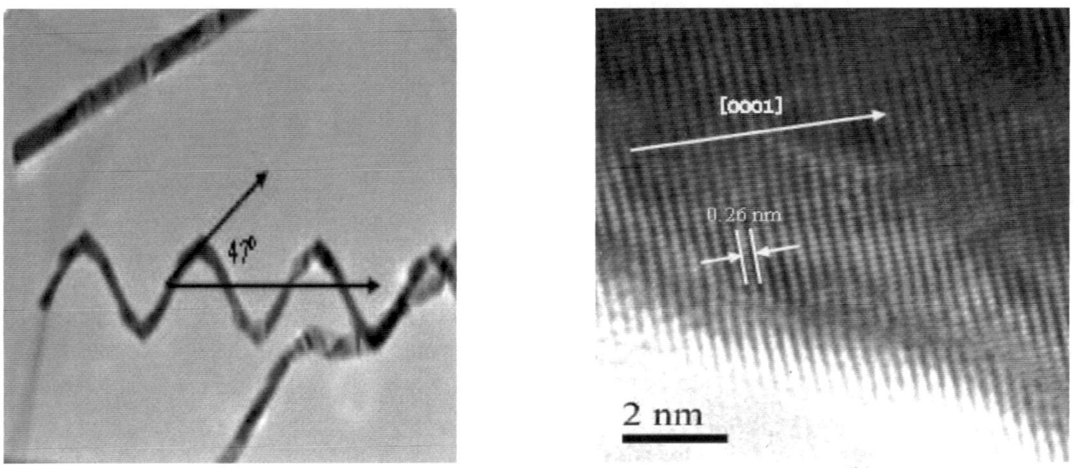

(a) (b)

Figure 3. (a). The structure of ZnO nanohelix under transmission electron microscope(TEM); (b). HRTEM images indicated at the connection between two borders on pieces near the outer surface.

An HRTEM image (Figure 3(b)) indicated at the connection between two border on pieces of the nanohelices shows the apparently separated interplanar spacing of 0.26nm, corresponding to the (0002) lattice spacing which establishes that the helical structure has a dislocation free single crystal structure and that the nanohelices as a whole have an axis of $<0001>$. Based on images of SEM and TEM the theoretical angle between the $<0\bar{1}11>$and [0001] directions is 47.25^0, it can be induced that the six pieces form the nanohelix structure grows in a spiral cycle along the six equivalent directions. Thus, the spiral surfaces of side should be composed of Zn^{2+} terminated $\{0\bar{1}11\}$, O^{2-}terminated $\{10\bar{1}\bar{1}\}$, and nonpolar $\{01\bar{1}2\}$ are in back or front. Every building block of the nanohelices is terminated by (0001) Zn-terminated surface. It has confirmed that the surfaces are chemical activity at the (0001) Zn-terminated, but the surfaces of $(000\bar{1})$ O-terminated and nonpolar surfaces are inert [9]. Therefore, (0001) the surface will be lead growth along the direction of uniform spiral. To investigate the growth conditions for ZnO nanohelices, we changed the growth time, pressure, gas flux and the growth temperature and produced a series of samples. The results show that the induction of Sb added the raw powder is very necessary for formation of ZnO nanomaterials. Without Sb, only nanowires with a growth direction of [0001] can be produced [10]. Introduction of Sb obviously decreases the evaporation temperature of ZnO and provides a special growth kinetics overcoming the surface energy barrier to form ZnO rectangular nanostick[11]. From the calculated results of the relevant change of electrostatic energy, the nanospiral is positive and favorable. It is worth noting that the ZnO nanohelices grow along the C-axis, and the length of obtained products could beyond several micrometers.

Figure 4. SEM images about SiO_2 nanowires in higher temperature zone.

The temperature of deposition is another important parameter influenced the morphology. W hat we discovered is that the ZnO nanohelices could only be fabricated at higher temp region about 1200^{0}C. At high temperature deposition region, thermal disturbing caused the formation of ZnO nanowires or SiO_2 nanowires (in Figure 4). According to the preceding information, a Vapor-Solid growth mechanism of Sb induced synthesis is put forward. The mechanism for th -e growth of nanohelix in detail is still under study.

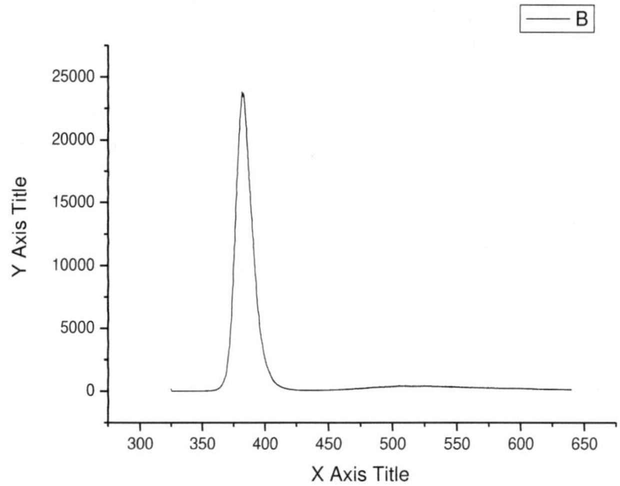

Figure 5. Room temperature PL curves of the as-synthesized ZnO helices.

Figure 5 shows the room temperature PL curves of the as-synthesized ZnO helices. What we analyzed is that the peak of near-band-edge ultraviolet (UV) at 384 nm is pretty strong compared with the green emission at 520nm, which is quite weak. It is usually believed that the oxygen deficiency in ZnO leads to the visible emission peak and the emission results from the radiative recombination of a photogenerated hole with an electron occupying the oxygen vacancy. The luminescence of the free exciton recombination causes the Ultraviolet emission [12]. For the PL spectrum of the nanohelices, an intense single ultraviolet emission peak is situated at 384 nm, which can be attributed to the free excitonic recombination emission, while the deep-level emission corresponding to oxygen vacancies or surface state emission is barely observed, being much weaker than previously reported [13].This fact of experiment shows the products' good crystalline nature.

4. Summary and conclusion

To summarize, through the way of thermal evaporation to get the uniform ZnO nanohelices was fabricated. The nanohelices have a uniform period, a faultless hexagonal cross-sectional area and an $\langle 0001 \rangle$ axial growth direction. It is necessary for the induced Sb in the source material to form the helical structure, which is energetically favorable. In the different deposition temperature zone, different morphologies of ZnO nanostructures show that another factor in growth mechanism is deposition temperature. Due to the perfect periodicity of helical structure along the axis, it is hoped that it can be used in a photonic band-gap material. In the helical structure, it is possible to find some interesting phenomenon related to Faraday rotation effect, which might be possible used for building pieces in photoelectric devices and optoelectronic coupling system also.

References

[1] Wang Z L 2009 *Mater. Sci. Eng. R.* **64** 33
[2] Pearton S J, Nortona D P and Ipa K 2005 *Prog. Mater Sci.* **50** 3
[3] Kong X Y, Ding Y, Yang R S and Wang Z L 2004 *Science.* ***303*** *2*
[4] Wang X D, Song J H and Wang Z L 2006 *Chem. Phys. Lett.* **424** 86
[5] Zhang Y H, Liu J, Liu T, You L P and Li X G 2005 *Cryst. Growth* **285** 54
[6] Yang R S, Ding Y and Wang Z L 2004 *Nano Lett.* **4** 7
[7] Bae S Y, Lee J Y, Jung H, Park J and Ahn J P 2005 *J. Am. Chem. Soc.* **127** 10802
[8] Gao H, Zhang X T and Zhou M Y 2006 *Sloid State Communications* **140** 9
[9] Baxter J B, Wu F and Aydil E S 2003 *Appl. Phys. Lett.* **83** 18
[10] Yang R S, Ding Y and Wang Z L 2004 *Nano Lett.* **4** 7
[11] Gao H, Zhang X T and Zhou M Y 2007 *Nanotechnology***18** 065
[12] Cheng C W, Xie E J and Liu B 2010 *Appl. Phys. Lett.* **96** 7
[13] Wan A H, Ji X X and lu L X 2012 *Chin. J. Low Temper. Phys.* **34** 2

IOP Conf. Series: Materials Science and Engineering **213** (2017) 012010 doi:10.1088/1757-899X/213/1/012010

Surface modification by deformational cutting for improving the bonding strength of polymers and polymer composite materials with metals and with each other

N N Zubkov[1,2] and I V Bezin[1]

[1]Bauman Moscow State Technical University, Moscow, Russia, 105005, 2-ya Baumanskaya st., 5/1

E-mail: zoubkovn@bmstu.ru

Abstract. This article covers the analysis of possibilities for the use of deformational cutting method as a tool for the creation of macroreliefs designed for increasing the strength of connections between materials which are characterized by a low adhesive quality. Surface reinforcement of carbon-fiber materials with sheet metal is also considered in this work. The results shown prove a major increase in strength values. All samples with macrorelief obtained by deformational cutting show high strength in exfoliation and shock tests with a changing damage mechanism from adhesion to cohesion.

1. Introduction

Metal alloys, polymeric and composite materials are widely used in industry. Each group has its significant advantages before other groups in terms of price, strength, hardness, wear resistance or other parameters. It would be reasonable to combine advantages of different types of structural materials in one prefabricated part structure while each component of an assembly unit is responsible for its own functionality. Assembly method is key factor in functionality, reliability and manufacturability of the prefabricated unit. Adhesive joints are able to connect various materials that differ from each other both in properties and thickness, and not to weaken the parts subject to connection. Sometimes bonding is the only possible way to connect the unit parts. The disadvantages of adhesive joints include low tear resistance, instability of physical and mechanical properties as well as the need for special treatment of the surface before bonding [1].

Modern engineering also sees a widespread use of laminated metal-polymer composites or metal-filled plastics - materials consisting of alternating thin sheets of metal alloys and polymer layers or polymer composite materials: organic, carbon, or fiberglass plastics (PCM). Such hybrid materials are distinguished by high strength and fracture toughness values achieved due to the fact that properties of their constituent components complement each other. Metal-polymer composites are important for applications, e.g. in aircraft structural elements, which makes it possible to reduce their weight and significantly improve the structural life [2-3]. Sports equipment (snowboards, skis) also has in its structure compounds of metal – carbon composite – antifrictional sliding polymer layer [4]. Reduction in the weight by 80 kg of new car models was achieved specifically due to a combination of adhesive joints of PCM and metals [5]. Not only adhesive joints are used for bonding of metals with polymer or polymer composite materials (PCM) or combinations thereof, but also joints obtained directly during the polymerization process of the polymer and/or PCM binder [6-8].

Modern adhesives and binders have good adhesion to PCM with a thermosetting matrix (glass-, carbon-, organic plastics, but the strength of their adhesive bonding to metals and thermoplastic polymers is insufficient [9-10]. If bonding or polymerization is made for connection, preliminary treatment of the surfaces by various methods is of particular importance. It plays a crucial role in ensuring durability of connections during operation. The objective of preliminary surface treatment is to create conditions for a physical and chemical interaction of surfaces to be bonded [11].

Currently, surfaces to be bonded are exposed to the procedures of degreasing, pickling, oxidation, phosphating, zinc plating, surface activation by electrical discharge, ultraviolet irradiation, flame treatment, etc. to ensure strength of joints "metal-PCM", "metal-polymer" and "polymer-PCM" [12]. More reliable compounds allow one to make the fixing microelements on metal via punching combs, pin milling or pressing, welding, and so forth (Figure 1) [13-15]. The topical issue is miniaturization of fasteners and ensuring the reproducibility of the required quality of their installation, which is quite a difficult technological challenge [9].

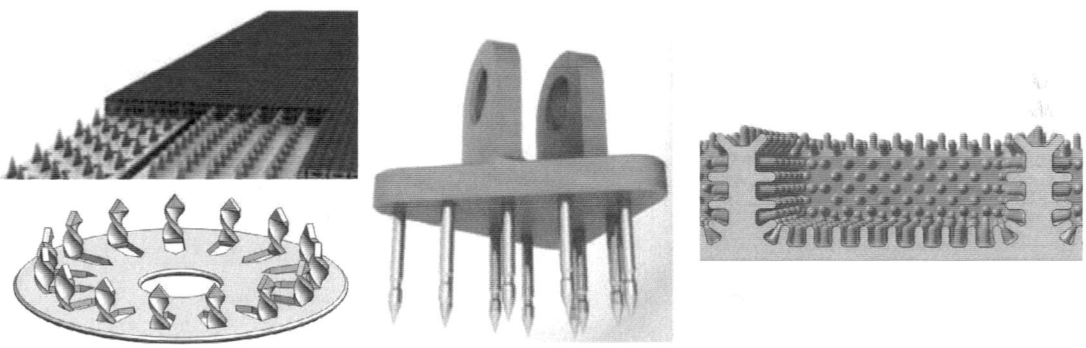

Figure 1. Fixing elements for connection of metal with PCM [6, 13].

2. Deformational cutting for surface macrostructuring.

An existing increase in strength and reliability of both glue conjunction and compound obtained via polymerization of the binder can be enriched via regular small pitch relief with a high height to pitch ratio. The creation of the deep micro or macroreliefs via deformational cutting (DC) is the most promising method.

The DC method is based on simultaneous cutting of the surface layers and their plastic deformation [16]. The cut layer is not removed completely from the blank, being mechanically connected with it from its narrow side. A set of the cut surface layers that retains the continuity of its connection with a blank forms a enhanced macrorelief in the form of finning or pin fins on the treated surface (Figure 2). DC technology boasts high performance and a wide range of sizes of the relief obtained and can be implemented both on specialized and unified metal-cutting equipment [17]. Turning machines are used for the processing of cylindrical surfaces and sheet bars using the mandrels. Planing and milling machines are used to process planes. DC technology allows obtaining a macrorelief only on ductile materials such as copper, aluminium and their alloys, titanium, steels, thermoplastic polymers, etc. Parameter which characterizes the plasticity of the material to the maximum extent is the relative elongation value δ_{10}. For the materials with δ_{10} <18% the cut layer does not remain in the fin form on the workpiece, but is separated in the form of chips [17].

IOP Conf. Series: Materials Science and Engineering **213** (2017) 012010 doi:10.1088/1757-899X/213/1/012010

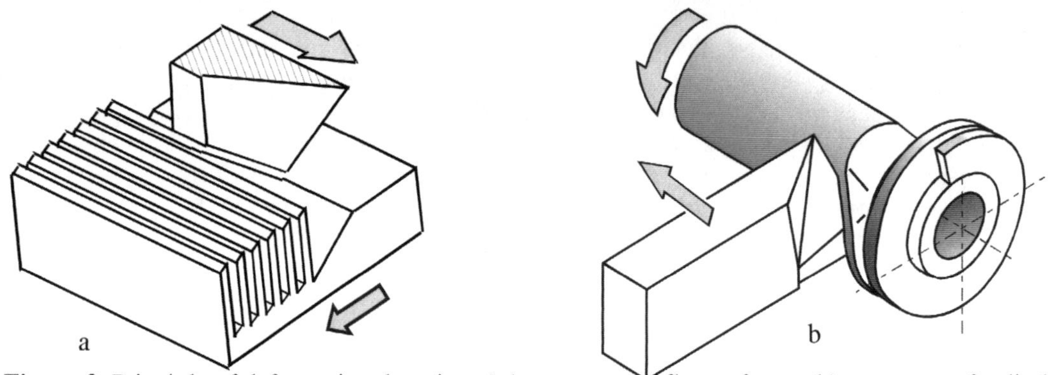

Figure 2. Principle of deformational cutting. (a) treatment of flat surfaces, (b) treatment of cylinder.

The typical profile of macrorelief in the form of finning which is obtained through DC technology and its geometric parameters are shown on Figure 3. For vertical fins, there is the following relationship between the parameters of the fins and technological processing parameters [17]. When turning is made, fin spacing S is determined by the tool feed based on one part revolution: $S=S_0$, for milling operations - feed per tooth: $S=S_{tooth}$, when shaping - feed per double stroke: $S=S_{fds}$

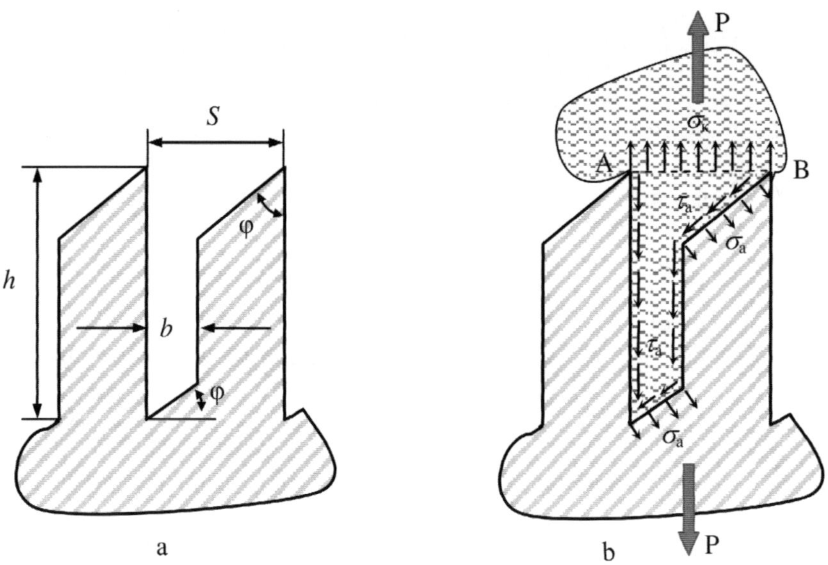

Figure 3. A typical profile and parameters of macrorelief in the form of fins obtained through DC technology (a) and layout of current stresses if adhesive joint is exposed to tearout forces (b).

The width of the interfin gap (b) is determined by the fin spacing S and undercutting angle φ according to the function as follows:

$$b=S\cdot(1-\sin\varphi) \quad (1)$$

where φ is the angle of undercutting.

Fin top angle is equal to the undercutting angle φ.

Let us consider an option of exposure of an adhesive joint to a uniform separation in an interfin gap. The strength of adhesive joint is caused both by the adhesion interaction of surfaces to be glued with an adhesive layer, and the cohesive strength of the adhesive itself. The ratio of the interfin gap width and height (fin height) is an important factor that enables maximum strength of the adhesive joint with the surface treatment using DC technology. The use of shallow grooves will not provide a reliable adhesive bond between the adhesive joint and cladding metal. The fins of excessive height will result in an increase in steel intensity of the cladding layer, while the production technology will become more difficult.

A reasonable option is an adhesive joint which functions with fully implemented values of adhesive permissible tensile strength σ_K(cohesive strength), i.e. when the failure occurs on the line A-B (Figure 3b) without detachment from the lateral sides of the interfin gap. Figure 3b shows with small arrows the tangents τ_a and normal lines σ_a of adhesion bond stress at the interface adhesive-metal. Inclined places of the interfin gap (triangular fin pointing and bottom of the interfin cavity) are exposed both to normal σ_a and τ_a tangential stress lines. Vertical walls are exposed to tangential stresses only. Let us take as an assumption a uniform distribution between normal and tangential stresses along the perimeter of interfin cavity. Such approach is justifiable for an estimate calculation [18], and is used for adhesive joints of small length [6], which is typical for interfin gap areas.

The condition of full implementation of adhesive cohesive strength means that the total shear and tensile strength of adhesive bonds along the perimeter of the interfin gap should be higher than the adhesive tensile strength along the line A-B. This position allows to minimize the macrorelief height while ensuring the maximum strength of the adhesive joint.

Let us consider the layout of adhesive interaction between the adhesive and metal along the perimeter of interfin gap on the area of the adhesive joint with a length equal to fin spacing. (Figure 3b). The force required for cohesive failure of adhesive over the area with the length equal to the fin spacing will amount to: $P=[\sigma_k]\cdot S$, where $[\sigma_k]$ is a permissible uniform tensile strength of the adhesive.

The total component of shear and normal stresses of adhesive bonds in the direction of tensile force with regard to the length of their impact on individual areas should be recorded as follows:

$$P=\frac{[\tau_a]\cdot S\cdot\sin\varphi}{\sin\varphi}\cdot\cos\varphi+\frac{[\sigma_a\cdot S\cdot\sin\varphi]}{\sin\varphi}\cdot\sin\varphi+[\tau_a]\cdot\left(h-\frac{S\cdot\sin\varphi}{tg\varphi}-\frac{S\text{-}S\cdot\sin\varphi}{tg\varphi}\right)+$$
$$+\frac{[\tau_a]\cdot(S-S\cdot\sin\varphi)}{\cos\varphi}\cdot\sin\varphi+\frac{[\sigma_a]\cdot(S-S\cdot\sin\varphi)}{\cos\varphi}\cdot\cos\varphi+[\tau_a]\cdot h. \quad (2)$$

where -$[\tau_a]$ is a shear strength of the adhesive bond of the adhesive and material to be glued,

$[\sigma_a]$ is the ultimate uniform tensile strength of the adhesive bond between the adhesive and material to be glued, h - fin height.

If we equate tensile strength of the adhesive over the area with the length equal to the fin spacing, with the sum of adhesive interaction components along the perimeter of interfin gap, we will receive the following after simplification:

$$[\sigma_k]\cdot S=[\sigma_a]\cdot S\cdot\sin\varphi+[\sigma_a]\cdot S\cdot(1-\sin\varphi)+2[\tau_a]\cdot h. \quad (3)$$

Based on this expression it is possible to obtain a minimum height of interfin gap for this finning spacing which ensures the absence of adhesion failure along its perimeter (i.e. ensuring maximum strength of adhesive joint based on its ultimate uniform tensile strength:

$$h\geq\frac{S\cdot([\sigma_k]-[\sigma_a])}{2[\tau_a]}. \quad (4)$$

As it can be seen from the expression, the macrorelief is required in the first place for adhesives with a significant difference between the cohesive and adhesive strength values. The greater the difference in these values, the deeper the interfin gap must be for this combination adhesive-metal.

The second important conclusion that can be drawn when analysing this formula, is the absence of undercutting angle therein which determines the ratio of the width of the interfin gap to its spacing: $b/S=1-\sin\varphi$, i. e. the strength of the adhesive joint is not affected by the interfin gap width (for this spacing and fin height).

3. Preparation of the surfaces for glueing by using the DC technology

Materials with a low adhesion were selected to assess the efficiency of use of macroreliefs obtained through deformational cutting in order to enhance the strength of adhesive joint: austenitic heat-resistant steel 316L, Teflon and ultrahigh molecular weight polyethylene (UHMW-PE).

Table 1 shows the parameters of samples with the macrorelief obtained via DC technology. Relief cross-sections are shown for samples No. 1-4 in the Table 1, top view of the structure is shown for the sample No. 5 on Figure 4a.

Macrorelief No. 1 on the steel samples was made on the strip with an initial thickness of 0.76 mm. Macrorelief (sample No. 2) was performed on one side of the Teflon cube having a side of 11 mm. Macrorelief on samples 3-6 was performed on a tape made of UHMW-PE with a width of 10 mm and a thickness of 0.8 mm. The tape was obtained through the planing of the end face of the sheet with the thickness 10 mm via free cutting method. When receiving the reliefs on the polymers, the peculiarities of their processing via DC method described in [19] were taken into account.

The macrorelief had inclination on all samples to increase strength of the adhesive joint. When putting samples No. 1 to the delamination test and during uniform tensile test of samples No. 2, no tearing of the adhesive from the interfin gap was recorded. The failure occurred along the adhesive. For samples made of UHMW-PE No. 3, even the use of inclined fin with large inclination angle of the fins of 32° during a delamination test was accompanied by the absence of cohesive failure. Epoxy adhesive "slipped out" even from the inclined interfin gap. The Figure 4b shows an area with epoxy glue following delamination of UHMW-PE, which remained on the carbon fibre and is a replica of the interfin gaps shape of the sample No. 3. This fact required the search for other macrorelief configurations on UHMW-PE which were implemented on samples No. 4 and 5. Interfin gap on the sample No. 4 has an expansion at the bottom part by forming a locked cavity for a cured adhesive. The width of the interfin gap is 130 μm in the expanded part and 60 μm in the narrow part. The principle of obtaining such subsurface cavities via DC technology is described in [20].

Macrorelief on the sample No. 5 was obtained through double finning of the tape made of UHMW-PE with the finning pitch on each pass of 0.3 mm and with the pass crossing angle 13°. After the processing the pin-type surface of the tape represented in its essence a variety of inclined "villi" with a length of 1.3 mm. The pin has in its cross section a rhombus shape with a side of 0.15 mm. The inclination of pins was 70° from the vertical. 1000 pins were arranged on every square centimeter. The principle of obtaining the pin structures is described in [21].

Table 1. Parameters of the macrorelief of samples to be glued and test conditions.

Sample No.	1	2	3	4	5
Material to be glued	Steel 316L	Teflon	UHMW-PE	UHMW-PE	UHMW-PE
Type of structure	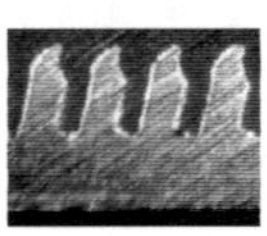	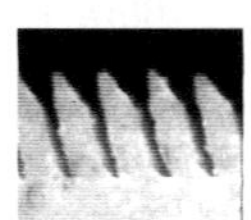	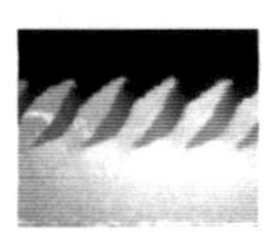	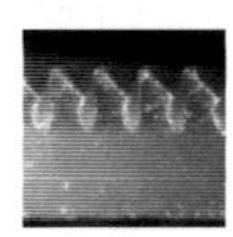	Figure 4a
Pitch, mm	0.4	0.5	0.4	0.3	0.3
Height of relief, mm	0.67	1.0	0.55	0.4	0.3
Interfin gap, mm	0.16	0.15	0.15	0.13/0.06	–
Fin, or pin fin inclination, deg	8	15	32	0	70

Material of counterpart	Carbon-fiber	Teflon	Carbon-fiber	Carbon-fiber	Carbon-fiber
Adhesive trade mark	Toolfusion 1A/1B (Airtech)	Russian epoxy binder. Specification 07510508.90-94	Araldite 2011 (Huntsman Corporation)	Araldite 2011 (Huntsman Corporation)	Araldite 2011 (Huntsman Corporation)

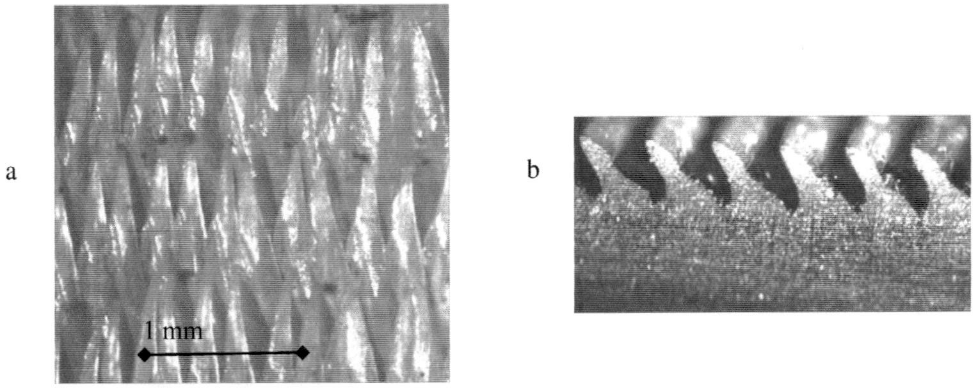

Figure 4. Micropin-type surface of UHMW-PE for bonding (a) and carbon fibre PCM view following delamination of finned tape UHMW-PE (b).

4. Test results

During the test of sample No. 1 (6 tests) using impact machine Walter+Bai AG PH150 with impact energy of 50 J, fracture energy An = 11.5 J was recorded, which resulted in the value of specific impact toughness An= 228 kJ/m^2. Samples with a cladding metallic layer without macrorelief exfoliated completely from carboplastic (Figure 5b), whereas the samples with microrelief exfoliated exclusively by the carboplastic layers (Figure 5c). Table 2 show that 53-60% increment in the value of specific impact strength of sample No. 1 is attained when compared with sheets without macrorelief [9].

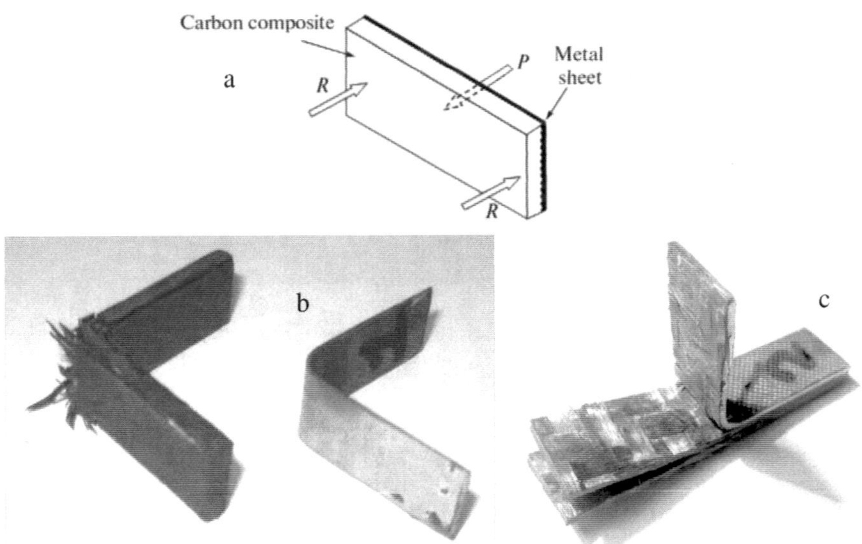

Figure 5. (a) Test scheme on impact-testing machine and samples after tests (b) without macrorelief on metal, (c) with macrorelief obtained via DC.

IOP Conf. Series: Materials Science and Engineering **213** (2017) 012010 doi:10.1088/1757-899X/213/1/012010

Table 2. Destruction energy of samples in tests on impact-testing machine [9].

Sample number	1	Cladding without relief	Without cladding
Destruction energy An (J),	11.5	7.2	4.1

Six tests were also carried out for the sample No. 1 to measure the peeling strength under the angle 90°. Test configuration is given on Figure 6a. The tests were conducted on a universal testing machine INSTRON 600-DX. The size of the contact conjunction area metal - carbon fibre was 25 x 100 mm. The samples were stretched at the speed of 25 mm/min, and an average breaking force was measured during the test. On the next phase, the force was converted into the line force, measured in Newton per 1 mm sample width (or in kN/m).

A multiple increase in the delamination force of the sheet metal having a macrorelief compared to cladding with smooth metal plates was recorded. Plates with no macrorelief flaked simultaneously over the entire bonding area (at force orders of magnitude. Cohesive failure occurred for finned samples, no tearing of the epoxy binder from macrorelief was observed (Figure 6). An average linear breaking force was within the range 4.4-9.1 kN/m [9].

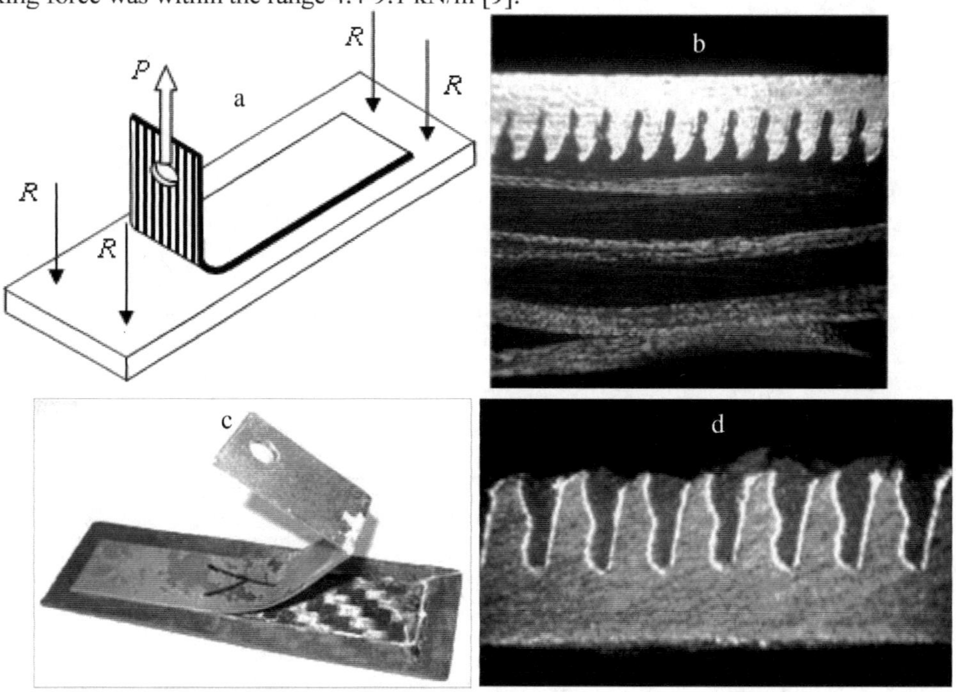

Figure 6. (a) scheme of 90° exfoliation tests. External view of samples after 90° exfoliation tests: (b) plated slab with macrorelief; (c) cross section of a plated layer with macrorelief after exfoliation.

The strength of adhesive joint is performing a uniform tensile of the samples No. 2 was measured when gluing the couple Teflon - Teflon. Teflon blocks were glued with sides having an applied macrorelief. Breakout force was recorded with the use of dynamometer model 9257 V made by "Kistler". When using EDP epoxy adhesive, a couple of glued samples withstood the greatest load: they had mutually perpendicular directions of microfins on the cube sides. The maximum tensile strength was 12 MPa. When bonding Teflon+Teflon by using hot-melt glue (ethylene vinyl acetate), tensile strength amounted to 5 MPa.

Strips made of UHMW-PE bonded with an epoxy adhesive Araldite 2011 were tested for delamination at an angle 90° from carbon fibre. Microfinning direction was along the strip being delaminated. The test results are shown in Table 3.

Table 3. Force "per unit of length" of 90° exfoliation of UHMW-PE from carbon-fiber.

	Surface treatment of UHMW-PE strip				
	DC treatment			Coarse-grained abrasive paper	Without surface treatment
	Sample number				
	3	4	5		
Exfoliation loading per unit length, kN/m	1.2	1.7	4.5	1.0	0.15

5. Discussion

When bonding metal with carbon fiber microfinning on the metal allowed improving the strength of the adhesive joint up to the carbon fibre matrix level. When testing Teflon for uniform separation, the maximum strength of adhesive joint amounted to 12 MPa, which is only two times lower than the tensile strength of Teflon (23 MPa). When testing for delamination of UHMW-PE from carboplastic, the relief in the form of microfinning did not produce significant results compared to the treatment by means of coarse-grained sand paper. Inclined fins did not provide a locked connection and an adhesive delamination occurred. Interfin gap with an expansion at the bottom part secured an anchor (lock-type) effect. Epoxy adhesive remained in the expansion of interfin gap, its failure occurred in the narrowest part having the width of 70 μm, which resulted in an increase in strength. However, for the sample No. 4, the total area of adhesive bridges subject to tension is five times less than the area of surface to be bonded, which did not allow to fully implementing the strength of epoxy adhesive. When testing UHMW-PE for delamination, bonding surface in the form of inclined micropins turned out to be the most effective in terms of strength. The peel strength value exceeded five times the similar strength for the surface treated by means of coarse-grained sandpaper.

Undoubtedly, the presented versions of macrorelief shapes and sizes are not optimal; however, the first results of strength tests prove the prospects of using the DC technology to enhance the strength of adhesive joints between materials with low adhesive properties.

6. Conclusions

1. Surface preparation of low-adhesive materials using DC method revealed to change from adhesion to cohesion destruction mode.

2. For all samples with macrorelief via DC, there was either interlayer destruction by carbon fabric layers or destruction of epoxy binder on the top of the fins.

3. With double-sided finning of metal sheets DC technology can also be used for the creation of multilayered composite materials based on thin metal sheets and adhesive prepregs – MPCM.

4. Microrelief obtained via deformational cutting exhibited metal–PCM contact strength in exfoliation to the carbon-fiber matrix strength.

5. In the shock tests for samples with microrelief on metal, there was detected an increase in the destruction energy to 60% in comparison with macrorelief free samples on metal.

6. For UHMW-PE only the micropinned surface showed sufficient result.

Acknowledgement
This work was supported by the Ministry of Education and Science of Russia.

References

[1] Adams R *et al* 1997 *Structural Adhesive Joints in Engineering* (London: Chapman & Hall) p 360

[2] Vlot A and Gunnink J 2011 *Fibre Metal Laminates*(New York: Springer Science & Business Media) p 532

[3] Kraemer B 2014 Kleben im Leichbau. Von der Materialherstellung bis zur Reparatur *Adhäsion Kleben und Dichten* **58**(10) 28-31

[4] Galetz M *et al* 2007 Carbon nanofibre reinforced ultrahigh molecular weight polyethylene for tribological applications *J. of Appl. Polymer Sc.* **104** 4173-81

[5] Lutz A and Schmaltloch S 2014 Composites strukturell kleben *Adhäsion Kleben und Dichten* **58**(9) 32-6

[6] Messler R 2004 *Joining of Materials and Structures from Pragmatic Process to Enabling Technology* (Burlington: Elsevier) p 816

[7] Troughton M 2008 *Handbook of Plastics Joining: A Practical Guide* (New York: William Adrew) p 600

[8] Dillard D 2010 *Advances in Structural Adhesive Bonding* (Burlington: Elsevier science technology) p 656

[9] Zubkov N *et al* 2016 Surface reinforcement of carbon composites with microstructural metal materials *Polymer Sc. Series D* **9**(1) 91-5

[10] Vermeeren C 2003 An Historic Overview of the Development of Fibre Metal Laminates *Appl. Compos. Mat.* **10** 189-205

[11] Marcus S *et al* 2006 Fehlervermeidung durch Inline-Monitoring des Oberflaechezustandes *Adhaesion Kleben und Dichten* **4** 20–7

[12] Brockmann W *et al* 2008 *Adhesive Bonding: Adhesives, Applications and Processes* (Hoboken: John Wiley & Sons) p 432

[13] Cherevashchenko A 2013 Structural technological solutions for metal–composite compounds operating under peeling stress *Probl. Proekt. Proizvod. Konstr.Letatel'nykh Appar.* **1**(73) 14-20

[14] Komarov G 2016 Trends in Joining Technology of Parts Made of Plastics and Joints with Them *Polymer Mat.* **10** 42-8

[15] Kunststoff 2015 Oberflächen passgenau modifizieren *Adhäsion Kleben und Dichten* **59**(12) 6

[16] Solovyeva L *et al* 2012 Novel Electrical Joints Using Deformation Machining Technology Computer Modeling *IEEE Transac. on Comp., Pack. and Manufact. Techn* **2**(10) 1711-7

[17] Zubkov N 2001 Features of realization of the method of deforming cutting *Tekhnol. Mashinostr.* **1** 19-26

[18] Chawla K 2012 *Composite Materials Science and Engineering* (New York: Springer-Verlag) p 542

[19] Zubkov N and Sleptsov A 2010 Production of slotted polymer filter tubes by deformational cutting *Russ. Engin. Res.* **30**(12) 1231-3

[20] Popov I *et al* 2013 Heat transfer during the boiling of liquid on microstructured surfaces: Part 1: Heat transfer during the boiling of water *Thermal Eng.* **60**(3) 157–65

[21] Zubkov N *et al* 2014 Using the pin structures of a new type for cooling of electronic equipment *Vestn. Mosk. Gos. Tekh. Univ. im. N. E. Baumana, Ser. Mashinostr.* **2**(95) 70-9

Effect of biobased fillers nature on biodeterioration of hybrid polyethylene composites by mold fungi

E E Mastalygina[1, 2, 3], A A Popov[1, 2] and P V Pantyukhov[1, 2]

[1]Department of Chemistry and Physics, Plekhanov Russian University of Economics, 36 Stremyanny Lane, Moscow, 117997, Russia
[2]Emanuel Institute of Biochemical Physics, Russian Academy of Sciences, 4 Kosygina Street, Moscow, 119334, Russia

E-mail: elena.mastalygina@gmail.com

Abstract. The paper is devoted to investigation of deterioration of natural fillers and polyethylene composites on their basis (polyethylene/filler=70/30) due to the action of mold fungi. The fillers chemical composition, dimensional parameters and biodegradability have been analyzed as factors exert a considerable impact on composite materials biodeterioration. It has been found that the principal factor determining the biodeterioration of polyethylene/filler composites by mold fungi is chemical composition of a filler and, in turn, its biodegradability. The excess of holocellulose content over lignin content and high protein content in a filler are able to induce biofouling of the polymeric composite materials. The presence of soluble and easy hydrolysed fraction in a filler increases its availability in a polymeric matrix. According to the study results, most effective natural fillers as additives stimulating polyethylene composites biodegradability are milled straw of seed flax and hydrolyzed keratin of bird's feather.

1. Introduction

Development of composite materials on the basis of traditional synthetic polymers and various additives initiating process of biodegradation is the most technologically and economically advantageous direction of biodegradable polymers creation. Oxo-additives, polymers synthesized from renewable vegetable raw materials by biotechnological methods and natural polymers are most often used as the additives initiating degradation of polyolefins in the environmental conditions.

Hybrid composite materials based on polyolefins with natural fillers such as starch [1], wood flour [2], natural rubber [3], chitosan [4], various natural fibers [5-8], etc. have prospects for wide use. Addition of various production wastes as natural fillers allows not only to provide biodegradability and to partially replace non-renewable petrochemical raw materials on renewable but also to reduce their cost [9]. In most cases, production of similar composites based on polyolefins and natural additives does not require cardinal change of production process that also causes profitability of their production.

The presence of a filler leads to existence of large surface contact area with polymeric matrix through which destroying agents can get into the polymer [10]. Thus, a natural additive acts as an initiator not only biological destruction but also polymer oxidizing. Besides, natural fillers being hydrophilic and easily assimilated by microorganisms promote increase in adhesion of microorganisms to polyolefins surface [11]. The result is an expedited attachment of microorganisms to the surface of composite material with subsequent biofouling. To a greater extent, properties and ability of hybrid composite materials based on polyolefins and natural fillers to be degraded and assimilated by microorganisms is defined by content and chemical composition of a filler, and also dimensional parameters and shape of its particles [12].

The main objective of this study was to investigate the influence of chemical composition and dimensional parameters of biofillers on biodegradability of polyolefin composites. Additionally the authors place emphasis on the correlation between ability to biodeterioration of individual fillers and composites on their basis.

2. Materials and methods

Objects of the research were fillers and composite materials on the basis of low-density polyethylene (LDPE) (TM 15803-020, JSC Neftekhimsevilen) with the following molecular mass characteristics: M_w=1.0×10^5 Da, M_n=1.5×10^4 Da, M_w/M_n=6.9 (GPC at (140±2)°C, solvent is 1,2,4-phenyl trichloride). Available natural raw materials, which mainly are waste of various productions, were used as fillers: powdered cellulose (hereafter cellulose) ("Politsell PTsS", TU 5410-029-32957739), native corn starch (starch) (GOST 51953), milled straw of seed flax (flax) (LM98 grade), milled chaff of winter wheat (wheat) (Krasnodarskaya-99 grade), wood powder of fir-tree breeds (wood), hydrolyzed keratin of bird's feather to oligopeptides (feather).

The filler content for all the composite materials based on LDPE was 30 wt.%. Obtaining of the composites was carried out by means of a Plasti-Corder mixer torque rheometer (Brabender type) in the atmosphere of gaseous argon at a temperature of (140±2)°C and with rotors speed of 30 rpm within 5 minutes. A weighted amount of LDPE was placed in the mixing chamber, 2 minutes later filler was added. After cooling, materials were crushed by means of a knife mill RM 120 (Vibrotechnik, LLC, Russia), and then subjected to pressing by a manual hydraulic press PRG-10 (VNIR, LLC, Russia) at a temperature of (140±2)°C and load of 70 kgf/m^2 on a cellophane substrate, followed by quenching in water at (20±2)°C. Finally, film samples with the thickness of (130±10) μm were obtained. The thickness of the films was measured by a micrometer screw gage.

The analysis of chemical composition was carried out by infrared spectroscopy on a FT-IR spectrometer Perkin Elmer Spectrum 100 (Germany) at a temperature of (22±2)°C in the range of wave numbers $4600 \leq \nu \leq 650$ cm^{-1} by a method of frustrated total internal reflection.

Thermal stability of fillers was determined by a method of thermogravimetric analysis (TGA) on a derivatograph Netzsch TG 209 F1 Iris (Germany) at a heating speed of 20°C/min in a temperature interval of 25-800°C.

Microscopic investigation was carried out using an optical microscope Axio Imager Z2m, Carl Zeiss (Germany) with Axio Vision software at magnification of 50x, 200x, 500x in transmitted and reflected light.

Mycological examination was carried out at the premises of the mycology and algology department of Lomonosov Moscow State University. Biodegradability of the fillers was estimated on water and agar media with 30 wt.% of a filler according to GOST 9.049 and ISO EN 846 with the use of *Aspergillus niger*, *Aspergillus terreus*, *Penicillium chrysogenum*, *Aspergillus flaus*, *Penicillum cyclopium* and *Paecilomyces variotii* test fungi cultures, which are the most universal decomposers of carboniferous substances. As control environments, Czapek's medium and wort agar on the basis of malt extract were used. The pH index was determined by means of pH chemical agent test strips.

Microbiological testing of the composite materials was carried out with *Aspergillus niger*, *Penicillium chrysogenum*, *Trichoderma viride* and *Paecilomyces variotii* test cultures according to GOST 9.049 and ISO EN 846. The incubation of the samples inoculated by spore suspension of micromycetes (1×10^{5}spor/ml) was carried out in the conditions of relative humidity more than 90% at (22±2)°C within 84 days with intermediate studies in 28 and 56 days. The fungal development on a sample surface was estimated by a rating scale according to ISO EN 846. The gain of fungal biomass was determined by the difference in weight of a sample with fungal mycelium and a cleared sample (washing in ethyl alcohol (95 vol.%)).

3. Results and discussion

The most important characteristics of fillers for polyolefins are decomposition temperature, particles shape and size, chemical composition, and biodegradation rate under the influence of microorganisms. The basic physical and chemical properties of the fillers under investigation are given in Table 1.

The fillers have various degree of dispersion and dimensional parameters of particles (length (L) and diameter (D), length to diameter ratio (L/D-ratio)) that can exert a considerable impact on materials structure, properties, and biodegradability [7]. Oblong soft fibers with a wide range of length ($L/D \approx 10$) are typical for the cellulose. Largely, particles of the milled flax straw have rigid oblong cylindrical units ($L/D \approx 8$). The particles size and shape of the milled wheat chaff and wood powder are similar with elliptical shape of particles ($L/D \approx 3$). The starch filler and hydrolyzed feather have small spherical particles of 10-20 μm with a tendency to large agglomerates formation (up to 250 μm). The presence of a soluble and easy hydrolyzed fraction in some of the fillers (flax, wheat and feather) could effect on the weight loss under soil medium resulting from removing of the available fraction.

Table 1. Properties of the fillers applied for polymeric composites.

Filler	Decomposition temperature (TGA), °C ($\Delta \pm 3$°C)	Dimensional parameters of particles, μm ($\Delta \pm 5$μm)	Easy hydrolysable fraction, wt.% ($\Delta \pm 5$wt.%)	Crystallinity index, D_{1423}/D_{895} (IR)	Holocellulose/ lignin, D_{1370}/D_{1510} (IR)	Lignin/ hemicellulose D_{1595}/D_{1645} (IR)
Cellulose	290	L=150-500 D=20-40 $L/D \approx 10$	5	0.76	-	-
Flax	240	L=100-400 D=25-50 $L/D \approx 8$	25	1.33	1.14	1.05
Wheat	270	L=80-100 D=20-40 $L/D \approx 3$	30	1.12	1.96	0.64
Wood	275	L=70-120 D=20-40 $L/D \approx 3$	10	0.87	1.05	1.45
Starch	290	L=10-20 D=10-15 $L/D \approx 1$	Gel formation	-	-	-
Feather	190	L=10-20 D=10-20 $L/D \approx 1$	55	-	-	-

One of the major factors determining a rate of biodegradation of synthetic polymers filled by natural components is the ability of microorganisms to use substances forming part of the fillers as a nutrient source [13]. For studying biological utilization rate of the fillers by microorganisms, the analysis of their chemical composition was carried out by means of IR-spectroscopy and TGA, as well as an assessment of fillers biofouling by mold fungi cultures.

According to TG and DTG-curves received by the TGA method the starch and hydrolyzed keratin fillers contain about 10 wt.% of easily volatiles as shown by an endothermic peak in the area of (70±15)°C [14]. The summarized peak of lignocellulose complex decomposition (hemicellulose, cellulose and lignin) with a maxima at temperatures of 307°C, 320°C and 345°C are observed on the DTG-curves of flax, wheat and wood, respectively. It is important to note that a decrease of thermal stability of cellulosic components compared to pure cellulose occurs due to the presence of water and hemicellulose in flax, wheat and wood [15]. In consequence of hemicellulose destruction in these fillers, non-flammable gases are evolved that is capable to accelerate the process of cellulose depolymerization [16]. Increasing the oxidation time of a carbon residue and share of a non-combustible ash content for the cellulosic fillers (10-13 wt.%) compared with pure cellulose (5 wt.%) indicates the presence of lignin. The existence of a narrow exothermic peak on the DTG-curve of the flax straw decomposition in the area of (450±5)°C can be referred to fats oxidation being its part [17].

In accordance with the results of IR-spectroscopy (ATR) of the powdered fillers, the peculiarities of chemical composition were determined (Figure 1). The ratio of absorption bands intensities of 1015 and 1047 cm^{-1}, which are responsible for amorphous and organized structures of starch, respectively [18], is equal to 1.65 that gives evidence of its poorly organized structure. The comparison of IR-spectra of the cellulosic fillers was carried out. From the ratio of band intensities of D_{1370}/D_{1510} [19] it was stated that the ratio of holocellulose/lignin increases in the series of wood-flax-wheat. The ratio of lignin and hemicellulose in the fillers was determined by the ratio of band intensity of 1595 cm^{-1} in xylans (fluctuation of C=O in ester groups) and band intensity of 1645 cm^{-1} in pectins, which are the basic components of hemicellulose [20]. It was shown that lignin share compared to hemicellulose content in the cellulosic fillers increases in the series of wheat-flax-wood. According to a ratio of the optical density values of D_{1423}/D_{895}, the crystallinity index of the fillers increases in the series of cellulose-wood-wheat-flax.

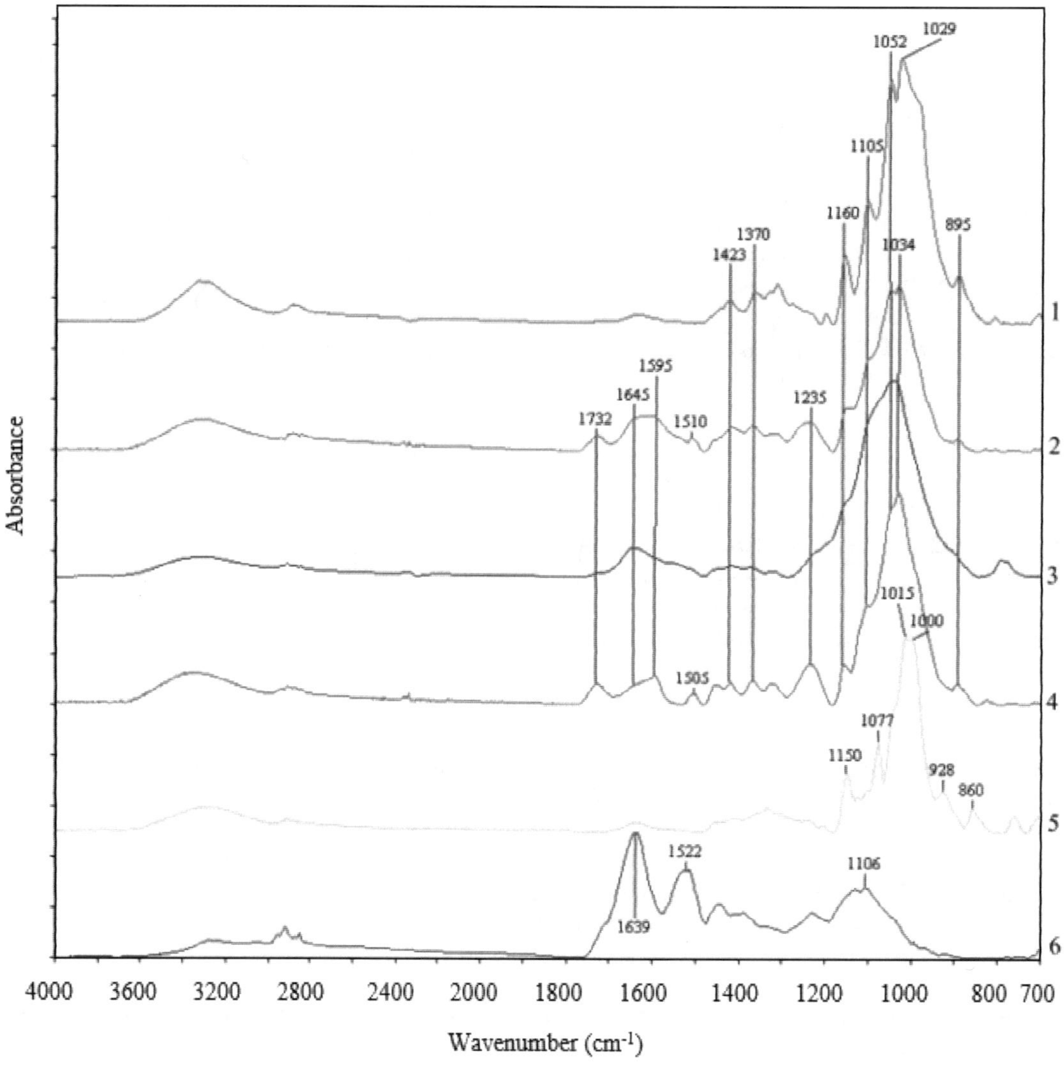

Figure 1. FTIR-spectra of the powdered fillers: cellulose (1), flax (2), wheat (3), wood (4), starch (5), feather (6).

According to the results of microbiological test of individual fillers, it was determined that mold fungi are capable to use all the tested fillers as a carbon source. However, the dynamics of micromycetes growth of on aqueous and agar media with different fillers under the identical conditions of incubation and the same filler content are not identical (Figure 2 (a)). The highest values of summarized biomass gain – (36±0.5)wt.% and (38±0.5)wt.%, respectively, were observed on the 14th day after test cultures inoculation of aqueous media with flax straw and hydrolyzed feather keratin. The wheat chaff and wood powder fillers were characterized by a biomass cumulative value of (27±0.5)wt.%. The lowest value of biomass gain from all the tested fillers was observed on media with cellulose and starch – (23.5±0.5)wt.% for the majority of test cultures. In case of cultivation of such cultures as *A. terreus*, *P. chrysogenum*, *A. flaus* and *P. cyclopium* on the hydrolyzed feather containing peptides, alkaline products of micromycetes metabolism were accumulated. The flax straw and hydrolyzed feather under the action of *A. niger* were characterized by the most intensive reduction of pH index.

Dynamics of mold fungi growth on agar media with different fillers has a complex character. As an example, growth curves of inoculated colony of *A. terreus* on agar media with 30 wt.% of fillers are given (Figure 2 (b)). A high intensity of a colony growth for the majority of mold fungi cultures was characteristic for media with hydrolyzed feather keratin (growth rate on this substrate was 3-4 mm/day) that is apparently connected with a high content of low molecular peptide components in it. The surface of agar media with cellulose and starch at the incubation period of 28 days was characterized only by separated areas of sporulation formation. The maximum colonization of agar media with introduction of cellulosic components was determined for cultures of *A. terreus* and *P. chrysogenum*. High rate of micromycetes growth was typical for medium with seed flax straw additive that could be caused by a high content of hydrolyzed holocellulose, and the presence of fat component. It should be noted that growth curves of the majority of cultures (*A. niger*, *A. terreus*, *P. cyclopium*, *P. variotii*) on media with wheat (Figure 2 (a) and (b)) had the existence of a long lag phase (up to 11 days) after that colony growth occured. The wood of fir-tree species also had reduced biodegradability, apparently, due to a high lignin content.

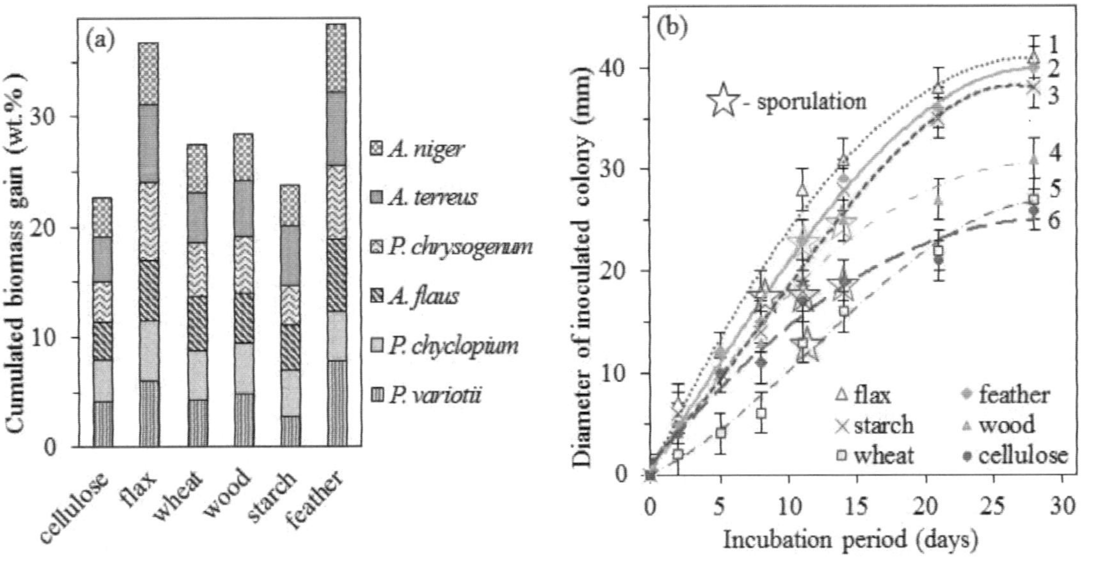

Figure 2. Cumulated biomass gain of mold fungi test cultures (wt.% of filler weight) on aqueous media with 30 wt.% of fillers on the 14th day after cultures inoculation **(a)**; Growth curves of inoculated colony of *Aspergillus terreus* on agar media with 30 wt.% of fillers: flax (1), feather (2), starch (3), wood (4), wheat (5), cellulose (6) **(b)**.

The fungal growth assessment of such test cultures as *A. niger*, *P. chrysogenum*, *T. viride* and *P. variotii* on film samples of LDPE/filler composites was carried out (Figure 3). It was experimentally shown that the film sample of neat LDPE is not nutrient medium for mold fungi (intensity of fouling is 0-1 point) (Figure 4 (a)). For all the LDPE/filler composites, it was observed the growth of fungal mycelium with sporulation development that argues for the content of enough nutrients, which provide fungal growth. The difference in the intensity of micromycetes growth on surface of the materials depending on the type of a filler was noted that is coherent with the results of biofouling assessment of individual fillers. Therefore, composite materials filled by cellulose contain nutrients, which provide low growth of mold fungi with fouling of less than 20% of the material surface (1-1.5 points depending on culture) (Figure 4 (b)). Filling of LDPE by starch also promoted insignificant increase in biodegradability of materials in comparison with neat LDPE (1.5-2.5 points depending on culture). At the same time the existence of additives like flax, wheat and wood flour in LDPE matrix promoted more intensive fungal growth (2.5-3.5 points) (Figure 4 (c)). The hydrolyzed bird feather containing proteinaceous component and having high ability to washing away from a polymeric matrix is easily available source of carbon for micromycetes. Therefore, the intensity of fungal growth on the LDPE/filler composites is connected with chemical nature and availability of a filler.

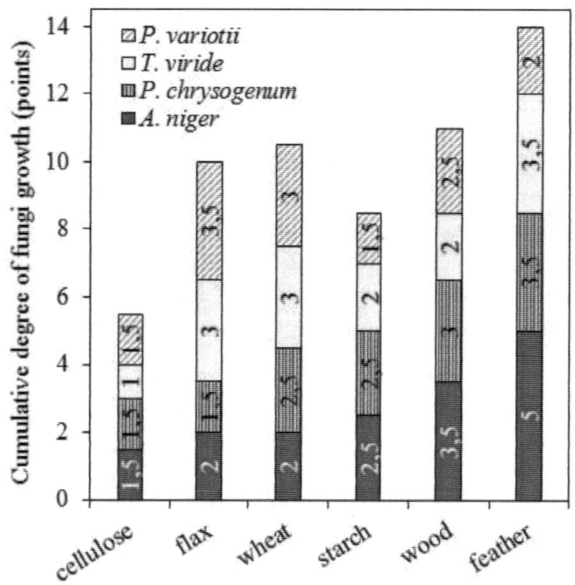

Figure 3. Dependence of cumulative degree of fungal growth on filler nature in the LDPE/filler composites.

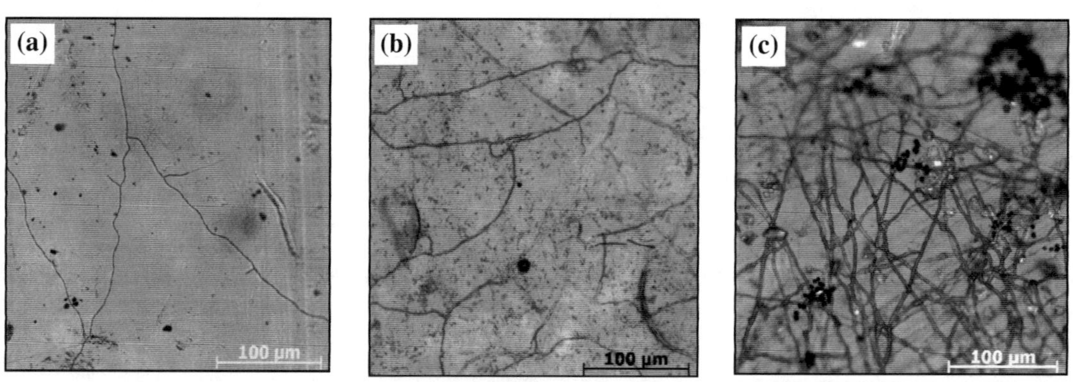

Figure 4. Microphotographs of LDPE (a), LDPE/cellulose (b), LDPE/flax (c) samples after 84 days from inoculation by spore suspension of *Aspergillus niger* culture (reflected light, 200x).

4. Conclusion

Novel hybrid polyethylene composites filled by natural powdered components (30 wt/%) have been investigated. Natural fillers and polymer composite materials on their basis have been destructed by mold fungi. The chemical compositions of the fillers, their dimensional parameters and biodegradability have been analyzed as factors exert a considerable impact on the biodegradability of composite materials. The data of biodegradability test for individual fillers is consistent with the data of biodegradability test for the LDPE/filler = 70/30 composites. It has been found that biodegradability of the composites under mold fungi action is preeminently affected by not materials structure, but chemical composition of a filler.

Protein content and presence of soluble and easy hydrolyzed fraction in the fillers are able to increase biodegradability of hybrid polyethylene composites. High rate of micromycetes growth on the PE/flax composites could be caused by the excess of hydrolyzed holocellulose content over lignin content, and presence of fat component. The intensity of biodegradability of the bird feather filler containing proteinaceous component and composites on its basis is twice as much as the other fillers. Easy hydrolyzed fraction (more than 10 wt.%) of fillers incorporated in the polyethylene matrix makes composites an easily available source of carbon for micromycetes. Higher L/D-ratio of filler particles could induce biofouling of the polymeric composite materials due to their accessibility.

According to the paper results, most effective natural fillers as additives stimulating polyethylene composites biodegradability are milled straw of seed flax and hydrolyzed keratin of bird's feather.

Acknowledgements

The authors express an acknowledgement to the Center of Shared Usage «New Materials and Technologies» of Emanuel Institute of Biochemical Physics and the Common Use Centre at the Plekhanov Russian University of Economics for provided equipment. The authors wish to acknowledge the assistance from Anastasiya Smirnova by the translation of this paper. This study was financed by a grant from the Plekhanov Russian University of Economics.

References

[1] Prachayawarakorn J and Pomdage W 2014 *Mater. Design* **61** 264
[2] Zykova A K, Pantyukhov P V, Kolesnikova N N, Popov A A and Olkhov A A 2015 *AIP Conf. Proc.* **1683** 020242 DOI: 10.1063/1.4932932
[3] Mastalygina E E, Varyan I A, Kolesnikova N N and Popov A A 2016 *AIP Conf. Proc.* **1736** 020097 DOI: 10.1063/1.4949672
[4] Matet M, Heuzey M C, Ajji A and Sarazin P 2015 *Carbohyd. Polym.* **117** 177
[5] Foulk J A, Chao W Y, Akin D E, Dodd R B and Layton P A 2006 *J. Polym. Environ.* **14**(1) 15
[6] Toupe J L, Trokourey A and Rodrigue D 2014 *Polym. Compos.* **35**(4) 730
[7] Pantyukhov P, Kolesnikova N and Popov A 2016 *Polym. Compos.* **37** 1461
[8] Kijenska M, Kowalska E, Pa1ys B and Ryczkowski J 2010 *Polym. Degrad. Stab.* **95** 536
[9] Arkatkar A, Arutchelvi J and Sudhakar M 2009 *The Open Environ. Eng. J.* **2** 68
[10] Rutkowska M, Heimowska A, Krasowska K and Janik H 2002 *Polish Journal of Environmental Science* **11**(3) 267
[11] Shah A A, Hasan F, Hameed A and Ahmed S 2008 *Biotechnol. Adv.***26** 246
[12] Mastalygina E E, Varyan I A, Kolesnikova N N, Monakhova T V, Karpova S G and Popov A A 2016 *AIP Conf. Proc.* **1783** 020150 DOI: 10.1063/1.4966443
[13] Ave´rous L and Le Digabel F 2006 *Carbohyd. Polym.* **66**(4) 480
[14] Mastalygina E E 2015 *Structure, Poperties and Biodegradability of Ternary Composites – Polypropylene, Low-density Polyethylene and Natural Fillers (Emanuel Institute of Biochemical Physics)* Ph.D. thesis http://ibcp.chph.ras.ru/docs/diss_mastaligina.pdf (*in Russian*)
[15] Niessen W R 2010 *Combustion and Incineration Processes: Applications in Environmental Endeneering* (Boca Raton: CRC Press) p768
[16] Poletto M, Zattera A J, Forte M M C and Santana R M C 2012 *Bioresource Technol.* **109** 148

[17] O'Brien R D 2008 *Fats and Oils: Formulating and Processing for Applications* (Boca Raton: CRC Press) p 680
[18] Delwiche S R, Pitt R E and Norris K H 1991 *Starch* **43** 415
[19] Fan M, Dai D and Huang B 2012 Fourier Transform Infrared Spectroscopy for Natural Fibres *Fourier Transform-Materials Analysis* ed S M Salih (Vienna: InTech) chapter 3 p 272
[20] Schwanninger M, Rodrigues J C, Pereira H and Hinterstoisser B 2004 *Vib. Spectrosc.* **36**(1) 23

IOP Conf. Series: Materials Science and Engineering **213** (2017) 012012 doi:10.1088/1757-899X/213/1/012012

Degradation kinetics of ethylene-octene copolymer/wood flour biocomposites in dependence to filler content

A K Zykova[1,2,3], P V Pantyukhov[1,2], T V Monakhova[2] and A A Popov[1,2]

[1]Department of Chemistry and Physics, Plekhanov Russian University of Economics, 36 Stremyanny Lane, Moscow, 117997, Russia
[2]Emanuel Institute of Biochemical Physics, Russian Academy of Sciences, 4 Kosygina Street, Moscow, 119334, Russia

E-mail: zykovaak@yandex.ru

Abstract. This article is focused on thermal oxidative degradation and biodegradation in soil of biocomposites based on ethylene-octene copolymer (EOC), filled by wood flour (from 30 to 70% wt.), in dependence to the filler content. The study of oxidative degradation of composites was carried out at two temperatures (80 and 130°C respectively). The induction period and the rates of oxidation were determined. It was concluded that as filler content raises, the induction period increases. It can be explained by the higher specific area of composites in comparison with pure EOC. However, high filled composites (60 and 70 % of the filler) are oxidized with a huge induction period because polyphenols in the filler inhibit the oxidation process. Biodegradation test under laboratory conditions was carried out to investigate the biodegradability of the material. Composites with lower filler content have lower weight loss rate. Small particles are capsulated by polymer and are isolated from moisture and microorganisms. On the other hand, at a high filling of the composite small particles stick together and act as large ones. Such filler agglomerates are connected with each other and allow microorganisms to penetrate into the composite. It was concluded as filler content raises the mass loss increases.

1. Introduction

New environmental legislation all over the world and a growing ecological awareness led to the development of eco-friendly composites – hybrid materials. It can be degraded in different natural medias, such as sea water [1] and soil [2] without environmental contamination. The total world production of plastics in 2004-2014 has changed from 225 m. tones to 311 m. tones and has a tendency to increase [3]. In Europe the total amount of polymer output for the same period has changed insignificantly from 60 m. tons to 59 m. tons. According to the new law [4], France banned the usage of fully plastic bags and serviceware. The authorities expect the reduction by 50% of waste in landfills by 2025.

Researches all over the world have been developing new types of materials, which have the same or better mechanical properties than common used polyolefins [5-6].

Biocomposites based on polyolefins and natural fillers are commonly used in different areas of economy, mainly in automotive (interior door linings and paneling, seat backrests) and construction industries (light structural walls, insulation materials, floor and wall coverings, geotextiles, thatch roofing) [7]. Biocomposites should have reduced price and can be partly or fully degraded in soil without harm to ecology.

Wood flour is widely used in composite engineering. This type of filler has several advantages such as high production output (>1.5 m. tons) [8] and low cost (0.2-0.4$/kg). A relatively high thermal stability (180-200°C) [9] allows compounding with traditional thermoplastics polyethylene (PE), polypropylene (PP), polystyrene (PS), polyvinyl chloride (PVC).

Synthetic biodegradable polymers such as polylactide acid (PLA) [10], polycaprolactone (PCL), polyhydroxybutyrate (PHB) [11], and polyhydroxyvalerate (PHV) can be used as polymer matrix in fully biodegradable biocomposites. However, they are still very expensive in comparison with the traditional polymers. Polyolefins such as polyethylene [12] and polypropylene [13] are widely used for partially biodegradable composites.

Some other polymer matrices are also examined for biocomposites development. For example, polyolefin elastomers (POEs), such as ethylene–octene copolymer (EOC) can be used as the polymer matrix in composites. Few studies have investigated the properties of composites based on POE. Most of these papers consider the mechanical, thermal and reological [14] properties of EOC–modified corn starch blends [15]. EOCs are known to have higher filler loading capacity than polyolefins. Some manufacturers (Ultrasoft Plastic LLC, Alexandrov, Vladimir region, Russia and Polymer Alliance LLC, Khimki, Moscow region, Russia) produce highly filled composites based on EOC.

There are a lot of methods, which examined the oxidative stability of polymers and composites, such as thermogravimetric analysis (TGA) and differential scanning calorimetry (DSC), isothermogravimetric analysis (IGA), thermal volatilization analysis (TVA), accelerated aging tests [16-17].

The objective of this study is to investigate the thermal oxidative degradation and biodegradation in soil of biocomposites based on EOC, filled by wood flour, in dependence to the filler content.

2. Materials and methods

Ethylene–octene copolymer (EOC) of Lucene 670 trademark manufactured by LG Chem (South Korea) was chosen as the polymer matrix. The properties of considered polymer, provided by LG Chem, are as follows: 1-octene content 35 % mol, melting point 57°C, density 0.870 g/cm^3, MFI (melt flow index) at 190°C 5 g/10 min, Mn/Mw= 32000/75000.

Birch wood flour (WF) was selected as the filler for the biocomposites. WF was produced according to GOST 16361-87 and referred to trademark № 120. WF was provided by «Novotop» Ltd. (Russia, Moscow). The chemical composition of the filler is presented in Table 1.

Table 1. Chemical composition of the filler.

Filler	Cellulose	Lignin	Hemicellulose	Proteins	Fats	Ash	References
Birch wood flour	37–56	18–22	25–40	0.3	0.4	0.4	[18-19]

The wood flour was sieved by a sieve set. The fraction 0–200 μm was chosen for the biocomposites preparation.

The components of the biocomposites were mixed in following proportions (pure EOC and wood flour respectively): 100/0, 70/30, 60/40, 50/50, 40/60, 30/70 (% wt.).

The compositions were mixed by heated mixing rolls UBL-6175BL (China) at T= 160°C for 5 min.

The mixtures were pressed by the manual hydraulic press VNIR PRG-1–10 (Russia) at 135°C under a load of 7 kN for 3 min with subsequent quick cooling. The obtained films were round, with a diameter of 7 cm. The thickness of the films varied from 100 μm for composites with 30% of the filler to 1000 μm for composites with 70% of the filler.

The study of oxidative degradation of composites was carried out at 80°C and 130°C and under the oxygen pressure of 500 mmHg in the system described and developed in Semenov Institute of

Chemical Physics of Russian Academy of Sciences [20]. This method simulates the process taking place in natural conditions during material degradation in the environment. Samples were placed in the vessels in which the oxygen is under the pressure. The vessels are placed in thermostat heated to a specific temperature. The intensity of the oxygen absorption by the material was measured in fixed intervals. Based on the data the kinetic curves of oxidation were plotted. According to the results, the influence of a filler (or compatibilizer) on oxidation rate can be seen. The total time of exposure at 80°C and 130°C was 1000 hours and 30 hours respectively. The rates and induction period of considered processes were determined for both temperatures. The rates were presented by two values. First value is characterized as the time (hours) for which the composite absorbs fixed amount of oxygen. The second value determines average rate of oxidation and was measured in mol/kg*s. It should be noted that the filler itself is oxidized insignificantly in compare with composite.

The biodegradation test under laboratory conditions was carried out in accordance with ASTM D 5988-96 [21] to investigate the biodegradability of the material. The soil was made of sand, garden soil, and horse manure taken in equal amounts. The resulting soil was held for 2 months at 20 ± 5°C with daily stirring and humidity maintained at 60%. The same soil was used in previous studies [22, 12, 23]. The film samples were immersed vertically in the prepared soil and held for 12 months. The samples were cleaned and dried to the constant mass. The difference in mass of the composites was measured.

The work was carried out using scientific equipment of the Center of Shared Usage «New Materials and Technologies» of Emanuel Institute of Biochemical Physics and the Common Use Centre of Plekhanov Russian University of Economics. This study was financed by a grant from the Plekhanov Russian University of Economics.

3. Results and discussion

Figure 1 shows the kinetic curves of oxygen absorption of pure EOC and composites based on EOC and WF.

(A) (B)

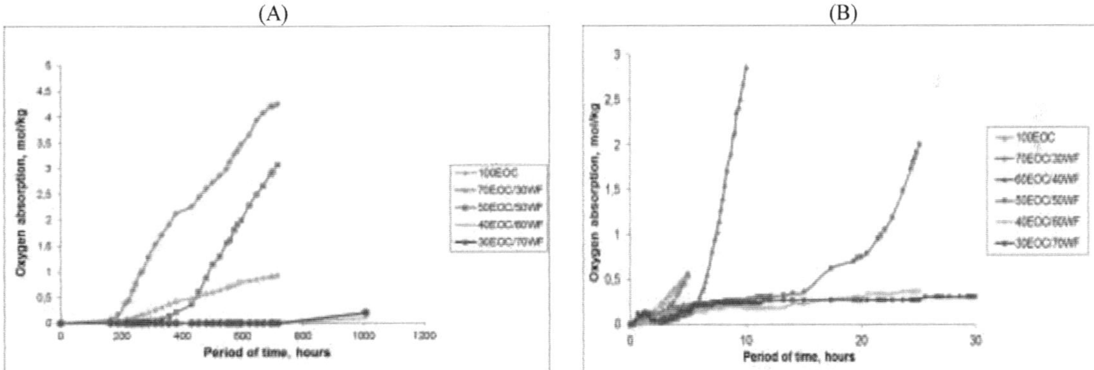

Figure 1. Kinetics of thermal oxidation of composites at 80°C (a) and at 130°C (b).

It is known that thermal-oxidative degradation of polyolefins in air is autocatalytic, i.e., the rate is slow at first but gradually accelerates to a constant value. [17, 20, 24]. The reactions are presented below according to the source [25].

The initiation steps in the radical-chain oxidation process are

$$RH \rightarrow R\cdot + H\cdot \tag{1}$$

$$RH + O_2 \rightarrow R\cdot + HOO\cdot \tag{2}$$

$$2RH + O_2 \rightarrow 2R\cdot + H_2O_2 \tag{3}$$

The chain propagation steps are

$$R\cdot + O_2 \rightarrow ROO\cdot \tag{4}$$

$$ROO\cdot + RH \rightarrow ROOH + R\cdot \tag{5}$$

The chain-branching steps are

$$ROOH \rightarrow RO\cdot + \cdot OH \tag{6}$$

$$ROOH + RH \rightarrow RO\cdot + R\cdot + H_2O \tag{7}$$

$$2ROOH \rightarrow ROO\cdot + RO\cdot + H_2O \tag{8}$$

The chain termination steps are

$$2R \rightarrow R_2 \tag{9}$$

$$R\cdot + ROO\cdot \rightarrow ROOR \tag{10}$$

$$2ROO \rightarrow ROOR + O_2 \tag{11}$$

EOC and composites based on EOC are oxidized with the induction period. It can be explained by high amount of polyphenols in WF. It was concluded that as filler content raises, the induction period increases. It should be noted that the role of antioxidant added during polymerization at the factory is insignificant, because the content of antioxidant in polymer is constant.

The rate of oxidation and induction period was determined. The results are presented in Table 3. According to the results, it was found that pure EOC has lower rate of oxidation than composites based on EOC. It can be explained by the higher specific area of composites in comparison with pure EOC. The comparison of pure EOC rate and oxidation rate of composites is not correct, because pure EOC is homogeneous system, and oxidation occurs only on the surface of polymer. Hence, composites are heterogeneous and have large (developed) interfacial contact area. A specific area of particles correlates with the interfacial contact area between the filler and EOC matrix, which is mostly prone to oxygen absorption. It was concluded that as filler content rises, the oxidation rate decreases.

Table 2. Oxidation rate and induction period of composites at 80 and 130°C and oxygen pressure 500 mmHg.

Composite	Oxidation rate, hours		Induction period, hours		Oxidation rate, mol/kg*s	
	80°C (0.2mol)	130°C (0.3mol)	80°C	130°C	80°C	130°C
100EOC	287	3	163	0.2	$5.97*10^{-7}$	$2.76*10^{-5}$
70EOC/30WF	287	4	163	0.2	$2.44*10^{-6}$	$5.19*10^{-5}$
60EOC/40WF	-	6	-	0.5	-	$1.3*10^{-4}$
50EOC/50WF	380	11	311	1	$3.43*10^{-6}$	$4.8*10^{-5}$
40EOC/60WF	1000	20	717	More than 25	-	-
30EOC/70WF	1000	25	717	More than 30	-	-

According to the results of biodegradation test under laboratory conditions (Figure 2), composites with lower filler content have lower weight loss. Pure EOC did not change its mass and it was concluded that pure EOC is not biodegradable polymer. Small particles in composites are capsulated by polymer and unavailable for penetration by microorganisms. On the other hand, large and long particles are connected with each other and allow microorganisms to penetrate into the composite. At a low filling of the composite the particles are separated and capsulated by polymer. At a high filling the small particles have the tendency to agglomerate and act as large ones or even as a network. It was concluded as filler content raises the mass loss increases. It is well correlates with previous studies [23]. At the initial stage of biodegradation there is no difference between composites with 60 and 70% of wood flour. Apparently, the quantity of wood flour at the surface is the same in these two composites, it should differ into the volume and it would be observed at the next stages of biodegradation process.

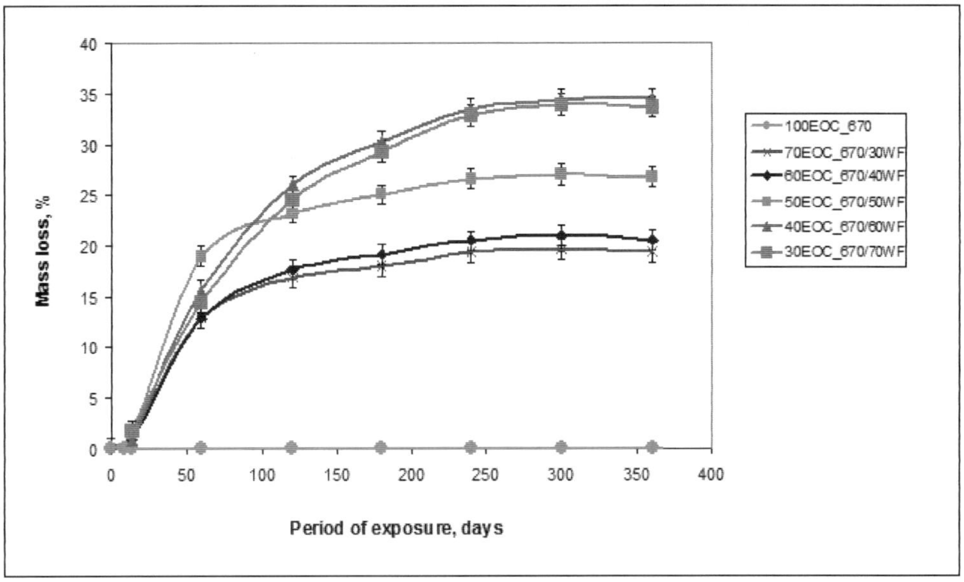

Figure 2. Kinetic curves of biodegradation of composites on recovered soil.

4. Conclusion

Biocomposites based on EOC and WF were investigated by two methods defining their degradation in the environment. At first stage of biodegradation process, high weight loss is observed cause of filler degradation at the composite surface. Later, when the microorganisms' enzymes have an effect on the composite surface, the microbial waste products initiate the oxidation. According to the data of thermal oxidation, it can be concluded that the addition of WF to the EOC matrix lead to the accelerated oxidative degradation of EOC in the composite. Thus, EOC in composite is stable on the air for some period of time (when it is used by consumer) but it degrades rather quickly after the induction period (when it is thrown out). Biodegradation test showed that WF biodegrades easily and makes composite partially biodegradable. Filler content in biocomposite correlates with the mass loss. Obtained materials are promising because they are much cheaper than fully biodegradable biocomposites but they have a good potential for degradation in the environment by both mechanisms – oxidative and biological degradation.

References

[1] Alvarez-Zeferino C J, Beltrán-Villavicencio M and Vázquez-Morillas A 2015 *Open J. Polym. Chem.* **5** 55-62
[2] Catto L, Montagna L S, Almeida S H, Silveira R M B and Santana R M C 2016 *Int. Biodeter. Biodegr.* **109** 11-22
[3] Plastics-the Facts 2015 An analysis of European plastics production, demand and waste data *Ital ian Association of Plastic Manufacturers* http://www.plasticseurope.org/Document/plastics---the-facts-2015.aspx

[4] Energy Transition to the Green Growth Act 2015 http://www.developpement-durable.gouv.fr/IMG/pdf/14123-8-GB_loi-TE-mode-emploi_DEF_light.pdf

[5] Bledzki A K and Faruk O 2003 *Appl. Compos. Mater.* **10** 365-79

[6] AlMaadeed M A, Nogellova Z, Micuzik M, Novak I and Krupa I 2013 *Mater. Design* **53** 29-37

[7] Sudell B C 2009 *Proc. Symposium on Natural Fibres Common Fund for commodities (Amsterdam)* P 144

[8] Oksman N K and Sain M 2008 *Wood-polymer Composites* (Boca Raton: CRC Press) p 366

[9] Poletto M, Zattera A J, Forte M M C and Santana R M C 2012 *Bioresource Technol.* **109** 148

[10] Podzorova M V, Tertyshnaya Yu V, Monakhova T V and Popov A A 2016 *Rus. J. Phys. Chem. B* **10**(5) 825-9

[11] Olkhov A A, Iordanskii A L, Staroverova O V, Gumargalieva K Z, Sklyanchuk E D, Gurev V V, Abbasov T A, Ishchenko A A, Rogovina S Z and Berlin A A 2016 *Fibre Chem.* **47** 348

[12] Pantyukhov P, Kolesnikova N and Popov A 2016 *Polym. Compos.* **37** 1461

[13] Mastalygina E E, Kolesnikova N N, Popov A A and Olkhov A A 2015 *AIP Conf. Proc.* **1683** 020143

[14] Fu X, Chen X, Wen R, He X, Shang X, Liao Z and Yang L 2007 *J. Polym. Res.* **14** 297-304

[15] Shang X Y, Fu X, Yang L S, Chen X D and Zhang M 2008 *Int. J. Polym. Mater.* **57**(4) 362-73

[16] Apostolov A and Djoumaliisky S 2014 *Mater. Sci. Appl.* **5** 579-91

[17] Mark J E 2007 *Physical Properties of Polymers Handbook* (USA: Springer) p 1047

[18] Nikitin V M, Obolenskaya A V and Schegolev V P 1978 *Chemistry of Wood and Cellulose* (Moscow :Wood Industry) p 420

[19] Fengel D and Wegener G 1989 *Wood: Chemistry, Ultrastructure, Reactions* (Berlin-New York: Walter de Gruyter) p 613

[20] Shlyapnikov Y A, Kiryushkin S G and Mar'in A P 1986 *Antioxidant Stability of Polymers* (Moscow : Chemistry) p 470 (*In Russian*)

[21] ASTM D5988-96 1996 *Standard Test Method for Determining Aerobic Biodegradation in Soil of Plastic Materials or Residual Plastic Materials After Composting* (West Conshohocken: ASTM International) www.astm.org

[22] Ottenbrite R, Huang S J and Park K 1996 *Hydrogels and Biodegradable Polymers for Bioapplications* (Washington: American Chemical Society) p 626

[23] Zykova A K, Pantyukhov P V, Monakhova T V, Kolesnikova N N, Popov A A and Ramos C C 2016 *AIP Conf. Proc.* **1783** 020239

[24] Emanuel N M and Buchachenko A L 1987 *Chemical Physics of Polymer Degradation and Stabilization* (The Netherlands: VNU Science Press) p 336

[25] Kuleznev V N and Shershnev V A 2014 *Chemistry and Physics of Polymers* (Russia: Lan') p 368 (*In Russian*)

IOP Conf. Series: Materials Science and Engineering **213** (2017) 012013 doi:10.1088/1757-899X/213/1/012013

Matching refractive indices of two fluids and finding interfacial tension for the purpose of fuel spray imaging

Y H Liang

Mechanical Engineering, University of Wisconsin Madison, WI 53706, USA

E-mail: liang46@wisc.edu

Abstract. This study attempts to prepare a fluid pair for use in spray dynamics investigations. Better understanding the behavior of fuel sprays is one of the things that can help improve the efficiency of internal combustion engines. To address the scattering issue in current imaging methods, the refractive index difference between the injected fluid and the medium that it is injected into is eliminated. Two immiscible fluids (sucrose solution and silicone oil) with the same refractive index was identified, their surface tension to build a model fluid engine system injection was also studied. At the same time, Weber number is found to help correct the difference. Results show that 63.7% mass sucrose solution has the same refractive index as silicone oil, and the sucrose solution/silicone oil interface has a surface tension of 0.08941 N/m, which is roughly four times larger than that of ethanol/air. This means using the sucrose/silicone oil fluid pair to model fuel spray will involve some adjustments to be accurate.

1. Introduction

It is completely necessary to effort towards more efficient vehicles. Even though the automotive industry is beginning to make the transition to electric and other alternative drivetrains, internal combustion engines are expected to play an important role in transportation and other fields until at least the next century. Fuel injection is a necessary function of most conventional internal combustion engines today. The spray of a fuel injector is an important subject to study because when it is better understood, it can be taken advantage of to create more optimal combustion. A detailed model of near nozzle sprays would serve to improve different aspects of the engine. Unfortunately, not enough is known about how fuel sprays behave to predict them from first principles [1-2]. If a simple, easily accessible visualization technique exists, much more experimental data can be gathered to help build accurate models of sprays. The main problem this study attempts to address is imaging the near nozzle spray of fuel injectors in a cost efficient manner [3-4]. Currently, there is a lack of understanding of the fundamentals of fuel injection. Imaging of fluid sprays have been attempted before. Argonne National Laboratory has the necessary X-ray equipment to reveal the structure of a fluid spray, but it is the only facility in the world that has such capability. Since 2005, it has been conducting a study to better understand fuel injection and sprays, and to develop improved spray models with annual budgets of about $1 million. A less costly way to study fuel sprays would benefit the industry in that it would make spray analysis more widely accessible, speeding up the development of more optimized engines. A very recent investigation into fuel spray imaging [5] makes use of a high speed camera with a framerate of 1,000,000 fps. Fuel is injected through a light source and the effects of fuel type and varying injection pressure on spray propagation is analyzed. It was discovered that injection pressure does not significantly affect the atomized region of a spray, and

IOP Conf. Series: Materials Science and Engineering **213** (2017) 012013 doi:10.1088/1757-899X/213/1/012013

there is a close correlation between viscosity and "mushroom length", the end region of a spray that resembles a mushroom [6]. The images created showed the space occupied by the spray in its entirety as a black shaded area, not revealing the structure of the spray or its breakup process. This is caused by the high optical density of the spray. The particle cloud in the spray is so dense that when light travels through it, scattering caused by refraction through each individual particle makes it impossible to clearly visualize the structure of the spray through this technique. If the phenomenon of refraction did not exist, this technique would produce images that showed a liquid core, which would be a small black shaded area in the middle; and a cloud of distinct fuel particles surrounding the core.
The study outlined by this article attempts to solve the scattering problem with this idea. Scattering is caused by a difference in index of refraction. By injecting a fluid into a medium of the same refractive index, scattering is eliminated. This study consists of identifying two immiscible fluids with the same refractive index, and finding their surface tension. Weber number can be calculated once surface tension is known, and that can be used to correct differences between our fluid pair and a real air/fuel spray.

2. Materials and method

2.1. Selecting fluid pair
The pair of fluids for this investigation was chosen based on two main criteria: immiscibility and matched (or the ability to match) refractive index. Cost and safety of handling would also be factors that would affect the final decision. All possible pairs are listed in Table 1. Glycerin (n=1.466) is immiscible with both amyl acetate (n=1.4017) and acetone (n=1.3588), benzene (n=1.5014) would have to be added to them to bring their refractive index up to that of glycerin; cyclohexane (n=1.4262) is immiscible with both ethanol (n=1.36) and methyl salicylate (n=1.523) mix, the mix can be mixed in varying ratios to achieve any refractive index between their own; 200 mPa.s silicone oil and sodium iodide solution pair is immiscible, the refractive index of a sodium iodide solution can be modeled as a linear function of its concentration, both of the two fluids are non-toxic, but iodide is costly. In the study, sucrose solution can also be considered instead of sodium iodide [7], as it is also immiscible with silicone oil and it can reach a maximum refractive index of 1.5 which is above the index of silicone oil. The major benefit of using sucrose solution is that it is very cost effective, so the silicone oil/sucrose solution pair proves most appropriate for the following investigation.

Table 1. Possible fluid pairs that have been considered.

Fluid pair (fixed fluid/adjustable concentration fluid)	Silicone oil /sodium iodide solution	Glycerin/amyl acetate or acetone	Cyclohexane/ethanol and methyl salicylate mix with Benzene added	Silicone oil/sucrose solution
Target Index (n)	1.453	1.466	1.4262	1.453

2.2. Index matching
The apparatus (Figure 1) made by our lab includes a container with four transparent windows and a frame to hold a motor for stirring the fluid mix. Fused silica was used as the viewing windows on the container, due to having the same refractive index (1.4585) as our target. There is an impeller driven by a motor reaching into the container for creating disturbances within the two fluids. A laser could shine through the fluid mixture via the windows, which could be index matched. The motors are mounted on an 80/20 extrusion frame, with mounting brackets that allow the motor positions, and in turn, positions of impellers within the container, to be adjusted. The mounting brackets each have a single straight slot for accurate and easily quantified adjustment of motor position. The "blender method" created by our lab is used to check whether two fluids have the same index as light scatters when it travels from one medium to another with a different refractive index.

Both fluids from a pair are poured into the apparatus with the adjustable index fluid at a high concentration (67.9%). The impeller was turned on while being careful not to exceed a speed that induces cavitation or pulls pockets of air into the mixture. If this happens, there will be unwanted refraction as the laser travels through these gaseous phases. The laser was turned on and positioned to travel right through the container windows, not hitting the impeller shaft or any remnants of sealant at the edges of the windows. Scattering was initialized by injecting two mL of silicone oil into the mixture. Observation and recording was done by visually inspecting the scattering pattern and capturing images of the wall on which light scattered. Steps above were repeated with different concentrations until light on the wall becomes focused, at which point the indices are matched. To confirm, concentrations below the matched index were also tested, to see that light is indeed more dispersed again. Three trials of this process have been done.

Figure 1. Index matching experimental setup with laser turned on.

2.3. Measuring surface tension between the two fluids

The pendant drop method is used to find surface tension between the two fluids. This method was chosen because it is a cheap and quick way to test for surface tension. A syringe with a needle that has been bent upwards is secured to the motor mounting frame. The container with 66.7% mass sugar solution is chosen and placed under the needle so that the needle opening is submerged by the sugar solution, concentration and surface tension has little relation. Droplets of the same size were produced by operating the syringe by a fluid dispenser. The droplet was illuminated using a flashlight.

A syringe needle is inserted into one fluid, and then a small amount of the other fluid is pushed out through the needle. A droplet will form on the tip of the syringe, and the geometry of the drop is indicative of the interfacial tension present between the fluids. The following equation [8]

$$\gamma = \frac{\Delta\rho\, g d_e^{\,2}}{H} \qquad (1)\ [8]$$

is used to find surface tension γ. Delta ρ is the density difference between the two fluids, d_e is the maximum width of the droplet, illustrated in Figure 2. H is a correction factor which is a function of the ratio of d_s/d_e, where d_s is the width of the droplet at a height of d_e from the tip of the droplet.

$$\text{Weber Number} = \rho\, v^2\, l / \sigma \qquad (2)\ [9]$$

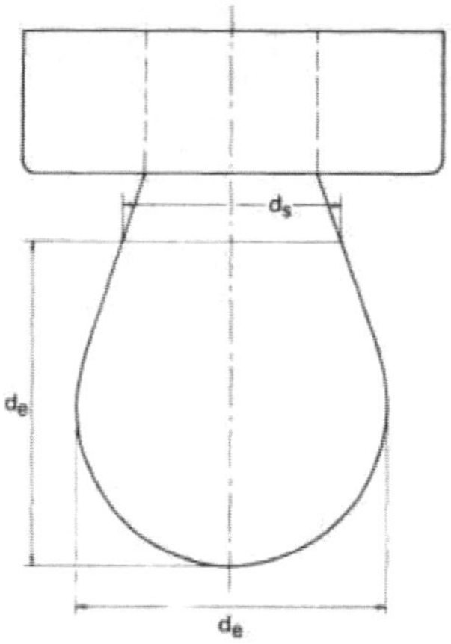

Figure 2. Method of measuring dimensions of droplet for surface tension analysis.

3. Results and discussion

3.1. Index matching of silicone oil and sucrose

Concentration of the sucrose solution is changed by adding water to the container. Since the experiment is based on refraction of light in different refractive indices, the sucrose solutions need to be completely free of solid sugar grains and other impurities. Initial testing has shown that sugar, especially when creating high concentration (50% mass+) solutions, dissolves very slowly. This is the reason a concentration that is above the expected matching point instead of a concentration that is below the expected matching point is used to start the process. Adjusting concentration by adding water takes much less time, therefore it expedites the experiment. In Figure 3, the two highlighted frames are laser outputs at the most closely matched points.

At the starting concentration of 67.9%, the laser exiting the container casts a scattered shade of green on the wall. Not much changes until the solution gets to 65.6%, where the lit region on the wall starts to become brighter and smaller. At 64.8%, a dot of laser becomes visible to the camera. As the concentration lowers further, the dot becomes small and focused. The laser is most focused at the 64.0% and the 63.5% concentrations. Three trials of this process have shown consistent results. Experimental results from the International Scale of Sucrose Solutions at 20 C show that a solution of roughly 65% would be index matched to the silicone oil used. This discrepancy possibly stems from different testing methods and different purities of sugar used.

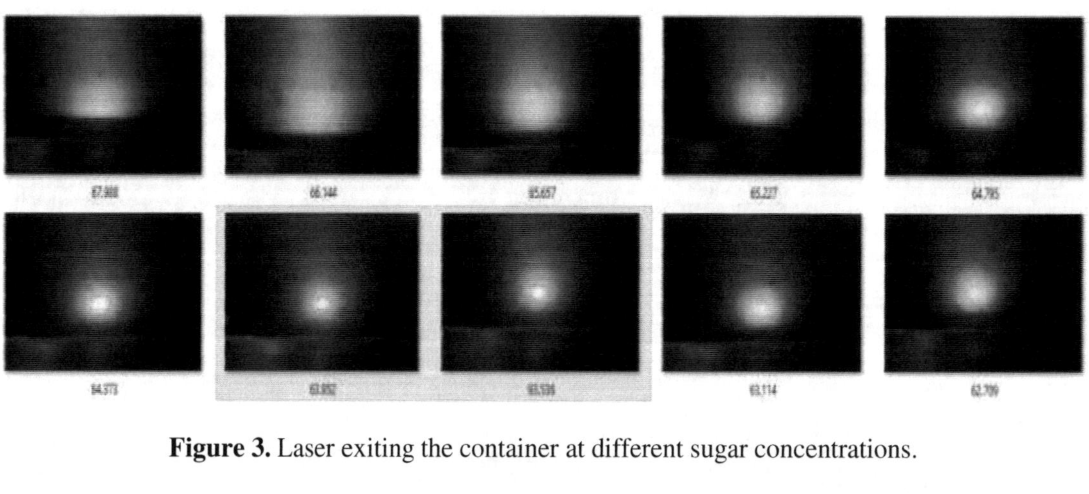

Figure 3. Laser exiting the container at different sugar concentrations.

3.2. Surface tension between the two fluids

The titration process that was carried out during laser scatter testing was also done with an oil droplet. A flashlight was directed at a viewing window on the container and droplets of silicone oil were produced inside various concentrations of sucrose solution. In comparison to laser, this appears to be a less straightforward way to check whether indices are matched, since with the laser method, there is always something to see (focused or scattered light); with a droplet, visual indication of when two fluids are matched is actually when nothing can be seen. The droplet of oil is visible in 67.9% sucrose solution. As the concentration decreases, the bubble is less visible, and it eventually disappears at around 62%. Figure 4 compares when the fluids are index matched and clearly unmatched.

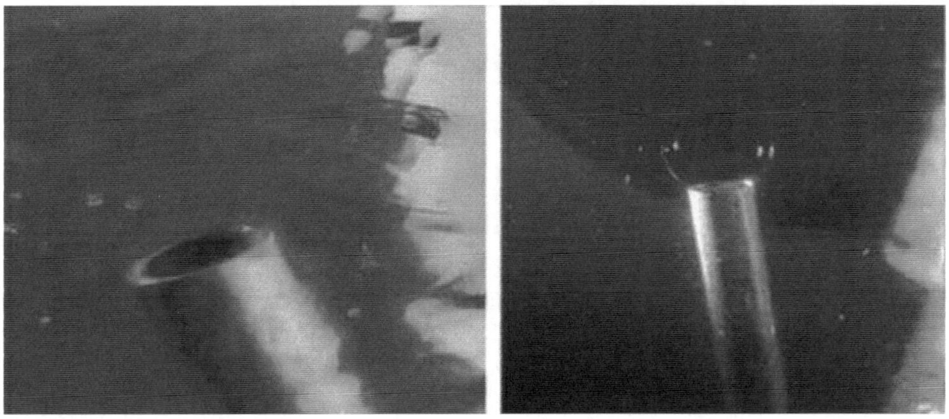

Figure 4. Comparison between silicon oil suspended in 62.65% (left) and 59.36% (right) sugar solution.

To calculate the Weber Number for the pseudo spray that is created by this fluid pair, we need to know the interfacial tension between the two fluids. Since the density of sucrose solution that is 1.32g/cm^3 (65% by weight) is higher than that of silicone oil (1.05g/cm^3), silicone oil will tend to rise to the top of the sucrose solution. This means the capillary needs to have an upwards exit in order for an oil droplet to form. A syringe with a bent needle is used. A picture of the droplet is taken through one of the windows on the container, and the lengths needed by the equation are found through deducing droplet dimensions from the known needle size.

Figure 5 is the picture used to extract droplet dimensions. Width of the widest point, de, and needle diameter were found in pixels. A conversion factor between pixel and meter was obtained by dividing the number of pixels in needle diameter by the actual needle diameter. Now that there is a factor with units of pixels per meter, all pixel values are turned into meters after being divided by this value. Needle diameter is 148 pixels, and its actual diameter is 0.002159 m. The conversion factor is 148px/0.002159m = 68550 [px/m]. This value will differ slightly between each picture depending on how much zoom there was when the picture was taken.

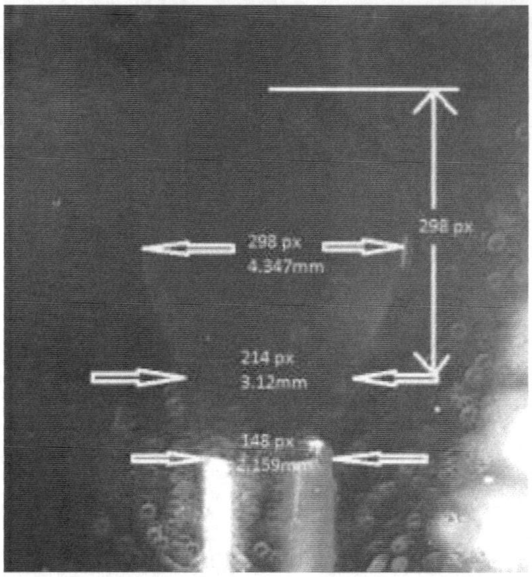

Figure 5. Silicon oil droplet with ds and de measured.

Knowing the length of d_e, one would find the length of d_s at width of the droplet that is d_e down from the tip. With the droplet shown in Figure 6, d_e is 298 pixels. 298 is added to 86, the Y coordinate of the tip, to arrive at where d_s should be measured. Since the Y axis is inverted in Microsoft Paint, adding to a Y coordinate means moving down the picture. d_s appears to be 214 pixels for this droplet. Table 3 converts these dimensions from pixel to meters, to be used in equation (1) to calculate surface tension.

Table 2. ds and de dimensions.

Dimension	Length [px]	Length [m]
de	298	0.004347
ds	214	0.00312

Equation (1) also requires the difference in density between the two fluids, gravitational acceleration, and a correction factor H. Although H can be derived from dimensions of the droplet, experimental data is available as a function of S, which is d_s/d_e. According to equation (1), the droplet in Figure 4 has a surface tension of 0.08941 [N/m]. As shown in table 3, the sucrose solution/silicone oil pair do not have similar surface tension and Weber number values with an actual fuel/air interface. Ethanol is used as an example.

Table 3. Comparison of surface tension and Weber number between different interfaces.

Interface	Surface Tension [N/m]	Weber Number
Sucrose Solution/Silicone Oil	0.08941	10943 V^2 L
Air/Ethanol	0.02250	35066 V^2 L

4. Conclusion

Matching the refractive indices of two fluids- one adjustable by concentration, one with a set index- by stirring a mixture of the two and shining a laser through the container has proved successful. Indication that they are matched comes from how the laser exits the container. When the adjustable fluid reaches a concentration where its refractive index matches that of the other fluid, the laser emerged the container as a focused beam, and that happened near the concentration at where they are expected to be matched. This study showed that 63.7% concentration has the best match. Compared to checking for the severity of laser scattering, matching the refractive index with an illuminated droplet of silicone oil inside sucrose concentrations of varying index was not as quick, or as simple to carry out. When moving on to the next concentration, the fluid would need to be stirred manually, detaching the droplet from the needle tip. Creating a new droplet requires a light press on the syringe plunger, which needs to be done very slowly and carefully. Using a new droplet for every concentration also takes away some degree of consistency.

Droplets however were useful in determining surface tension. The surface tension between two fluids is 0.0891 N/m. This knowledge can be used to correct fuel spray behavior predictions, as surface tension and Weber number between actual fuel/air interfaces are not quite the same, which is roughly four times larger than that of ethanol/air. This means using the sucrose/silicone oil fluid pair to model fuel spray will involve some adjustments to be accurate.

Acknowledgements

The author hereby expresses his gratitude to Jaal Ghandifor for his comment and suggestion on the research. Mr. Varun Ramesh's continued assistance throughout the project is greatly appreciated. The author would also like to acknowledge University of Wisconsin's Engine Research Center for providing this unique research opportunity and financial support.

References

[1] Kentaro H S and Mitsuhisa I Y 2012 Measurements of droplets spatial sistribution in spray by combining focus and defocus images *16th Int. Symp. on Applications of Laser* pp 9-12
[2] Kawaguchi T and Kobayashi M 2002 Measurement of spray flow by an improved interferometric laser imaging droplet sizing (ILIDS) system *Laser Techniques for Fluid Mechanics* 209-220
[3] Zhang Y Y, Yoshizaki T and Nishida K 2015 Imaging of droplets and vapor distributions in a diesel fuel spray by means of a laser absorption-scattering technique *Appl. Optics* **33** 6221-29
[4] Sean P D, Jason M P and Terence E P 2015 Ballistic imaging of diesel sprays using a picosecond laser: characterization and demonstration *Appl. Optics* **54** 1743-50
[5] Ding H C , Wang Z M and Xu H 2016 Initial dynamic development of fuel spray analyzed by ultra high speed imaging *Fuel* **169** 99-110
[6] Rajat S and Kenneth T C 2015 Surrogate immiscible liquid pairs with refractive indexes matchable over a wide range of density and viscosity ratios *Physics of Fluids* **27** 087103
[7] Frederiksen, S. "Manual for Sugar Solution Prism." 2004:Web.www.fredenksen.eu
[8] Merzkirch w 1987 *Flow Visualization* 2nd ed (New York: Academic) pp 138-204
[9] Evans G M, Jameson G J and Atkinson B W 1992 Prediction the bubble size generated by a plunging liquid jet bubble column *Chem. Eng. Sci.***47** 3265-72

Molecular dynamic simulations of selective self-diffusion of $CH_4/CO_2/H_2O/N_2$ in coal

Y Song[1,2], B Jiang[1,2,3] and F L Li[1,2]

[1]School of resources and earth science, China University of Mining & Technology, Xuzhou 221116, China
[2]Key Laboratory of Coal bed Methane Resource & Reservoir Formation Process, Ministry of Education, Xuzhou 221008, China

E-mail: jiangbo@cumt.edu.cn

Abstract. The self-diffusion coefficients (*D*) of $CH_4/CO_2/H_2O/N_2$ at a relatively broad range of temperatures(298.15~ 458.15K)and pressures (1~6MPa) under the NPT, NPH, NVE, and NVT ensembles were obtained after the calculations of molecular mechanics(MM), annealing kinetics(AK), giant canonical Monte Carlo(GCMC), and molecular dynamics (MD) based on Wiser bituminous coal model (WM). The *D*s of the adsorbates at the saturated adsorption configurations are $D_{CH4}<D_{CO2}<D_{H2O}<D_{N2}$(NPT, 298.15K, 0.1MPa). The diffusion activation energy (*E*) is E_{H2O} (1.07kJ/mol)<E_{N2}(1.82kJ/mol)<E_{CO2} (2.94kJ/mol)<E_{CH4}(10.88kJ/mol), indicating that CH_4 diffusion requires to overcome the highest adsorption energy barrier and N_2 and H_2O to the lowest. The order of different ensembles is D_{N2} (NVE)< D_{N2} (NVT)≈D_{N2} (NPH)≈D_{N2} (NPT) (T<418K) and D_{N2} (NVE) is remarkable higher than other ensembles when T>418K. The average swelling ratios manifest as H_2O (14.7~35.18%)>CO_2 (13.38~32.25%)>CH_4 (15.35~23.71%)> N_2 (11.47~22.14%) (NPH, 1~6MPa). There exits differences in *D*, *swelling ratios* and *E* among various ensembles, indicating that the selection of ensembles has an important influence on the MD calculations for self-diffusion coefficients.

Keywords: Molecular dynamic; Selective self-Diffusion; $CH_4/CO_2/H_2O/N_2$; Wiser model

1. Introduction

Mainly existing as adsorption state, coalbed methane (CBM) is first desorbed from the micro-pores surface [1-2] then diffuses to the macro-pore and fracture systems. Finally, CBM enter the gas phases. The yields of the CBM well are determined by the micro-pore diffusion ability of the main components [3]. The investigations on the self-diffusion characteristics of CBM main components (CH_4/H_2O) and driving gases (CO_2/N_2) are the basis to reveal mass transport diffusion coefficients and Fick diffusion coefficients, which is of significant practical and theoretical importance in CBM development and CO_2 storage.

The investigations on organic–inorganic nanocomposites and the microscopic interactions are of great importance for advanced materials exploration [4]. Due to the remarkable electrical and magnetic properties, organic nanoparticles have received an increasing interest in the literature, which open a great potential for both fundamental and application device [5-6]. As a porous composite, coal has nano-pores in are multiscale ranging from millimeter to angstrom. In nano-pores of >15nm and

2~15nm, the diffusion types are dominant by Fick diffusion and transitional diffusion respectively. Knudsen diffusion is significant in micro-pores of 1.2~2.0nm [7]. Suffered from strong adsorption affinity in molecule scale micro-pores, gas mainly exists as adsorbed phase and only surface diffusions of adsorption phase and solid solution diffusions occur [8-9], which obeys the molecular sieve principles and controlled by geometric shapes of adsorbates, as well as the micro-pore structures of adsorbents. The self-diffusion coefficients of gases in coal macro-molecule can be obtained through molecule simulation. Hu et al. [10] simulated the adsorption and diffusion of CH_4 and CO_2 in coal by GCMC and molecule dynamic (MD) and the results show that CH_4 has higher diffusion velocities and lower saturated adsorption capacities than CO_2 at the same temperature. Based on Wiser model and Einstein equation, Zhao et al. [11] investigated the self- and transport- diffusion characteristics of CO_2/CH_4 in coal under NPT ensemble and demonstrated that the simulated self- and transport- diffusion coefficients are both consistent with the experimental results based on GCMC and MD, indicating that the self-diffusion coefficients can effectively characterize the diffusion characteristics of CO_2/CH_4 in micro-pores of coal. Xiang et al. [8] calculated the self-diffusion coefficients of $CO_2/CH_4/H_2O$ in Yanzhou coal based on the isothermal absorption curves of ternary mixtures (CO_2:CH_4:H_2O=1:1:1) coupled with Einstein equation under NVT ensemble and revealed that the self-diffusion coefficients of CO_2, CH_4, and H_2O are 3.35×10^{-12}, 1.42×10^{-11} and $1.86\times10^{-11} m^2s^{-1}$ respectively at 298.15K. Niekerk et al. [12] studied the non-bonding's key roles in the swelling mechanisms between coal and pyridine, methyl pyrrole and CS_2/NMP solvents through MD. Hu et al [13] analyzed diffusion coefficients of one-component in CO_2-CH_4 mixture based on MD calculations and pointed that the self-diffusion coefficients of CO_2 and CH_4 decrease with gas concentration but increase with temperature (268~428K).

The apparent diffusion coefficients of gases in coal can also be obtained through experiments. Cai et al [14] analyzed the effects of nano-pore structures on adsorptions and diffusions of adsorbates based on high-pressure mercury intrusion and low-pressure liquid nitrogen adsorption. Dutka et al. [15] investigated the replacement processes of CO_2 for CH_4 at the CH_4's saturated adsorption states based on the experiment methods and acquired the replacement rates at the various pressures. Coupled with low-pressure liquid nitrogen adsorption and isothermal adsorption experiments, Ji et al. [16] extracted the small organic molecule from coal by microwave-assisted method and studied the effects of small organic molecules on gas diffusions in coal. Saghafi et al. [9] applied the newly self-developed instruments to explore the self-diffusion coefficients of CO_2 (1.18×10^{-10}~$10.2\times10^{-10} m^2s$) and CH_4 (1.23~$4.88\times10^{-10} m^2s$) in Sydney basin coal ($R_{o,max}$=0.66~1.45%). Staib et al. [17] analyzed the effects of pressures and concentrations on CO_2's diffusions in Australia bituminous coal through the experimental kinetic adsorption data based on unipore, bidisperse, and Fickian diffusion-relaxation model (FDR). By gravimetric experimental, Delphine et al. [18] reported the adsorption kinetics of CO_2 and CH_4(0.1~5MP, 283.15~333.15K)and demonstrated that the diffusion coefficient was estimated at about 10^{-12} m^2s^{-1} and the activation energy of CO_2 adsorption was about 18 kJ mol^{-1} compared with CH_4 (34 kJ mol^{-1}).

As mentioned above, domestic and overseas studies on gases diffusion in coal are mainly concentrated on certain single ensemble during molecule simulations. The effects of various ensembles on the self-diffusion coefficients of $CH_4/H_2O/CO_2/N_2$ in coal at a relatively broad temperatures and pressures have minimally been reported. Actually, self-diffusions are originated from the atoms' incessant generation and recombination caused by the vacancies and clearances among atoms in the crystal. Hence, self-diffusions are born in the randomized Brownian motions from one location/site to another, which is independent of the concentration gradient and promoted by the thermal vibration generated from the migration of atoms. Therefore, the self-diffusion coefficients are the basis to investigate the mass transport/Fick diffusion coefficients [11], [13]. As the thermal motions are significantly affected by the temperature, pressure and enthalpy and each system in the ensemble have the identical structures and macro-restrictions, thus the self-diffusion coefficients of $CH_4/H_2O/CO_2/N_2$ in coal should be influenced by the identical structures and macro-restrictions in different ensembles. Micro-canonical ensemble (NVE) ensemble is an isolated conservative statistical ensemble, in which the atom number

(*N*), pore volume (*V*) and energy (*E*) all stay constant during molecule simulation. Canonical ensemble (NVT) ensemble is in thermal equilibrium condition with the virtual caldarium and has the constant *N*, *V* and temperature (*T*). Isothermal-isobaric ensemble (NPT) is similar to NVT ensemble in the constant temperature method and implemented by regulating the velocity or constrained force of the systems. Because of $H=E+PV$, the pressure and enthalpy in constant pressure and enthalpy (NPH) ensemble maintain constant during the molecule simulation. Hence, the self-diffusion coefficients of $CH_4/H_2O/CO_2/N_2$ in various ensembles should exit differences and the analysis on these differences is the basis to further understand the diffusion mechanisms and tp accurately conduct the molecular simulation.

Therefore, the energy-optimized configuration were obtained through molecule mechanics (MM), MD to optimize the geometric configuration and annealing kinetics (AK) to overcome the molecular structure energy barrier based on WM model in periodic boundary conditions (PBC). Then the saturated adsorption configurations of WM for $CH_4/H_2O/CO_2/N_2$ were calculated through GCMC. Finally the self-diffusion coefficients of $CH_4/CO_2/H_2O/N_2$ in WM were computed through MM and MD under NPT, NPH, NVE and NVT ensembles respectively. Coupled with slit-shape pore models, Steele potential function, and cavity theory, the diffusion mechanisms of $CH_4/CO_2/H_2O/N_2$ in WM and the effects of temperature (298~458K) and pressures (1~6MPa) were revealed in NPT and NPH ensembles, as well as the corresponding diffusion activation energy (*E*). The study is of significant practical and theoretical importance for comprehensive and profound recognitions on efficiently promoting the CBM recovery, as well as perfecting CO_2-ECBM theory, and developing the geological disposal technologies of CO_2 in deep un-minable coalbeds [19-20].

2. Model and calculation methods

2.1. Calculation methods and parameter settings

2.1.1. MM and MDs. The charge distributions of the molecular force field were obtained through the charge equilibrium method in MM and MD calculation in Materials studio8.0 [11], [21]. The $E_{van\ der\ Waals}$ and $E_{electrostatic}$ were calculated based on the atom base method with a cut off distance of 12.5 Å and an initial charge of 0 (Convergence:5.0e-4) in the dreiding force field. The parameters of MM were set as follows: smart minimizer method iterative steps, 50000; RMS force standard, 0.1 kcal/mol·Å; and energy difference, 0.001 kcal/mol. MD were run in a NPT ensemble with the following parameters: initial velocities, random; temperature, 298K; pressure, 0.1 MP a; time; 0.1fs; total simulation time, 2000 ps; number of steps, 50000; thermostat, nose; barostat, Berendsen; and energy deviation, 50000.0 kcal/mol.

Three-dimensional PBCs and relaxations were applied to ensure the energy minimization and geometry optimization of the WM and $CH_4/H_2O/CO_2/N_2$. Then, MD calculations of 1500 ps were conducted under NPT, NPH, NVE, and NVT ensembles to reach the thermodynamic equilibrium states. The atoms' initial velocities were randomly given according to the Boltzmann distribution. The temperatures and pressures were controlled through the Nose [22] and Berendsen methods in a dreiding force field [23]. The effects of various ensembles, gas species, temperatures and pressures on gas diffusion activities are discussed based on the simulated results.

2.1.2 Self-diffusions. The mean square displacement (MSD) is the deviation degree of the spatial position at t from the initial position at t=0. As a result of the special chemistry environment created by the abundance of aliphatic side chains and surface oxygen-containing functional groups in WM, the porosity zigzag factors of diffusion particles would be changed, resulting in the delayed diffusion of the particles due to their adaptation to pore morphology [21], [24]. Hence, the abnormal diffusion existed before the self-diffusion of $CH_4/CO_2/H_2O/N_2$ on the surface of WM. When the anomalous diffusion occurs [7-8], [25]:

$$\left\langle \sum_{i=1}^{N} [r_i(t) - r_i(0)] \right\rangle^2 \propto t^m \tag{1}$$

where t is the time from 0 to the anomalous diffusion, ps; $r(t)$ is Cartesian coordinate tensor of gas molecules at t moment, Å; and m is the fitting constants, $m \neq 1, 2$ (Einstein diffusion).

When Einstein diffusion occurs, it follows Einstein's diffusion law [26]. The diffusion coefficient of the adsorbate is proportional to the MSD of the mass centre of the diffusion particles. The equation is:

$$D = \lim_{t \to \infty} \frac{1}{6Nt} \left\langle \sum_{i=1}^{N} [r_i(t) - r_i(0)] \right\rangle^2 \tag{2}$$

where N is the number of diffusion particles.

The adsorbates diffusing are an activation process. According to the Arrhenius equation, the reaction rate constants have an exponential relationship with the activation energy (E):

$$k = A \bullet e^{-E/RT} \tag{3}$$

where A is the pre-exponential factor, s^{-1}; E is the apparent activation energy of the reactants, kJ/mol; R is the universal gas constant, 8.314 J/(mol • K); and T is the reaction temperature, K.

The adsorption pores are constituted by micro-pores (<2 nm) and ultra-micro-pores (<1 nm). The mass transfer of $CH_4/CO_2/H_2O/N_2$ in nanopores contains two stages: (1) continuous diffusion or viscous flow, caused by the collision among the adsorbates; (2) the migration in micro-pores [7], dominated by the adsorption forces (dispersion force and induced force) and Knudsen diffusion, resulting from the collision and surface diffusion between adsorption layers, and configurational diffusion in the ultra-micro-pores. For the slit-shape pores (Figure 1), when the pore width is L, the distance between the adsorbate molecule and pore well is z. Then, the potential can be calculated through the Steele 10-4-3 potential function [27]:

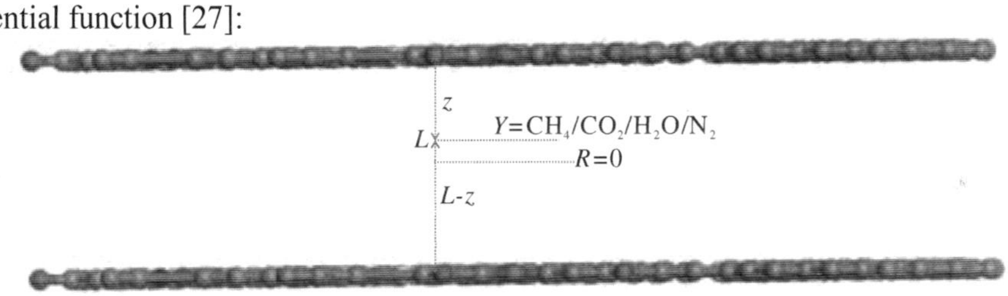

Figure 1. Slit-shape pore models schematic.

$$\Phi(z) = \psi(z) + \psi(L - z) (0<z<L) \tag{4}$$

$$\psi(z) = 4\pi\rho_{atoms}\varepsilon_{gs}\sigma_{gs}^2 \cdot \left[\frac{\sigma_{gs}^{10}}{5(z)^{10}} - \frac{1}{2}\sum_{i=1}^{4} \frac{\sigma_{gs}^4}{\{z + (i-1)\cdot\sigma_{ss}\}^4} \right] \tag{5}$$

where $\varepsilon_{gs}=(\varepsilon_{gg}*\varepsilon_{ss})^{0.5}$, $\sigma_{gs}=(\sigma_{gg}+\sigma_{ss})/2$, and ρ_{atoms} is the number of molecules per unit lattice, 114 nm^{-3}; σ_{gg}, and σ_{ss} are the effective distances of adsorbate and adsorbent, respectively, 0.3758, 0.335 nm for CH4, 0.330, 0.400 nm for CO2, and 0.364, 0.368 nm for N2; εgs is the well depth of cross potential and k is the Boltzmann constant, 1.381×10-23J/K, εgg/k and εss/k are 148.1 and 24.0K, respectively, for CH4; and(εgg,εss) and(σgg,σss)are the characteristic parameters of the Lorentz-Jones potential between C and CH4/N2/CO2 on the coal matrix surface.

2.2. Coal molecule model

Mathews et al. [24], [28] conducted comprehensive reviews on domestic and overseas coal molecule models and revealed that at least 134 molecular level coal models, ranging from low to high rank coals, have been put forward between 1942 and 2012, such as the Wender lignite and brown coals models [29], the Given bituminous model [30], the Shinn subbituminous model [31], the Wiser bituminous model [32] and Vishnyakov anthracite model [33]. Wiser bituminous model can reflect the modern concepts of coal model [21], [28]; thus, this article adopts the Wiser model (Figure 2a). Figure 2b shows the final stable Wiser bituminous model after MM, MD and AK calculations, whose corresponding lattice parameters are $a=b=c=15.4$ Å and $\alpha=\beta=\gamma=90°$. As the optimization process goes on, the energy change and energy gradient both decrease with the increasing optimization steps (Figure 3a), indicating that the energy compositions of WM gradually tended to be stable. $E_{\text{total energy}}$ (720.898 kcal/mol) was dominated by $E_{\text{non-bond}}$ (393.609 kcal/mol, $E_{\text{van der Waals}}$=393.752 kcal/mol). E_{Valence} (327.290 kcal/mol) mainly consisted of E_{angel} (132.089 kcal/mol) and $E_{\text{inversion}}$ (111.133 kcal/mol). The energy composition indicates that $E_{\text{van der Waals}}$ has an important role on structural stability under PBCs. The increase of E_{angel} and $E_{\text{inversion}}$ reveal the increases of the crooked degrees in the final structure (Figure 3b, Table 1). Belonging to ultra-micropores, the average pore diameter of WM is 0.5 nm. The microporous porosity is 9.7%, which is within the scope of 6%~14% reported by Brochard et al [34]. The flow chart illustrating the molecule model and diffusion coefficient calculation method is shown in Figure 3c.

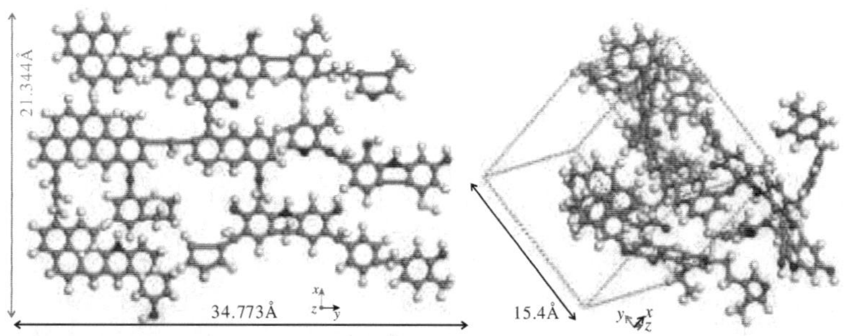

Figure 2. WM plane structure (a) and the optimal configuration (b).

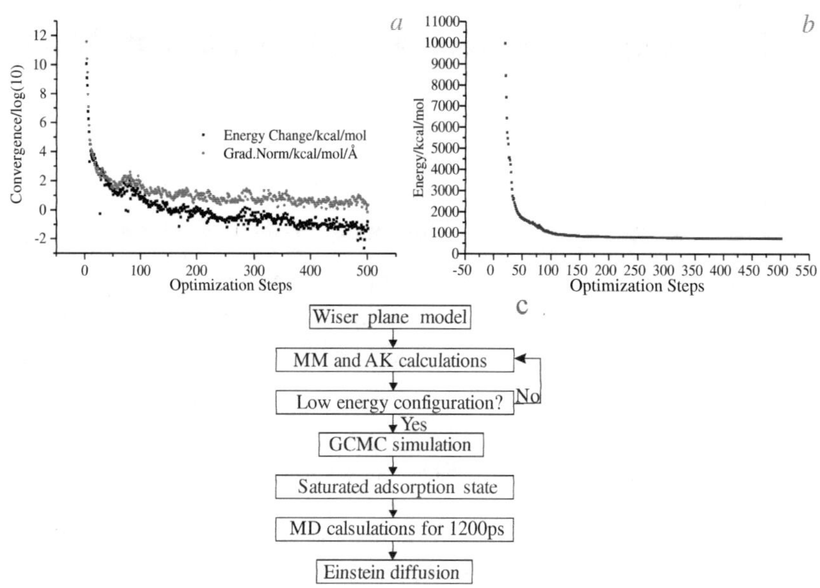

Figure 3. Energy change during the optimization process (a), energy composition of the optimal configuration (b), and flow chart illustrating the molecule model and diffusion coefficient calculation method (c).

Table 1. Energy composition after optimization.

$E_{Total\ energy}$/ kcal/mol		720.898		
$E_{valence}$/ kcal/mol	E_{Bond}/ kcal/mol	E_{Angle}/ kcal/mol	$E_{Torsion}$/ kcal/mol	$E_{Inversion}$/ kcal/mol
327.29	81.904	132.089	111.133	2.164
$E_{Non\text{-}bond}$/ kcal/mol	$E_{Hydrogen\ bond}$/ kcal/mol	$E_{van\ der\ Waals}$/ kcal/mol	$E_{Eelectrostatic}$/ kcal/mol	
393.609	-0.144	393.752	0	

GCMC is applicable to the microscopic particle adsorption simulations of fixed chemical potential and variable particle number. Hence, it is a powerful tool for analysing the adsorption characteristics and mechanisms of porous materials [8], [11]. GCMC simulation was conducted using 5 optimal configurations as adsorbents under PBCs (Figure 3b). The calculation of saturated adsorption capacity was calculated using an isotherm adsorption task in a dreiding force field. The potential constitution of a dreiding force field is as follows [35]:

$$E_{total}=E_{valence\ energy}+E_{non\text{-}bond\ energy} \tag{6}$$

$$E_{valence\ energy}=E_{bond}+E_{angel}+E_{torsion}+E_{inversion} \tag{7}$$

$$E_{non\text{-}bond\ energy}=E_{van\ der\ Waals}+E_{electrostatic}+E_{hydrogen\ bond} \tag{8}$$

The simulated temperature and pressure were set at 298~458K (temperature interval: 10K) and 1~6 MPa, respectively. The initial configuration was the lowest energy of the WM stereoscopic model after MM, MD and AK optimizations. Then, and thereafter, each configuration was determined by the energy change, based on the Metropolis rules [36], to accept or reject the generation, disappearance, translation and rotation of an adsorbate, which ensured the adsorption compound in the lowest energy state and the then output of the new configuration. The convergence criterion was 5×10^{-4}. The Ewald sum method was used for electrostatic action, with a precision of 4.186×10^{-3}kJ/mol. To ensure the balance of the system, 2×10^7 GCMC steps were adopted. The former 10^7 steps were used to reach the balance state and the latter 10^7 steps were to calculate the statistical averages. The final result was the average of three parallel run processes, with a variation coefficient of less than 0.01.

In the process of gradually increasing the adsorbate, the E_{Bond}, E_{Angle}, $E_{Torsion}$ and $E_{Inversion}$ of the adsorption complex was 0 and released heat, resulting in the gradual decreasing of $E_{no\text{-}bond}$. $E_{no\text{-}bond}$ decreased to a minimum and then gradually increased with the increasing of the adsorbates, indicating the adsorbate was no longer absorbed and begun to repel. The adsorption numbers of the adsorbates at the lowest E_{total} was the saturation adsorption capacity per WM. Finally, they reached saturated states after absorbing 30 CH_4s, 42 CO_2s, 56 H_2Os, and 17 N_2s per 5 WMs. The saturated adsorption configurations are shown in Figure 4a~d. The difference in the adsorbate molecules numbers indicates that the adsorption capacity of coal for different gases is: $H_2O> CO_2> CH_4> N_2$. Taking the saturated adsorption configurations as the initial structures, the MD for 1500ps was computed.

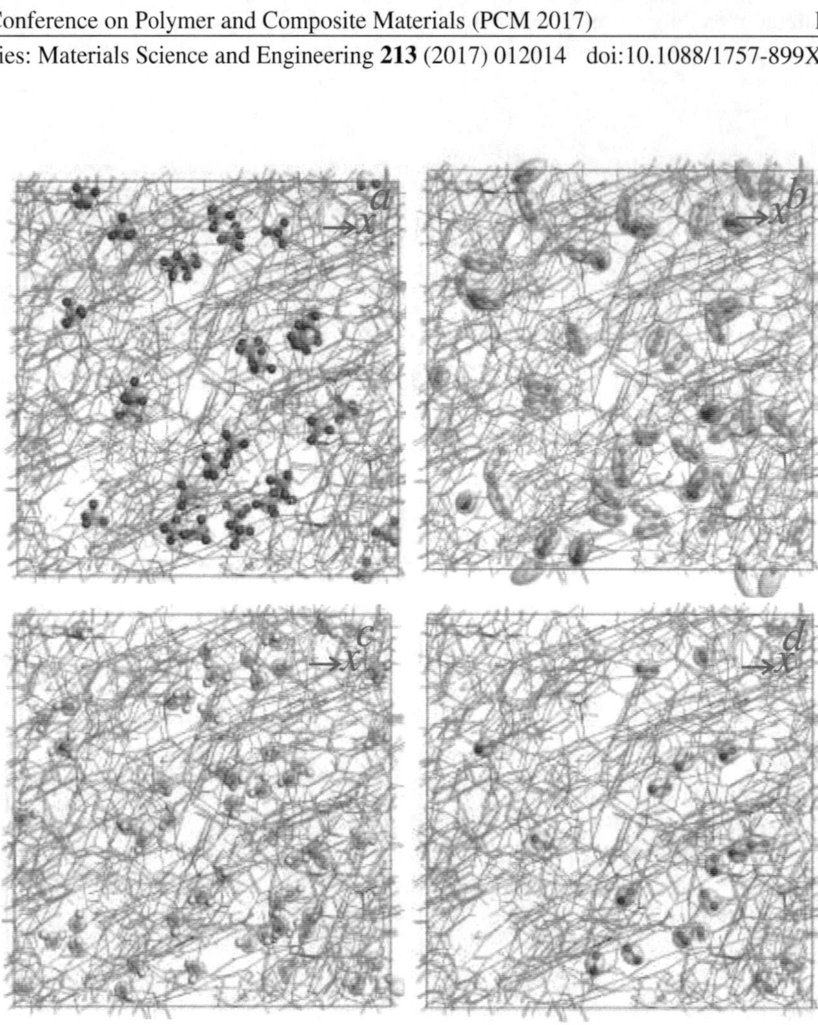

Figure 4. Saturated adsorption configurations of CH_4 (*a*), CO_2 (*b*), H_2O (*c*), and N_2 (*d*).

3. Results and discussions

3.1. Simulation reliability demonstration

To approach the Einstein diffusions, the MD conclusions were conducted for quite a long time (1500 ps), until reaching the linear region of MSD vs *t* curves to ensure the kinetic equilibrium of the WM-adsorbate systems, based on the saturated adsorption configurations. The abnormal diffusion and Einstein diffusion are characterized by slopes of 0.5 and 1, respectively, in the final log(MSD)~log(*t*)curves [10-11].

As is shown in Figure 5a~d, the required times to reach the Einstein diffusion of the adsorbates manifests as t_{CO2}(200 ps)<t_{N2}(250 ps)<t_{H2O}(600 ps)<t_{CH4}(650 ps)at 298.15K and 0.1 MPa under the NPT ensemble. The self-diffusion coefficients of CH_4, CO_2, H_2O, and N_2 were further obtained through equation (2), and the order is D_{CH4} ($4.7*10^{-9}$ m^2/s) <D_{CO2} ($6.9*10^{-9}$ m^2/s) <D_{H2O} ($14*10^{-10}$ m^2/s) <D_{N2} ($1.8*10^{-10}$ m^2/s) at 298.15K. Hu et al. [13] calculated the self-diffusion coefficients of CH_4 and CO_2 based on the MD methods and D_{CO2} (2.51 $*10^{-10}$) and D_{CH4} ($2.41*10^{-10}$ m^{2}s^{-1}) are both slightly higher than the values obtained in this paper at the same temperature. The reasons for the lower self-diffusion coefficients are that Hu investigated mixture systems (25 CO_2 and 25 CH_4 molecules) in unsaturated configurations. Saghafi et al. [9] investigated the diffusion coefficients of CH_4 (1.23~4.88×10^{-10} m^{2}s^{-1}) and CO_2 (1.18×10^{-10}~10.2×10^{-10} m^2 s^{-1}) in coal ($R_{o,max}$=0.66~1.45%) from the Sydney basin, New South Wales, Australia, based on a gravimetric method and the results are close to the diffusion coefficients in this paper, indicating the rationality of the model and method in this paper. Xiang et al. [8] analysed the isothermal adsorption curves and self-diffusion coefficients of CO_2, CH_4, and H_2O in coal ($C_{222}H_{185}N_3O_{17}S_5$, $R_{o,max}$=0.60%) at 298.15K under the NVT ensemble. The similar results of D_{CH4} (3.35×10^{-12} m^2/s)<D_{CO2} (1.42×10^{-11} m^2/s)<D_{H2O} (1.86×10^{-11} m^{2}s^{-1}) was obtained.

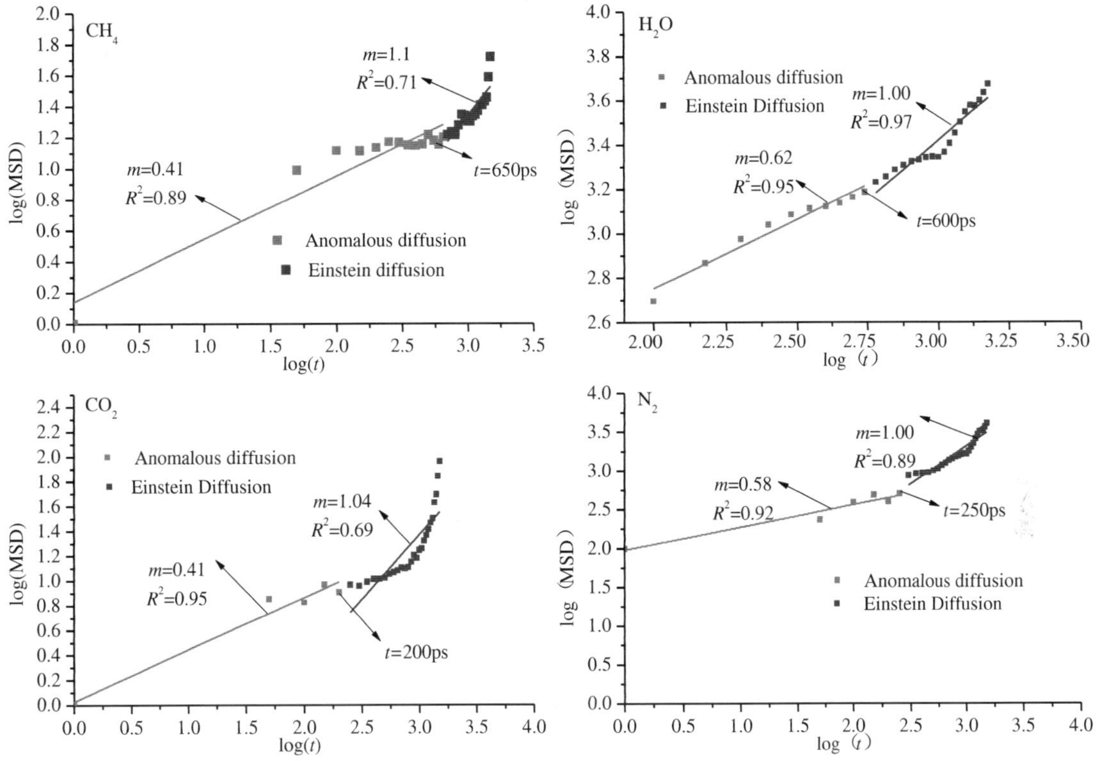

Figure 5. Log(MSD)~log(t) curves of CH_4(*a*), H_2O(*b*), CO_2(*c*) and N_2(*d*).

3.2. Self-diffusion mechanisms

The distributions of the interaction potentials between CH4/CO2/N2 and pore wells in micro- and ultra- micro-pores in WM are shown in Figure 6a, b, as well as the lowest potential energy distribution (Figure 6c). Figure 6a and Figure 6b shows a normalized pore wall in WM macromolecule interaction potential along the dimensionless distance from the pore center. When the pore half width is larger than 0.7 nm, enhancement of the negative potential in the center of the pore begins and a double minimum of the potential develops near the two pore walls. This gives the onset of the overlapping of the potential fields of the opposing walls. Strong enhancement of potential minimum begins when the pore half width is lower than 0.7 nm and larger than 0.5 nm. The double minimum is replaced by a single minimum in the pore. The potential reaches its negative minimum at pore center. Further decrease of the pore width causes a significant positive increase of the potential and the potential becomes positive when the pore half width is lower than 0.5nm and larger than 0.2nm.

For CH4, when L≥1.5 nm, the maximum Φ at z=0 is larger than 0, indicating that CH_4 in the pore centre are not affected by the potential energy field of the pore walls. The diffusions are dominant by the gas phases' Knudsen diffusions and adsorbed phases' surface diffusions [37]. When 0.7 nm<L≤1.5 nm, the maximum Φ at the z=0 is negative, and there exist two lowest points, which are higher than the adsorption surface free energy. The energy required to desorb CH4 from the adsorption sites to the gas's phases of the pore centre decreases with decreasing L. The corresponding diffusive types are activation desorption and Knudsen diffusion along the pore centre in these pores [38]. When 0.5 nm<L≤0.7 nm, there exists only one negative maximum value at z=0, and the corresponding adsorption potential is the lowest. In these pores, the adsorption phases do not exist and the configuration

diffusion dominates, which needs to overcome the adsorption energy barrier. In pores of 0.2 nm⩽L⩽0.5 nm, the adsorption potential increases and becomes positive. The adsorption potential minimums locate at z=0 and other locations are repulsive forces, indicating that it is difficult for CH4 to enter pores of 0.2 nm≤L≤0.5 nm (Figure 6a, b).

The lowest potentials at different diameters in Figure 6a, b were extracted and the adsorption potential distributions of CH4/CO2/N2 were obtained. As is shown in Figure 5c, CH4, CO2, and N2 have highest adsorption energy at the pore widths of 0.72, 0.67, and 0.65 nm, respectively, half of which are equal to the gases' diameters, indicating that pores with a half-width (radius) close to the adsorbates' diameters are the most beneficial to adsorption. The optimum aperture of CH4 in the Steele potential analysis are close to the adsorption equilibrium distances (3.6~4.0 Å) in our previous work of molecular simulation of CH4 absorbing on coal vitrinite [35]. Shieh et al. [39] considered that the aerodynamic diameters of CH4/CO2/N2 are close to their respective molecular sieve dimensions and are critical for gas diffusion in porous media through investigations on the gases' permeability, diffusivity and solubility in poly (4-vinylpridine). The adsorption potential order manifests as CO2 (24.28 kJ/mol)>CH4 (22.26kJ/mol)>N2 (17.27kJ/mol), indicating that the adsorption affinity of coal-CO2 is the strongest, and second strongest for CH4. The highest adsorption potentials are about two times larger than those in pores >1.5 nm for CH4, CO2, and H2O (CH4: 11.28kJ/mol; CO2: 12.42kJ/ mol; N2: 8.79kJ/mol).

CH4, CO2 and N2 cannot enter pores with widths <0.56 nm, and CO2 and N2 can enter pores <0.62 nm, indicating that the minimum pore widths that can be entered by CO2 and N2 are lower than that of CH4. From the above analysis, the diffusions of gases in micro-pores are controlled by the adsorption energy and the molecular dynamics diameters. CO2 a has larger adsorption energy and lower aerodynamic diameter, therefore it is easier to diffuse in micro-pores, followed by N2, meanwhile indicating that differences exist for various gas types in pore tortuosity factors due to the different diameters. The larger the gas diameter, the higher the pore tortuosity factors, which was also revealed in polymer (4-vinylpridine) and other porous medias [39]. The adsorption potential of CH4 is higher than CO2 only in pores of 0.72~0.91 nm, which may be attributed to limitations of the 10-4-1 model. In addition, the adsorption force of CO2 is larger than CH4 and N2 in almost all of the microporous segments, and has the lowest aerodynamic diameter. Hence, CO2 can enter smaller micro-pores than CH4, leading to advantages in the competitive adsorption and selective self-diffusion among the main components of CBM.

The cell volume of coal molecule is constituted by the occupied volumes and free volumes. The size and shape of the free volumes have a significant effect on the adsorbates' diffusion activities in coal macro-molecules, which provide both the necessary activity spaces of coal macro-molecules and the diffusion spaces of the adsorbates. Gas molecules are constantly moving in the coal molecules. The magnitude of the displacement is small and can be regarded as vibration in most cases. Occasionally, a big jump to another area such as a cavity, to continue small amplitude vibrations occurs. When small gas molecules vibrate to a certain site where the coal macromolecular structure is or has been twisting or rotating, then small gas molecules may jump to the neighboring cavities and continue vibrating in the cavities. Figure 7a~d show the diffusion trajectories and displacement -t curves of $CH_4/CO_2/H_2O/N_2$ in coal from the initial position(x_0, y_0, z_0)=(0, 0, 0)at 298.15K and 0.1 MPa under an NPT ensemble.

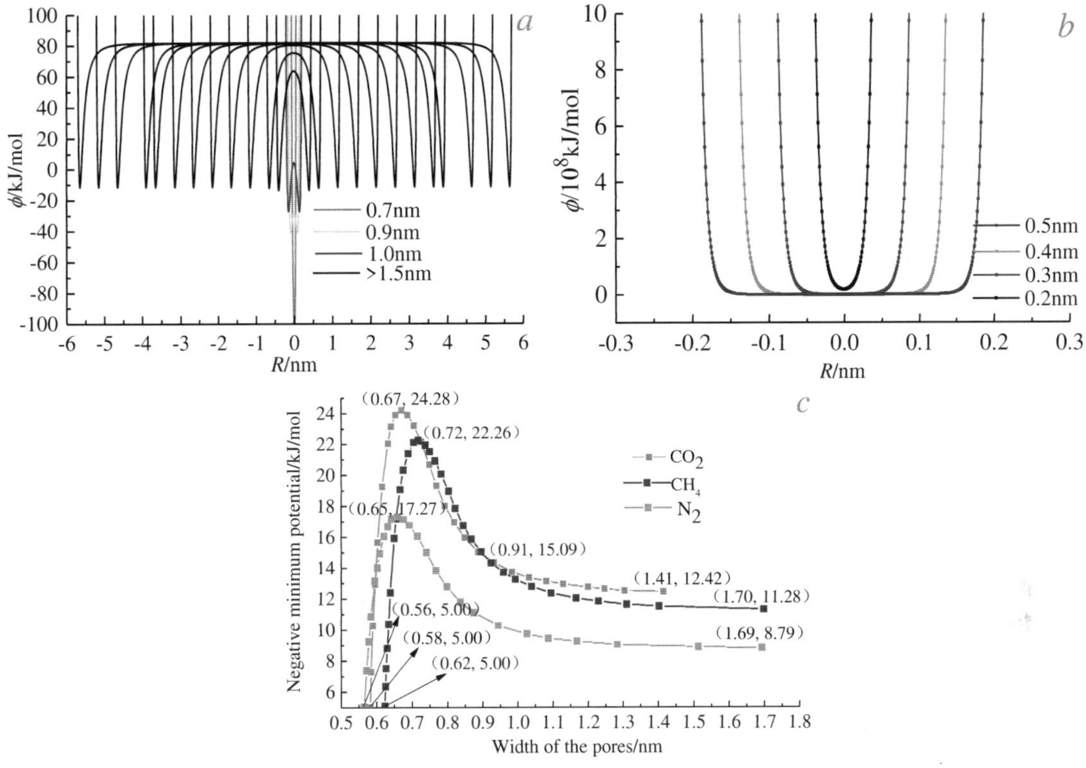

Figure 6. Potential distributions of CH_4 in slit-shape pores of different widths in coal (*a*, *b*). Lowest potential of CO_2/CH_4/H_2O in slit-shape pores of different widths (*c*).

The aerodynamic equivalent diameters (σ) of CH_4, CO_2, H_2O, and N_2 manifest as σ_{CO2} (0.33)<σ_{H2O} (0.35)<σ_{N2} (0.37)<σ_{CH4} (0.38 nm). As is shown in Figure 7a~d, the distance of a single jump of N_2 and H_2O are significantly higher than those of CH_4 and CO_2 in the 1500 ps period. The calculated result shows that each transition distance of N_2 and H_2O are > 0.5 nm. Because of the weak adsorption capacity of N_2 on WM, the jumps among cavities are dominated by configurational diffusion, while activation desorption almost did not exist and was relatively stationary for the rest of the time. Two jumps among cavities occurred within a relatively long time (848 and 1350 ps) and the jump distances of the 848 and 1350ps were 3.36 and 9.88 Å, respectively (Figure 7d), which is related to the smaller lowest accessible pore width and the weak adsorption affinity on coal. The displacement-*t* curve of N_2 indicates the rapid diffusion behavior inside the coal macromolecule.

The displacements of each H_2O jump are larger than 1.3 nm, and the diffusion proceeds in a stepwise manner. The displacements generally do not increase within 252~550, 850~1005, and 1300~1400ps, during which the diffusion manner manifests as internal vibration within a certain cavity (Figure 7c). When *t*=550, 1005, and 1400ps, the displacements rapidly increase as the diffusion processes continue, indicating the jumps among the adjacent cavities. The stepwise characteristics resulted from the strong adsorption forces and significant heterogeneities of the coal surface for H_2O; especially the interactions between H_2O and oxygen-containing functional groups, which are stronger than those of CO_2 and CH_4 [8]. The internal vibration within a certain cavity is caused by the Debye forces and London dispersion forces from the preferential adsorption locations/sites, such as aromatic structures and heteroatom rings. Whereas the jump of large displacements (>10 nm) among the adjacent cavities in activation desorption and surface diffusion manger originated from the strong affinity of neighboring locations/sites where the coal molecule structure is or has been twisting or rotating, leading to surface diffusion from relatively weak adsorption sites to strong adsorption sites. The diffusion speeds and jump distances of N_2 are lower than those of H_2O, which is related to the strong hydrogen bonding between H_2O and coal. At certain adsorption sites, H_2O will jump for a large displacement when affected by the hydrogen bonding of the adjacent adsorption sites.

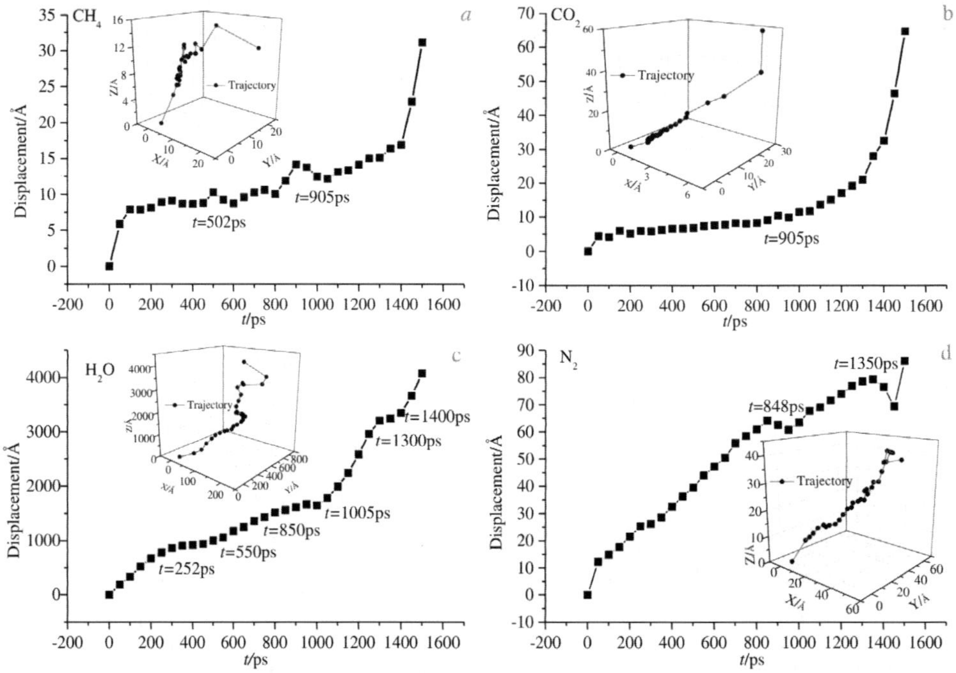

Figure 7. Displacement-*t* curves and diffusion trajectories of different gases.

The displacement-*t* curve of CH_4 manifests as a long time period, staying in a single cavity in the abnormal diffusion stages, and the small jumps of 1.2 and 4.5 Å only occur at 502 and 905ps, respectively (Figure 7a). Combined with the Steele potential function results, CH_4 has the lowest adsorption energy and highest molecule diameter among $CH_4/CO_2/H_2O/N_2$; hence, the required pore width to generate jumps among the adjacent cavities is the largest. The time to reach Einstein diffusion is also the longest (650ps) and the magnitude of the movement is small in most cases. Caused by the highest kinematic viscosity (17.0710) in contrast to CO_2 (8.369) and N_2 (15.753 mm^2/s), the diffusion type of CH_4 is dominated by Knudsen diffusion and surface diffusion, and the displacement increases slowly at Einstein diffusion stages. The motion amplitude of CO_2 is even in a smaller variation range than CH_4, and a jump with a displacement of >1 nm only occurs at *t*=905ps (Figure 7a). The small displacement variation range is created by the high molecule weight and large dynamic viscosity (14.932μPa·s) compared with CH_4 (11.067), N_2 (13.805) and H_2O, leading to a smaller mean free walk. In addition, CO_2 has a strong affinity with coal. Affected by the smaller mean free walk and strong affinity, the diffusion velocity is slow at the anomalous diffusion stage. However, the rapid increase of displacement at Einstein diffusion stage is controlled by its lowest kinematic viscosity. Similar to CH_4 in diffusion behaviour, the vibration displacements of CO_2 are <0.5 nm. The vibration of CO_2 also belongs to motion in a single cavity. The molecule weight of CO_2 is far higher than CO_2 and the occasional jumps among neighbouring cavities will not occur for CO_2. As analysed above, the self-diffusion mechanisms of CH_4, CO_2, H_2O, and N_2 in coal are profoundly influenced by molecular diameter, molecule weight, viscosity and adsorption affinity with coal. At anomalous diffusion stages, the diffusions are mainly dominated by the molecular diameter and adsorption affinity. While at the Einstein diffusion stages, the diffusions are dominated by the molecule weight and kinematic viscosity.

3.3. Effects of temperature

The relationships between the self-diffusion coefficients and temperatures under NPT, NPH, NVE, and NVT ensembles are shown in Figure 8a, c, d and e. The pressure of NPH and NPT ensemble are both 1 MPa. The self-diffusion coefficients of CH_4, CO_2, and N_2 gradually increase with increasing temperature. When the temperature is < 368K, 368~428K, and 438~458K, the self-diffusion coefficients manifest as D_{CH4}(0.1~2.24*10^{-9} m^2/s)<D_{CO2}(0.69~1.3*10^{-9} m^2/s)<D_{N2}(1.8~4.8*10^{-9} m^2/s), D_{CO2}(1.4~1.6*10^{-9} m^2/s)<D_{CH4} (1.9~ 4.0*$10^{-9}$$m^2$/s) <$D_{N2}$(4.8~6.8*$10^{-9}$$m^2$/s) and D_{CO2} (1.6~2.0*10^{-9} m^2/s)<D_{CH4}(9.6~9.6*10^{-9} m^2/s)<D_{N2} (10.4~29.6*10^{-9} m^2/s), respectively. D_{H2O} (0.1~58.08*10^{-9} m^2/s) is higher than D_{CH4}, D_{CO2}, and D_{N2} at each temperature during 298~ 458K, and fluctuation increases with increasing temperature. The higher D_{H2O} is related to light molecule weight and small molecular diameter.

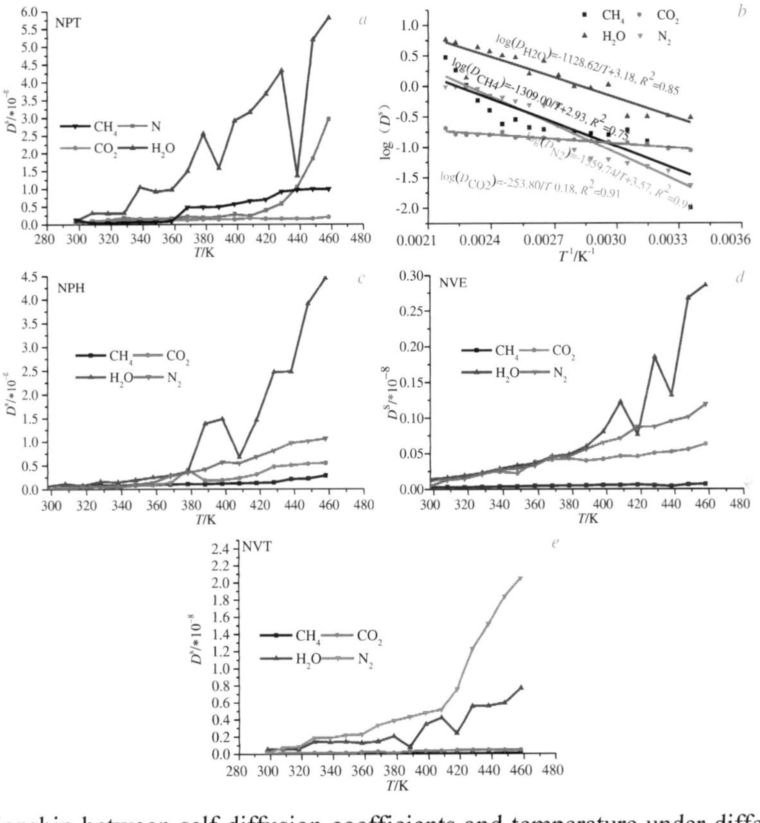

Figure 8. Relationship between self-diffusion coefficients and temperature under different ensembles (a, c, d, e) and the activation energy calculation method (b).

The self-diffusion of adsorbates in adsorbents is a process of activation. The diffusion reactivity is in accordance with Arrhenius law. As is shown in Figure 8b, log(D) has a good linear relationship with $1/T$ (R^2=0.7523~0.9157) (Figure 8b), which is also consistent with equation (3). Activation energy (E) can be obtained through the slopes and manifests as E_{H2O}(1.07)<E_{N2}(1.82) <E_{CO2}(2.94) <E_{CH4}(10.88kJ/mol); the corresponding pre-exponential factors are 24.20, 19.08, 0.83, and 18.75s^{-1}, respectively, indicating that the Einstein diffusion of CH_4 is the most difficult to occur, as CH_4 has a higher adsorption energy barrier to overcome than CO_2, H_2O and N_2. The Einstein diffusion of CO_2 is easier to occur than CH_4, which is related to its lowest kinematic viscosity combined with the Steele 10-4-1 potential function results. The results also revealed that CO_2 should be injected effectively into the coal bed with high speed to form competitive adsorption and selective self-diffusion with CH_4 in order to replace the CH_4 during E-CBM engineering. The adsorption capacity of N_2 is lower than CO_2 and CH_4 [40]; when the temperature is higher than 208K, D_{N2} is higher than D_{CH4}, indicating that the displacement mechanism of N_2 is different from CO_2. The rapid diffusion of N_2 in a coal reservoir can effectively decrease the partial pressure of CH_4, which could further influence the isothermal absorption curve and reduce the methane adsorption capacity. E_{H2O} and E_{N2} are both the lowest, and the Einstein diffusions of H_2O and N_2 are the easiest to occur, which is also consistent their highest self-diffusion coefficients.

Under the NPH ensemble, the self-diffusion coefficient order is $D_{CH4}(0.01*10^{-9})<D_{CO2}(0.04*10^{-9})<D_{H2O}(0.06*10^{-9})<D_{N2}(0.08*10^{-9}\ m^2/s)$ at 298K, which is similar to the NPT ensemble. The self-diffusion coefficients during 298~458K overall manifest as $D_{CH4}(0.01\sim2.8*10^{-9})<D_{CO2}(0.04\sim5.5*10^{-9})<D_{N2}(0.06\sim\ 44.5*10^{-9})<D_{H2O}(0.08\sim10.6*10^{-9}\ m^2/s)$. The self-diffusion coefficients of CH_4, CO_2, and N_2 gradually increase with increasing temperature, while D_{H2O} fluctuation increases and is sensitive to temperature (Figure 8c). The self-diffusion coefficients of $CH_4/CO_2/H_2O/N_2$ under the NVE ensemble are the lowest compared with other ensembles. D_{CH4} $(0.03\sim0.07*10^{-9}\ m^2/s)$ is even lower than $CO_2/H_2O/N_2$ during 298~458K, and gradually increases with increasing temperature. During 298~378K, $D_{H2O}(0.14\sim0.48\ *10^{-9})$is slightly higher than $D_{CO2}(0.12\sim0.42*10^{-9})$ and $D_{N2}(0.04\sim0.45*10^{-9}\ m^2/s)$. When the temperature is 458K, the self-diffusion coefficients are $D_{CO2}(0.39\sim0.62*10^{-9})<D_{N2}(0.55\sim1.18*10^{-9})<D_{H2O}(0.59\sim2.89*10^{-9}\ m^2/s)$(Figure 8d). While under the NPH ensemble, the self-diffusion coefficients order is $D_{CH4}(0.01\sim0.06*10^{-9})<D_{CO2}(0.07\sim\ 0.45*10^{-9})<\ D_{H2O}(0.54\sim7.67*10^{-9})<D_{N2}(0.10\sim20.4*10^{-9}\ m^2/s)$. The self-diffusion coefficients of N_2 and CH_4 gradually increase and those of CO_2 and H_2O jump increase with increasing temperature (Figure 8e).

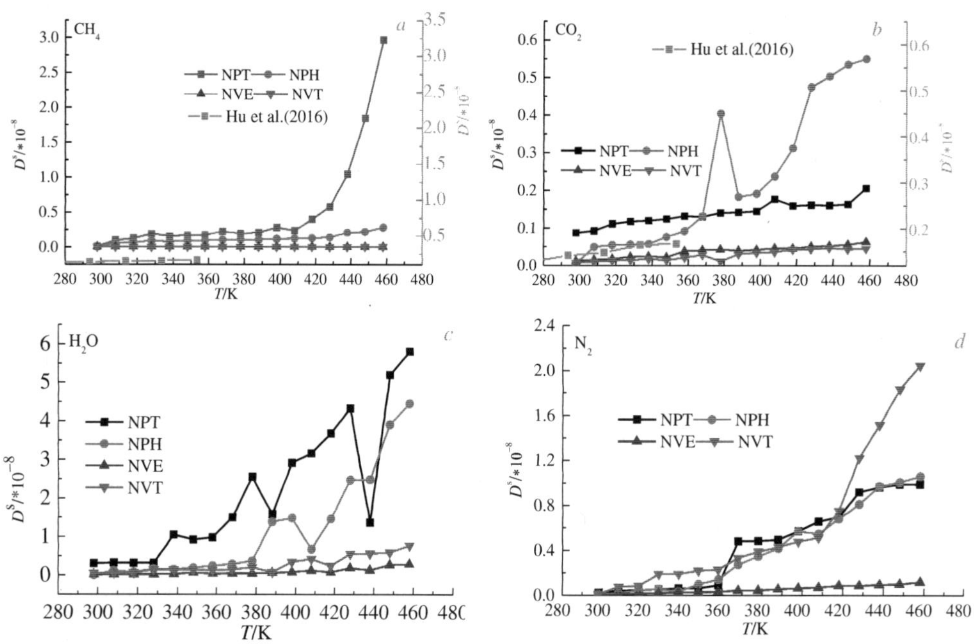

Figure 9. Relationships between self-diffusion coefficients and temperatures under the same ensemble.

D_{CH4} gradually increases with increasing temperature independent of the ensembles. However, the D_{CH4}, which increases slowly with increasing temperature under NVT and NVE ensembles, increases slowly when the temperature is lower than 418K and rapidly when the temperature higher than 418K in NPT and NPH ensembles. At the same temperature, D_{CH4} manifests as D_{CH4} (NVE)$\approx D_{CH4}$ (NVT)$<D_{CH4}$ (NPH)$<D_{CH4}$ (NPT) (Figure 9*a*). D_{CO2} leaping increases under NPH and NPT ensembles and gradually increases under NVT and NVE ensembles. D_{CO2} manifest as D_{CO2} (NVT)$<D_{CO2}$ (NVE)$<D_{CO2}$ (NPH)$<D_{CO2}$ (NPT) (298K$\leqslant T \leqslant$368K)and D_{CO2} (NVT) $<D_{CO2}$ (NVE) $<D_{CO2}$ (NPH)$<D_{CO2}$ (NPT)(378K$\leqslant T \leqslant$458K) at the same temperature (Figure 9*b*). The calculated results of Hu et al. (2016) are close to the magnitudes of CO_2 and CH_4 in this paper, but the values are significantly lower than the values in this paper. The lower values are because the initial configuration (5Wiser model+25CH_4, 5Wiser model+25CO_2) before MD calculations has not yet reached a saturated adsorption state, still subjected to strong adsorption forces when diffusing in coal. D_{H2O} all show an increase with the temperature independent of the ensembles (Figure 9*c*). D_{N2} gradually increases with temperature in various ensembles, and the amplitude becomes significant with increasing temperature when the temperature higher than 358K. The D_{N2} of NVT, NPH, and NPT ensembles are close and obviously higher than those of the NVE ensemble. Unlike other gases, D_{N2} of the NVT ensemble is significantly higher than other ensembles when the temperature is higher than 418K (Figure 9d). The diffusion coefficient differences of various ensembles are essentially created by macroscopic conditions, such as temperature, pressure, and enthalpy, as well as the thermodynamic properties of various gases. The NPT ensemble has the highest diffusion coefficients for CH_4, CO_2, H_2O, and N_2, which is caused by the violent Monte Carlo movements from one potential energy surface to another. Originating from random impulse force [41]; NPH ensemble has the constant pressure and enthalpy, leading to second highest diffusion coefficients. D_{CH4}, D_{H2O}, and D_{N2} are the lowest in the isolated and conservative NVE ensembles, and D_{CO2} is the lowest in NVT ensembles of constant kinetic energy, which is controlled by macroscopic conditions and thermodynamic properties.

3.4. Effects of pressure

3.4.1. NPH ensemble. The self-diffusion coefficients of gas molecules at 298K and in the NPH ensemble decrease with increasing pressure (Figure 10a). At higher pressures, the interactions between gas molecules are stronger and the gas density is also higher, leading to a slower diffusion velocity, which is also the reason for CBM's low diffusion coefficients at a deep buried depth [13]. Skoulidas et al. [42] also found this rule by analysing the self-diffusion and mass-transfer coefficients of small-molecule gases in metal-organic framework materials (MOE) through the MD method. When the pressure is lower than 1.68 MPa and higher than 1.68 MPa, the self-diffusion coefficients manifest as $D_{H2O}>D_{CO2}>D_{N2}>D_{CH4}$ and $D_{H2O}>D_{CH4}>D_{CO2}>D_{N2}$, respectively, showing that higher reservoir pressure is favourable to the replacement of CH_4 by CO_2 and N_2. In contrast to the unipore model of Clarkson et al. [43] and the bidisperse model of Cui et al. [7], the diffusion coefficients in the bidisperse model are obtained through the apparent micro-pore diffusivity (s^{-1}) multiplied by the path length (m). The simulation results of CH_4 and CO_2 in this paper agree with those of the unipore and bidisperse models in the order of magnitude, which further proves the rationality of the MD simulation method. However, the bidisperse model involves pressures below 3 MPa, and D_{CH4} and D_{CO2} first increased and then decreased with increasing pressure, during which CO_2 and CH_4 have the largest diffusion coefficients at 0.75 MPa and 1.25 MPa, respectively. The pressure in this paper ranges from 1 MPa to 6 MPa, and the trend in this interval is in consistent with the trends of Cui's bidisperse model and unipore model at 0.5-6.5 MPa. Unlike the diffusion coefficients of temperature-dependence, the diffusion coefficients of various gases are $D_{CH4}(0.256*10^{-10}$ $m^2/s)<D_{N2}(0.296*10^{-10}$ $m^2/s)<D_{CO2}(0.4433*10^{-10}$ $m^2/s)$ $<D_{H2O}(2.271*10^{-10}$ $m^2/s)$ when the pressure is lower than 1.75 MPa. The diffusion coefficient of H_2O is the largest among CH_4, CO_2, H_2O, and N_2, which is related to its largest kinetics viscosity and lighter molecule weight. By ^{13}C NMR and FT IR, Xiang et al. (2014) calculated the self-diffusion coefficients of $CH_4/CO_2/H_2O$ in the self-building Yanzhou coal macromolecular model at 298K and the results manifested as $D_{CH4}(0.335*10^{-12}$ $m^2/s)$ $<D_{CO2}(0.142*10^{-11}$ $m^2/s)$ $<D_{H2O}(0.186*10^{-11}$ $m^2/s)$, which are close to the simulated values, in order of magnitude, in this paper. When the pressure is higher than 3 MPa, D_{CH4} diffusion coefficients are larger than D_{CO2} and D_{H2O},

indicating that higher reservoir pressure is conductive for promoting the diffusion coefficients of CH_4. The diffusion coefficient changes of CH_4 are caused by the lighter molecule weight and larger molecular free length at higher pressure, when the temperature is constant.

Molecular swelling of various degrees will occur during the interaction processes of adsorption and diffusion between coal molecules and gases to form a more stable configuration [19],[17], [44]. Xiang et al. holds that the valence electron energy of the coal molecule decreases after absorbing CH_4, CO_2, and H_2O, indicating that the "strain coal" originating from pressure action in the geological conditions may be the results of molecular rearrangement after swelling effects during the interactions between fluids and coal. Gas molecules in the coal diffusion process can also cause swelling phenomenon; the *swelling ratio* is defined as [11]:

$$\textit{swelling ratio} = (V_i - V_0)/V_0$$

where V_i and V_0 are the volumes occupied by the molecules before and after diffusion, respectively; A^3.

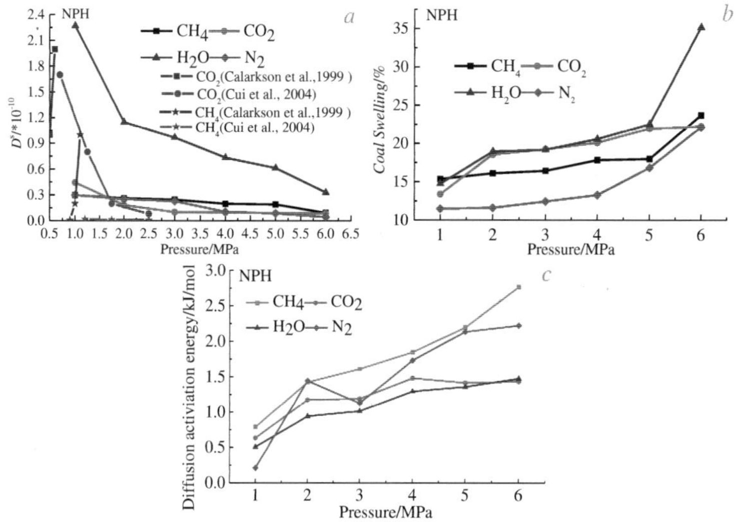

Figure 10. Relationships between the *D* (*a*), *swelling ratio* (*b*) and *E* (*c*) with the pressure under the NPH ensemble.

With the increasing pressure, the *swelling ratio* of gas molecules increases gradually through either the adsorption or diffusion process. During 1~6 MPa, the average *swelling ratios* of different gases are shown as: H_2O(14.7~35.18%)>CO_2(13.38~32.25%)>CH_4(15.35~23.71%)>N_2(11.47~22.14%) (Figure 10b). The increase trend of *swelling ratios* with pressure becomes significant at high pressures, which are in consistent with Busch's findings. The swelling action can shrink the micro-pore volume and change the width of the fracture to change the reservoir diffusion coefficient and permeability. For H_2O and CH_4, when the pressure is less than 5 MPa, the *swelling ratio* is low, the dominant diffusion type is surface diffusion of the adsorption phase, and the diffusion activation energy increases with the increasing pressure to cause surface diffusion to overcome the higher molecular structural energy barrier (Figure 10c), leading to a gradual decrease in the diffusion coefficients. When the pressure is higher than 5 MPa, the molecular *swelling ratio* of coal increases sharply, leading to rapidly narrowed micro-pore width, and the dominant diffusion type is configuration diffusion; while the diffusion energy barrier also gradually increased, resulting in a rapid decrease in diffusion coefficients. For CO_2, when the pressure is less than 2 MPa, the *swelling ratio* increases greatly, and the main diffusion type is surface diffusion; when the pressure is higher than 2 MPa, the *swelling ratio* and diffusion activation energy increases slowly, and the dominant diffusion types are surface diffusion and Knudsen diffusion. The diffusion type of N_2 is mainly Knudsen diffusion of gas phases, and surface diffusion of the adsorbed phases rarely exists, which leads to the lowest *swelling ratio* and the lowest activation energy (Figure 10c). Therefore, the required energy barrier to overcome for Knudsen diffusion is also the lowest among $CH_4/CO_2/H_2O/N_2$.

The WM-N_2 diffusion system has the lowest *swelling ratios*, which indicates that the decrease in the throat radius of the micro porous pores caused by the coal molecular swelling is also the lowest, so that the diffusion of N_2 in the coal molecules maintains relatively wide pore throats, which can efficiently reduce the partial pressures of CH_4 and the micro-crack reductions caused by CH_4-coal's swelling effects. However, the WM-H_2O diffusion system molecular swell has an adverse effect on the pore throat radius, and the *swelling ratio* of the system is the highest, leading to a significant reduction of the gas inlet radius in coal micro-pores. The *average swelling ratio* of the WM-CO_2 system is higher than that of the WM-CH_4 system under different pressures, which indicates that the pore throat radius of the WM-CO_2 system is smaller than that of the WM-CH_4 system, so that the pressure required for decreasing the diffusion coefficient is lower [11]. Siemons et al. [45] analysed the rate of sorption of CO_2 under various pressures and various grain size fractions using volumetric experiments and found at that at low pressures (<1 MPa), fast gas diffusion(surface diffusion)is the prevailing mechanism for sorption, whereas at higher pressures, the slow diffusion process controls the gas uptake by the coal.

As analysed above, the WM's molecular-level micro-pores (<2 nm) contain two transmission mechanisms: (1) diffusion from one end to the other end in ultra-micro-pores: as the pore sizes are close to the diameter of the adsorbates in magnitude, only the gases with diameters smaller than the pore sizes can enter the ultra-micro-pores, while the larger gas molecules could not. (2) surface diffusion: the adsorbates of gas phase jumps from one certain adsorption site to neighbouring sites [7]. Both adsorption mechanisms have to overcome the energy barrier on the surface of the coal matrix. Hereby, the corresponding activation energy of diffusion of $CH_4/CO_2/$ H_2O/N_2 in the coal is calculated under the NPH ensemble. Under the pressure of 1-6 MPa, the activation energy of diffusion is shown as follows: E_{H2O}(0.51~1.48kJ/mol)<E_{CO2}(0.63~1.43kJ/mol)<E_{N2}(0.21~2.22kJ/mol)<E_{CH4}(0.78 ~2.76kJ/mol). CH_4 has the largest E and needs to overcome the highest energy barrier. E_{CO2} and E_{H2O} are relatively low and they could overcome the energy barriers easily, which is also consistent with the higher self-diffusion coefficients of CO_2 and H_2O.

3.4.2. NPT ensemble. The changing laws of the self-diffusion coefficients with the pressure under the NPT ensemble are similar to the NPH ensemble. When the pressure is lower than 2 MPa, the self-diffusion coefficients manifest as D_{CH4} (1.57~1.65*10^{-10})<D_{N2} (2.15~5.12*10^{-10})<D_{CO2} (1.46~1.91*10^{-10})<D_{H2O} (5.49~7.44*10^{-10} m^2/s). However, when the pressure is higher than 3 MPa, D_{CH4} is higher than D_{N2} and D_{CO2} and lower than D_{H2O} (Figure 11a). The self-diffusion coefficients of the NPT ensemble are higher than those of the NPH ensemble at each pressure, which is consistent with the temperature effect. Compared with the results of Zhao et al. [8] by WM, this paper's results and Zhao's results are both in the 10^{-10} magnitude order. Zhao demonstrated that the self-diffusion coefficients of CO_2 and CH_4 are the highest at 1.5 MPa and 1.2 MPa, respectively, slighter higher than the pressure peak value of 1 MPa in this paper. The simulated diffusion coefficients of pressure-dependence in this paper are of a same trend to the results of the unipore model of Clarkson et al. [43], the bidisperse model of Cui et al. [7] and Zhao at a pressure range of 0~3 MPa. The simulated diffusion coefficients of pressure-dependence in this paper are higher than the bidisperse model by a magnitude order. The deviations are caused by the differences in pore structure and chemical structures of the bidisperse model and WM. Coupled with surface functional groups and heteroatomic rings, the WM's pore structure in this paper is mainly constituted by ultra-micro-pores (<1 nm) and micro-pores (<2 nm). While the bidisperse model is mainly made up of micro-pores (<2 nm) and aromatic nuclei as the basic structure units (BSUs), in which the adsorption capacities for CH_4 and CO_2 are stronger than WM (adsorption capacities of functional groups are aromatic structure> heteroatom rings>oxygen functional groups; Song et al., [35]. The swelling ratios and E of pressure-dependence under the NPT ensemble gradually increase with increasing pressure (Figure 11b, c). Differences in the swelling ratios of the NPH and NPT ensembles exist at the same pressure and temperature, indicating that the swelling ratios are also influenced by macro-restrictions. The average swelling ratios of the NPT ensemble for different gases manifest as CH_4 (18.95~24.80%)>H_2O (16.12~31.73%)>CO_2 (15.21~24.32%)>N_2 (13.17~18.59%) (Figure 11b).

The *swelling ratios* are of the same trend but significantly higher than Zhao's results, indicating that the diffusions in saturated adsorption configurations will cause significantly higher *swelling ratios* than the unsaturated adsorption state. Because of the differences in the thermodynamic properties of the gases and the macro-restrictions of the ensembles, the *swelling ratios* of various ensembles are close, but differ in the order of various gases. The results in this paper also demonstrate the importance of the ensemble selection during molecule simulation.

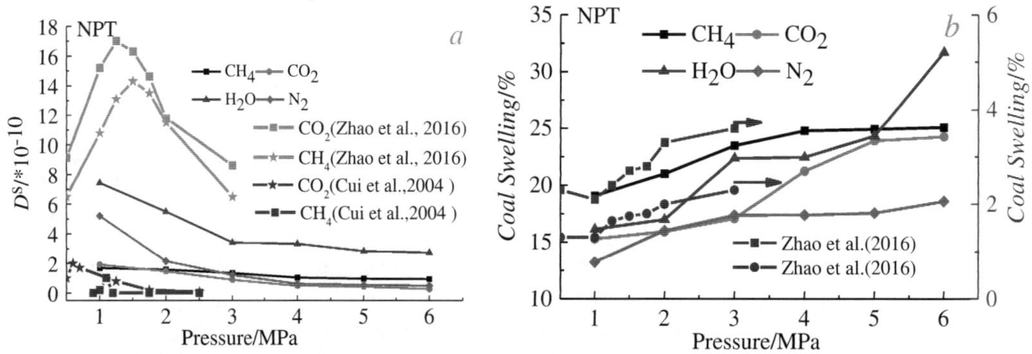

Figure 11. Relationships between the self-diffusion coefficients (*a*), *swelling ratios* (*b*) and pressure in coal under NPT ensemble.

4. Conclusions

The self-diffusion coefficients (D) of $CH_4/CO_2/H_2O/N_2$ at 298.15~458.15K and 1~6MPa under the NPT, NPH, NVE, and NVT ensembles were obtained after the calculations of MM, AK, GCMC, and MD based on WM. The adsorption capacities of 5 WMs for $CH_4/CO_2/H_2O/N_2$ manifest as H_2O(56)>CO_2(42)>CH_4(30)>N_2(17)under PBCs at under 298.15K.

The Ds of the adsorbates in saturated adsorption configurations are $D_{CH4}(0.47*10^{-9})$ $<D_{CO2}(0.69*10^{-9})$ $<D_{H2O}(1.4*10^{-9})<D_{N2}(1.8*10^{-9}m^2/s)$ (NPT, 298.15K, 0.1MPa).The diffusion type of N_2 is dominant by configurational diffusion in micro-pores. H_2O diffuses in micro-pores in a stepwise way(252~550, 850~1005, 1300~1400ps)due to the strong heterogeneity of adsorption capacity on coal surface for H_2O. For CH_4, Knudsen- and surface diffusions occur in a single cavity occasionally (498, 897ps) associating with a skip between two cavities during the abnormal diffusion stages, resulting a relatively long period of time staying in a single micro-pore cavity. The displacements of CO_2 manifest almost unchanged in the abnormal diffusion stages and rapid increasing in the Einstein diffusion stages respectively.

D_{H2O} lies between 0.1~58.08*$10^{-9}m^2/s$ when 298≤T≤458K and larger than D_{CH4}, D_{CO2} and D_{N2} at each temperature. The diffusion activation energy (E) is E_{H2O} (1.07)<E_{N2} (1.82)<E_{CO2} (2.94kJ/mol)<E_{CH4} (10.88kJ/mol), indicating that CH_4 diffusion needs to overcome the higher adsorption energy barrier than N_2 and H_2O. D_{CH4} increases slowly when T<418K and rapidly when 418~458K. The D_{CH4} order of different ensembles is D_{CH4}(NVE)≈D_{CH4}(NVT)< D_{CH4}(NPH)< D_{CH4}(NPT)at the same temperatures. D_{CO2} shows a jump increase in NPH and NPT ensembles and gradual increase in NVE and NVT ensembles. The D_{CO2} orders of different ensembles are D_{CO2} (NVT)<D_{CO2} (NVE)<D_{CO2} (NPH)<D_{CO2} (NPT) (298K≤T≤368K) and D_{CO2} (NVT)<D_{CO2} (NVE)<D_{CO2} (NPH)<D_{CO2} (NPT) (378K≤T≤458K) at the same temperatures. D_{H2O} all displays a jumpily increasing under each ensemble with the order of D_{H2O} (NVE)<D_{H2O} (NVT)<D_{H2O} (NPH)<D_{H2O} (NPT)($T\neq$ 438K). D_{N2} increases with the increasing temperatures and the amplifications become significantly when T>358K. The order of different ensembles is D_{N2} (NVE)<D_{N2} (NVT)≈D_{N2} (NPH)≈D_{N2} (NPT) (T<418K) and D_{N2} (NVE) is remarkable higher than other ensembles when T>418K. The self-diffusion coefficients of $CH_4/CO_2/H_2O/N_2$ gradually decrease whereas swelling ratios of coal matrix gradually increase with the increasing pressure (NPH, NPT) when T>418K. The average swelling ratios manifest as H_2O (14.7~35.18%)>CO_2 (13.38~32.25%)>CH_4 (15.35~23.71%)>N_2 (11.47~22.14%) (NPH, 1~6MPa). There exits differences in D, *swelling ratios* and E among various ensembles, indicating that the selection of ensembles has an important influence on the MD calculations for self-diffusion coefficients.

Acknowledgments
This work is supported by the National Natural Science Foundation of China (No. 41430317), Major Projects of National Science and Technology (2016ZX05044001-02), and Research and Development Plan of Postgraduate Research in Jiangsu Province (KYLX16_0550, KYLX16_0551).

References
[1] Kang Y, Huang F, You L, Li X and Gao B 2016 Impact of fracturing fluid on multi-scale mass transport in coalbed methane reservoirs *Int. J. Coal Geol.* **154** 123-35
[2] Tian H, Li T, Zhang T and Xiao X 2016 Characterization of methane adsorption on overmature Lower Silurian–Upper Ordovician shales in Sichuan Basin, Southwest China: coalbed methane recovery *Int. J. Coal Geol.* **152** 15-24
[3] Li Z, Liu D, Cai Y and Shi Y 2016 Investigation of methane diffusion in low-rank coals by a multiporous diffusion model *J. Nat. Gas Sci. Eng.* **33** 97-107
[4] Ridene R, Mastrour N, Gamra D and Bouchriha H 2015 Energetic behavior of excitons in hybrid organic–inorganic parabolic quantum dots and its electric field dependence *Int. J. Mod. Phys. B* **29**(30) 1550211
[5] Mastour N and Bouchriha H 2016a Effect of CdSe nanoparticles on the fluorescence spectra of conjugate polymer P3HT: An experimental and theoretical study *Phys. Lett. A.* **380**(45) 3866-9
[6] Mastour N, Ridene R and Bouchriha H 2016b Charge and non-radiative energy transfer in P3HT-CdSe nanocomposite *Int. J. Hydrogen Energ.* http://dx.doi.org/10.1016/j.ijhydene.2016.10.138
[7] Cui X, Bustin R M and Dipple G 2004 Selective transport of CO_2, CH_4, and N_2 in coals: insights from modeling of experimental gas adsorption data *Fuel* **83**(3) 293-303
[8] Xiang J, Zeng F, Liang H, Li B and Song X 2014 Molecular simulation of the CH4/CO2/H2O adsorption onto the molecular structure of coal *Sci. Chi. Earth Sci.* **57**(8) 1749-59
[9] Saghafi A, Faiz M and Roberts D 2007 CO_2 storage and gas diffusivity properties of coals from Sydney Basin, Australia *Int. J. Coal Geol.* **70**(1) 240-54
[10] Hu H, Li X, Fang Z, Wei N and Li Q 2010 Small-molecule gas sorption and diffusion in coal: Molecular simulation *Energy* **35**(7) 2939-44
[11] Zhao Y, Feng Y and Zhang X 2016 Molecular simulation of CO_2/CH_4 self- and transport diffusion coefficients in coal *Fuel* **165** 19-27
[12] Van Niekerk D and Mathews J P 2011 Molecular dynamic simulation of coal–solvent interactions in Permian-aged South African coals *Fuel Process. Technol.* **92**(4) 729-34
[13] Hu H, Du L, Xing Y and Li X 2017 Detailed study on self-and multicomponent diffusion of CO2-CH4 gas mixture in coal by molecular simulation *Fuel* **187** 220-8
[14] Cai Y, Liu D, Pan Z, Yao Y, Li J and Qiu Y 2013 Pore structure and its impact on CH4 adsorption capacity and flow capability of bituminous and subbituminous coals from Northeast China *Fuel* **103** 258-68
[15] Dutka B, Kudasik M, Pokryszka Z, Skoczylas N, Topolnicki J and Wierzbicki M 2013 Balance of CO_2/CH_4 exchange sorption in a coal briquette *Fuel Process Technol.* **106** 95-101
[16] Ji H, Li Z, Yang Y, Hu S and Peng Y 2015 Effects of organic micromolecules in coal on its pore structure and gas diffusion characteristics *Transport porous Med.* **107**(2) 419-33

[17] Staib G, Sakurovs R and Gray E M A 2013 A pressure and concentration dependence of CO_2 diffusion in two Australian bituminous coals *Int. J. Coal Geol.* **116** 106-16
[18] Busch A, Gensterblum Y, Krooss B M and Littke R 2004 Methane and carbon dioxide adsorption–diffusion experiments on coal: upscaling and modeling *Int. J. Coal Geol.* **60**(2) 151-68
[19] Busch A and Gensterblum Y 2011 CBM and CO2-ECBM related sorption processes in coal: a review *Int. J. Coal Geol.* **87**(2) 49-71
[20] An H, Wei X R, Wang G X, Massarotto P, Wang F Y, Rudolph V and Golding S D 2015 Modeling anisotropic permeability of coal and its effects on CO2 sequestration and enhanced coalbed methane recovery *Int. J. Coal Geol.* **152** 15-24
[21] Xin H, Wang D M, Qi X Y, Qi G S and Dou G L 2014 Structural characteristics of coal functional groups using quantum chemistry for quantification of infrared spectra *Fuel Process Technol.* **118** 287-95
[22] Yu C F, Chen K L, Cheng H C and Chen W H 2016 A study of mechanical properties of multi-layered graphene using modified Nosé–Hoover based molecular dynamics *Comp. Mater. Sci.* **117** 127-38
[23] Takanohashi T, Iino M and Nakamura K 1998 Simulation of interaction of coal associates with solvents using the molecular dynamics calculation *Energy Fuel* **12**(6) 1168-73
[24] Mathews J P, van Duin A C T and Chaffee A L 2011 The utility of coal molecular models *Fuel Process Technol.* **92**(4) 718-28
[25] Tambach T J, Mathews J P and van Bergen F 2009 Molecular Exchange of CH_4 and CO_2 in Coal: Enhanced Coalbed Methane on a Nanoscale *Energy Fuel* **23**(10) 4845-7
[26] Einstein A 1905 On the movement of small particles suspended in a stationary liquid demanded by the molecular-kinetic theory of heart *Ann physic-berlin* **17** 549-60
[27] Steele W A 1974 The interaction of gases with solid surfaces *Pergamon* 251-2
[28] Mathews J P and Chaffee A L 2012 The molecular representations of coal–a review *Fuel* **96** 1-14
[29] Wender I 1976 Catalytic synthesis of chemicals from coal *Catalysis Rev.-Sci. Eng.* **14**(1) 97-129
[30] Given P H 1960 The distribution of hydrogen in coals and its relation to coal structure *Fuel* **39**(2) 147-53
[31] Shinn J H 1996 Visualization of complex hydrocarbon reaction systems *ACS, Division of Fuel Chem.* **41** 510–5
[32] Wiser W H 1984 Magnetic Resonance *Conversion of bituminous coal to liquids and gases: chemistry and representative processes (Vol 124)* (Netherlands: Springer) **124** pp 325-50
[33] Vishnyakov A, Piotrovskaya E M and Brodskaya E N 1998 Capillary condensation and melting/freezing transitions for methane in slit coal pore *Adsorption* **4**(3-4) 207-24
[34] Brochard L, Vandamme M, Pellenq R J M and Fen C T 2012 Adsorption-induced deformation of microporous materials: coal swelling induced by CO_2–CH_4 competitive adsorption *Langmuir* **28**(5) 2659-70
[35] Song Y, Zhu Y M and Li W 2017 Macromolecule simulation and CH_4 adsorption mechanism of coal vitrinite *Appl. Surf. Sci.* **396** 291-302
[36] Metropolis N, Rosenbluth A W, Rosenbluth M N, Teller A H and Teller E 1953 Equation of state calculations by fast computing machines *J. Chem. Phys.* **21**(6) 1087-92
[37] Burggraaf A J 1999 Single gas permeation of thin zeolite (MFI) membranes: theory and analysis of experimental observations *J. Membrane Sci.* **155** 45-65
[38] Xiao J and Wei J 1992 Diffusion mechanism of hydrocarbons in zeolites-I. *Theory Chem. Eng. Sci.* **47**(5) 1123-41
[39] Shieh J J and Chung T S 1999 Gas permeability, diffusivity, and solubility of poly (4-vinylpyridine) film *J. Polym. Sci. Poly.* **37**(20) 2851-61
[40] Wang H, Fu X, Jian K, Li T and Luo P 2015 Changes in coal pore structure and permeability during N 2 injection *J. Nat. Gas Sci. Eng.* **27** 1234-41
[41] Andersen H C 1980 Molecular dynamics simulations at constant pressure and/or temperature *J. Chem. Phy.* **72**(4) 2384-93

[42] Skoulidas A I and Sholl D S 2005 Self-diffusion and transport diffusion of light gases in metal-organic framework materials assessed using molecular dynamics simulations *J. Phy. Chem. B.* **109**(33) 15760-8

[43] Clarkson C R and Bustin R M 1999 The effect of pore structure and gas pressure upon the transport properties of coal: a laboratory and modeling study. 2. Adsorption rate modeling *Fuel* **78**(11) 1345–62

[44] Charrière D, Pokryszka Z and Behra P 2010 Effect of pressure and temperature on diffusion of CO2 and CH4 into coal from the Lorraine basin (France) *Int. J. Coal Geol.* **81**(4) 373-80

[45] Siemons N, Wolf K H A and Bruining J 2007 Interpretation of carbon dioxide diffusion behavior in coals *Int. J. Coal Geol.* **72**(3) 315-24

IOP Conf. Series: Materials Science and Engineering **213** (2017) 012015 doi:10.1088/1757-899X/213/1/012015

Influence of drill helical direction on exit damage development in drilling carbon fiber reinforced plastic

Y Bai[1], Z Y Jia[1], F J Wang[1,2], R Fu[1], H B Guo[1], D Cheng[1] and B Y Zhang[1]

[1]School of Mechanical Engineering, Dalian University of Technology, Ganjingzi, Dalian, Liaoning Province 116024, China

E-mail: wfjsll@dlut.edu.cn

Abstract. Drilling is inevitable for CFRP components' assembling process in the aviation industry. The exit damage frequently occurs and affects the load carrying capacity of components. Consequently, it is of great urgency to enhance drilling exit quality on CFRP components. The article aims to guide the reasonable choice of drill helical direction and effectively reduce exit damage. Exit observation experiments are carried out with left-hand helical, right-hand helical and straight one-shot drill drilling T800S CFRP laminates separately. The development rules of exit damage and delamination factor curves are obtained. Combined with loading conditions and fracture modes of push-out burrs, and thrust force curves, the influence of drill helical direction on exit damage development is derived. It is found that the main fracture modes for left-hand helical, right-hand helical, and straight one-shot drill are mode Ⅰ, extrusive fracture, mode Ⅲ respectively. Among them, mode Ⅲ has the least effect on exit damage development. Meanwhile, the changing rate of thrust force is relative slow for right-hand helical and straight one-shot drill in the thrust force increasing phase of stage Ⅱ, which is disadvantaged for exit damage development. Therefore, straight one-shot drill's exit quality is the best.

1. Introduction

Carbon fiber reinforced plastic(CFRP) composites have been widely adopted for manufacturing aircraft and spacecraft structural parts because of their superior properties such as excellent damage tolerance, high strength-to-weight and stiffness-to-weight ratios[1]. Although CFRP composites are produced to near-net shape, machining is still inevitable in the assembly process. Among machining processes, drilling is one of the most frequently used to make holes for screws, rivets and bolts, and often a final operation during aerostructures assembly [1-2].

CFRP composites have strong anisotropy, lamination property and low layer-to-layer bond strength at the macro level. Meanwhile, the thrust force produced in drilling is quite large, so it is easy to generate delamination. In addition, other damage, such as cracks and burrs, often occurs because of fiber bending when it is difficult to cut off fiber. Damage reduces the service life of CFRP components substantially [3-4]. For example, as much as sixty percent of composite components failure is caused by delamination [5-6].Damage can be classified as exit damage, entry damage and hole wall damage according to the damage position. Among them, exit damage is the most serious which can decrease the strength, stiffness and bearing capacity of laminates to a large extent [7]. It is known that drill geometry affects the contact state between tool and workpiece directly and is one of the most important factors affecting exit damage [8].

Drill geometry can be categorized to drill bit geometry and drill flute geometry. Many researchers have made great efforts to study the effect of drill bit geometry on exit damage. Hocheng et al. [4, 9-13] investigated the exit damage in drilling CFRP using twist drill or the drill with special drill bit geometry, such as saw drill, candle stick drill and step drill, and built the corresponding critical thrust force model of push-out delamination. The results indicated that push-out burrs and push-out delamination caused by one-shot drill are weaker than those caused by twist drill. Jain and Yang in Reference [14], and Tsao and Hocheng in Reference [15] found that the lower point angle of twist drill, the lower push-out delamination. Chisel edge had an important effect on push-out delamination, and derived the optimal range of chisel edge with respect to drill diameter. Durão et al. [16] pointed out that decreasing twist drill's helix angle would contribute to increasing the thrust force, torque and push-out delamination. Above all for the final exit damage, in order to make further optimization on drill bit geometry and decrease drilling damage, some researchers experimentally studied the formation and development of exit damage as well. Dipaolo et al. [17] photographed the integrated development process of exit damage, and significant damage types of each drilling-out stage were observed. Correlations between the average drill forces and damage development, relations between instantaneous forces and damage development in different fiber cutting angle were shown respectively. However, there is still a lack of study that explains the influence of drill flute geometry on exit damage development at present. Drill helical direction is one of the most important components of drill flute geometry, and most of drills adopt right-hand helical flutes while few of them have the left-hand helical or straight flutes [18-19]. Drill helical direction affects drill cutting angle and chip flowing directly, and further changes drilling forces and temperature, which has a great effect on exit quality. Therefore it is important to study the influence of drill helical direction on exit damage development.

This article conducts drilling experiment using right-hand helical, left-hand helical, and straight one-shot drill to drill T800S CFRP laminates separately. Based on the experimental results, the development rules of exit damage are shown drilling with three kinds of one-shot drills. Combined with the loading conditions and fracture modes of push-out burrs, and thrust force analysis, the influence of drill helical direction on exit damage development is derived, which guides the reasonable choice of drill helical direction.

2. Exit observation experiments using one-shot drills with different helical directions

2.1. Experimental drills

The paper conducts exit observation experiments with three kinds of one-shot drills whose helical directions are right-hand helix, left-hand helix and straight respectively. Drills are made of K44UF tungsten carbide without coating which have two primary cutting edges, four secondary cutting edges, and four minor cutting edges as illustrated in Table 1. The helix angles of right-hand helical one-shot drill and left-hand helical one-shot drill are both 6° , while the helix angle of straight one-shot drill is 0° . Other dimensions of three one-shot drills are identical whose specific values are shown in Table 2.

Table 1. One-shot drills with different helical directions.

Left-hand helical one-shot drill	Right-hand helical one-shot drill	Straight one-shot drill

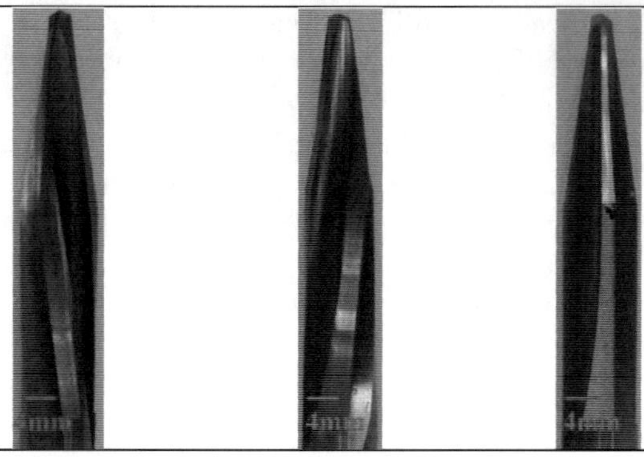

Table 2. Specific dimensions of one-shot drills.

Diameter (mm)	Primary point angle (°)	Secondary point angle (°)	Rake angle of primary cutting edges (°)	Helical rake angle (°)	Chisel edge length (mm)	Axial length of primary cutting edges (mm)	Axial length of secondary cutting edges (mm)
7.94	120	18	5	10	0.41	0.57	18.8

2.2. Experimental materials and equipment

In order to guide drilling of aviation parts preferably, the workpiece used in the experiments is multi-directional CFRP composites and laminated by 20 layers of P2352 prepreg in the fiber orientation of $[(-45/0/45/90)_2/0/0]_s$. The reinforced fiber is Toray-T800S and the fiber volume content is 60%. The mechanical properties of the composites are listed in Table 3. After curing under high temperature and pressure conditions, the workpiece has the dimensions of 150×20×4 mm.

Table 3. Mechanical properties of composites.

Longitudinal tensile strength (MPa)	Transverse compressive (MPa)	Longitudinal Young's modulus (GPa)	Transverse Young's modulus (GPa)	Poisson's ration
2842	165.6	160	8.97	0.28

A series of drilling operations were carried out on the GONA 5-axis machining center. In order to observe conveniently, the spindle adopted angle head to change vertical drilling into horizontal drilling. High-speed camera was used to photograph the drilling-out process of one-shot drills whose frame rate was set as 1000fps. A Kistler force sensor, 9257B dynamometer, was fixed on the platform of the machining center to support the clamping system and measure the thrust force during drilling. The kistler amplifier and data acquisition were employed to read and save data from the force sensor in a sampling frequency of 3 KHz, and experimental equipment is shown in Figure 1. CFRP laminates were observed on Keyence digital microscope to measure the final exit quality, and the observation rate was 20. Drilling parameters of 3000rpm in spindle speed, 150mm/min in feed rate, 25mm in clamping length and 30mm in drilling-in length were adopted in the experiments, and drilling operations were performed under dry condition.

Figure 1. Experimental equipment.

3. Results and discussion

3.1. Analysis of the loading conditions and fracture modes of push-out burrs

The paper divides one-shot drills' drilling-out process into three stages as shown in Figure 2. The chisel edge and primary cutting edges drill out of workpiece in stage Ⅰ. During this stage, the chisel edge, primary cutting edges and secondary cutting edges, or primary cutting edges and secondary cutting edges participate in drilling. Stage Ⅱ is the process of secondary cutting edges drilling-out gradually. Secondary cutting edges drill alone at the beginning, and then minor cutting edges drill into the workpiece. The process of minor cutting edges drilling-out is defined as stage Ⅲ which only has minor cutting edges drilling. Because of the difference of three one-shot drills' helical direction near secondary cutting edges and minor cutting edges, the influence of drill helical direction on exit damage still exits. Among them, the position containing helical direction acts on exit directly in stage Ⅱ and stage Ⅲ, which has a significant effect on exit damage.

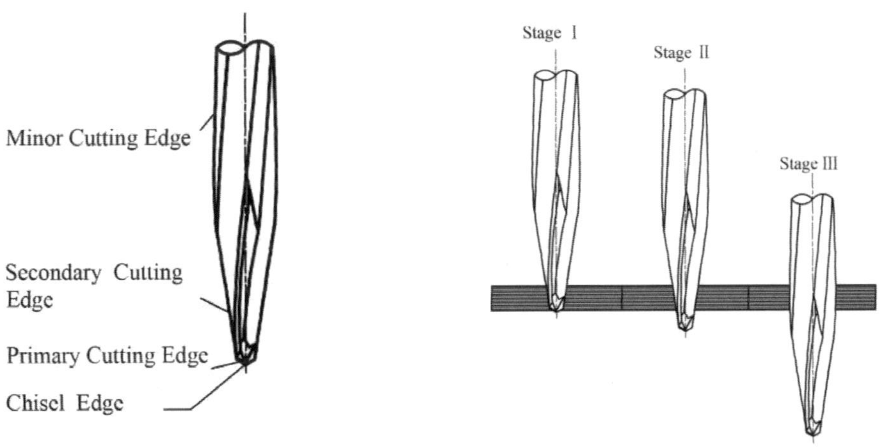

(a) Schematic diagram of one-shot drill (b) Schematic diagram of drilling-out stages

Figure 2. Division of drilling-out stages.

Feed rate is far less than cutting speed, so the instantaneous relative motion between one-shot drill and exit of CFRP laminates can be approximated to rotary motion. It can be seen that feed motion has little effect on exit loading conditions. Considering the comprehensive effect of relative motion and drill helical direction, the resultant force for a point A on laminates' exit can be decomposed into the thrust force F_z aligned with the drill axis, the lateral force F_x along the radial direction, and the tangential force F_y that is perpendicular to both in stage II, as in Figure 3. At the same time, Figure 4 shows that the resultant force for a point B on laminates' exit can be decomposed into the thrust force F_z aligned with the drill axis and the tangential force F_y corresponding to the tangential direction in stage III. The thrust force F_z produced by right-hand helical drill still points entry from exit, while it has opposite direction for left-hand helical drill and straight drill. In addition, the thrust force F_z generated by left-hand helical drill is higher than that generated by straight drill.

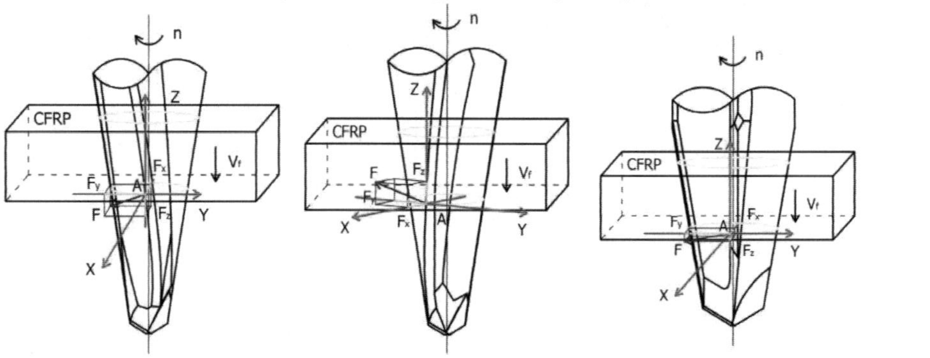

(a) Left-hand helical one-shot drill (b) Right-hand helical one-shot drill (c) Straight one-shot drill

Figure 3. Exit loading conditions in stage II.

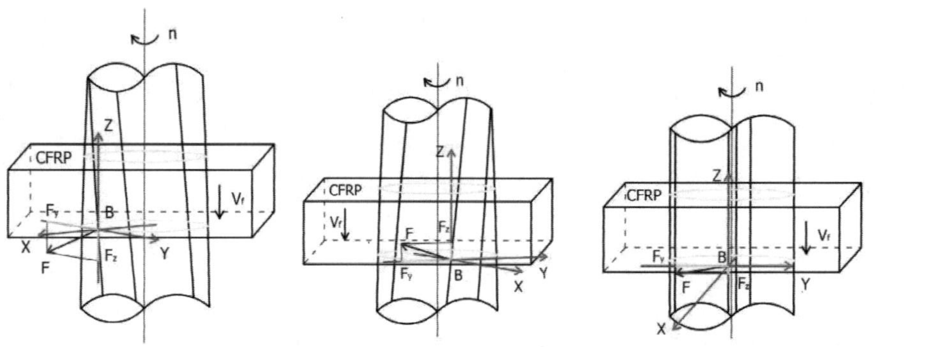

(a) Left-hand helical one-shot drill (b) Right-hand helical one-shot drill (c) Straight one-shot drill

Figure 4. Exit loading conditions in stage III.

For left-hand helical one-shot drill and straight one-shot drill, the thrust force F_z stretch burr to exit at the point, which causes the fracture of mode I, and the tangential force F_y shears exit burr, which contributes to the fracture of mode III. However, mode I of left-hand helical one-shot drill is more significant compared to that of straight one-shot drill because left-hand helical drill has higher thrust force F_z. The current research showed mode I had more serious effect on exit damage than mode III [20]. Consequently, straight one-shot drill is more superior in exit quality than left-hand helical one-shot drill. Moreover, exit burrs is extruded into laminates and sheared along the tangential direction drilling with right-hand helical one-shot drill. The fracture modes are mode III and extrusive fracture.

Figure 5 and Figure 6 show that left-hand helical drill tears burrs outside laminates, straight drill mainly shears burrs along the tangential direction whose tearing effect is weak, and right-hand helical drill extrudes burrs into laminates and shears them along the tangential direction. The experimental results correspond to the above analysis of loading conditions and fracture modes of push-out burrs.

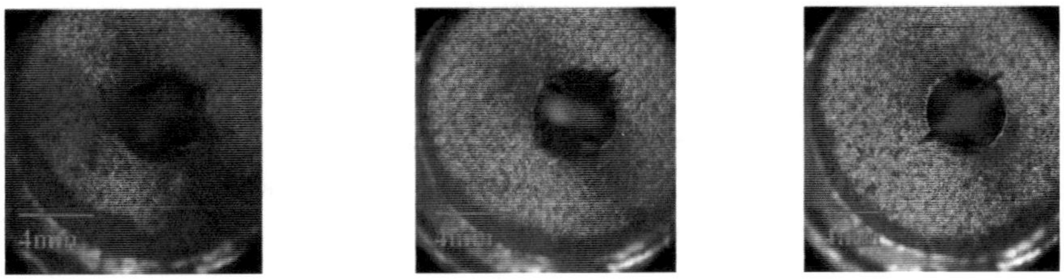

(a) Left-hand helical one-shot drill (b) Right-hand helical one-shot drill (c) Straight one-shot drill

Figure 5. CFRP laminates exit in stage II.

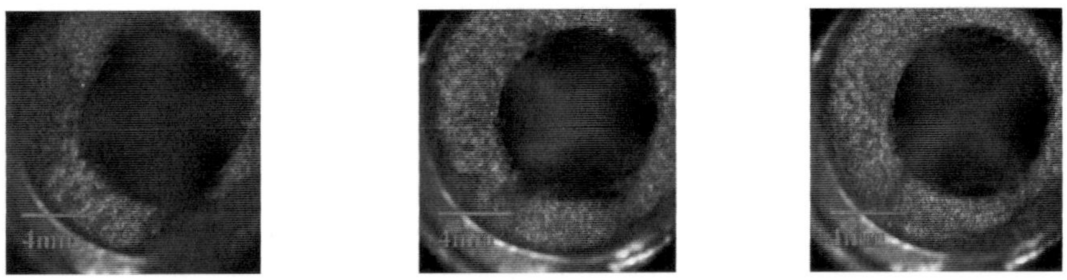

(a) Left-hand helical one-shot drill (b) Right-hand helical one-shot drill (c) Straight one-shot drill

Figure 6. CFRP laminates exit in stage III.

3.2. Thrust force

The data obtained from drilling is filtered in the software with a low pass filter set to 10Hz, because the axial motion is a continuous low frequency process. The obtained typical thrust force curves are shown in Figure 7. It can be seen that thrust force curves of three one-shot drills are almost same. The main cutting edges enter the workpiece gradually in drilling-in stage, which leads to the thrust force increasing rapidly. The maximum value of thrust force occurs when the chisel edge is close to the bottom of laminates. The thrust force decreases quickly in stage I, and is found to: first decrease, then increase and decrease again in stage II. During stage III, it is very low and maintains a constant value basically. However, the thrust force changing rate of left-hand helical drill is much higher than that of right-hand helical drill and straight drill in the increasing phase of stage II. Besides that, in stage III, the thrust forces of left-hand helical drill, right-hand helical drill and straight drill are positive value, negative value and nearly zero separately.

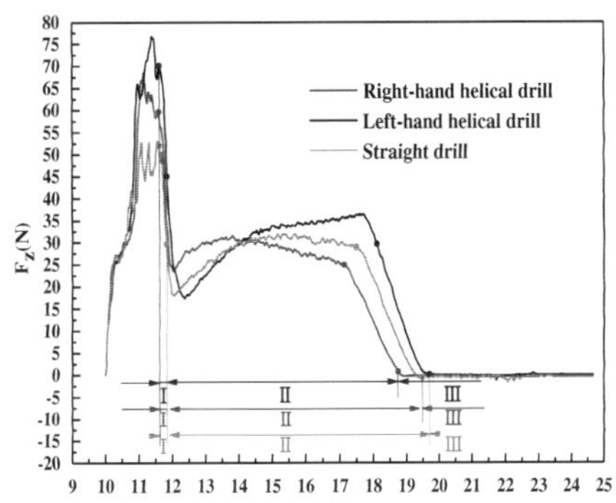

Figure 7. Typical thrust force curves of three one-shot drills.

3.3. Exit damage

The experimental results indicate that the development rules of exit damage are fundamentally uniform for three drills. A longitudinal bulge along fiber orientation is observed at laminates exit when the chisel edge is close to the bottom of laminates, whose main damage type is delamination. In the stage of chisel edge drilling-out, a longitudinal crack along fiber orientation produces at laminates exit, which has the damage of delamination and cracks in Figure 8. Primary cutting edges drill exit fiber and generate massive burrs because of their weak removal capacity in the stage of primary cutting edges drilling-out. During this stage, burrs can't be removed timely, which makes further efforts to aggravate cracks and delamination, and cracks mainly grow along fiber orientation in Figure 9. In stage II, plenty of push-out burrs are removed rapidly, meanwhile cracks and delamination aggravate at first. Subsequently, secondary cutting edges drill few remaining burrs corresponding to fiber orientation, damage decreases constantly, and cracks develop along fiber orientation as well in Figure 5. Minor cutting edges drill push-out burrs, and burrs, delamination and cracks develop slowly in stage III, as shown in Figure 6. It is clear that the recovery capacity of stage II is higher and stage II has a significant influence on the final exit quality.

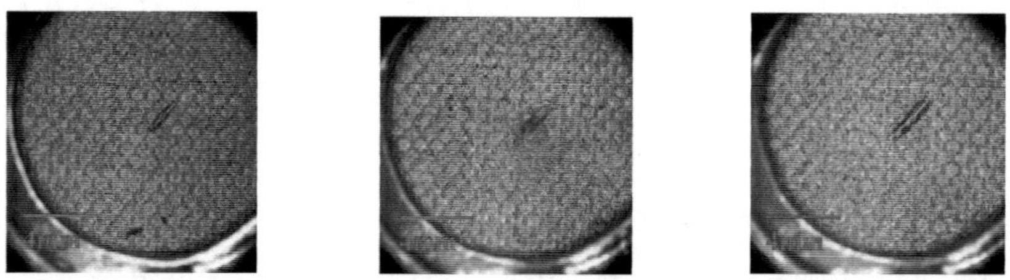

(a) Left-hand helical one-shot drill (b) Right-hand helical one-shot drill (c) Straight one-shot drill

Figure 8. CFRP laminates exit in stage of chisel edge drilling-out.

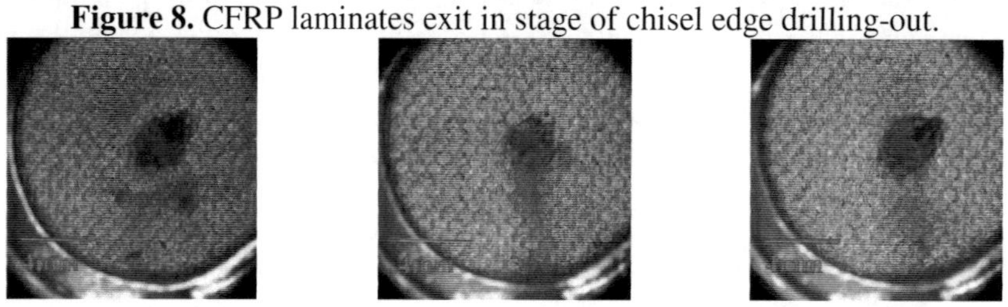

(a) Left-hand helical one-shot drill (b) Right-hand helical one-shot drill (c) Straight one-shot drill

Figure 9. CFRP laminates exit in stage of primary cutting edges drilling-out.

The main damage types are delamination, burrs and cracks for three drills drilling CFRP laminates from above analysis. Among them, delamination effects bearing capability of parts greatly, so the paper adopts it as the basis of evaluating quality. There are many ways to measure and evaluate push-out delamination at present, and this article chooses the widely applicable and convenient delamination factor as evaluation basis. The delamination factor was defined as the ratio of the maximum diameter, D_{max}, of the delamination area to the hole normal diameter, D_{norm}, described as the following equation, and illustrated in Figure 10.

$$F_d = D_{\max} / D_{norm} \quad (1)$$

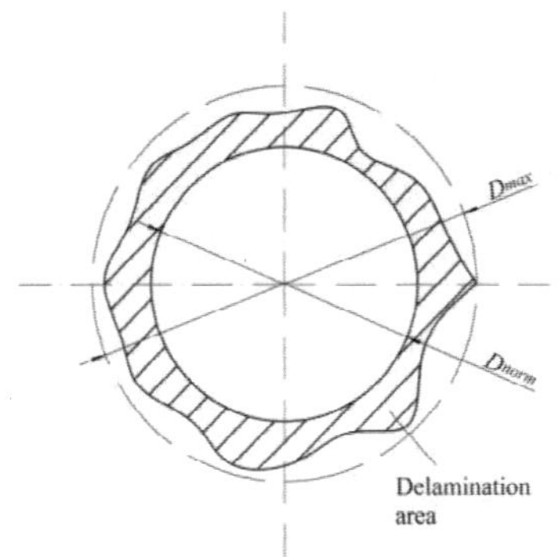

Figure 10. Schematic diagram of delamination factor.

Delamination factor curves of three drills are all found to: first increase, then decrease quickly and increase slowly again in stage II, while they increase slowly as the increasing of drilling-out length in stage III as shown in Figure 11. It agrees with above analysis of exit damage development. Maximum values of delamination factor are 2.071, 2.262, 2.217 for left-hand helical drill, right-hand helical drill and straight drill respectively in stage II. When secondary cutting edges drill out completely, values of delamination factor are about 1.282, 1.365, 1.181 for left-hand helical drill, right-hand helical drill and straight drill separately. It can be calculated that reduction values of delamination factor are 0.789, 0.897, 1.036 for left-hand helical drill, right-hand helical drill and straight drill respectively, so secondary cutting edges' recovery ability is the best for the straight drill compared with other two drills. The delamination factor changes a little and has less effect on exit damage in stage III. The final holes' delamination factor are 1.379, 1.432, 1.185 for left-hand helical drill, right-hand helical drill and straight drill respectively in Figure 12. In conclusion, the drilling quality of straight drill is better than that of left-hand helical drill and right-hand helical drill.

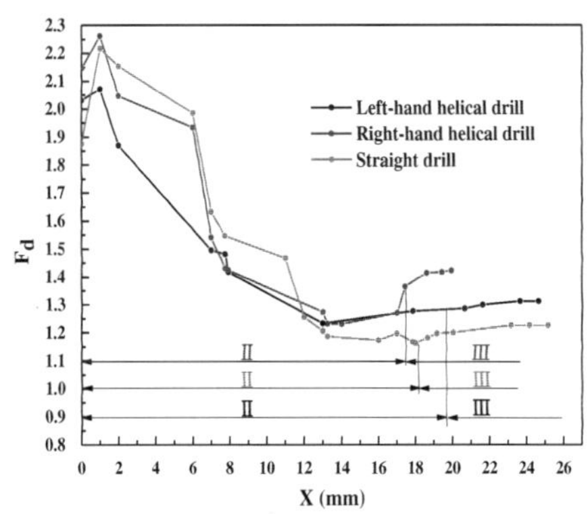

Figure 11. Typical delamination factor curves of three one-shot drills.

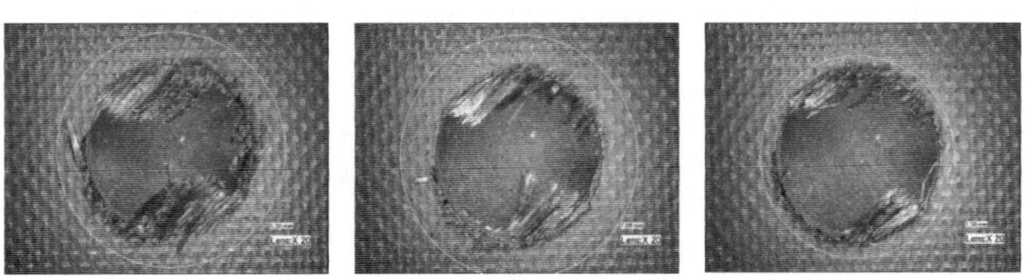

IOP Conf. Series: Materials Science and Engineering **213** (2017) 012015 doi:10.1088/1757-899X/213/1/012015

(a) Left-hand helical one-shot drill (b) Right-hand helical one-shot drill (c) Straight one-shot drill

Figure 12. The final exit quality of three one-shot drills.

Figure 13 shows that the decreasing phase of delamination factor curve corresponds to the increasing phase of thrust force curve in stage II. The increasing rate of thrust force for left-hand helical one-shot drill is high, and the drill's damage recovery capacity is the worst. For right-hand helical one-shot drill and straight one-shot drill, their increasing rate of thrust force is very low and basically a constant, but their damage recovery capacity is high. Therefore it can be concluded that the damage recovery capacity is irrelevant to thrust force's value but is relevant to thrust force's changing rate. Meanwhile the thrust force's changing rate for right-hand helical one-shot drill and straight one-shot drill is almost same, but the damage recovery capacity of straight one-shot drill is higher than that of right-hand helical one-shot drill. The reason may be that extrusive fracture is beneficial to damage development compared with mode III.

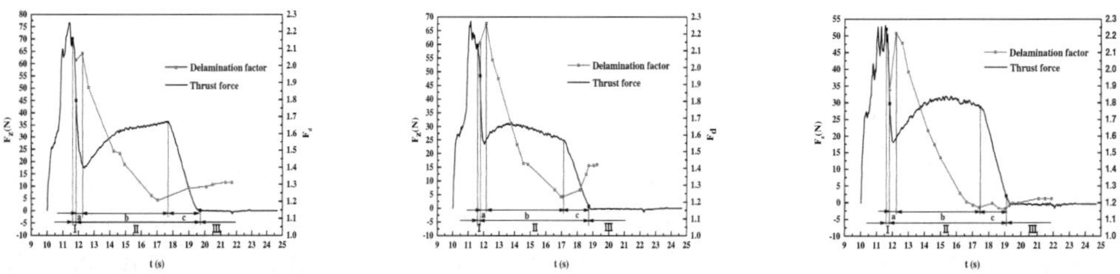

(a) Left-hand helical one-shot drill (b) Right-hand helical one-shot drill (c) Straight one-shot drill

Figure 13. Thrust force and delamination factor contrastive curves of three one-shot drills.

4. Conclusions

(1) In stage II and stage III, left-hand helical one-shot drill tears burrs outside laminates, which mainly causes mode I. Straight one-shot drill mainly shears burrs along the tangential direction whose tearing effect is weak, and mode III is the major fracture mode. Right-hand helical one-shot drill extrudes burrs into laminates and shears them along the tangential direction, which mainly generates extrusive fracture. Among them, mode III has the least influence on exit damage development.

(2) The thrust force changing rate of left-hand helical one-shot drill is much higher than that of right-hand helical one-shot drill and straight one-shot drill in the increasing phase of stage II. The damage recovery capacity is irrelevant to thrust force's value, but it is relevant to thrust force's changing rate.

(3) Development rules of exit damage are fundamentally uniform for three drills. Delamination factor curves of three drills are all found to: first increase, then decrease quickly and increase slowly again in stage II. Secondary cutting edges' recovery ability and the drilling quality are the best for the straight one-shot drill compared with other two drills.

Acknowledgments

This work is supported by National Natural Science Foundation of China-Joint Funds of Liaoning Province, No. U1508207; National Natural Science Foundation of China, No.51575082; National Key Basic Research Program of China (973 Program), No. 2014CB046503; National Innovative Research Group, No.51621064; Education Ministry's New Century Excellent Talents Supporting Plan Program, No.NCET-13-0081.

References

[1] Che D, Saxena I, Han P, Guo P and Ehmann K F 2014 Machining of Carbon Fiber Reinforced Plastics/Polymers: A Literature Review *J. Manuf. Sci. Eng.* **136**(3) 034001

[2] Melentiev R, Priarone P C, Robiglio M and Settineri L 2016 Effects of Tool Geometry and Process Parameters on Delamination in CFRP Drilling: An Overview *Procedia CIRP* **45** 31-4

[3] Teti R 2002 Machining of composite materials *CIRP Annals-Manufacturing Technology* **51**(2) 611-34

[4] Hocheng H and Dharan C K H 1990 Delamination during drilling in composite laminates *Journal of Engineering for Industry* **112**(3) 236-39

[5] Wong T L, Wu S M and Croy G M 1982 An analysis of delamination in drilling composite materials *14th National SAMPE Technical Conference* **47** 471-83

[6] Stone R and Krishnamurthy K 1996 A neural network thrust force controller to minimize delamination during drilling of graphite-epoxy laminates *Int. J. Mach. Tools Manu.* **36**(9) 985-1003

[7] Gururaja S and Ramulu M 2009 Modified exit-ply delamination model for drilling FRPs *J. Compos. Mater.* **43**(5) 483-500

[8] Caprino G and Tagliaferri V 1995 Damage development in drilling glass fibre reinforced plastics *Int. J. Mach. Tools Manu.* **35**(6) 817-29

[9] Hocheng H and Tsao C C 2003 Comprehensive analysis of delamination in drilling of composite materials with various drill bits *J. Mater. Process. Technol.* **140**(1) 335-9

[10] Xu J, An Q, Cai X and Chen M 2013 Drilling machinability evaluation on new developed high-strength T800S/250F CFRP laminates *Int. J. Precis. Eng. Man.* **14**(10) 1687-96

[11] Grilo T J, Paulo R M F, Silva C R M and Davim J P 2013 Experimental delamination analyses of CFRPs using different drill geometries *Composites Part B* **45**(1) 1344-50

[12] Davim J P and Reis P 2003 Study of delamination in drilling carbon fiber reinforced plastics (CFRP) using design experiments *Compos. Struct.* **59**(4) 481-7

[13] Piquet R, Ferret B, Lachaud F and Swider P 2000 Experimental analysis of drilling damage in thin carbon/epoxy plate using special drills *Composites Part A* **31**(10) 1107-15

[14] Jain S and Yang D C H 1993 Effects of feedrate and chisel edge on delamination in composites drilling *Journal of Engineering for Industry* **115**(4) 398-405

[15] Tsao C C and Hocheng H 2003 The effect of chisel length and associated pilot hole on delamination when drilling composite materials *Int. J. Mach. Tools Manu.* **43**(11) 1087-92

[16] Durão L M P, Gonçalves D J, Tavares J M R, Albuquerque V H C, Vieira A A and Marques A T 2010 Drilling tool geometry evaluation for reinforced composite laminates *Compos. Struct.* **92(7)** 1545-50

[17] DiPaolo G, Kapoor S G and DeVor R E 1996 An experimental investigation of the crack growth phenomenon for drilling of fiber-reinforced composite materials *Journal of Engineering for Industry* **118**(1) 104-10

[18] Fernandes M and Cook C 2006 Drilling of carbon composites using a one shot drill bit. Part II: empirical modeling of maximum thrust force *Int. J. Mach. Tools Manu.* **46**(1) 76-9

[19] Fernandes M and Cook C 2006 Drilling of carbon composites using a one shot drill bit. Part I: Five stage representation of drilling and factors affecting maximum force and torque *Int. J. Mach. Tools Manu.* **46**(1) 70-5

[20] Sereshk M R V and Bidhendi H M 2017 The Contribution of Different Fracture Modes on Drilling Delamination Crack Propagation *J. Manuf. Sci. Eng.* **139**(1) 011013

IOP Conf. Series: Materials Science and Engineering **213** (2017) 012016 doi:10.1088/1757-899X/213/1/012016

New multifunctional zirconia composite nanomaterials – from electronics to ceramics

I Danilenko[1,2], O Gorban[1], A Shylo[1], L Akhkozov[1], M. Lakusta[1] and T Konstantinova[1]

[1]Donetsk Institute for Physics and Engineering NAS of Ukraine, Kiev, Nauki av, 46, 03028, Ukraine

E mail: igord69@ukr.net

Abstract. In the present study the multifunctional composite materials on zirconia based for different applications were synthesized, studied and analyzed. It has been shown that the using of liquid solutions reagents instead of powders mixing allows entering in the matrix material the oxide dopants in supersaturated concentrations. The nanoparticles synthesized during this investigation contain a potential for the formation of various structures. By the controlling of heating regimes during nanopowders firing or sintering processes it become possible to control the diffusion processes on the particles volume and boundaries. In this study was found the decomposition of initial supersaturated solid solution during heat treatment leads to formation of multilevel nanocomposite structure in the ceramic matrix with enhancing mechanical, electrical, optical and catalytic properties. It was established that the modification of zirconia even a slight amount of aluminum oxide, zinc oxide leads to a significant changing in the properties of nanopowders and ceramics.

1. Introduction.

It is known, the properties of nanomaterials are very different from the properties of ordinary materials [1-2]. Consequently, during transformation from nanostructure to microstructure, for example due to heating, the properties of material change. Thus, the controlling of the treatments regimes allows regulating the structural transformations in nanomaterials and adjusts the properties of material to the exploitation conditions. At the same time we continued work in nanomaterials region. This conception is perfectly matched to the conception of multifunctional materials.

The conception of multipurpose materials has been appearing during last several years [3-4]. Multifunctional materials include materials that have the ability to perform multiple functions through judicious combinations of structural properties and at least one additional functional capability as dictated by the system application requirements. This direction is favorably for industry and it is compete with task oriented material.

Zirconia is well known as a multifunctional material with wide range of applications and as a bulk material used many years ago. This material combines the high fracture toughness and biocompatibility, ionic conductivity and chemical stability. It is used in energetic, technical ceramics, optics, catalyst, etc. [5-7]. Zirconia based nanomaterials and nanocomposites are perfect match for demonstration of the universality of material. Electrical conductivity, catalytic and optical properties of zirconia based nanocomposites have due to the peculiarities of nanoparticle formation and surface properties of nanoparticles. So, controlling the structure of composites and their chemical composition allow demonstrating the different properties of material. We think that the key parameter is the preparation of metastable supersaturated solid solutions in nanodimentional state and its decomposition during heating. This process opens a new possibility in creation of different types of inclusions: clusters, grains, boundary enrichments and interfaces. This diversity of forms can be realized by different physical and chemical synthesis methods (laser and microwave synthesis, CVD and PVD methods) as well as by using the simple wet chemical methods.

Critical factors in the commercialization of advanced materials are the starting powders and the method of their formation. The cheap coarse starting powders and, consequently, high sintering temperature often decrease the level of functional properties of materials, e.g. mechanical strength, density, magnetization, etc. In case of composite materials the amount of critical factors increased in several times, because creation of uniform microstructure of multicomponent composite materials by traditional mixing process is very hard. However now, this method is the most commonly used in the industry. So, we have supposed, there is one way for the cheap improving the properties of composite ceramic materials for the industry at the present time. This is using of mixing of liquid solutions of reagents instead of powders mixing. It is wet chemical methods, for example sol-gel [8-9], colloidal [10], hydrothermal [11] or precipitation [12-13] techniques. In all these techniques the precipitation of insoluble phase is occurs. The using of mixing of reagents liquid solutions instead of powders mixing allows entering in the matrix material the oxide dopants in supersaturated concentrations. The decomposition of initial supersaturated solid solution during heat treatment can lead to formation of multilevel nanocomposite structure in the matrix with enhancing mechanical, electrical, optical and magnetic properties.

In this study we have analyzed the structures and functional properties of different types of zirconia composite materials obtained by precipitation technique and compared these results with structure and properties of materials, obtained by traditional technology. Also we have shown that the traditional ceramic technology processes to impose greater restrictions on the chemical composition of materials and parameters of technological process.

2. Experimental

Zirconia (3Y-TZP) and zirconia based composites ZrO_2-3mol%ZnO, ZrO_2 -3mol%Y_2O_3+n%Al_2O_3, ZrO_2+4.4 mol%Y_2O_3 (4.4YZrO_2) doped by $CuCl_2$ and MnS nanopowders were synthesized with a co-precipitation technique using $ZrOCl_2$*nH_2O, $Y(NO_3)_3$*nH_2O, $AlCl_3$*6H_2O and $ZnCl_2$ salts. The amount of alumina was varied from 0.5 to 5 wt%. The amount of ZnO was varied from 3 to 50 wt%. All used chemicals were of high chemical purity. The technological aspects of the precipitation process were described in [14-16]. After washing and filtration, the hydrogel was dried in a microwave furnace with an output power of 700 W and at a frequency of 2.45 GHz. The dried zirconium hydroxides and composites were calcined in a resistive furnace at different temperatures with a dwelling time of 2 h. This variant was named CT and this abbreviation marked in powder name. After the 4.4YZrO_2 powder had been calcined at 700°C for 2 h, it was additionally doped by $CuCl_2$ or MnS (1 wt %) powder and then calcined at 800°C for 120 min in a laboratory furnace again [16].

In order to compare our results with traditional mixing techniques an appropriate amount of commercially available α-Al_2O_3 or ZnO powder was mixed with obtained 3Y-TZP nanopowders in distilled water. The α-Al_2O_3 particle size distribution was wide enough with an average size at 1 μm (manufacturer's data). ZnO powders were synthesized by technique which was mention above and after calcination at 500, 700, 1000°C the average particle size was 45, 80 and 250 nm, respectively. These mixtures were planetary-milled (MSK-SFM-1 MTI Corp., USA) at 400 rpm for 10 h using YSZ milling balls. This variant was named BM and this abbreviation marked in powder name.

The powders and sintered specimens were characterized using XRD analysis (Dron-3) with Cu-Kα radiation for crystallite sizes and quantitative phase analyses with a proven method [17]. Particle sizes of different calcined powders were estimated by transmission electron microscopy (JEM 200, Jeol, Japan).The optical properties of ZrO_2-ZnO nanopowders were measured on Cary 5000 UV-Vis-NIR spectrometer (Agilent Technologies, USA). The chemical composition of synthesized materials was checked by energy dispersive spectroscopy (EDS) analysis (Oxford Instruments). The fracture toughness values were calculated by Niihara equation for Palmquist type cracks. The microstructures of the ceramics were studied by scanning electron microscopy (JSM 6490LV Jeol) after polishing of the surfaces as well as fractured surfaces.

3. Results and discussion

3.1. Formation of structure and properties in ZrO_2-3mol%ZnO system

The dynamic reflectance spectroscopy (DRS) results of mixed ZrO_2-3mol%ZnO composite have shown the traditional results. The peak near 380 nm is corresponded to ZnO and peak near 220 nm correspond to ZrO_2 (Figure 1). With increasing of calcining temperature and consequently with increasing ZnO particle size, the reflection peak near 380 nm, which was corresponded to ZnO, decreased for all values of ZnO concentrations (Figure 1a).

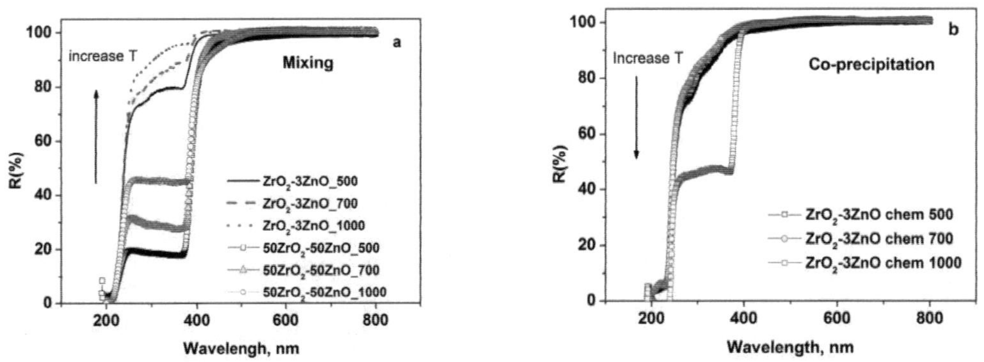

Figure 1. The DRS spectra of ZrO_2-ZnO nanopowders obtained by mixing technique - a) and co-precipitation – b) and calcined at different temperatures: 500 (black), 700 (red) and 1000°C (green).

In case of the ZrO_2-3mol%ZnO powder synthesized by co-precipitation technique the reflection near 380 nm, which correspond to ZnO reflection, did not fixed (Figure 1b) after calcination at temperatures 400-700°C. After calcination at 1000°C the reflectance peak near 380 nm appeared, but the absorption level in this case was in ten times higher than in powders obtained by mixing technique. It is shown the absorption level of 50% ZnO in mixed sample equivalent of 3mol% ZnO in sample obtained by precipitation. So, the opposite behavior of optic properties from method of powder preparation was demonstrated. In case of mixing technique – increasing ZnO particle size leads to decreasing the reflectance level, but in case of co-precipitation technique – increasing of ZnO particle size leads to strong increasing the reflectance level. Consequently, distribution of ZnO in ZrO_2 in this case is different in comparison with powders obtained by mixing technique. The typical TEM images of ZrO_2 - ZnO particles are depicted on Figure 2.

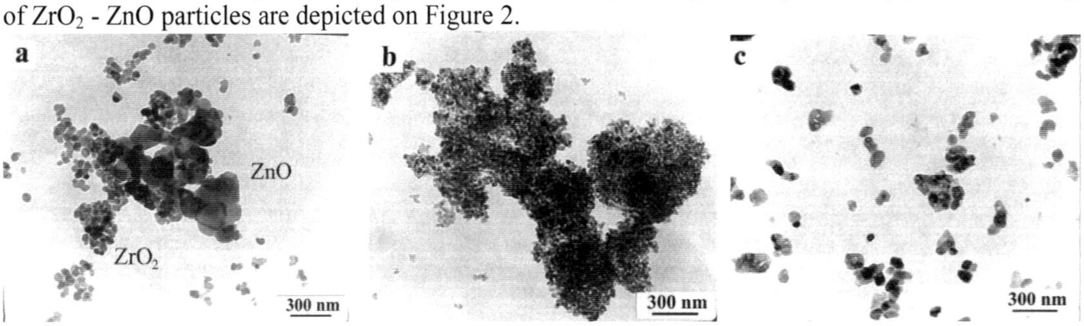

Figure 2. TEM images of ZrO_2-ZnO nanopowders, obtained by mixing - a) and co-precipitation – b, c) techniques and calcined at 700°C – a, b and 1000°C – c).

The mixture of small ZrO_2 particles and big ZnO particles obtained by mixing technology is shown on Figure 2a. The TEM images of powders obtained by co-precipitation technique are shown on Figure 2b and Figure 2c. The nanopowders ZrO_2-ZnO demonstrate the agglomeration but have a uniform structure of ZrO_2-ZnO nanoparticles with average size 20-30 nm after calcination at 700°C. Increasing the calcination temperature up to 1000°C leads to formation of ZrO_2 particles with elliptical shape ZnO inclusions (5-7 nm). The TEM and DRS data can be confirmed and explained by XRD results.

The XRD data of pure ZrO_2 and ZrO_2 – ZnO mixes are typical and commonly known [12, 18]. The pure ZrO_2 was presented by monoclinic structure $P2_1/c$, ZnO by wurtzite structure $P6_3mc$. The XRD spectra of ZrO_2 – ZnO mixes were presented by the monoclinic phase of ZrO_2 and ZnO wurtzite phase for all samples (Figure 3a). Thus, it was not found the formation of solid solution ZnO in ZrO_2 for all ZnO concentrations, which were used in this study. The increasing of intensity peaks is due to the increasing of particles size during heating. In case of the ZrO_2-ZnO powder synthesized by co-precipitation technique and calcined at temperatures 400-700°C XRD analysis was not found peaks which correspond to ZnO. After calcinations at 1000°C the absorption peak appeared and ZnO was fixed by XRD (Figure 3b). At the same time the tetragonal phase of ZrO_2, which was formed at low firing temperatures transforms to monoclinic phase at 1000°C.

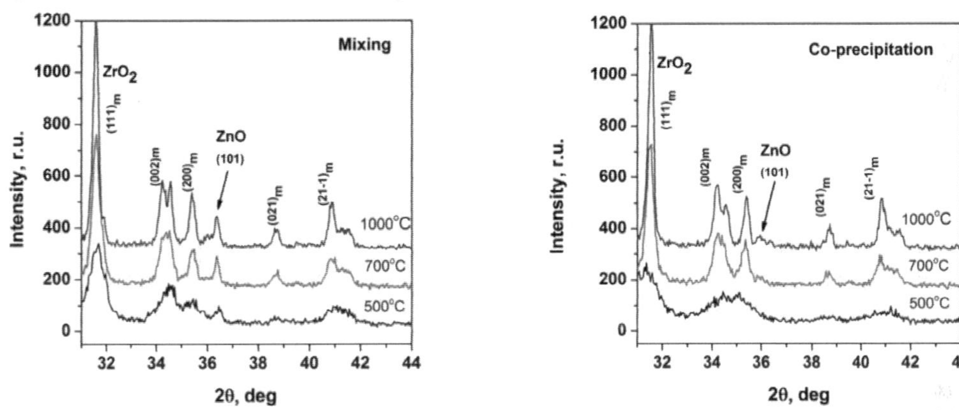

Figure 3. The XRD spectra of ZrO_2 – ZnO powders synthesized by mixing - a) and co-precipitation – b) technique.

In case using the co-precipitation technique it is possible to create conditions (at low firing temperatures) when the nonequilibrum supersaturated solid solutions exists. We have supposed the ZnO concentration was low (near 3-5 wt%) and Zn atoms were located in nonequilibrum solid solution in ZrO_2 lattice and the absorption spectrum was determined by ZrO_2, the reflection, which conforms to ZnO was not fixed and ZnO can influence on changing in long wave tails in absorption spectra only. When the firing temperature increased the ZnO diffused from ZrO_2 particles the absorption which was corresponded to ZnO appeared. The XRD analysis data detected the changing of phase composition and lattice parameters ZrO_2 and ZnO lattices in composites with low ZnO amount during increasing firing temperature. During heating at low temperatures the tetragonal phase of ZrO_2 was formed. After increasing the firing temperatures up to 800-1000°C the tetragonal phase transformed to monoclinic phase and XRD analyses fixed ZnO as separated phase.

3.2. Formation of structure and properties in ZrO_2-3mol%Y_2O_3 – Al_2O_3 system

The analogical situation in formation of supersaturated solid solutions of Al^{3+} ions in ZrO_2 lattice was found in ZrO_2-3mol%Y_2O_3–Al_2O_3 composite system. In this case the formation of complex structure of composite was demonstrated not only in structure of nanopowders but in the structure and properties of sintered ceramics. The formation of complex structure of Al_2O_3 inclusions in zirconia matrix during sintering leads to increasing the K_{1C} and wear resistance of zirconia ceramics in 1.5-2 times [19]. At the same time the density and strength values remains at the same level.

The composite powders obtained by ball-milling 3Y-TZP nanopowder (particle size 32 nm) with commercial α-Al_2O_3 powder and ZrO_2-3mol%Y_2O_3 – Al_2O_3 nanopowder are shown on Figure 4.

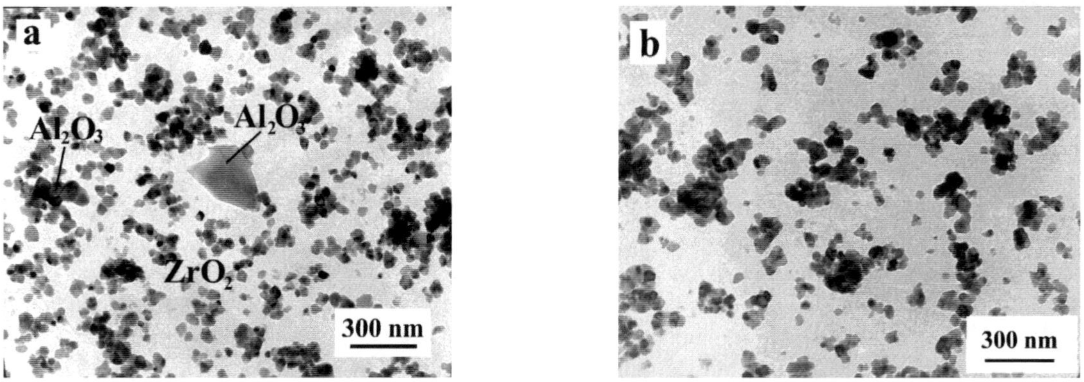

Figure 4. TEM structure of the ball-milled 32 nm 3Y-TZP powder with 2w% a-Al_2O_3 – a) and 3Y-TZP-2w% Al_2O_3 obtained by the co-precipitation technique – b). Calcination temperature was 1000°C.

After milling, the particle sizes of ZrO_2 did not change. From XRD data (peak on 43.36°) was identified the α-Al_2O_3 in the composite powder that was obtained with the ball-milling technique (Figure 5). According TEM and XRD data the mean particle size of ZrO_2-3mol%Y_2O_3 – Al_2O_3 nanopowders obtained by the co-precipitation technique decreased from 32 to 19.3 nm with increasing concentrations of Al_2O_3 from 0 to 5 wt%, respectively. In contrast with previous data, in the nanocomposite powder obtained by the co-precipitation technique, the (101) α-Al_2O_3 reflex has not been found during powder characterization (Figure 4). The XRD investigation also determined that the lattice parameters (***a***, ***c***) and lattice volume (***V***) of T-ZrO_2 changed from ***a*** = 0.50934nm, ***c*** = 0.51682 nm, ***V*** = 0.13408 nm^3 for matrix 3Y-TZP powder to ***a*** = 0.509162 nm, ***c*** = 0.516377 nm, ***V*** = 0.13386 nm^3, for 3Y-TZP-2wt% Al_2O_3, in powders obtained by co-precipitation technique. In case of 3Y-TZP-2wt% Al_2O_3, obtained by ball-milling technique such changes have not been found.

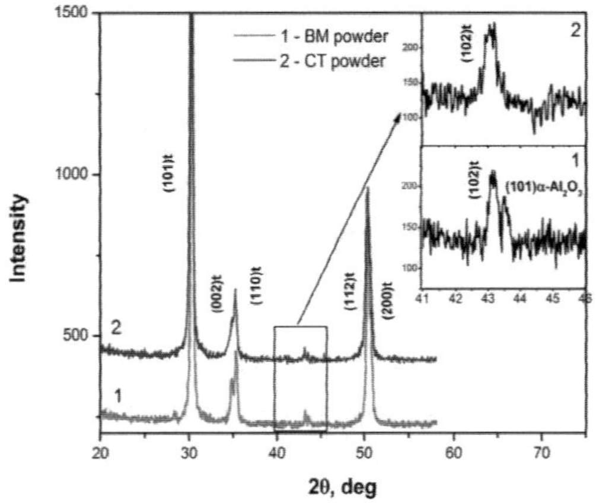

Figure 5. XRD data of 3Y-TZP-2wt% Al_2O_3 powders. 3Y-TZP-2wt% Al_2O_3 composite obtained by the ball-milling technique – 1 and co-precipitation technique – 2). In the inserts, the regions 40 – 46° are shown.

In case of using co-precipitation method, Al^{3+} ion are located in zirconia lattice. Al^{3+} solubility in zirconia is very low, and the active diffusion of ions Al^{3+} from the volume to the grain boundary or triple junctions takes place during sintering. The process of grain growth during sintering of co-precipitated nanopowders is competing with Al^{3+} segregation. Some inclusions can be captured by growing grains and turn into intertype inclusions. An amount of alumina can remain on sub-grains and grains boundaries. The final structure consists of matrix grains and sub-grains with or without alumina segregations, inter- and intratypes of inclusions. The crack propagation from such a structure should be hampered. In the case of ball-milling powder preparation method, the initial Al_2O_3 particles are located only on zirconia grain surfaces and structure changes took place near the its starting location, so intragranular alumina inclusions in zirconia grains during sintering could not be formed in appropriate amount as well as Al^{3+} segregations, because the solubility of Al^{3+} ions in ZrO_2 is too low.

Thus, the increase in the K_{IC} value of zirconia ceramics with a small amount of alumina sintered from nanopowders obtained using co-precipitation techniques can be conditioned through a series of processes for composite structure formation during precipitation, crystallization, and sintering of nanopowders. These processes strongly differ from structure formation processes in composites produced from ball-milled powders.

3.3. Other examples of application of ZrO_2 NPs

The formation of different states on NPs surfaces during heating of nanopowders in presence of impurities is one of the promising methods for creation of selective properties of nanopowders. For example the sensitivity of ZrO_2 nanoparticles to humidity increased with increasing of calcination temperature up to 700°C (Figure 6).

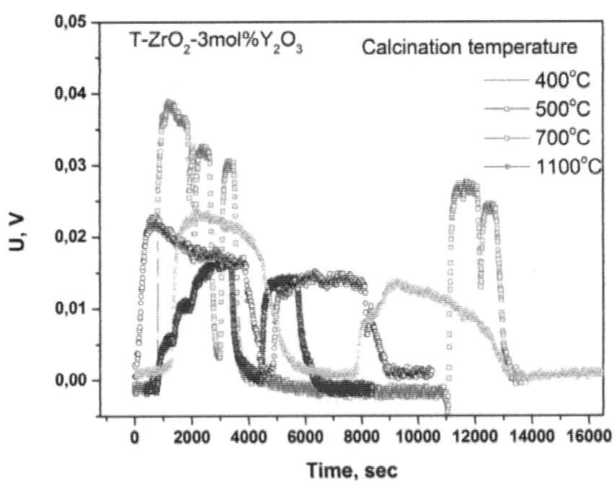

Figure 6. The electric potential of ZrO_2 samples, prepared from different type of nanopowders.

The electric potential is generated on zirconia NPs surface according to processes adsorption of water molecules and its decomposition on NPs surface. In this process the crucial point is the formation of surface centers of certain type. It was found that the calcination in the presence of residual Cl^- atmosphere leads to formation of basic Bronsted type surface centers. Increasing of calcining temperature up to 700°C leads to removing Cl^- ions from NPs surface and formation of basic Bronsted centers. Further increasing of calcining temperature results in annealing of defects and decreasing the effectiveness the conversion humidity to electricity.

The analogical influence of annealing of ZrO_2 nanopowders with small amounts of $CuCl_2$ or MnS leads to the changing in optical properties of matrix material (Figure 7).

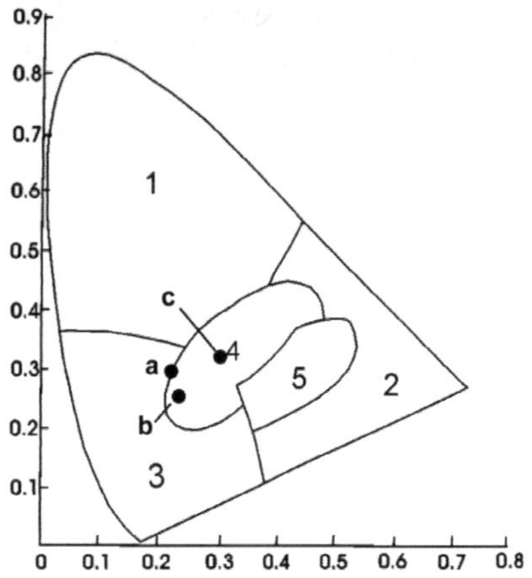

Figure 7. Representation of the PL spectrum of the 4.4Y-ZrO_2 powder in color coordinates: (1) green, (2) red, (3) blue, (4) white, and (5) pink. (a) 4.4Y-ZrO_2 : 1% $CuCl_2$, (b) 4.4Y-ZrO_2 : 3% $CuCl_2$, and (c) 4.4Y-ZrO_2 : 1% MnS. Figure was taken from [8].

The annealing of ZrO_2+4.4mol%Y_2O_3 with $CuCl_2$ or MnS leads to an increase in the intensity of the photoluminescence (PL) spectrum in the range 350–550 nm, where this spectrum is likely to be a superposition of individual PL bands with $\lambda_{max} \approx 388$, 420, 455, and 497 nm. The appearance of PL in the 4.4YZrO_2 nanopowder may be due to the temperature-induced changes in its oxygen sublattice. These results demonstrate that 4.4Y-ZrO_2:(1%)MnS can be used to producing a single-component nanophosphor with a continuous radiation spectrum and with color coordinates (X = 0.3, Y = 0.31) corresponding to white radiation.

4. Conclusions

In the present study has been shown that zirconia materials in nanostructured state has a wide spectrum of properties, which was impossible to achieve in traditional microscopic state. This technology approaches lead to creation of nonequilibrum supersaturated solid solutions, which were partially or fully destroyed during calcination and sintering process. The process of dopant migration from grain volume to the grain surface leads to creation of complex multilevel composite structures such as ion segregation on zirconia grain boundaries, clusters, intra- and intercrystalline inclusions in zirconia grains. These processes strongly differ from structure formation processes in composites produced from industrial ball-milled powder.

Acknowledgments

This work was supported by the grant AZ 90355 of the VW foundation program "Trilateral Partnerships" and projects 690968 NANOGUARD2AR 691010 HUNTER of H2020-MSCA-RISE-2015 Programs.

References

[1] Guo D, Xie G and Luo J 2014 *J. Phys. D: Appl. Phys.* **47** 013001
[2] Chen X and Mao S 2007 *Chem. Rev.* **107** 2891
[3] Srinivasan M, Rajabi M and Mousa S 2015 *Nanomaterials* **5** 1690
[4] Sahay R, Reddy V and Ramakrishna S 2014 *Int. J. Mech. Mater. Eng.* **9** 25
[5] Lopez-Esteban S, Rodriguez-Suarez T, Esteban-Betego F, Pecharroma C and Moya J 2006 *J.Mater. Sci.* **41** 5194
[6] Vasylkiv O, Sakka Y and Skorokhod V 2003 J. *Am. Ceram. Soc.* **86** 299
[7] Naglieri V, Palmero P, Montanaro L and Chevalier J 2013 *Materials* **6** 2090
[8] Suciu C, Hoffmann A and Kosinski P 2008 *J. Mater. Proc. Tech.* **202** 316
[9] Liu T, Fang J, Li S, Wang C and Ji C 2015 *J. Ceram. Soc. Jap.* **123** 554
[10] Vasylkiv O and Sakka Y 2001 *J. Am. Ceram. Soc.* **84** 2489
[11] Lin J and Duh J 1997 *J. Am. Ceram. Soc.* **80** 92

[12] Chatterjee M, Ray J, Chatterjee A and Ganguli D 1992 *Ceram. Int.* **18** 337
[13] Oliveira A and Torem M 2001 *Powder Technol.* **119** 181
[14] Danilenko I, Konstantinova T, Volkova G and Glazunova V 2013 *Mater. Charact.* **82** 140
[15] Konstantinova T, Danilenko I, Glazunova V, Volkova G and Gorban O 2011 *J. Nanopart. Res.* **13** 4015
[16] Bacherikov Y, Optasyuk S, Konstantinova T and Danilenko I 2007 *Tech. Phys.* **52** 747
[17] Garvie R and Nicholson P 1972 *J. Am. Ceram. Soc.* **55** 303
[18] Štefanic G, Music S and Ivanda M 2009 *J. Mol. Struct.* **924-926** 225
[19] Danilenko I, Konstantinova T, Volkova G, Burkhovetski V and Glazunova V 2015 *J. Ceram. Sci. Tech.* **6** 191

Core/shell nanocomposite for biomedicine and environmental safety

O Gorban[1,3], I Danilenko[1], N Mitina[2], S Gorban[1], A Zaichenko[2] and T Konstantinova[1]

[1]Department of Materials Science and Engineering, Donetsk Institute for Physics & Engineering NAS of Ukraine, Nauki av., 46, Kyiv, Ukraine
[2]Lviv Polytechnic National University, S. Bandera St., 12, Lviv, Ukraine

E-mail: oxanag1@ukr.net

Abstract. The two different routes for creation Cu(II) modified zirconia were seen. FTIR and ESR spectroscopy were investigated the influence of synthesis route on features of formation of such structure. It is shown that for both cases the interaction between copper (II) and Zr-OH and Zr=O is occurred and these centres have paramagnetic nature. When the Cu-contained polymer is used for shell formation, only one kind of paramagnetic centres of copper (II) is formed, but when the Cu-contained inorganic complex is used for shell formation two kinds of copper (II) paramagnetic centres are formed.

1. Introduction

Task-oriented nanocomposites of core/shell structure (NC) and surface-modified nanoparticles are promising materials for using in biomedical and environmental safety application [1]. These materials must to have pre-determined properties there are catalytic or photocatalytic, antibacterial or fungicide properties [2-3]. CuO, Ag or Au particles are “the oldest” bactericide agents, but in high concentrations the ions of these elements show a high toxicity to living organism [4]. It is also necessary noted that these materials are very expensive. It leads to need developing of a new composite materials in which such bactericide active ions are on nanoparticles surface [5].That approach allows decreasing of amount of active elements in nanoparticles with safety of their high bactericide properties but to decrease toxicity to living organism and cost of material [5]. The specific features of composite materials are the crucial influence of the interface on their structural and functional properties [6-7]. Efficiency of surface functionalized material in these applications has been achieved due to its high dispersity and homogeneity of components distribution in the complex system [5]. Cu(II) – functionalized zirconia is promising material for creation functional ceramic and inorganic-polymeric composite [8-9]. Composite systems may be inorganic/inorganic and inorganic/organic nature. These composite structures may be different nature, for example, oxide nanoparticles with surface is enriched by Cu (II) complexes or oxide core/Cu-contained polymeric shell [8,10]. The state of copper (II) on surface is a main moment for functionality of such composite structures.

Some routes of creation of functional structures are used in material science technology [11-13]. In practice, the impregnation of Cu salts in oxide surface is a more often used method in catalysts chemistry for creation or enhance catalytic systems [12]. However in this approach often synthesized

IOP Conf. Series: Materials Science and Engineering **213** (2017) 012017 doi:10.1088/1757-899X/213/1/012017

the structure with separated chemical components or another words substrate-carrier with high dispersity of active copper oxide components [13].

But for some applications the metal ions linked with oxide surface structure are needed [14]. It may be realized due to simultaneously precipitation of oxide matrix and metal hydroxocomplexes and for this case the oxide nanoparticles with copper ions enriched surface is formed [15]. In another approach the creation of oxide core/Cu(II) contained shell structures may be realized modifying of oxide nanoparticles surface by specially developed Cu(II) contained oligomer molecule [16]. It is especially important for forming of polymer compatible nanomaterials. For case of core/shell structure and as core and as shell together determines of functionality of such system [6]. It is noted that for both cases the copper (II) must be linked with zirconia surface.

In this work we show that two routes may be effective use for creation of CuO surface enriched zirconia particles and zirconia core/Cu-contained polymeric shell structure and the linking of copper (II) in these systems will be the same. Preservation of copper (II) in ionic or chelate forms is very important moment for keeping of antibacterial, catalytic and optic properties of composite materials.

2. Experimental

Two routes for creation of Cu(II) surface functionalized zirconia were used. First route based on wet co-precipitation technology and only inorganic components are used. Next procedure was used for ZrO_2-3 mol. % Y_2O_3 core/Cu-contained shell system creation: the mix of 2 mol/L of zirconium chloride [$ZrOCl_2$] and 0.044 mol/L of yttrium nitrate [$Y(NO_3)_3$] salts and 0.09 mol/L of $Cu(NO_3)_2$ in distilled water was used as initial salt solution. Co-precipitation was carried out using 6 wt% solution of ammonia at pH 10-11, and then was precipitate washing to pH 7. The precipitate was drying under microwave irradiation (MW, frequency is 2.45 GHz and power is 700W). Cu(II) functionalized ZrO_2-3 mol. % Y_2O_3 NPs (ZrO_2-Y_2O3-Cu(II) complex) was obtained.

Second route based on wet co-precipitation method for oxide core creation with follow modification of Cu-contained polymer. Next procedure was used for ZrO_2-3 mol.% Y_2O_3 core creation: the mix of 2 mol/L of zirconium chloride [$ZrOCl_2$] and 0.044 mol/L of yttrium nitrate [$Y(NO_3)_3$] salts in distilled water was used as initial salt solution. Co-precipitation was carried out using 6 wt% solution of ammonia at pH 10-11 and then the precipitate was washed to pH 7. Precipitate was drying under microwave irradiation (MW) with power of 700 W and frequency of 2.45 GHz. ZrO_2-3 mol. % Y_2O_3 crystalline NPs (ZrO_2-Y_2O_3, zirconia nanoparticles) was synthesized by calcining the dried precipitate. The annealing temperature was 500°C. Coordinated Cu^{2+} complex of copolymer of vinyl acetate, 5-tertbutylperoxy-5-methyl-1-hexene-3-yne and maleic acid (OMC) was used for obtaining functional shell via its irreversible adsorption on nanoparticle surface. Complex structure presents in Figure 1.

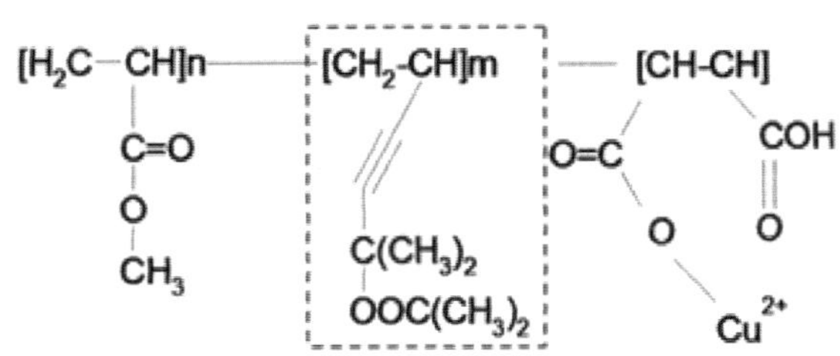

Figure 1. Structure of OMC. The blue square shows the 2-tertbutylperoxy-2-methyl-5-hexene-3-yne fragment (follower BEP).

Fourier transform infrared spectroscopy (FTIR spectroscopy) was used to study the initial materials and composite structures. The FTIR spectra of NPs were registered within the range of 375-4000cm^{-1} with resolution of 4cm^{-1}. FTIR spectrometer "Bruker Tensor 27" was used for measurement.

Electron spin resonance spectra (ESR spectra) of xerogels were obtained at room temperature. The spectrometer CMS8400 with a high-frequency modulation at 9 GHz is used. The ESR measurements were carried out in presence of calibration sample – Mn(II) in cubic MgO. The fifth CTC line of Mn(II)/MgO was chosen as a reference of the paramagnetic centers position. The relative measurement error did not exceed 20%.

3. Results and discussion

According to the presented synthesis routes for the first case the Cu(II) are adsorbed on nanoparticles of zirconium hydroxide hydrogel in form complexes with different structures. There are copper ammonia complexes $[Cu(NH_3)_m(H_2O)_n]^{2+}$ or copper hydroxide complexes $[Cu(H_2O)_m(OH)_k]^{2-k}$ or mix copper ammonia hydroxide complexes $[Cu(NH_3)_m(OH)_n]^{2-n}$ and also the different associates of their complexes may be formed. Realization of determined structure of Cu(II) complex are depended on pH solution and concentration of copper salts and ammonium hydrate [17].

The identification of Cu(II) complex structure which may be realized in conditions of synthesis at pH 8 was carried out by FTIR spectroscopy, see Figure 2. FTIR spectrum shows the presence the set of absorption bands for valence vibrations of O-H bond in hydroxyls and water molecule (3546cm^{-1} and 3426cm^{-1}) and for valence vibrations of N-H bond of ammonia ligand (3125cm^{-1} and 3050cm^{-1}) in equatorial and axial positions in a complex, respectively. The set of adsorption bands at 1354cm^{-1}, 1384cm^{-1}, 1422cm^{-1}, 1635cm^{-1}, 1763cm^{-1} in region of deformation vibrations of FTIR spectrum indicates on the presence of ligands with NH and OH bonds in Cu(II) complex and on residual amount of NO_3 groups. Thus, peaks at 1635cm^{-1}, 1422cm^{-1}, 1354cm^{-1} identify NH_3 ligands, the peak at 1384cm^{-1} indicate on residual amount of NO3 groups, peak at 1763cм$^{-1}$ is summary from oscillation of both NH_3 and NO_3 groups. Peak at 1048cm^{-1} indicates on Cu-(OH) or Cu-(OH_2) groups in complex structure and peak at 677cm^{-1} corresponds to librational vibration of water ligands. The band at 858cm^{-1} corresponds to pendulum oscillations of NH_3. The band at 511cm^{-1} corresponds to vibrations of Cu-O bond and vibrations of Cu-N occur at 456cm^{-1} and 431cm^{-1} frequencies.

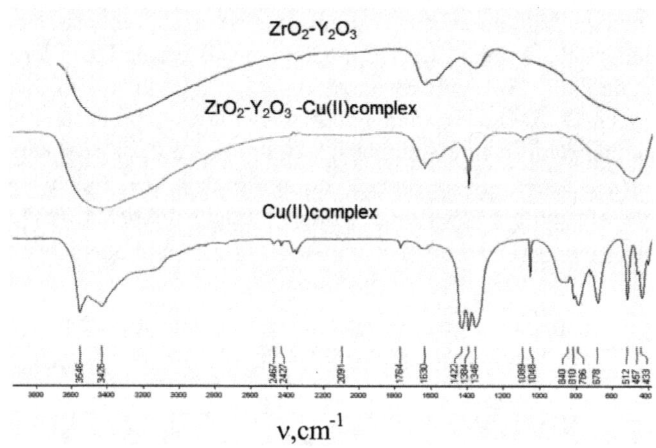

Figure 2. FTIR-spectrum of synthesized systems: Cu(II) complex, ZrO_2-Y_2O_3 and precipitated ZrO_2-Y_2O_3-Cu(II) complex.

FTIR spectra of ZrO_2-Y_2O_3 amorphous NPs show the presence of surface hydroxyls and water molecules with different extents of connectivity. The wide peak at 3434cm^{-1} corresponded to terminal hydroxyl and low frequency shoulder corresponds to two and three coordinate hydroxyls and water molecules. The peak at 1635cm^{-1} indicated physically adsorbed water. The spectra of investigated systems contained an adsorption band at 1093cm^{-1}. The presence of this adsorption band in the FTIR spectrum of oxide was found to be associated with the appearance of surface cation-oxygen (or cation-hydroxyl group) vibrations of different strengths [18].

For Cu(II) complex modified ZrO_2-Y_2O_3 in region of valence vibrations of O-H and N-H bonds FTIR spectrum shows the wide bands with central frequency at 3429cm^{-1}. This adsorption band is shifted in low frequency region on compared with pure Cu(II) complex. FTIR spectrum of this system shows the change of intensities of bands of deformation vibrations of N-H bonds (in range of 1300-1400cm^{-1}). It may be evidence changing in Cu(II) surrounding ligands at adsorption of complex Cu (II) on the surface of ZrO_2-Y_2O_3 nanoparticles. FTIR spectra of Cu(II) modified ZrO_2-Y_2O_3 shows a wide non-divided band at 400-1000cm^{-1} in range of lattice vibrations. The qualitative picture of FTIR spectrum of Cu(II) modified ZrO_2-Y_2O_3 may be corresponded to replace some NH_3 ligands from Cu(II) complex as result of interaction Cu(II) complex with surface oxygen-contained groups ((Zr=O or ZrOH) and oxygen vacancies.

In the second case the ZrO_2-Y_2O_3 core is first formed and only then copper-contained polymer shell is instilled to it. Synthesis ZrO_2-Y_2O_3 may be divided to same stage. On first stage the amorphous product is formed. On the second stage the heat treatment of amorphous NPs at 500°C leads to forming tetragonal crystalline NPs with sizes about 10nm. Figure 3 shows the FTIR spectrum of ZrO_2-3 mol.% Y_2O_3 NPs. FTIR spectra of crystalline NPs shows the presence of surface hydroxyls and water molecules with different extents of connectivity. The peak at 3692cm^{-1} corresponded to terminal hydroxyl groups that were bases. The existence of a wide band at 3440cm^{-1} indicated the presence of two and three coordinate hydroxyls and water molecules that were acids. The peak at 1635cm^{-1} indicated physically adsorbed water. The spectra of investigated systems contained an adsorption band at 1093cm^{-1}. The presence of this adsorption band in the FTIR spectrum of oxide was found to be associated with the appearance of surface cation-oxygen (or cation-hydroxyl group) vibrations of different strengths [18].

Figure 3 shows the FTIR spectrum of OMC. As one can see, bands on the FTIR spectrum ranged from 2900–2930cm^{-1} and were assigned to aliphatic CH_n group vibrations.

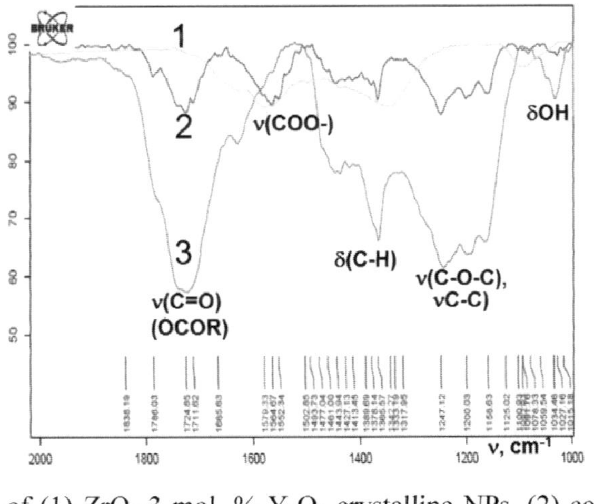

Figure 3. FTIR spectra of (1) ZrO_2-3 mol. % Y_2O_3 crystalline NPs, (2) composite structure ZrO_2-3 mol.%Y_2O_3/OMC, (3) OMC.

The presence of weak-intensity peak in range of 1900–2000cm^{-1} may be corresponding to C≡C bond vibrations. The peak at frequency of 1728cm^{-1} and shoulder at 1784cm^{-1} correspond to vibrations of the C=O bond of carbonyl groups of different fragments of OMC molecules (Figure 1). The peak at 3486cm^{-1} was assigned to the associated OH bond of the carbonyl groups. The peak at 1633cm^{-1} was assigned to the carboxyl ion bound in the chelated complex by Cu^{2+} ions. Bands at 1436cm^{-1} and 1364cm^{-1} correspond to anti-symmetric and symmetric vibrations of aliphatic CH_2 or CH_3 groups, respectively. The wide subminimum in range of 1100–1300cm^{-1} (peaks at 1241cm^{-1}, 1199cm^{-1} and

1165cm^{-1}) were ascribed as vibrations of C-O-C and C-C skeleton bonds. The peak at 1032cm^{-1} corresponded to a vibration of an OH bond of the carboxylic groups bound to Cu^{2+} ions. Figure 3 shows the FTIR spectrum of composite structures of ZrO_2-3 mol.% Y_2O_3 NPs core/OMC shell. The composites were obtained using 3 wt.% OMC solution. Vibrations of C=O, CH3, and CH2 of functional groups and C-C and C-O-C skeletons of OMC were good appeared in the FTIR spectra. The presence of adsorption band at 1727cm^{-1} corresponds to ν(c=O) of carbonyl groups of OMC fragments and also the new band at 1565cm^{-1} is appearance. It corresponds to carboxylic groups that are linked with surface. Also the disappearance of adsorption band at 1086 cm^{-1} (surface Zr-O bonds) is observed in FTIR. The change in band intensity at 1089cm^{-1} in the spectra of the modified system was used to estimate the extent of coverage of the organic phase surface. The qualitative picture allows to propose that the interaction OMC with NPs are realized as Cu coordination with Zr=O or Zr-OH. Figure 4 shows the possibility mechanism of interaction of zirconia nanoparticles with OMC.

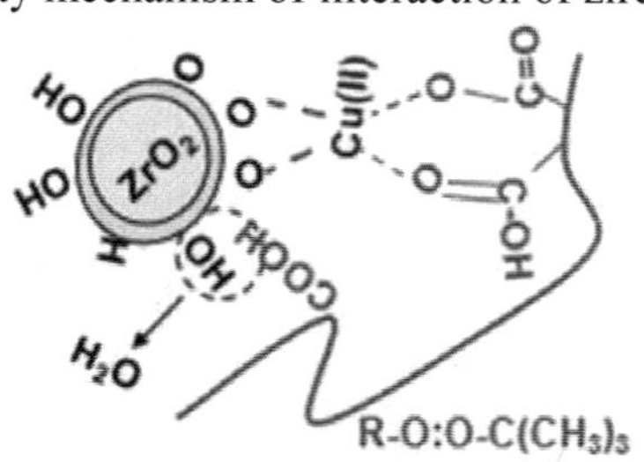

Figure 4. Possible mechanism of interaction OMC with zirconia surface.

Accordingly to noted features of FTIR spectra of zirconia/OMC structure, in particular the shifting of band from 1093cm^{-1} to 1086cm^{-1} (vibrations of surface Zr=O(OH) groups for zirconia and zirconia/OMC) and appearance a new band at 1565cm^{-1} (vibrations of C=O carboxylic ions), coordination OMC on zirconia surface occurs by linked carboxylic groups and Cu(II) coordination with Zr=O(OH) surface groups.

Figure 5 shows the ESR spectra of Cu(II) modified zirconia which were synthesized by different routes. In ESR spectra of both systems the broad signal are observed.

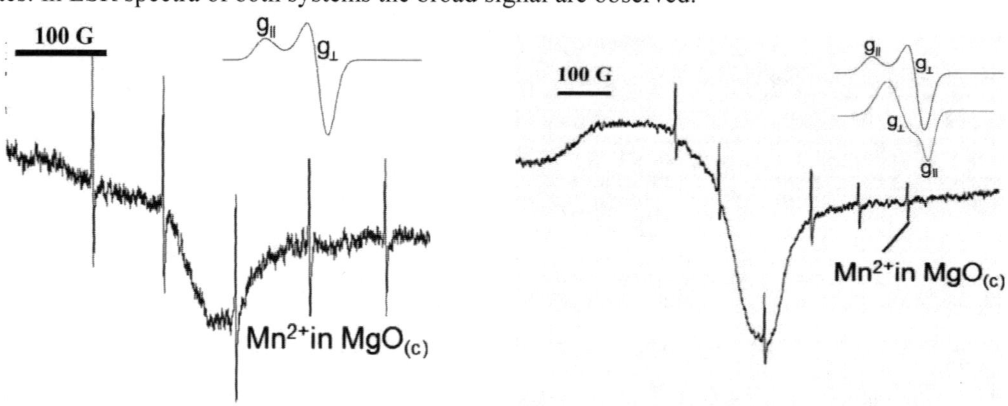

a b

Figure 5. ESR spectra of a) OMC modified ZrO_2-3 mol.% Y_2O_3 NPs, b) Cu(II) modified ZrO_2-3 mol.% Y_2O_3 NPs.

Early [19] we show that the ESR signal for pure Cu(II) complex has a complex characters and shows the presence of two kinds of Cu(II) paramagnetic centers ($g_{\parallel}$=2.25, $g_{\perp}$=2.08 and $g_{\perp}$=2.17, $g_{\parallel}$=2.05). For Cu(II) modified ZrO_2-3 mol.% Y_2O_3 also two kinds of Cu(II) paramagnetic centers ($g_{\parallel}$=2.24, $g_{\perp}$=2.06 and $g_{\perp}$=2.15, $g_{\parallel}$=2.03) are observed in ESR spectra. The ESR spectrum of ZrO_2-3 mol% Y_2O_3 NPs which is modified by OMC show the broad signal with g-factors $g_{\|}$=2.17, $g_{\perp}$=2.06, which may be corresponded the coordination of Cu(II) on NPs surface as chelate complex.

4. Conclusion

Thus for both routes of formation of Cu(II) modified ZrO_2-Y_2O_3 and ZrO_2-Y_2O_3/OMC-shell structures the Cu(II) is linked with zirconia surface via surface oxygen-containing groups (Zr=O or ZrOH). The features of formation of Cu(II) modified ZrO_2-Y_2O_3 in presented synthesis conditions led to form surface Cu(II) paramagnetic centers in different environment and as result with different degree of linking with zirconia surface. In case formation ZrO_2-Y_2O_3/OMC-shell only one kind of Cu(II) paramagnetic is formed and oligoperoxide shell additional contains the functional peroxide groups as centers of active oxygen forms. The proximity of the values of the g-factors of Cu(II) centers for both kinds of formed structures indicates on approximately equal environment of Cu(II) centers that is linked with zirconia surface.

Using of different synthesis routes may be effective key to controlling of the surface state of composite nanoparticles and kind of their active surface centers. These structures are proposed as active material with antibacterial and catalytic properties in systems of air purification.

Acknowledgments

Work was partially supported by the projects 690968 NANOGUARD2AR of H2020-MSCA-RISE-2015 Programme and by the projects 47/16-H of "Fundamental problems of the novelty of nanomaterials and nanotechnologies.

Reference

[1] Ratnasamy P, Srinivas D, Satyanarayana C V V, Manikandan P, Senthil K R S, Sachin M, Vasudev N and Shetti 2004 *J. Cat.* **221** 455-65
[2] Gawande M B, Goswami A, Asefa T, Guo H Z, Biradar A V, Peng D L, Zboril R and Varma R S. 2015 *Chem. Soc. Rew.* **44** 7540-90
[3] Shokufeh A 2015 *Res. J. Chem. Environ.* **19**(8) 28-32
[4] Bondarenko O, Juganson K, Ivask A, Kasemets K, Mortimer M and Kahru A 2013 *Arch. Toxicol.* **87**(7) 1181-200
[5] Dhanalekshmi K I and Meena K S 2014 *Spectrochemica Acta: Part A. Molecular and biomolecular spectroscopy* **128** 887-90
[6] Samzadeh-Kermani A, Mirzaee M and Ghaffari-Moghaddam M 2016 *Adv. Biol. Chem.* **6** 1-11
[7] Levitin E Y, Kryskiv O S and Scoryk M 2015 *Scripta Scientifica Pharmaceutica* **2**(1) 53-9
[8] Palza H, Quijada R and Delgado K 2015 *J. Bioactive Comp. Polym.* **1** 1-15
[9] Actis L, Gaviria L, Guda T and Ong J L 2013 *J. Korean Assoc. Oral Maxillofac. Surg.* **39** 43-54
[10] Samson K, *et al.* 2014 *ACS Catal* **4**(10) 3730-41
[11] Liu X, Bai S, Zhuang H and Yan Z 2012 *Frontiers of Chem. Sci. and Eng.* **6** 47-52
[12] Pakharukova V, Moroz E M, Zyuzin D A, Dolgikh L Yu and Strizhak P E 2015 *J. Phys Chem C* **119** 28828-35
[13] Basahel S N, Mokhtar M, Alsharaeh E H, Ali T T, Mahmoud H A and Narasimharao K 2016 *Catalysts* **6** 57
[14] Wang G, Ji J, Zhang X, Zhang Y, Wang Q, You X and Xu X 2014 *Sci. Rep.* **4** 5480
[15] Chen C, Ruan C, Zhan Y, Lin X, Zheng Q and Wei K 2014 *Int. J. Hydrogen Energy* **39** 317-24
[16] Zaichenko A, Mitina N, Kovbuz M, Artym I and Voronov S 2000 *J. Polym. Sci., Part A: Polym. Chem.* **38** 516-27
[17] Anufrienko V F, Zenkovec G A, Schutilov R A, Gavrilov V Yu, Vasenin N T, Schubin A A, Ismagilov Z R and Parmon V N 2011 *Phys. Chem.* **440**(5) 651-4
[18] Davudov A A and Anufrienko B F 1984 *IR Spectroscopy in Oxide Surface Chemistry* (Novosibirsk: Nauka)
[19] Gorban O, Gorban S, Zarechnaya O, Kharchenko M and Konstantinova T 2016 *Nanophysics,*

Nanophotonics, Surface Studies and Applications (Cham: Springer International Publishing AG) pp 543-9

IOP Conf. Series: Materials Science and Engineering **213** (2017) 012018 doi:10.1088/1757-899X/213/1/012018

Morphological and thermal properties of PLA/OMMT nanocomposites prepared via vane extruder

Y Luo[1], H Y Liu[1], G Z Zhang[1] and J P Qu[1,2]

[1]National Engineering Research Center of Novel Equipment for Polymer Processing, The Key Laboratory of Polymer Processing Engineering of Ministry of Education, South China University of Technology, Guangzhou 510640, China

E-mail: jpqu@scut.edu.cn

Abstract. Polylactide/Organo-Montmorillonite (PLA/OMMT) Nanocomposites were prepared by melting extrusion using a novel vane extruder (VE), which can induce global elongational flow. In the study, the influence of different concentrations of the OMMT on the morphological and thermal properties were investigated. The morphology and structure of the nanocomposites were evaluated using Fourier Transform Infrared Spectroscopy (FTIR), the X-ray diffraction (XRD) and transmission electron microscopy (TEM) respectively, whereas the thermal behaviors and thermal stabilities were characterized using differential scanning calorimetry (DSC) and thermal gravimetric analysis (TGA) respectively. The results illustrate that PLA/OMMT nanocomposites displayed clear intercalation and/or exfoliation structures. Interestingly, increasing the clay content did not lead to the agglomeration of OMMT layers. Moreover, the presence of nanoclay decreased the enthalpy of crystallization of PLA/OMMT composites. Also, the melting temperatures of the nanocomposites were reduced by the addition of nanoclay.

1. Introduction

Recently, biodegradable and biocompatible polymers have attracted tremendous attention both from biomedical and ecological possibility [1]. PLA, one of such polymer, has superb properties such as biocompatibility, thermal plasticity, high strength and stiffness, so there is a large prospect of developing PLA as ideal substitutes for the petrochemical plastics in polymer applications [2-3]. However, its drawbacks, such as low heat distortion temperature and inferior hydrophobicity, limit its use in various end-use applications [4]. In spite of the dissatisfaction, many considerable attempts have been carried out to enhance these properties [5-6].

In recent decades, the incorporation of nanoclay into the PLA matrix has received considerable interest as one of the most universally approved and feasible measures, since significant improvements in properties are observed at lower clay content. Besides that, clay materials are easily obtainable, environmentally harmlessly. Forming a kind of delamination and/or exfoliation structure of nanoclay in PLA matrix is necessary for preparing superior nanocomposites. Because the synergistic effects of the nanoscale structure and the interaction of the fillers and polymer matrix perhaps can improve some properties. Several processing methods have been considered to prepare PLA-layered silicates for the superstructure and micromorphology of PLA in situ intercalative polymerization [7], solvent intercalation [8-9] and direct melt intercalation [5-6, 10-11] using a polymer mixer or screw extruder. Ogata et al. [8] first prepared PLA organically modified clays blends by the solvent-cast method.

Sinha Ray et al. [11] prepared PLA/clay nanocomposites with a twin screw extruder and the intercalated structure of silicate layers of the clay was observed, it was revealed that clay randomly distributed in the PLA matrix. Okamoto [12] studied a range of PLA-based systems differing from the OMMT in the aspect ratio of the nature of the organic modifier. The results indicated that PLA/clay could possess an excellent enhancement in physical as compared with the pure PLA matrix.

However, all these improvements are based on classic screw extruder and there are some disadvantages of polymer composites prepared by a conventional extruder which is based on the shear flow. In polymer processing, the shear flow is less efficient than the elongation flow. More specially, according to some researchers, it is relatively effective to crush the melt drops under elongation flow rather than under shear flow [13-14]. Given that the incorporating of strong elongation flow has a good prospect on nanocomposite processing, a novel vane extruder (VE) will particularly be used in this study. Qu. [15] designed the VE, which can induce a global elongation flow. Moreover, the vane extruder shows a short thermo-mechanical history and wide suitability in polymer processing, hence the vane extruder can reduce particle size and macromolecule chain broken. From our previous experiments, finer dispersion particles and well mechanical properties have been obtained using VE because of its positive displacement-type characteristics [16-18], and the comparison between VE and traditional extruder has been researched in our previous works [16,18-19].

The aim of this study is to evaluate the effectiveness of OMMT concentration on dispersion, morphological, and thermal properties of PLA/OMMT composites prepared by elongation flow. In this way, it can offer a theoretical background to understand the nanocomposite formation in VE.

2. Experimental

2.1. Materials

Polylactide (PLA 3051D, density=1.24g/cm^3) was obtained from NatureWorks. Its glass transition temperature (T_g) and melting temperature (T_m) were 62℃ and 164℃, respectively. The melt flow index of the PLA was 25g/10 min (2.16 kg load, 210℃). Montmorillonite clay (1.34 TCN, Nanocor density = 2.00g/cm^3) was modified with methyl dihydroxylethyl hydrogenated tallow ammonium.

2.2. Preparation of PLA/OMMT nanocomposites

PLA/OMMT nanocomposites were prepared by a novel vane, whose schematic diagram is showed in Figure 1. Clearly, a number of groups of Vane Plasticating and Conveying Units (VPCUs) constitute the vane extruder. One group of VPCUs is made up of a rotor, a stator, two baffles and four vanes. Obviously, there are four cylindrical chambers between two baffles along the axial direction of the rotor. The polymer will be fed into the first VPCU and subsequently discharged into the next VPCU via the hole in the baffle, as the volume of chamber ranges from the maximum to minimum periodically. Simultaneously, materials get the necessary heat from the external heating and particulate deformation. It indicates that an elongational flow is induced by the volume transportation. Three different formulations including 2.5wt%, 5wt% and 7.5wt% OMMT fillers in the PLA matrix were fabricated by VE. Neat PLA pellets were prepared under the same conditions to get a reference material (the same thermal history). Before compounding, 8 hours' drying process was carried to PLA and OMMT in a vacuum oven at 80℃. Then PLA and OMMT clay were put into a dry plastic zip-lock bag, rolling over the bag to mix the materials and afterward feeding into the VE for melt-processing. The temperature setting of four zones from the hopper to die were 175-180-195-195℃, the rotor speed was set at 60rpm. The extruded composites were cooled naturally and subsequently pelleted by a granulator. The test samples for TEM were compression molded in a hydraulic press at 200℃ for 30 min, which were in the form of 150×80×4mm^3. The samples for XRD, DSC and TG characterization were obtained from the fresh extruder at the exit of the die. The films samples (0.4mm thickness) for FTIR test were hot compressed from the extruder molded plaques.

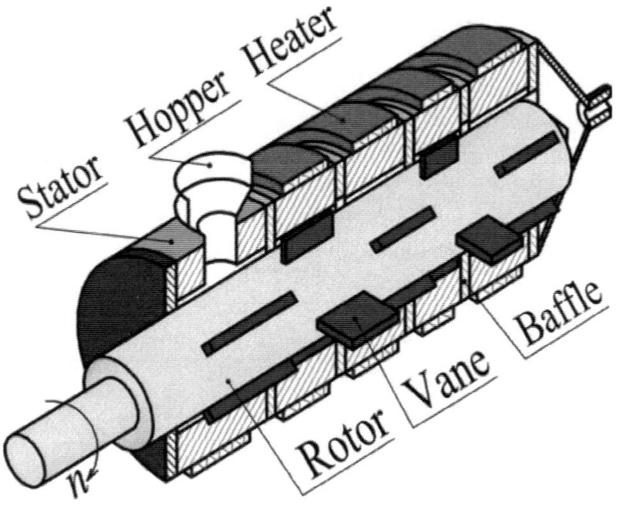

Figure 1. Schematic diagram of the vane extruder.

Table 1. Main parameters of vane extruder.

Main parameter	Value(mm)
Number of VCPUS	19
Radius of rotator	40
Internal radius of stator	48
Aspect ratio of rotator axis	10
Eccentricity distance	4

2.3. Characterization

X-ray diffraction (XRD) experiments were conducted to measure the d-spacing of the OMMT clay and the clay in the PLA/OMMT nanocomposites on a Germanic diffractometer (D8 ADVANCE), operating at the X-ray radiation (wave strength λ= 0.15418 nm, 40 KV and 40 mA). The mass of each sample include pure OMMT and PLA/OMMT nanocomposite with 2.5wt%, 5wt% and 7.5wt% are kept the same for measuring the d-spacing of OMMT layers in samples. The measurement was performed at a 2θ range of 2-30°, a scanning step of 0.02°, and a scanning rate of 2°/min.

Transmission electron microscopy (TEM: A Philips TECNAL 10) was used to characterize the dispersion of OMMT in the PLA/OMMT at an accelerated voltage of 120kV. The ultrathin sections for the TEM characterization were obtained from the samples by a Powertome X (Boeckeler Instrument) cryo-ultramicrotome.

Fourier Transform Infrared Spectroscopy (FTIR) analysis was conducted using an American instrument (Nexus 670, Nicolet Company). The spectra from 500-4000 cm^{-1} were recorded with an absorption of 32 scans.

Differential scanning calorimetric (DSC) measurements were carried out via a Netzsch DSC204 (Germany) under the nitrogen atmosphere. All test samples of 5-10 mg were heated from room temperature to 230℃ at a rate of 10℃/min and maintained 10 min, and then cooled -70℃, hold for 10 min. Lastly, these specimens were heated to 230℃.

Thermo-gravimetric analysis (TGA) of the samples were performed on an instrument (TG209, NETZSCH) at a heating rate of 20℃/min under a flow of N_2, from the room temperature to 700℃. TGA analyses were repeated in the air.

3. Results and discussion

3.1. Interactions between PLA and nanoparticles

The FITR spectrum of PLA and PLA/OMMT nanocomposites was presented in Figure 2. It is noteworthy that the 1000-3130cm^{-1} spectrum is omitted because there is no change in the range. The peaks at 3700-3400cm^{-1} are ascribed to the OH stretching of the –CH (CH_3)–OH end group of PLA. The peaks at 3700-3590cm^{-1}, 3590-3300cm^{-1} are assigned to the free hydroxyl and hydroxyl. The intensity of the free hydroxyl peak enhances with the loading of silicate clay, and the peak position shifts to lower wavenumbers. Interestingly, the position of associating hydroxyl do not change, and the intensity increases with the OMMT loading. The strong interactions between the hydroxyl of PLA end groups and OMMT lamella surfaces and/or the hydroxyl groups of the surface active agent in the OMMT fillers have caused above phenomenon.

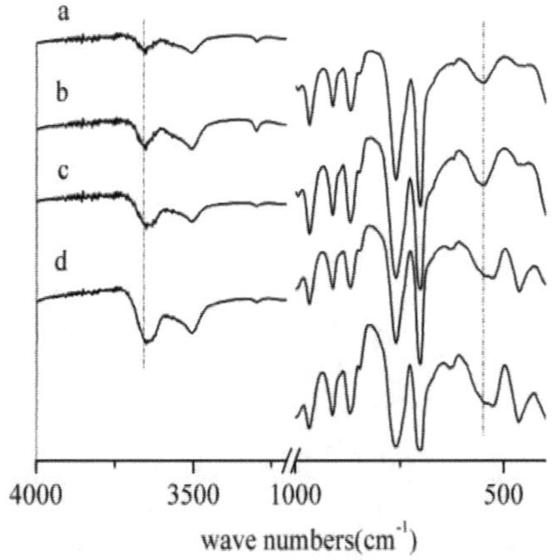

Figure 2. FITR spectrum of Pure PLA and PLA/OMMT nanocomposites, (a) Pure PLA, (b) PLA/OMMT (2.5wt %), (c) PLA/OMMT (5wt %), (d) PLA/OMMT (7.5wt %).

3.2. Interactions between PLA and nanoparticles

3.2.1. X-ray Diffraction (XRD). Figure 3 presents the XRD patterns of montmorillonite clay and PLA/OMMT nanocomposites. In Figure 3, log scale was used to compare the peaks near 5 degree more clearly. It indicates that there exists OMMT nanofiller intercalation and/or exfoliation in diverse PLA/OMMT nanocomposites. The order diffraction (001) of neat OMMT clay is registered at $2\theta = 4.84°$, which means a basal spacing of 1.83nm, according to Bragg's law ($\lambda = 2$ d sin θ; λ is the X-ray wavelength; d is the layer spacing). In the PLA/OMMT nanocomposites, 2θ of the (001) diffractions shifts to a higher angle for the composites of 2.5wt% (5.34°), 5wt% (5.05°) and 7.5wt% (5.00°) and the corresponding gallery spacing decreases than that of neat OMMT, as displayed in Table 2. This change is attributed to the elongational flow field which can break the homogeneous OMMT layers [20]. Also, the strong interplay between the PLA and the modified OMMT layer is another factor. The preservation of the peak (001) in the blends reveals that a portion of OMMT with the lamellar structure still maintains after the melt blending process despite part of OMMT may be intercalated and/or exfoliated in diverse PLA/OMMT blends.

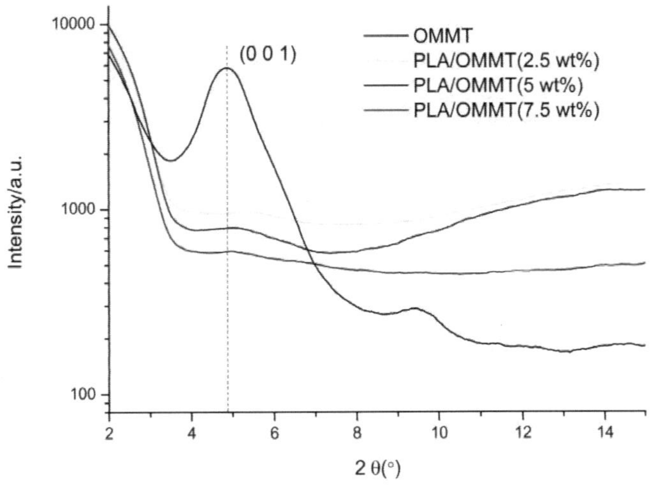

Figure 3. XRD patterns for OMMT and PLA/OMMT nanocomposites.

Table 2. D-spacing of OMMT and PLA nanocomposites.

Samples	2 θ(°)	d-spacing (Å)
OMMT	4.84	18.3
PLA/OMMT (2.5 wt %)	5.34	16.5
PLA/OMMT (5 wt %)	5.05	17.4
PLA/OMMT (7.5 wt %)	5.00	17.7

3.2.2. Transmission Electron Microscopy (TEM). Figure 4 displays the micrographs of PLA/OMMT nanocomposites. In Figure 4, the light regions are identified as PLA matrix, and the deep linear objects are piles of OMMT crystal lamellar. As showed in the Fig 4a-c, portions of the OMMT platelets successfully intercalate into the PLA/OMMT compounds. It indicates that the intercalated and/or exfoliated structure coexisted. Moreover, at high particle concentration (7.5wt%), the PLA matrix penetrates the OMMT layers, promoting the exfoliation of OMMT, as complete separation of the montmorillonite nanoclay by PLA chains. Interestingly, there is no agglomeration in the composites (Figure 4c), which corresponded to the XRD analysis. It is indicated that there are presentable dispersion and distribution of OMMT nanofiller in PLA matrix at various contents, which can be ascribed to the high pressure stressed of the polymer melt during processing in VE and the ammonium surfactant in the OMMT particle.

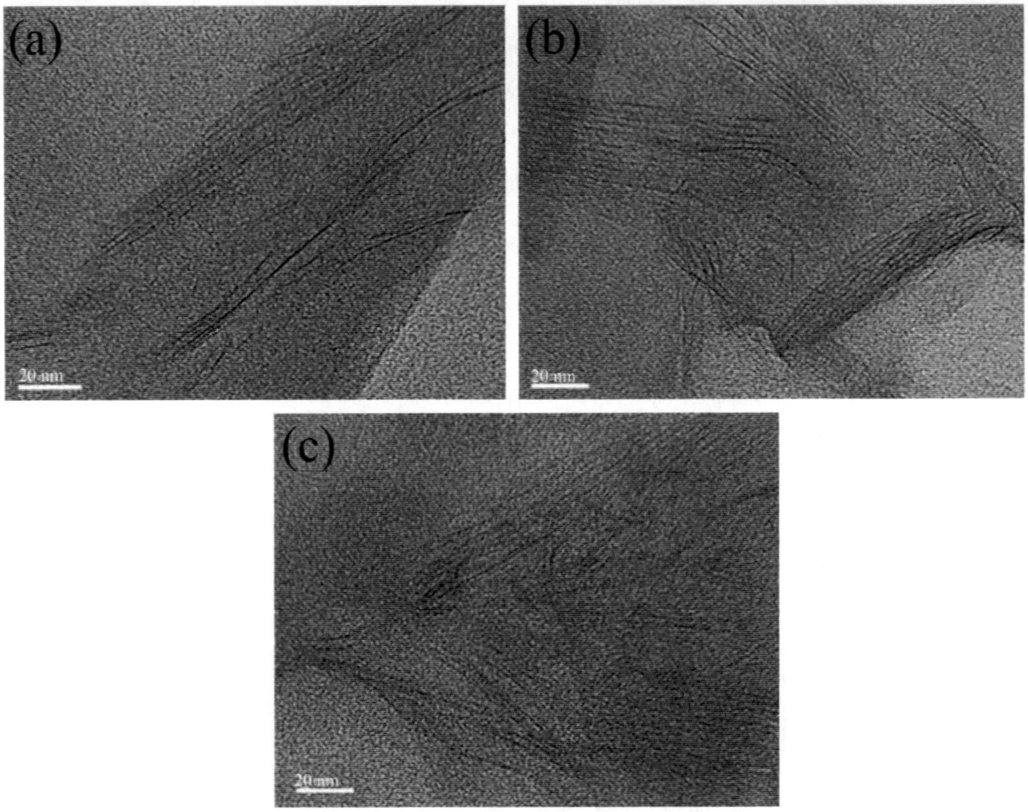

Figure 4. TEM micrographs of PLA nanocomposites with OMMT: (a) 2.5wt% OMMT, (b) 5wt% OMMT, (c) 7.5wt% OMMT.

3.3. Thermal analysis

3.3.1. Crystallization and melting behavior. The DSC thermograms of PLA and PLA/OMMT blends in the first cooling scan after rapid heating at 10°C/min are presented in Figure 5. All thermodynamic parameters can be calculated from the curves, as showed in Table 3. It can be found that the increasing of OMMT concentration decreases the glass transition of PLA (62.4°C) from Table 3. In Figure 5, we can see that the crystallization peak of PLA is found at the peak temperatures about 118.9°C in the crystallization curve. However, the crystallization temperatures of PLA/OMMT nanocomposites decreases at different extent compared to neat PLA. These findings confirm that when rapidly cooled from the molten state, PLA/OMMT nanocomposites are inadequate for crystallization. Moreover, the enthalpy of crystallization of PLA matrix decreases with the loading of OMMT nanofillers. It can be considered that there is a good interplay between PLA matrix and OMMT fillers and large surface areas and volume ratios of the OMMT clays [21]. DSC melting thermograms of PLA and its nanocomposites in the second heat scan after rapidly cooling from the molten situation are presented in Figure 6. In Figure 6, a melting peak at 153°C (T_{m1}) with a shoulder at 163.0°C (T_{m2}), and a cold crystallization peak as well are observed. However, the addition of OMMT transforms these peaks into two separated melting peaks and there are no cold crystallization peaks. Since the imperfect crystals have sufficient time to melt and recombine into crystals with higher structure perfection, and remelt at a higher temperature [21]. Simultaneously, the relatively small melting peak overlap with the crystallization peak, thus the melting of polymer is took place in parallel with the stacking and crystallization of molecular chain. The energy released by the crystallization has been counterbalanced, because the melting rate is greater than the rate of crystallization, and the energy required for melting is greater than the energy released by the crystallization. Thus it is observed of this independent bimodal melting peak in the melting curve of PLA/OMMT nanocomposites. Additionally, the incorporation of OMMT clay induces a clear decrease of the melting temperature of neat PLA (153°C), implying the OMMT fillers can promote the crystallization ability of PLA matrix. Especially, the largest degree of decline of the melting temperature of PLA is occurred in the PLA/OMMT blend with 2.5wt% nanofillers owing to the stronger interaction between organic surfactant of OMMT and the functional groups of PLA chains.

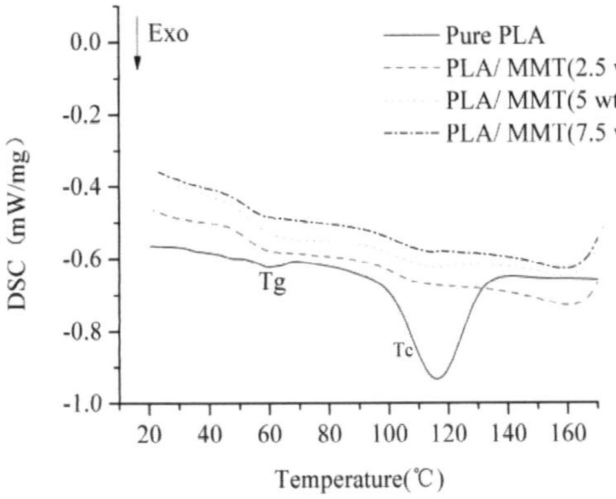

Figure 5. Crystallization curves of PLA and PLA/OMMT nanocomposites.

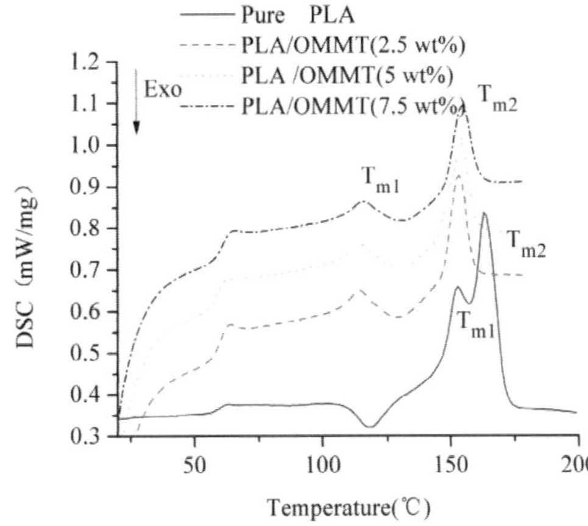

Figure 6. Melt curves of PLA and PLA/OMMT nanocomposites.

Table 3. Characteristic thermal properties of PLA and PLA/OMMT nanocomposties.

OMMT (wt %)	T_g(°C)	T_c(°C)	T_{m1}(°C)	T_{m2}(°C)
0 (Pure PLA)	62.4	118.9	153.0	163.3
2.5	60.8	110.1	113.9	153.5
5	57.6	115.2	114.2	154.7
7.5	56.3	112.7	115.2	154.8

Tg: the glass transition temperatures, Tc: cold crystallization temperature, Tm1, Tm2: the melting temperatures of PLA.

3.3.2. Thermal Gravimetric Analysis (TGA). Thermo-gravimetric analysis was made to study the thermal degradation characteristics of PLA/OMMT blends. Generally, the nanocomposite has most of weight losses in the region of 300-600°C. Figure 7 presents the TGA curves of PLA and PLA/OMMT nanocomposites. The initial degradation temperatures (T_d, at 10% weight loss), the degradation temperature at 50% weight loss ($T_{50\%}$) and the weight of the residue at 600°C (ω_t^{600}) are shown in Table 4. From the curves and the table, the T_d of the PLA/OMMT blends increases slightly with the loading of OMMT, suggesting that the incorporation of silicate nanofiller into PLA matrix can cause an enhancement of the thermal stabilities of the polymer matrix. The ω_t^{600} increases gradually with the increasing clay loadings of OMMT. This result can be explained by the good thermal stability of OMMT and to the interaction between the polymer matrix and the clay particles. Moreover, this changing tendency of the ω_t^{600} with the OMMT concentration is ascribed to the high heat resistance exerted by the OMMT clay itself. Similar phenomenon was also reported by some researchers [22-23]. In addition, the $T_{50\%}$ increases gradually with the clay content with a maximum increase of 29°C from the value of neat PLA (373°C for the 5wt% OMMT and of 26°C, 23°C for the 2.5wt%, 7wt% respectively. With increasing of OMMT concentration, a slight decrease has been eyed in the nanocomposites based on 7.5wt% clay loadings. Such phenomenon can be explained by that increasing the incorporation of OMMT silicate nanofiller into PLA matrix results in relatively more exfoliated particles and enhances the thermal stabilities of the PLA/OMMT blends.

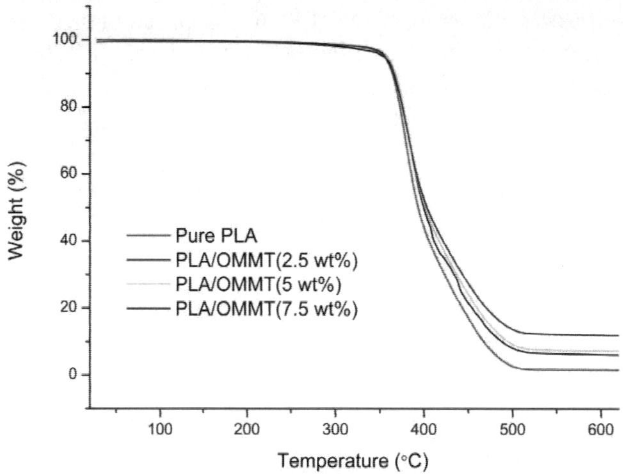

Figure 7. TGA thermograms for PLA and PLA/OMMT nanocomposites.

Table 4. TGA for PLA and PLA/OMMT nanocomposites.

OMMT (wt %)	T_d(°C)	$T_{50\%}$ (°C)	$\omega_t^{600}(\%)$
0 (Pure PLA)	365	373	0.22
2.5	369	399	6.49
5	372	402	7.53
7.5	368	396	12.20

4. Conclusions

PLA/OMMT nanocomposites with different OMMT content were successfully prepared by melt-blend using a novel VE. The effect of OMMT concentration on morphological, thermal properties of the PLA/OMMT blends has been discussed. Morphological studies show that intercalation/ exfoliation structure in the PLA matrix was revealed, and good dispersions of OMMT nanoparticles were observed in all samples with various OMMT concentrations. The results can be explained by the high elongation stressed of the polymer melt in vane extruder. Also with the loading of OMMT increases, the crystallization ability of the blends enhanced. Furthermore, it is indicated that the incorporation of OMMT clay enhanced the thermal stability of PLA/OMMT nanocomposites, thus, providing the possibility of application of PLA/OMMT composites at a higher temperature than pure PLA.

Acknowledgments

We acknowledge the National Natural Science Foundation of China (Grant No. 51435005, 51505153), PhD Start-up Fund of Natural Science Foundation of Guangdong Province, China (Grant No. 2016A030310429), the Special-funded Program on National Key Scientific Instruments and Equipment Development of China (Grant No. 2012YQ230043) and Science and Technology Program of Guangzhou, China (Grant No.201607010240) for financial support.

References

[1] Bordes P, Pollet E and Avérous L 2009 *Prog. Polym. Sci.* **2** 125
[2] Rasal R M and Hirt D E 2009 *Macromol. Biosci.* **9** 989
[3] Zhang Z and Feng S S 2006 *Biomaterials* **27** 4025
[4] Rasal R M, Janorkar A V and Hirt D E 2010 *Prog. Polym. Sci.* **35** 338

[5] Nam P H, Fujimori A and Masuko T 2004 *J. Appl. Polym. Sci.* **93** 2711
[6] Pluta M, Galeski A, Alexandre M, Paul M A and Dubois P 2002 *J. Appl. Polym. Sci.* **86** 1497
[7] Cao H, Wang P, Yuan W and Lei H 2010 *J. Appl. Polym. Sci.* **115** 1468
[8] Ogata N, Jimenez G, Kawai H and Ogihara T 1997 *J. Polym. Sci., Part B: Polym. Phys.* **35** 389
[9] Wu T M and Wu C Y 2006 *Polym Degrad Stabil.* **91** 2198
[10] Jonoobi M, Harun J, Mathew A P and Oksman K 2010 *Compos. Sci. Technol.* **70** 1742
[11] Nam J Y, Ray S S and Okamoto M 2003 *Macromolecules.* **36** 7126
[12] Ray S S and Okamoto M 2003 *Prog. Polym. Sci.* **28** 1539
[13] Tokihisa M, Yakemoto K, Sakai T, Utracki L A, Sepehr M, Li J and Simard Y 2006 *Polym. Eng. Sci.* **46** 1040
[14] Scott C E and Macosko C W 1991 *Polym. Bull.* **26** 341
[15] Qu J P 2008 *C.N. Patent* 200,810,026,054.X
[16] Jia S K, Qu J P, Wu C R, Liu W F, Chen R Y, Zhai S F, Huang Z and Chen F Q 2013 *Langmuir* **29** 13509
[17] Qu J P, Zhang G Z, Chen H Z, Yin X C and He H Z 2012 *Polym. Eng. Sci.* **52** 2147
[18] Qu J P, Zhao X Q, Li J B and Cai S Q 2013 *J. Appl. Polym. Sci.* **127** 3923
[19] Jia S, Qu J P, Zhai S, Huang Z, Wu C R, Chen R Y and Feng Y H 2014 *Polym. Compos.* **35** 884
[20] Ray S S, Okamoto K and Okamoto M 2003 *Macromolecules* **36** 2355
[21] Jiang L, Zhang J and Wolcott M P 2007 *Polymer* **48** 7632
[22] Chang J H and An Y U 2002 *J. Polym. Sci. B Polym. Phys.* **40** 670
[23] Chang J H, An Y U and Sur G S 2003 *J. Polym. Sci. B Polym. Phys.* **41** 94

IOP Conf. Series: Materials Science and Engineering **213** (2017) 012019 doi:10.1088/1757-899X/213/1/012019

Surface analysis of Fe-Co-Mo electrolytic coatings

G Sh Yar-Mukhamedova[1,3], N D Sakhnenko[2], M V Ved'[2], I Yu Yermolenko[2] and S I Zyubanova[2]

[1]Physics Department, Al-Farabi Kazakh National University, IETP, 71 Al-Farabi avenue, Kazakhstan
[2]National Technical University "Kharkiv Polytechnic Institute", Kirpychova, 2, Kharkov, 61002, Ukraine

E-mail: gulmira-alma-ata@mail.ru

Abstract. Coatings Fe-Co-Mo with a composition of 47 at.% iron, 28 at.% Cobalt and 25 at.% Molybdenum were deposited from citrate electrolyte using pulse electrolysis mode. Scanning electron and atomic force microscopy have established the surface morphology and topography. It was identified the parts with a globular structure which have an average size of 0.2-0.5μm and singly located sharp grains. Within the same scan area sites with developed surface were detected the topography of which is identical to the crystal structure of cobalt with the crystallites size of 0.2–1.75μm. The parameters Ra and Rq for parts with different morphology as well as average characteristics of coatings demonstrated the low roughness of the surface. It is found that the coercive force of Fe-Co-Mo films is 7-10 Oe, which allow us to classify the Fe-Co-Mo coatings as soft magnetic materials.

1. Introduction

Surface of machine and mechanism parts in the operation primarily exposed for external factors influence that cause wear of friction parts, the emergence of fatigue cracking, crushing and at the end - corrosion destruction. One of the common surface protection methods is electroplating wherein coatings by binary and ternary alloys of iron subgroup metals with refractory components are of great practical interest [1-2]. Such coatings are distinguished due to combination therein functional properties that exceed the corresponding parameters for alloying metals [3]. The complex implementation in thin layers of wear and corrosion resistance [4], catalytic [5] and magnetic properties [6] combined with high microhardness makes such coatings universal and allows to significantly expanding their application area. It was shown prospects for usage of Co-Mo systems in the production of magnetic head elements for recording and reproducing information [6].

The functional properties of the coatings are structurally dependent therefore composition and surface morphology will predetermine the performance behavior of materials. The roughness and surface friction are the main characteristics of the coatings quality. Previously, it was shown [7] that binary and ternary coatings obtained in a pulsed mode are characterized by high microhardness and wear resistance due to their smooth surface. The roughness of electrolytic deposits can be significantly decreased using pulse mode. Since the surface structure of the electrolytic alloys causes the properties and application of coatings, study of their morphology and topography remains relevant. In spite of a sufficient number of works devoted to the binary alloy Fe (Co, Ni)-W (Mo) [8–10] matters concerning the topography and morphology of ternary coatings as well as their influence on the properties requires detailed study.

The aim of this work is to study the surface morphology and magnetic properties of Fe-Co-Mo ternary coatings.

2. Experiment

The Fe-Co-Mo coatings were deposited onto a mild steel substrate from citrate electrolyte containing, g/dm^3: $Na_3C_6H_5O_7 \times 2H_2O$ – 95–100; $Fe_2(SO_4)_3 \times 9H_2O$ – 30–45; $Na_2MoO_4 \times 2H_2O$ – 15–25; $CoSO_4 \times 7H_2O$ – 30–45; Na_2SO_4 – 15–45; H_3BO_3 – 6. Pretreatment of samples includes mechanical polishing, polishing, degreasing, chemical etching in a mixture of the 50 % nitric and 50 % sulfuric acids, thorough washing with distilled water and drying. Electroplating was performed by unipolar pulse current with an amplitude of 3–4 A/dm^2 in the working range of the pulse t_{on} 10–20 ms and pause duration t_{off} 10–50 ms. The ratio of the cathode to the anode area was 1 : 5, volume current density was kept at the level 2 A/dm^3. The thickness of the coatings is controlled by the deposition time and is 8–10 μm.

The chemical composition of the coatings was determined by X-ray fluorescence method using a portable spectrometer "SPRUT" with a relative standard deviation of 10^{-3}–10^{-2}. The error at determining the content of the components is ± 1 wt.%. To analyze the phase the energy-dispersive X-ray spectroscopy was performed using an electron probe micro analyzer Oxford INCA Energy 350 integrated into the SEM system.

The surface morphology of Fe-Co-Mo thin films was studied by an atomic force microscopy AFM using NT–206 microscope. The tapping mode was conducted to measure samples surface morphologies. Scanning was performed by using the contact probe CSC-37 with a cantilever lateral resolution of 3 nm [11]. And the scan sizes were fixed at 39.9×39.9 μm and 10.0×10.0 μm and the height of the surface relief was recorded at a resolution of 256×256 pixels. For each sample a variety of scans were obtained at random locations on the surface of Fe-Co-Mo thin films. In order to analysis the AFM images all image data were converted into Surface Explorer software. The root mean square (R_q), mean particle height and its distribution, surface skewness and particle diameter were determined [12].

Magnetic properties of Fe-Co-Mo coatings of thickness 8 μm were studied using a vibrating magnetometer with fields up to 20 kOe at room temperature. Accuracy of magnetic characteristics measurement is ± 2%. The coercive force H_c and the saturation field H_s were defined by hysteresis loops measured in the fields applied to the coating plane. The saturation magnetization I_s and the saturation induction B_s were estimated by hysteresis loops obtained for the magnetization of specimens by the film plane normal.

3. Results and discussion

The analysis of SEM images (Figure 1) shows the globular structure of Fe-Co-Mo coatings surface wherein agglomerates of larger size 2–5μm are formed from smaller spheroids of size 0.2-0.5μm. It was shown [1, 3, 7] that globular structure of the surface is caused by the refractory metals presence in the alloy. Such composition and character of the surface it was found to be favorable for increasing microhardness, corrosion resistance, and catalytic activity of the material [13-14].

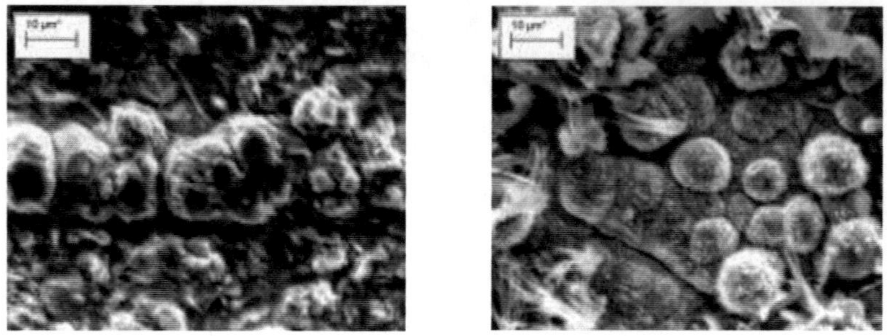

Coatings composition, at.%

Fe – 46.7; Co – 28.9; Mo – 24.4   Fe – 46.1; Co – 29.1; Mo – 24.8

a   *b*

Figure 1. SEM image and composition of Fe-Co-Mo coatings. Thickness 8 μm; pulse current density *i*, A/dm^2: a – 3; b – 4.

Topography maps of Fe-Co-Mo coatings reflect that their surface includes the parts A and B of different morphology as one can see from Figure 2. The torsion of measuring console and substantial changes in skin friction at these sites within the same scan area indicate non homogeneity of the material surface [12]. In this regard, the more detail study of the morphology of parts A and B is of great importance.

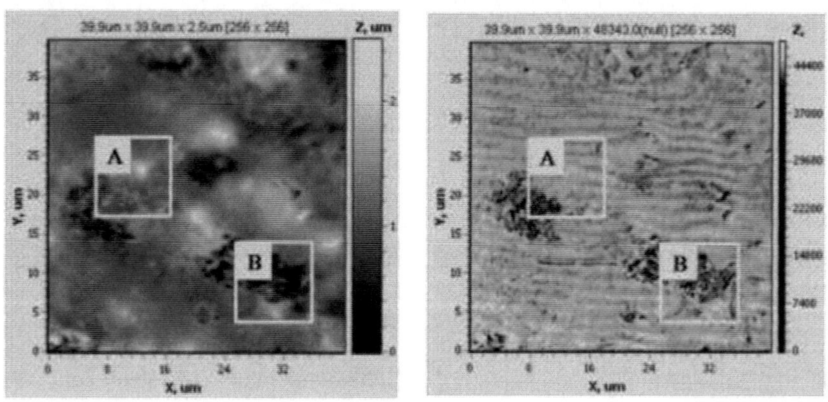

a   b

Figure 2. 2D-topography (a) and 2D-torsion (b) maps of the surface for Fe-Co-Mo coating deposited in pulse mode. Scan area AFM 39.9×39.9μm.

Figure 3 represents 3D- and 2D maps of the surface and cross section profile between markers 1 and 2 for different sites of coating Fe-Co-Mo. The part A differs by a surface consisting of spheroids (globules) 0.2–0.5 μm in size (Figure 3 *a*) and singly located cone-shaped hill with a base diameter of about 3 μm and height of 2.5 μm (Figure 3 *a*, *c*). As appears from 2D- and 3D- topography maps of the surface (Figure 3 *a*, *b*) the cone-shaped hills are formed of the smaller spheroids. The analysis of histogram of distribution the inclination angle to the surface normal of 17.5° indicates a predominance of spheroids on the surface (Figure 4*a*).

The part B is characterized by more developed surface compared with a part A (Figure 3 *d*). The morphology of the surface similar to pure cobalt [5] with sufficiently sharp hills alternating by valleys is visualized on the 2D- and 3D- maps of the site B (Figure 3 *d*, *f*). The cross section profile of coating between markers 1 and 2 allows determining the average size of agglomerates on the surface (Figure 3 c, g). This site B is characterized by a smaller difference in the size of the agglomerates of 0.5–1.75 μm and the heights of the hills and valleys of 0.5–0.7 μm. Histogram of distribution the inclination angle to the surface normal in this site demonstrates the uniform distribution of the sharp hills of different height (Figure 4*b*).

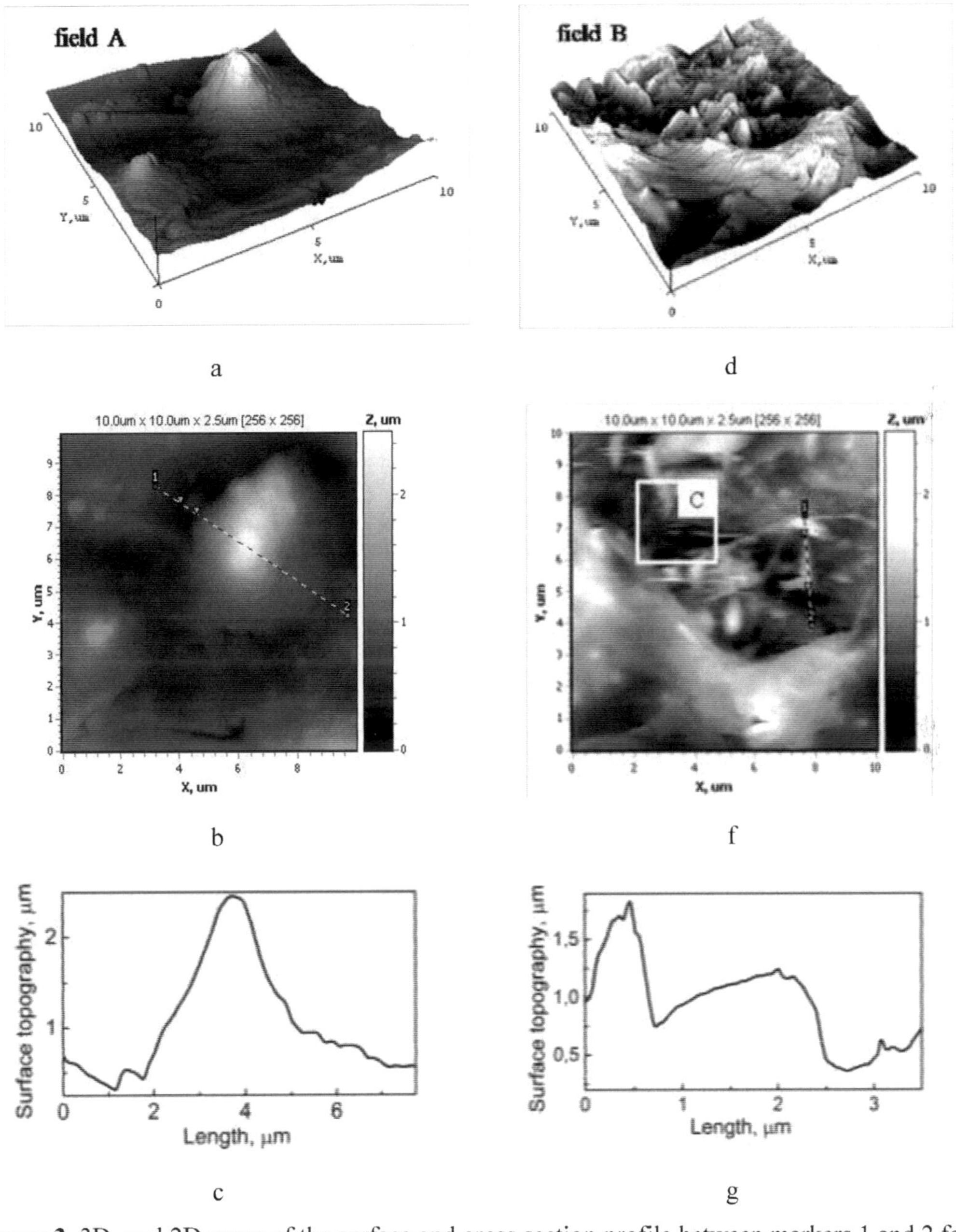

Figure 3. 3D- and 2D maps of the surface and cross section profile between markers 1 and 2 for site A (a, b, c) and site B (d, f, g) of coating Fe-Co-Mo. Scan area AFM 10.0×10.0μm. Thickness is 8μm.

IOP Conf. Series: Materials Science and Engineering **213** (2017) 012019 doi:10.1088/1757-899X/213/1/012019

Micro pores (site C Figure 3*f*) are visualized at the coating surface in the area adjacent to the site B. Observed surface and volume heterogeneity causes anisotropy of the coating which is a prerequisite for the formation of different magnetic phases [6, 15]. The parameter R_q for part A and part B was defined as 0.35 and 0.30 respectively, reflecting the greater roughness of the part A caused by higher hills. However the distinction in the parameters R_q for parts with different morphology does not affect the average roughness of the coatings surface which is R_a=0.25. As follows from the values of R_a and R_q the surface has 8–9 grade of roughness.

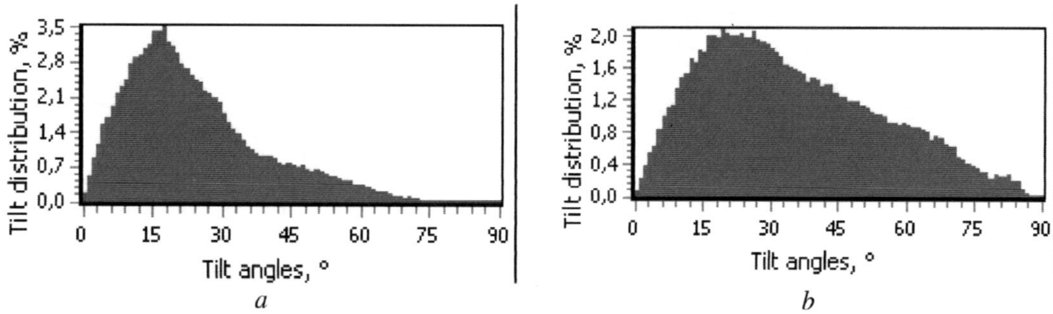

Figure 4. Histogram of the angles to the surface normal distribution for the respective topography sites A (*a*) and B (*b*).

As one can see from Figure 5 the shape of hysteresis loop in the saturation interval is smoothed that indicates the presence of an amorphous structure in the electrolytic deposits [10]. At the same time we observe the saturation of magnetization as well as demagnetization which is stepwise that confirms the presence of two magnetic phases in the coating (Figure 2).

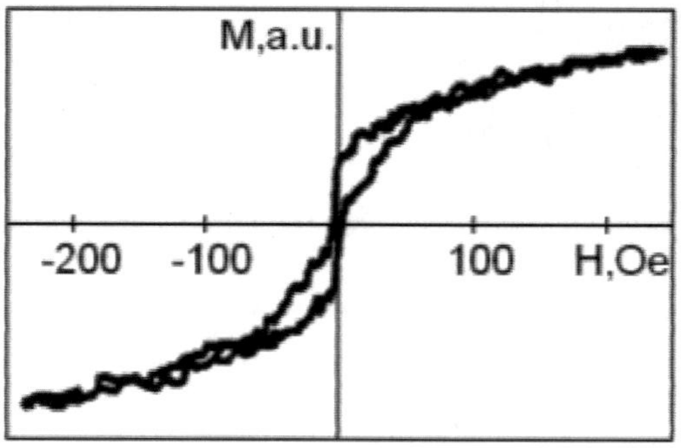

Figure 5. Hysteresis loop of Fe-Co-Mo coating.

It is found that the coercive force of Fe-Co-Mo films is 7-10 Oe, which allow us to classify the Fe-Co-Mo coatings as soft magnetic materials.

4. Conclusions

(i) Coatings Fe-Co-Mo containing 47 at.% iron, 28 at.% cobalt and 25 at.% molybdenum deposited by pulse current amplitude of 3–4 A/dm^2 are characterized by the surface containing different sites with globular structure alternating by sharp hills. The surface of Fe-Co-Mo coatings has 8–9 grade of roughness.

(ii) The observed anisotropy of the coating caused by both surface and volume heterogeneity is a prerequisite for the formation of different magnetic phases.

(iii) The coercive force of Fe-Co-Mo films is 7-10 Oe, which allow us to classify the coatings as soft magnetic materials.

References

[1] Tsyntsaru N *et al* 2009 *Powder Metall. Met. Ceram.* **48** 419
[2] Karakurkchi A, Ved' M, Yermolenko I and Sakhnenko N 2016 *Surf. Eng. Appl. Electrochem.* **52** 43
[3] Tsyntsaru N, Cesiulis H, Donten M, Sort J, Pellicer E and Podlaha-Murphy E 2012 *Surf. Eng. Appl. Electrochem.* **48** 491
[4] Karakurkchi A, Ved' M, Sakhnenko N and Yermolenko I 2015 *Russ. J Appl. Chem.* **88** 1860
[5] Glushkova M, Bairachna T, Ved' M and Sakhnenko M 2013 *MRS Proc.* **1491** 18
[6] Gomez E, Pellicer E and Valles E 2001 *J. Electroanal. Chem.* **517** 109
[7] Karakurkchi A, Ved' M, Sakhnenko N, Yermolenko I, Zyubanova S and Kolupayeva Z 2015 *Funct. Mater.* **22** 181
[8] Tsyntsaru N, Belevskii S, Volodyna G, Bersirova O, Yapontseva Yu, Kublanovskii V and Dikusar A 2007 *Surf. Eng. and Appl. Electrochem.* **5** 9
[9] Bobanova Zh, Petrenko V, Volodyna G, Grabko D and Dikusar A 2011 *Surf. Eng. and Appl. Electrochem.* **47** 17
[10] Dragos O, Chiriac H, Lupu N, Grigoras M and Tabakovic I 2016 *J. Electrochem. Soc.* **163** 83
[11] Burnham N and Colton R 1989 *J. Vac. Sci. Technol.* **4** 2906
[12] Labardi M, Allegrini M, Salerno M, Fredriani C and Ascoli C 1994 *Appl. Phys.* **59** 3
[13] Yar-Mukhamedova G, Ved' M, Sakhnenko N, Karakurkchi A and Yermolenko I 2016 *Appl. Surf. Sci.* **383** 346
[14] Yar-Mukhamedova G, Ved' M, Sakhnenko N and Koziar M 2017 *Appl. Surf. Sci.* https://doi.org/10.1016/j.apsusc.01.196
[15] Das K, Bhattacharyya A, Kulkarni R, Dhar S and Thamizhavel A 2014 *Phys. Rev. B* **89** 134418

Mixed alumina and cobalt containing plasma electrolytic oxide coatings

G Sh Yar-Mukhamedova[1,3], M V Ved'[2], A V Karakurkchi[2] and N D Sakhnenko[2]

[1]Physics Department, Al-Farabi Kazakh National University, IETP, 71 Al-Farabi avenue, Kazakhstan

[2]Physical Chemistry Department, National Technical University "Kharkiv Polytechnic Institute", 2 Kyrpychova Street, Kharkiv,61002, Ukraine

E-mail: gulmira-alma-ata@mail.ru

Abstract. Principles of plasma electrolytic oxidation of the AL25 aluminum alloy in diphosphate alkali solutions containing cobalt(2+) cations are discussed. It has been established that a variation in the concentration of the electrolyte components provides the formation of mixed-oxide coatings consisting of the basic matrix materials and the cobalt oxides of different content. An increase in the cobalt oxide content in the coating is achieved by the variation in electrolysis current density as well as the treatment time due to both the electrochemical and thermo-chemical reactions at substrate surface and in spark region. Current density intervals that provide micro-globular surface formation and uniform cobalt distribution in the coating are determined. The composition and morphology of the surface causes high catalytic properties of synthesized materials, which confirmed the results of testing in model reaction CO and benzene oxidation as well as fuel combustion for various modes of engine operation.

Keywords: AL25 alloy; Cobalt oxides; Plasma electrolytic oxidation; Mixed oxide system; Catalytic activity

1. Introduction

Environmental and technical problems accompanying the operation of internal combustion engines (ICE) are associated with the incomplete combustion of fuel and, accordingly, the inevitable formation of toxic components (CO, NOx, and soot). One of the perspective ways to improve the performance of ICE can be the use of catalysis in situ in the combustion chamber [1]. Catalytic processes are characterized by lower activation energies and consequently temperatures of the fuel burning start as well as its higher completeness [2]. These lead to a decrease in the maximum pressure in the combustion chamber and, hence, in the severity of engine operation and to the minimization of emissions of hazardous substances with exhaust gases. The first paragraph after a heading is not indented.

The currently available hydrocarbon fuel combustion catalysts are divided into two groups: catalysts based on noble metals (most commonly Pd and Pt) and transition-metal oxides (Mn, Co, Fe, etc.), particularly those with a complex composition. At high temperatures, significant advantages are shown by a family of catalysts based on d-metals in particular cobalt oxides, which exhibit thermal lability and high oxygen affinity [3]. Non stoichiometric cobalt oxides not only provide the thermal stability of the catalysts but also lead to a significant improvement of its reactivity.

In terms of rational designing and effective technology, it seems appropriate to deposit a catalytic layer directly on the surface of ICE pistons manufactured using AL25 alloys with an optimal combination of physic-mechanical and performance properties [4]. The most promising method for the formation of thin layers on aluminum alloys, in particular, silumins, is plasma electrolytic oxidation (PEO). This method provides the incorporation of catalytically active components into an alumina matrix owing to the implementation of both electrochemical and thermo-chemical reactions in high-energy modes [5-6]. However, it should be taken into account that the PEO of silumins is complicated because these materials contain a large number of alloying components and inter-metallic compounds with different chemical properties of oxides. Previously, we have shown PEO expediency of the AL25 alloys in a diphosphate solution with the addition of manganese salts to produce mixed manganese and aluminum oxides of different compositions [7]. On this basis, the aim set in this paper is a study of the electrolyte composition and the PEO modes for the formation of a mixed oxide layer with high cobalt content on the surface of the AL25 alloy.

2. Experimental

The studies were conducted using pure aluminum and AL25 alloy samples with a working surface of 2.0cm^2. The concentration alloying elements in a substrate, mass %: Si – 11.0-13.0, Cu – 1.5-3.0, Ni – 0.8-1.3, Mg – 0.8-1.3, Mn – 0.3-0.6, Fe<0.8, Zn<0.5, Ti<0.2, Cr<0.2, Pb<0.1. The surface preparation included sequential procedures of mechanical grinding, etching in an alkaline solution, thorough washing with distilled water, and drying. Electrolyte solutions for the pretreatment and PEO were prepared from certified reagent-grade reactants and distilled water.

The PEO of samples was conducted using a B5-50 stabilized power supply that maintained voltage up to 300V. PEO was implemented in an electrolytic cell under conditions of forced cooling of the electrolyte to a temperature of 20–30°C at a current density varied in a range of i=2–10 A·dm^{-2} up to final formation voltage 180–235V; duration of oxidation process was to 40–60 minutes. PEO was implemented from working electrolytes composition, g·dm^{-3}: $K_4P_2O_7$ – 65.0–300.0; $CoSO_4$ – 7.5–46.5.

The chemical composition of the coatings was determined by X-ray photoelectron spectroscopy on an INCA Energy 350 energy dispersive spectrometer; X-ray radiation was excited via exposing the samples to a 15-keV electron beam. In addition, X-ray fluorescence analysis was conducted using a SPRUT portable spectrometer with a relative standard deviation of 10^{-3}–10^{-2}; the error in determining the component content was ±1 mass %. The surface morphology of the coatings was studied using a Zeiss EVO 40XVP scanning electron microscope (SEM). Images were recorded by the registration of secondary electrons (BSEs) via scanning with an electron beam; this mode made it possible to study the topography with a high resolution and contrast ratio. The surface roughness was evaluated by the contact method on 10 x 10 x 2 mm samples with an NT-206 scanning probe AFM microscope (lateral and vertical resolutions 2 and 0.2 nm, respectively; 1024 x 1024 scanning matrix, CSC cantilever B as probe, probe tip radius 10 nm).

The catalytic properties of the resulting oxide systems were studied in the oxidation of carbon(II) oxide to carbon(IV) oxide and in the flameless oxidation of benzene with air [8].

3. Results and discussion

The results of experimental studies found that PEO process of alloy AL25 in diphosphate electrolytes containing cobalt(II) is a classic case. It means that the voltage chronograms are divided into four sections: pre-spark, spark, microarc and arc. The pre-spark section is almost linear, which is explained by the growth of barrier alumina film and a corresponding increase in the resistance. In the transition to the sparking section the U, t dependence becomes non linear, and the voltage growth slows down significantly, apparently due to the inclusion of the electrolyte components into the coating. This is particularly evident in the formation of coatings in cobalt-containing solutions when blue freeforms begin to appear in the anode areas, primarily occurring at sharp edges and in defect surface sites. Simultaneous progress of electrochemical oxidation and thermal decomposition of electrolyte components implemented in the mode of PEO contributes high adhesion of an oxide coating to the substrate and incorporation of catalytic component – cobalt oxides. Composition of multi component oxides depends both on the concentration ratio of the electrolyte components and the oxidation current density. Thus, the oxide system is logically enriched with cobalt by increasing its content in the solution, but the electrolyte stability as well as service time is reduced. At the same time, increasing the diphosphate concentration in the bath is not desirable because it leads to a decrease of PEO efficiency. This is due to the formation of stable complex ions $[Co(P_2O_7)_2]^{6-}$ in excess of the ligand, which is why the introduction of cobalt into the anode material is difficult.

It was also established that the cobalt content in mixed oxide coatings increases when PEO is performed in the incident power mode. Starting current density i_1 is higher for forming barrier aluminum oxide film on the AL25 substrate, which is accompanied by oxidation of the alloying components and particularly silicon with soluble substances forming and promotes homogenization of the alloy surface [9]. After the process enters a mode of stable arcing current density for the second stage i_2 is reduced to ensure a formation of the mixed oxide layer. The current density for the first PEO stage in cobalt-diphosphate solution does not exceed the interval 8-10 $A \cdot dm^{-2}$, while the second stage is advisable to hold at current density of 2.5-3.0 $A \cdot dm^{-2}$. Arcing voltage U_i of PEO process in electrolyte 200 $g \cdot dm^{-3}$ $K_4P_2O_7$ and 30 $g \cdot dm^{-3}$ $CoSO_4$ is 115-120 V, and the resulting formation voltage U_f is 140–160 V. It should be noted that all voltage parameters grow with the electrolyte dilution and with increasing current density. The cobalt content in the mixed oxides increases with the current density and uniformity of its distribution on the surface is improved [10].

Surface morphology and composition of PEO coatings doped with cobalt oxides both depends on the concentration of electrolyte and oxidation conditions – current density, voltage and time of formation (Figure 1). The incorporation of the cobalt in oxide systems affects the formation of separate spheroid islets of characteristic blue-purple color. During the treatment the number of islet structures significantly increases, some the islets grow together, agglomerated and form a ceramic like mosaic patterns. The higher content of cobalt the more developed but at the same time the less uniform is the surface of coating (Figure 1 a, b). Formed oxide systems with cobalt content 8–30 at.% and low concentration of impurities have high developed micro globular surface, which is essential for the high catalytic properties of the material [5]. Analysis of the composition of oxide systems indicates non stoichiometric ratio of cobalt and oxygen which is favorable for the implementation of catalytic processes involving oxygen.

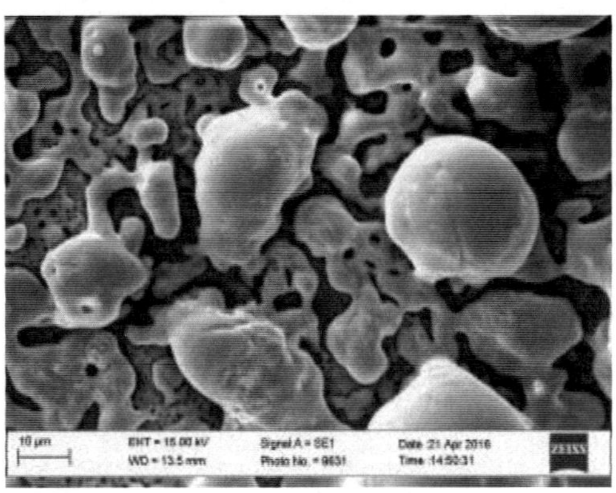

Al – 48.7; Co – 23.6; O – 25.9; Si – 1.8

a

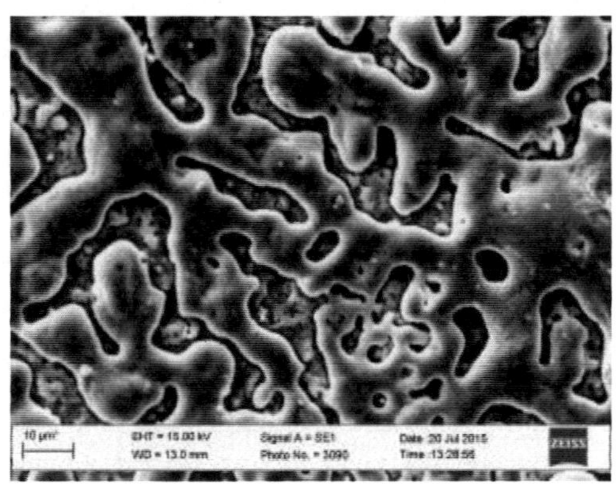

Al – 33.3; Co – 8.7; O – 55.2; Si – 2.8

b

Figure 1. Surface morphology and composition (at.%) of mixed oxide systems $Al | Al_2O_3 \cdot CoO_x$ obtained in PEO two stage mode (i_1=8 $A \cdot dm^{-2}$, i_2=2.5 $A \cdot dm^{-2}$) in diphosphate electrolytes (200 $g \cdot dm^{-3}$ $K_4P_2O_7$) with $CoSO_4$ concentration, $g \cdot dm^{-3}$: a – 30; b – 10. Magnification 2000.

IOP Conf. Series: Materials Science and Engineering **213** (2017) 012020 doi:10.1088/1757-899X/213/1/012020

It was established by AFM analysis that the surface topography of Al | $Al_2O_3 \cdot CoO_x$ system with cobalt content of 8–12at.% (Figure 2) is characterized by the high roughness and large grain size. Mixed oxide coatings are composed from the set of spheroid grains with an average size of 1-2μm which gather and form the parts of smooth surface with the spherical grain agglomerates of 6-8μm (Figure 2a). The surface roughness may be characterized with the spread of grain size on the cross section between markers 1-2 (Figure 2b). In addition, the surface topography of oxides is characterized by rather large difference between hills and valleys of 0.3μm up to 1.2μm and alternating the spheroid and torus-shaped structures (Figure 2c).

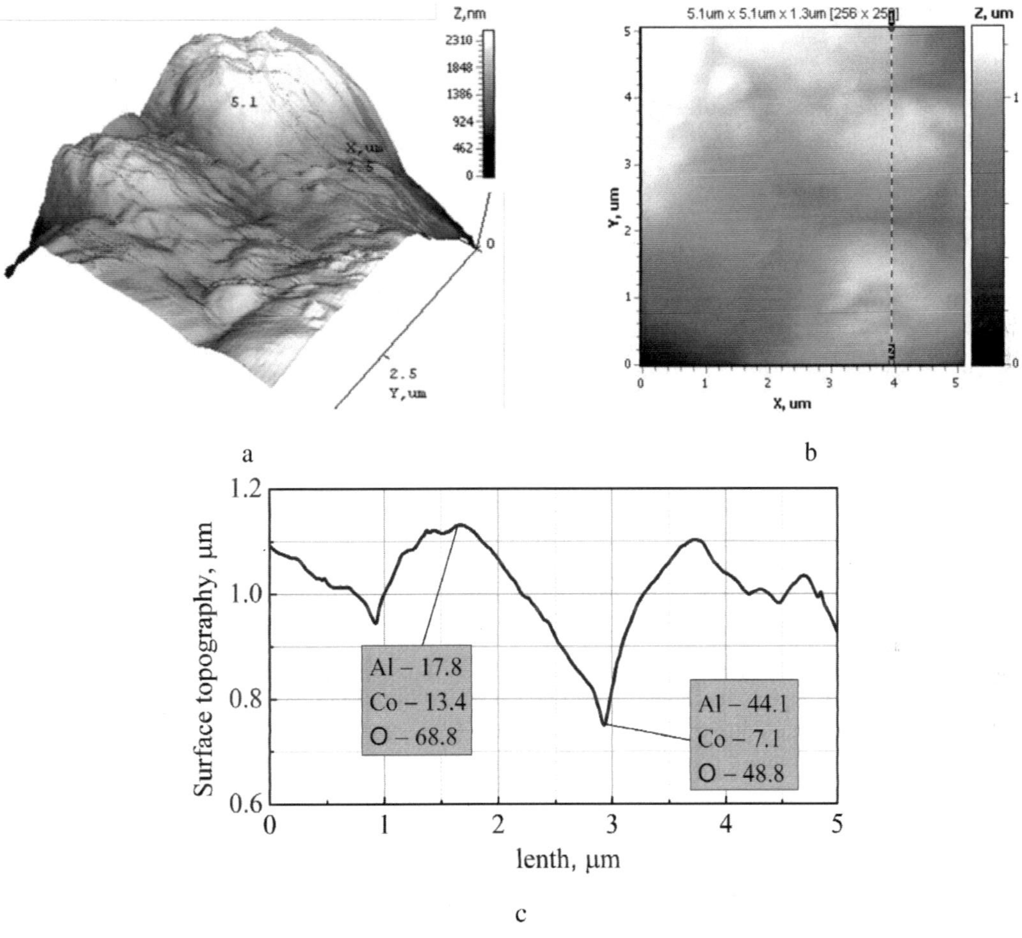

Figure 2. 3D- (a) and 2D (b) maps of the surface and cross section profile (c) between markers 1 and2 for mixed oxide system Al | $Al_2O_3 \cdot CoO_x$ with cobalt content 8.7 at.%. Scanning area 5 x 5 μm.

Testing of the synthesized oxide systems in the model reaction of carbon(II) oxide oxidation to CO_2 showed that the ignition temperature is 230 °C and the complete conversion is achieved at 305 °C which is comparable with the indicators of platinum catalysts 217 °C and 297 °C respectively [7]. Completeness of model reaction of flameless oxidation of benzene at Al | $Al_2O_3 \cdot CoO_x$ catalyst is also higher than at palladium-containing ones. For instance the degree of benzene oxidation, which is determined by the amount of resulting CO_2, for example, at a temperature of 300 °C, is 5.0 times higher than on contacts $SiO_2 \cdot Pd$ (with Pd content about 5 wt.%). Tests of materials based on mixed oxide systems in model reactions of CO and benzene oxidation have shown that their catalytic activity is not inferior to the contacts that contain noble metals. This can be attributed to several factors: firstly due to a high affinity of non-stoichiometric cobalt oxides to oxygen and the corresponding decrease in the activation energy of destruction double bonds O=O [11-12]. Secondly, the developed micro globular surface of oxide systems is characterized by a large number of catalytic centers and provides an increase in the number of elementary collisions and interactions in multicomponent systems.

Pilot tests of mixed oxides deposited on the surface of combustion chambers piston of ICE confirm the change in fuel consumption and NO_x emissions at forced diesel engine operating modes (Table 1). Analysis of data shows that the use of a catalytic coating based on manganese oxide compared to serial piston coated with $Al_2O_3{\cdot}SiO_2$ virtually unchanges hour fuel consumption (Table 1). However, the catalytic coating based on cobalt oxide can reduce diesel fuel consumption within 1–3%. Catalytic coating based on manganese oxide has almost no effect at NO_x emissions (Table 1), but cobalt oxide system can reduce NO_x emissions by 4–5% at loads when the emission of nitrogen oxide is most significant namely at crankshaft torque 40–70 N·m or loads level at 50–100 %. This is due to lower ignition temperature of the air-fuel mixture on $Al_2O_3{\cdot}CoO_x$ catalyst composed of non-stoichiometric oxides and their more developed surface comparing with $Al_2O_3{\cdot}MnO_x$ system.

Table 1. Indicators for fuel combustion in the chamber of ICE.

Composition of catalytic layer	NO_x emission (ppm)					Fuel consumption ($kg{\cdot}h^{-1}$)				
	Crankshaft torque (N·m)									
	15	30	40	55	70	15	30	40	55	70
$Al_2O_3{\cdot}SiO_2$	600	1,200	1,750	2,200	2,400	1.25	1.60	2.00	2.4	3.00
$Al_2O_3{\cdot}MnO_x$	590	1,200	1,730	2,160	2,370	1.25	1.61	1.99	2.38	2.98
$Al_2O_3{\cdot}CoO_x$	575	1,150	1,660	2,090	2,280	1.23	1.57	1.96	2.34	2.90

Thus we can conclude that the varying conditions of the PEO process allows formation of oxide coatings of different composition and morphology that are expected to affect their functional properties including catalytic activity in the process of burning fuel in internal combustion engines, and reduce the toxicity of emissions improving fuel efficiency and environmental performance of engines.

4. Conclusions

(i) Mixed oxide systems $Al_2O_3{\cdot}CoO_x$ were obtained on Aluminum alloy AL25 by plasma-electrolytic oxidation in a diphosphate electrolyte added with cobalt(II) sulfate. It was shown the composition and surface morphology of oxides strongly depends on the bath concentration and PEO regimes. It was established non stoichiometric ratio of cobalt and oxygen in deposits as well as that the cobalt content in mixed oxide increases when PEO is performed in the incident power mode.

(ii) Obtained mixed oxide systems have developed micro globular surface containing torus-shaped structures which roughness is caused by cobalt content. Such morphology is characterized by a large number of catalytic centers.

(iii) Synthesized oxides exhibit catalytic properties not inferior to the contacts that contain noble metals and can be recommended for use in the neutralization of gas emissions systems, as coatings for ICE combustion chamber for the removal of toxins of various natures.

References

[1] Gardiner W C 1984 *Combustion Chemistry* (New York: Springer-Verlag)

[2] Stiles A B 1987 *Catalyst Supports and Supported Catalysts: Theoretical and Applied Concepts* (Stoneham, MA: Butterworth)

[3] Glushkova M, Bairachna T, Ved' M and Sakhnenko M 2013 *MRS Proc.* **1491** 18

[4] Glazoff M V, Zolotorevsky V S and Belov N A 2007 *Casting Aluminum Alloys* (Oxford: Elsiever)

[5] Tyrina L M, Rudnev V S, Ustinov A Y, Lukiyanchuk I V, Nedozorov P M, Chernykh I V and Dmitrieva E E 2014 *Prot. Met. and Phys. Chem. Surf.* **50** 508

[6] Yar-Mukhamedova G 1999 *Mater. Sci.* **35** 598
[7] Parfenov E V, Yerokhin A, Nevyantseva R R, Gorbatkov M V, Liang C J and Matthews A 2015 *Surf. Coat. Tech.* **269** 2
[8] Sakhnenko N D, Ved' M V, Androshchuk D S and Korniy S A 2016 *Surf. Eng. and Appl. Electrochem.* **52** 145
[9] Sakhnenko N D, Ved' M V, Vestfrid Yu V and Stepanova I I 1996 *Russ. J. Appl. Chem.* **69** 1346
[10] Yar-Mukhamedova G 2000 *Mater. Sci.* **36** 598
[11] Ved' M V, Sakhnenko N D, Bogoyavlenska O V and Nenastina T O 2008 *M Mater. Sci.* **44** 79
[12] Sakhnenko N, Ved' M, Karakurkchi A and Galak A 2016 *East Europ. J. Enterpr. Technol.* **81** 37

IOP Conf. Series: Materials Science and Engineering **213** (2017) 012021 doi:10.1088/1757-899X/213/1/012021

Microwave plasma assisted process for cleaning and deposition in future semiconductor technology

S Altmannshofer[1,2], J Boudaden[1], R Wieland[1], I Eisele[1] and C Kutter[1]

[1]Fraunhofer Research Institution for Microsystems and Solid State Technologies, Munich, Germany

E-mail: stephan.altmannshofer@emft.fraunhofer.de

Abstract. The epitaxial growth of silicon layers is an important step in the fabrication of semiconductor devices. For conventional silicon epitaxy, high temperatures, up to 900 °C are necessary. However, in future, semiconductor technology epitaxy processes at lower temperatures are required to increase the integration density. The goal of this study was to investigate microwave plasma assisted processes for the selective removing of thin silicon oxide, the cleaning of silicon surfaces and the depositing of high quality silicon films. The main focus was to apply these processes for low temperature epitaxy. All processes, such as oxide removal, cleaning and deposition, were done in one chamber and with microwave plasma assistance. In order to remove silicon dioxide, the etching behavior of hydrogen, fluorine, and hydrogen/fluorine plasma was studied. It was shown, that with hydrogen/fluorine plasma, the best selectivity of oxide to silicon was reached. The deposition process of silicon was studied by growing μc-Si films. The process was characterized and optimized by spectral ellipsometry. After a successful characterization of all process steps, silicon epitaxy layers have been grown with in-situ removal of native oxide and in-situ surface cleaning. The temperature for all process steps was reduced below 450 °C.

1. Introduction

Silicon based epitaxy is a key process in semiconductor technology. Modern devices such as SiGe heterojunction bipolar transistors (HBTs) [1] or strained silicon MOS transistors [2] are based on epitaxy processes. Conventional chemical vapor deposition (CVD), which is used in industry, needs temperatures higher than 700°C. However, lower deposition temperatures are required due to the reduction of the structures` size, sharp doping profiles, and increased thickness of strained layers,.

Presently, high temperatures are still required not only for the epitaxial deposition, but also for the preceding cleaning State-of-the-art for pre-epitaxial cleaning is a high temperature annealing in hydrogen atmosphere, the so call "hydrogen bake" [3]. Therefore in order to realize a low temperature epitaxy, both the cleaning as well as the deposition must be conducted at low temperatures.

For low-temperature pre-epitaxial cleans, different hydrogen plasmas were investigated [4]. Usually, an additional ex-situ removal of the native oxide with HF-acid is required [5]. Nishino et al. [6] describe the selective etching of silicon dioxide to silicon in a NH3/NF3 plasma. This process is not widely investigated and until now not used for pre-epitaxial cleaning.

Plasma enhanced chemical vapor deposition (PECVD), which is used to grow amorphous [7] or microcrystalline (μc) silicon films [8], was also investigated for epitaxial applications [8]. However, an in-situ simultaneous surface cleaning and epitaxial growth at low-temperature, without any additional wet-clean step, not yet, to our knowledge, is published.

In this paper, we describe a microwave-plasma enhanced chemical vapor deposition (MECVD) system for silicon epitaxy with in-situ cleaning steps. For the removal of the native oxide, the etching behavior of hydrogen, fluorine and hydrogen/fluorine plasma was systematically studied. Especially the influence of the plasma on the etching behavior was investigated. For the deposition process, the growth of μc-Si is studied to find out ideal process conditions for high crystallinity films. Finally, we will report in-situ surface cleaning followed by growing a silicon epitaxial layer, without any additional wet-cleaning steps.

2. Experimental

The plasma system is a prototype, built by Muegge GmbH. The 2.45 GHz microwave plasma source is attached on top of an aluminum reactor (Figure 1). The source is designed for a power from 600 W up to 3000 W and the typical process pressure is between 150 mTorr and 1000 mTorr. The system is suited for 200 mm wafers. The distance between the floating substrate heater and the plasma source can be varied from 20 mm to 170 mm. This is an extremely important feature because it allows investigating the influence of different plasmas species, i.e. ions, electrons and radicals. If the substrate is next to the plasma, radicals, ions and electrons reach the surface. However, if the substrate is far away from the plasma source, only the radicals reach the substrate surface. A detailed description of the system is given in [9].

For all experiments, p-type (100) Si wafers were used. For the etching experiments, 15 nm thermal oxide was grown and patterned by lithography followed by a wet-chemical etching. This allowed measuring both, the silicon and silicon dioxide etching rate at the same time. The silicon dioxide thickness was measured by spectral ellipsometry and the silicon thickness by atomic force microscopy (AFM) step measurements. The μc-Si films have been deposited on samples having a 100 nm thermal oxide. The characterization was done by spectral ellipsometry. For the epitaxial deposition standard (100) Si wafer without any pre-treatment were used. The crystalline quality was determined using Raman spectroscopy and electron channeling pattern (ECP).

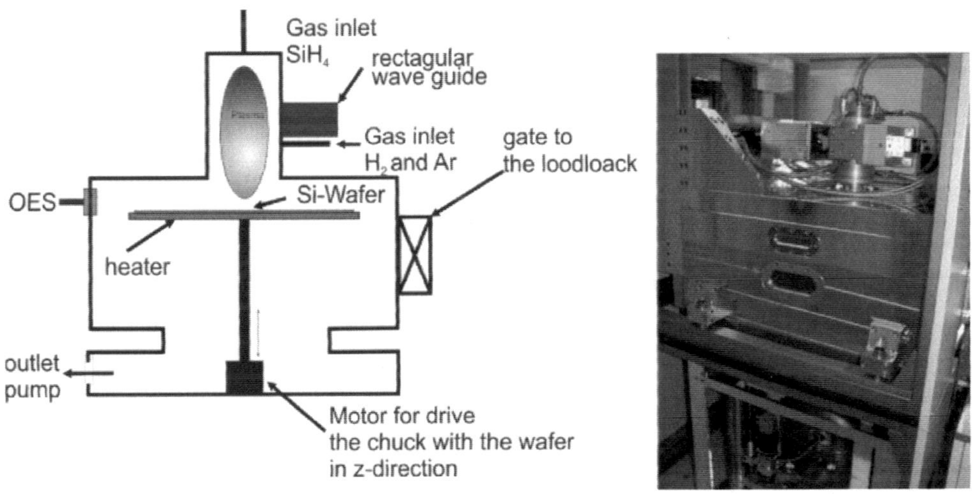

Figure 1. Scheme and photo of the microwave Plasma system.

3. Results and discussion

3.1. Hydrogen plasma etching of silicon and silicon dioxide

At first, the etching behavior of silicon as well as of silicon dioxide in the microwave plasma was investigated. For stable plasma conditions, the hydrogen was diluted in argon. The hydrogen to argon

IOP Conf. Series: Materials Science and Engineering **213** (2017) 012021 doi:10.1088/1757-899X/213/1/012021

flow ratio was 1 to 6. The substrate was heated up to 400 °C and a microwave power of 1.5 kW was applied.

The dependency of the etching rate on the distance, between the surface and the plasma source was investigated in order to understand the influence of ions, electrons, and radicals on the reaction kinetics. For these experiments, the pressure was kept at 200 mTorr. The results are shown in Figure 2. Etching is only observed at a distance less than 70 mm, and quite a different etching behavior for silicon dioxide and silicon can be observed. The silicon dioxide etching rate is almost constant in the range between 15 mm and 70 mm. The silicon etching rates are much higher between 15 mm and 40 mm and drop drastically beyond this distance. An explanation for this behavior can be given by the different influence of plasma species depending on the distance to the microwave source. In the region directly in the vicinity of the plasma, ions, electrons and radicals are available. Impinging ions break up the chemical bonds at the surface and a chemical etching with hydrogen radicals is possible. Because the Si-Si bond is much weaker than the Si-O bond, the difference in etching rate can be explained. If the distance increases the density of ions decreases rapidly while the high energy electrons are still present. Therefore at a distance between 40 mm and 70 mm mostly electrons reach the substrate surface. While electrons hitting the semiconducting silicon surface are neutralized by recombination processes, the isolating silicon dioxide will be negatively charged by them. This in turn leads to an increased ion bombardment of the silicon oxide compared to silicon. This explains why the etching rate of silicon dioxide does not drop as much as that of the silicon. Beyond 70 mm no charged particle reach the surface, and only hydrogen radicals are remaining, which are not able to initiate etching process [10]. This proves that between 20 mm and 70 mm a physical impact of charged particle dominates.

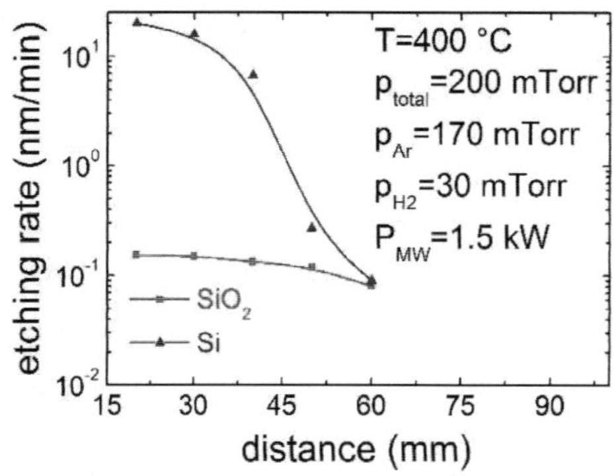

Figure 2. Etching rate of silicon dioxide (left axis, red) and silicon (right axis, blue) as a function of distance to the plasma source.

After these basic investigations the etching rate of Si and SiO2 as a function of the working pressure for a minimum distance of 20 mm has been investigated (Figure 3). It can be seen, that both etching rates decrease by increasing pressure. At pressures above 400 mTorr no etching rate can be determined. Normally, if the pressure increases, the radical density increases. This should lead to an increasing etching rate. However, it was shown above, that hydrogen radicals on their own do not etch silicon or silicon dioxide. If the pressure increases, the ion energy decreases because of the decrease of the mean free path. Therefore it can be concluded, that the limiting factor for hydrogen plasma etching is the ion energy, and not the radical density.

In conclusion it can be stated that the etching of silicon and silicon dioxide by hydrogen radicals is only successful, if an addition physical impact exists. The limiting factor for the chemical etching is the energy of ions, which are able to crack the chemical bond at the surface.

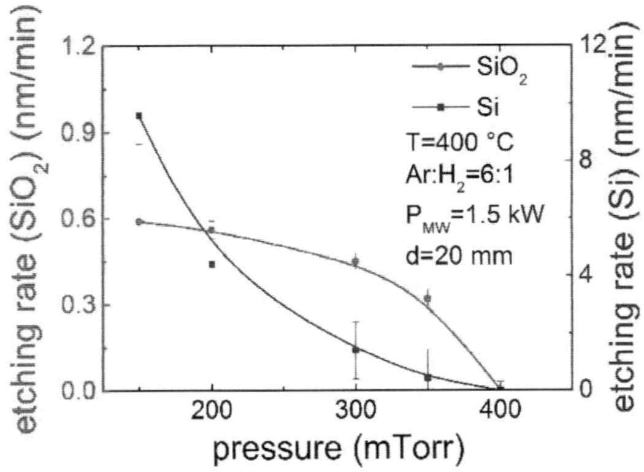

Figure 3. Etching rate of silicon dioxide (left axis, red) and silicon (right axis, blue) as a function of working pressure.

3.2. Fluorine plasma etching of silicon dioxide

Fluorine radicals are known to etch silicon as well as silicon dioxide [11]. Typical precursors usually are CF_4 [12] or NF_3 [13], but molecular F_2 can also be used [14]. The flourine etching is a typical dry chemical etching, which should only be limited by the radical density. The etching behavior of silicon dioxide in a F_2/He and a NF_3/Ar microwave plasma was investigated. The explored parameters were working pressure and distance between the substrate and plasma source. The total gas flow was fixed at 840 sccm, the power was kept at 1.5 kW and a temperature of 400 °C had been chosen.

The results are shown in Figure 4 and 5. In contrast to hydrogen etching for both gas chemistry the etching rate increases with increasing working pressure. This proves, the radical density and not the physical impact is the limiting factor for the chemical reaction.

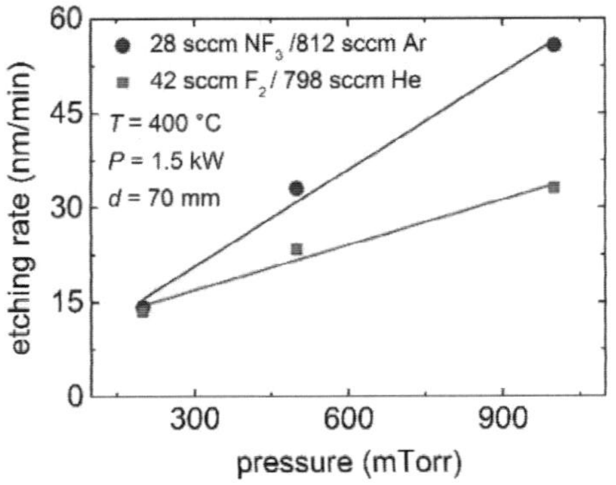

Figure 4. Etching rate of silicon dioxide in a NF_3 and F_2 microwave plasma as a function of working pressure.

For the investigation of the effect of distance, the pressure was fixed at 200 mTorr. As can be seen from Figure 5 two different etching regions are found. In both regions, the etching rates decrease linearly, but with different slopes. For a distance between 20 mm and 70 mm the etching rates decrease rapidly. Apparently the impact of ions and electrons enhances the etching rate. However, it can be clearly seen that even at a distance beyond 70 mm, where only radicals are present a significant etching rate exists, i.e. pure chemical etching takes place. In this region, the decrease of the etching rate can be explained by the reduced radical density. The etching behavior of NF_3 compared to F_2 is very similar. This suggests that in both cases F-radicals are the main etching species. The results show that the physical impact can enhance the etching, but the limiting factor is the radical density.

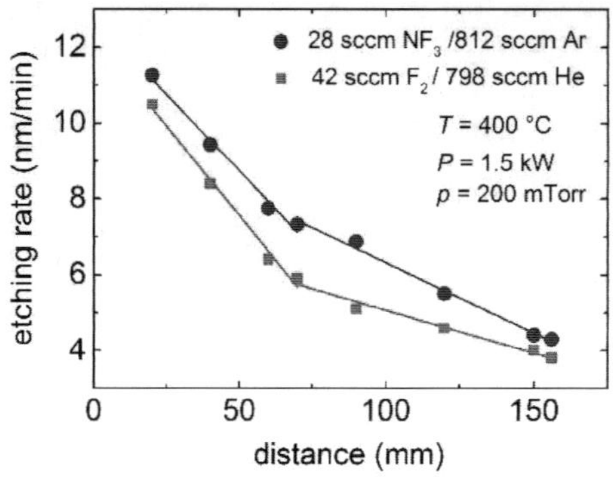

Figure 5. Etching rate of silicon dioxide as a function of distance between the substrate and the plasma source using NF_3 and F_2 microwave plasmas.

3.3. *Hydrogen/Fluorine plasma etching*

As shown above, hydrogen, as well as fluorine plasma, can be used to etch silicon dioxide. However, the selectivity to silicon is not very good. The main reason is that the Si-O binding energy is significantly higher as compared to the Si-Si bond. It was shown that in a NH_3/NF_3 plasma, a selective etching of silicon dioxide over silicon can be realized [6]. The microwave plasma system was investigated to realize the selective remove of oxide with a H_2/NF_3-discharge. These experiments were done at a distance of 70 mm. This means that the hydrogen radicals do not etch. The power and the working pressure were 1.5 kW and 200 mTorr, respectively.

The etching rates of silicon and silicon dioxide were investigated in dependency on the hydrogen dilution. The results are shown in Figure 6. At a low dilution a high etching rate of silicon is observed. Besides, the selectivity is very low due to the presence of many F-radicals. With increasing hydrogen dilution the silicon etching rate decreases much faster than the etching rate of silicon dioxide and an increase of the selectivity up to around 10 is observed. This can be explained by a reaction of fluorine with hydrogen to form HF, which only etches silicon dioxide.

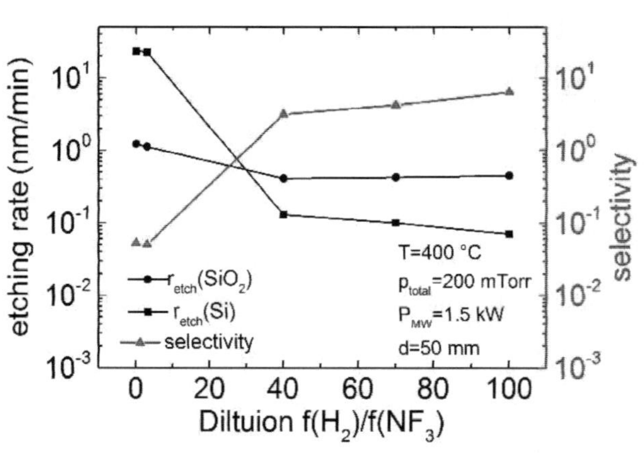

Figure 6. Etching rate of silicon dioxide in a NF_3 and F_2 microwave plasma as a function of distance from plasma source.

3.4. *Deposition of μc-Si layers*

To investigate the growth conditions of silicon films in the microwave plasma, microcrystalline silicon (μc-Si) was grown and investigated by spectral ellipsometry. For the characterization of the

IOP Conf. Series: Materials Science and Engineering **213** (2017) 012021 doi:10.1088/1757-899X/213/1/012021

crystallinity, a Bruggeman-effective-medium-approximation-model (BEMA) [15] was developed. The deposition was done from a silane/hydrogen/argon-plasma at a substrate temperature of 400 °C. It was found that high hydrogen dilution, low working pressure and high microwave power favor crystalline growth. The hydrogen flow and the silane flow were 1400 sccm and 4 sccm, respectively. To stabilize the plasma, 400 sccm Ar were added.

To investigate the effect of the plasma on the growth, the film properties in dependency on the distance were studied. Figure 7 summarizes the obtained deposition rate and the crystallinity. The deposition rate decreases exponentially with increasing distance because the number of available particles decreases. It also is obvious that the crystallinity is high and nearly constant, if the distance is below 50 mm. If the distance is further increased, the crystallinity decreases rapidly. At a distance of 70 mm and more, the film becomes mostly amorphous. These results prove that an additional physical impact of ions is necessary to achieve the growth of μc-Si with a high crystallinity.

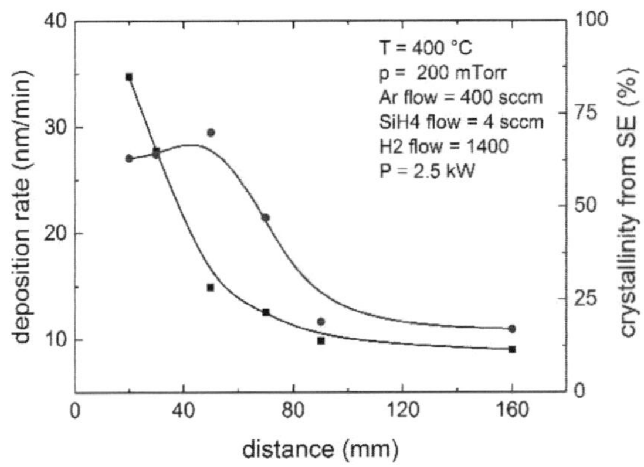

Figure 7. Deposition rate (left axis, black) and crystallinity (right axis, blue) determined by spectral ellipsometry as a function of distance from plasma source.

3.5. Silicon epitaxy at 450°C

Finally, the results of etching and μc-Si growth have been combined in order to achieve silicon homoepitaxy at low temperatures. At first the aim was to realize an in-situ removal of the native oxide followed by an epitaxial deposition of silicon layer. A NF_3/H_2 plasma etching was carried out to remove the native oxide from the silicon substrate. For this process, the substrate was positioned at a distance of 70 mm and the H_2 to NF_3 flow ratio was 100 to 1. In these conditions, a high etch selectivity of silicon dioxide to silicon was obtained. The power was 1.5 kW, the pressure 200 mTorr and the substrate temperature 450 °C. After the etching, the deposition was immediately done. During the deposition process, the distance was reduced to 40 mm, because the growth of μc-Si shows, that for high crystallinity, an additional impact of ions is required. Pressure and temperature were kept constant. The power was increased to 2.5 kW. The deposition was done with 4 sccm silane and 1400 sccm hydrogen. The film was characterized by electron channeling pattern (ECP) and Raman spectroscopy. Figure 8 shows the Raman spectra and the ECP results. The spectrum shows only the peak of the silicon crystal at 520 cm^{-1}. In the ECP picture a clear (100) pattern can be observed. This proves that epitaxial silicon was successfully grown at a low temperature with an in-situ low temperature removal of the native oxide.

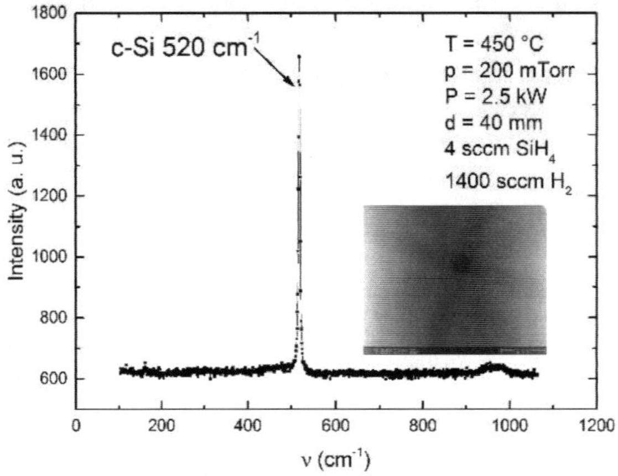

Figure 8. Raman spectrum and electron channeling pattern (ECP) of an epitaxial silicon film grown at 450°C on an in-situ NF_3/H_2 cleaned sample.

4. Conclusion

It has been shown that "Microwave Enhanced Processes" at temperatures as low as 450°C can be used for numerous applications in semiconductor technology.

The results, obtained from the present work, can be summarized as follows:

1. In a hydrogen microwave plasma silicon and silicon dioxide can only be etched if a physical impact by ions or electrons exists.

2. In a NF3 or F2 microwave plasma, the etching of silicon dioxide is mainly limited by radicals. The physical impact plays a secondary roll.

3. Both hydrogen and fluorine plasma show a higher etching rate of silicon compared to silicon dioxide.

4. In a NF3/H2 microwave discharge, a selective removal of native oxide is achieved, if the substrate is at a distance where only radicals are available and a high hydrogen dilution is used.

5. The selective etching of silicon dioxide to silicon can be explained by the formation of HF in the vapor phase.

6. μc-silicon films on silicon dioxide can only be grown, if an addition physical impact exists. Deposition with radicals leads to amorphous growth.

7. Epitaxial silicon films can be grown at 450°C with an in-situ removal of the native oxide. Additional ex-situ cleaning steps are not required.

References

[1] Heinemann B *et al* 2014 *IEEE EDL* **8** 814–6
[2] Oh J 2012 *Microelectron. Eng.* **97** 26–8
[3] Carroll M S, Sturm J C and Yang M 2000 *J. Electrochem. Soc.* **174** 4652-9
[4] Ramm J, Beck E, Züger E, Dommann A and Pixley R E 1992 *Thin Solid Films* **222** 126-31
[5] Anthony B, Hsu T, Breaux L, Quian R, Banerejee S and Tasch A 1990 *J. Electron. Mater.* **19** 1027-32
[6] Nishino H, Hayasaka N and Okano H 1993 *J. Electron. Mater.* **74** 1345-8
[7] Takeda T, Tanaka K, Inoue H, Hirayama M, Tsumori R, Aharoni H and Ohmi T 2007 *Jpn. J. Appl. Phys.* **46** 2542–53
[8] Müllerová J, Šutta P, van Elzakker G, Zeman M and Mikula M 2008 *Appl. Surf. Sci.* **254** 3690-5
[9] Altmannshofer S, Eisele I and Gschwandtner A 2016 *Surf. Coat. Technol.* **304** 359-63

[10] Aßmuth A, Stimpel-Lindner T, Senftleben O, Bayerstadler A, Sulima T, Baumgärtner H and Eisele I 2007 *Appl. Surf. Sci.* **253** 8389-93
[11] Kang S C, Kim D J, Hwang J Y, Yun Y B, Lee N E, Jang Y C and Bae G H 2007 *Thin Solid Films* **515** 4945-9
[12] Kastenmeier B E E, Matsuo P J, Beulens J J and Oehrlein G S 1996 *J. Vac. Sci. Technol. A* **14** 2802-13
[13] Matsuo P J, Kastenmeier, B E E, Oehrlein G S and Langan J G 1999 *Vac. Sci. Technol. A* **17** 2431-7
[14] Yun Y B, Park S M and Lee N E 2008 *J. Korean Phys. Soc.* **53** 2386-90
[15] Losurdo M, Rizzoli R, Summonte C, Cicala G, Capezzuto P and Bruno G 2000 *J. Appl. Phys.* **88** 2408

IOP Conf. Series: Materials Science and Engineering **213** (2017) 012022 doi:10.1088/1757-899X/213/1/012022

Effect of initiators on synthesis of poly(L-lactide) by ring opening polymerization

D Pholharn[1], Y Srithep[2,3] and J Morris[2]

[1]Department of Chemistry, Faculty of Science and Technology, Rajabhat Maha Sarakham University, Mahasarakham 44000, Thailand
[2]Manufactuing and Materials Research Unit, Faculty of Engineering, Mahasarakham University, Mahasarakham 44150, Thailand

E-mail: yottha.s@msu.ac.th

Abstract. We studied the effect of several aliphatic alcohols, including1-dodecanol, 1-octanol and methanol,. as initiators on synthesis of poly(L-lactide) (PLLA) by ring opening polymerization. The reaction starts with L-lactide monomer and uses stannous octoate as catalyst. Fourier transform infrared spectroscopy and X-ray diffraction analysis verified that PLLAs were produced successfully. Weight, number average molecular weight and polydispersity index of PLLAs were measured by gel permeation chromatography. The PLLA initiated by methanol (PLLA-Meth) presented the highest molecular weight and yield percent. From differential scanning calorimetry, PLLA-Meth showed the highest melting temperature at ~167°C, crystallization temperature at 110°C and degree of crystallinity 80%. The thermal stability was assessed by thermogravimetric analysis: this confirmed that PLLA-Meth was superior with the highest degradation temperature compared to PLLA initiated by other initiators. We concluded that methanol was the most appropriate initiator for PLLA synthesis by ring opening polymerization.

1. Introduction

Poly(lactic acid) or polylactide (PLA) can be derived from natural sources such as cassava, corn and sugar beet. It is a sustainable development plastic which is biodegradable and biocompatible [1]. So it has extensive medical applications including sutures, bone fixation material and drug delivery by microspheres. PLA has stereoisomers: poly(L-lactide) (PLLA), poly(D-lactide) (PDLA) and poly(DL-lactide) (PDLLA). Readily available from natural sources, PLLA has been studied extensively because of its excellent biocompatibility and mechanical properties [2].

PLA can be synthesized in two ways: the first is direct polycondensation of lactic acid and the second is ring opening polymerization of lactide. In direct condensation, solvent is used and higher reaction times are required. The resulting polymer has low molecular weight and poor mechanical properties. Ring opening polymerization (ROP) of the lactide needs a catalyst and results in PLA with controlled molecular weight, depending on the monomer used and reactions conditions[2-3]. Several different types of metal catalysts, e.g. zinc (II) and titanium (IV) have been evaluated, but tin (II)2-ethylhexanoate or stannous octoate ($Sn(Oct)_2$) has been widely used because of high reaction rates, solubility in the monomer melt and the ability to produce high molecular weights [4]. An initiator is important for ring opening polymerization. Duda and Penzek described a polymerization scheme based on the insertion-coordination mechanism as shown in Figure 1. In lactide polymerization catalyzed by stannous octoate (a), a hydroxyl compound (alcohol) is required as an initiator. The alcohol first reacts with stannous octoate to form a tin alkoxide bond by ligand exchange (b). Then, an exocyclic carbonyl oxygen of the lactide temporarily coordinates with the catalyst tin atom in the alkoxide form (c), enhancing the nucleophilicity of the initiator's alkoxide group and also the lactide carbonyl group's electrophilicity. Then the acyl-oxygen bond (between the carbonyl group and the endocyclic oxygen) of the lactide breaks (d), opening the lactide chain allowing insertion into the tin-oxygen bond (alkoxide) of the catalyst (e). Repeated addition of additional lactide molecules into the tin-oxygen bond propagates forming the polymer (f) [5-6].

However, there are no reported studies of the effect of different alcohols used as initiator for ring opening polymerization. Therefore, we assessed the effect of initiator on yield percent, molecular weight, thermal properties and thermal stability of PLA (specifically PLLA)polymerized by ring opening with stannous octoate as a catalyst.

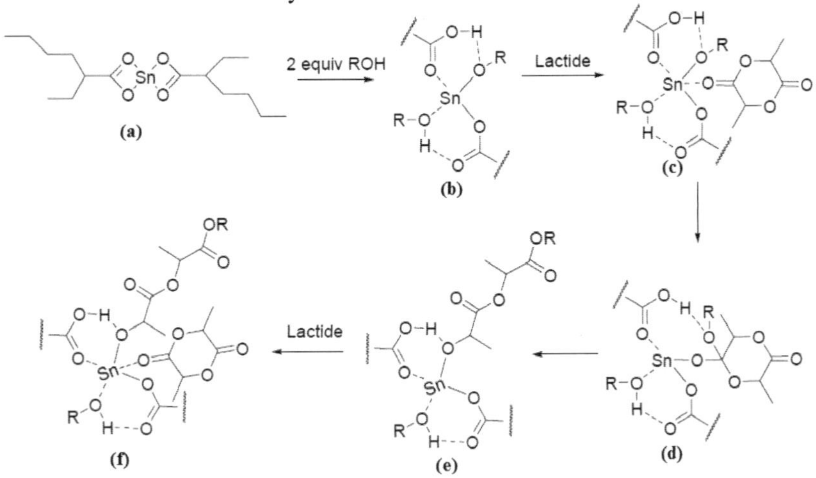

Figure 1. Coordination-insertion mechanism of $Sn(Oct)_2$ catalyzed polymerization of L-lactide (taken from [5]).

2. Experimentals

2.1. Materials

98% L-lactide was donated by Purac (Thailand) Ltd. For ring opening initiators, AR grade methanol 99% was purchased from Carlo Erba,1-octanol 99% from Panreac Sintesis and 1-dodecanol 98% from Across Co. Ltd. Stannous octoate ($Sn(Oct)_2$) was supplied by Sigma Aldrich. Solvents, including toluene and chloroform, were obtained from RCI Lab scan.

2.2. Synthesis of Poly(L-Lactide) by ring opening polymerization

PLLA was prepared from L-lactide. Before using, it was dried in a vacuum oven at 45°C for 12h. Poly(L-lactides) (PLLA) were synthesized by bulk L-lactide ring opening polymerization in a nitrogen atmosphere at 165°C for 4 h, initiated with methanol, 1-octanol and 1-dodecanol.$Sn(Oct)_2$ 0.1mol% was used as a catalyst. We varied temperature and synthesis time and found an optimum matching previous reports [1-3, 7]. After incubation, the product was dissolved in chloroform and removed to purify by precipitation in methanol. The precipitate was filtered under vacuum. Finally, PLLA was dried in a vacuum oven at 50°C for 24 h. The synthesis scheme of the PLLA by ring opening polymerization is shown in Figure 2. The samples were labelled PLLA-*initiator* where the *initiator* was Dodec, Oct and Meth, standing for 1-dodecanol, 1-octanol and methanol, respectively.

IOP Conf. Series: Materials Science and Engineering **213** (2017) 012022 doi:10.1088/1757-899X/213/1/012022

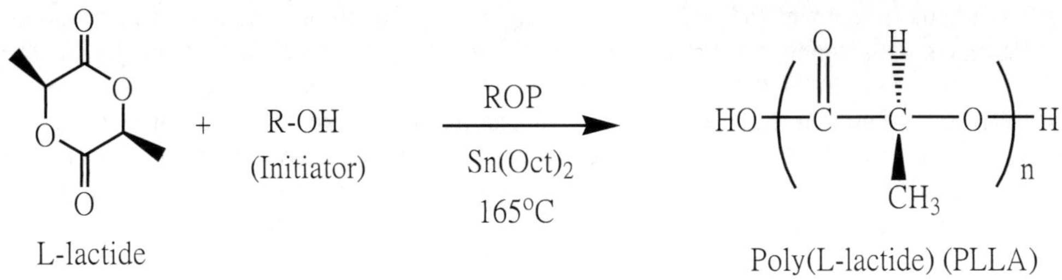

Figure 2. Synthesis of PLLA by ring opening polymerization.

2.3. Measurements and characterization

*2.3.1. Fourier transform infrared spectroscopy (FTIR).*The chemical structure of L-lactide and PLLAs were verified by FTIR spectra from 500 to4,000 cm^{-1} (2 cm^{-1} resolution) recorded with OPUS 7.0 software on a Bruker Tensor 27 FTIR spectrometer, using the Attenuated Total Reflection (ATR) mode.

2.3.2. X-ray diffraction analysis (XRD). To confirm the crystal structure, X-ray diffraction (XRD) spectra were recorded on a Bruker/D8 Advance Bruker BioSpin AG (Karlsruhe, Germany). After synthesis, samples were ground and placed on the XRD platform. The range of 2θ from 2° to 40° in the refraction mode was scanned at 2°/min.

2.3.3. Gel permeation chromatography (GPC). The M_n, M_w and PDI of PLLAs were measured by GPC. About 7 mg of samples were dissolved in 3 ml tetrahydrofuran (THF) via continuous stirring in a water bath at 60°C for 1 h. GPC columns were packed with Styragel HR5E 7.8 × 300 mm column (molecular weight resolution range: 2,000-4,000,000). A Waters 2414 refractive index (RI) detector was used. The GPC columns were eluted using tetrahydrofuran at a flow rate of 1.0 mL/min at 40°C and monodispersed polystyrenes were used for calibration.

2.3.4. Differential scanning calorimetry (DSC). Thermal histories of the samples were analyzed by DSC (Perkin-Elmer, model DSC 4000). Approximately 5 mg of samples were placed in aluminium pans and subjected to two heating scans from -10 to 200°C at 10 degC/min. T_m was determined from the second scan peak at maximum value of the endotherm peak. After the first heating scan, the sample was kept at 200°C for 3 min and then cooled to -10°C at 10 degC/min, to determine the crystallization temperature (T_c). Results from the initial heating cycle in DSC experiments were discarded because they included the previous thermal history. The melting temperature (T_m) and apparent melting enthalpy (ΔH_m) were determined from the second heating scan. The degree of crystallinity $\chi_c(\%)$ was estimated from the first heating cycle using

$$\chi_c(\%) = \frac{\Delta H_m}{\Delta H_m^o} \times 100 \quad (1)$$

where ΔH_m is the heat of fusion of PLA and $\Delta H_m^o$is theoretical melting enthalpy of perfect PLA crystal (93 J/g) [8]

2.3.5. Thermogravimetric analysis (TGA). Thermal stabilities of L-lactide and PLLAs were investigated using a TGA 4000 (Perkin-Elmer) under a constant nitrogen flow. About 10 mg of samples was heated from 25 to 800°Cat 20 degC/min.The weight loss was recorded and normalized against the initial weight.

3. Results and discussions

3.1. Chemical structure

Figure 3(a) and (b) shows the FTIR spectra of lactide monomer and PLLAs. PLLAs were initiated with 1-dodecanol (PLLA-Dodec), 1-octanol (PLLA-Oct) and methanol (PLLA-Meth): their spectrum showed peaks corresponding to asymmetric (2,995 cm-1) and symmetric –CH stretching (2,948 cm^{-1}). We also observed a band shift of the C=O stretch in the lactide at 1,727 to 1,752 cm^{-1}in the PLLA as show in Figure 3(b). This, accompanied by a difference in the peak intensity caused by changing of the environment of molecules in the polymer chain, confirms polymerization [3, 9].This confirms that the synthesis conditions and all initiators used were able to synthesize PLLA by ring opening.

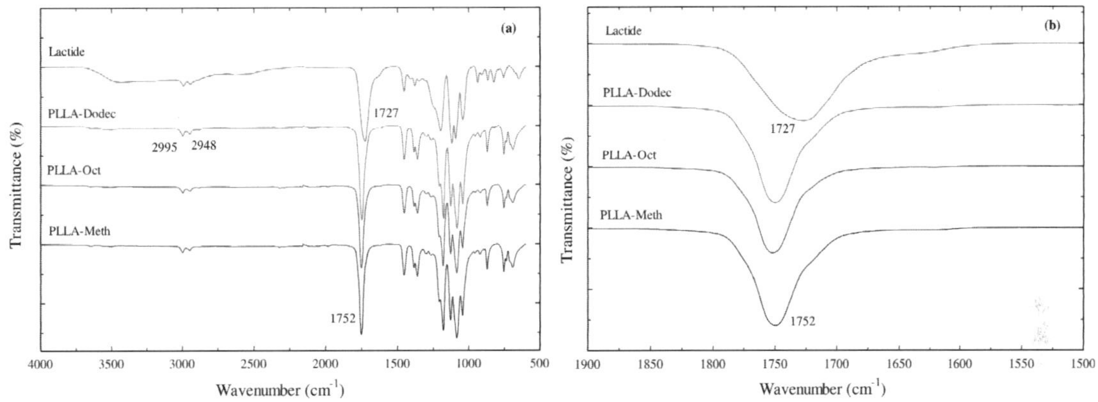

Figure 3. FTIR spectra of lactide and PLLAs.

3.2. Crystallize analysis

X-ray scattering was used to probe the crystallinity of lactide and PLLA and to confirm PLLA formation when using different initiators. The diffraction traces of the lactide and PLLAs are shown in Figure 4. The lactide shows several important diffraction peaks at 8.7°, 13.3 and 14.3°. After ring opening polymerization, the lactide polymerized to PLLA which presented three strong peaks at around 16.7°, 19.2° and 22.4°[8]. For all PLLAs samples, we clearly see all the diffraction peaks corresponding to PLA without lactide diffraction peaks. This confirms that all initiators were effective at initiating the reaction of lactide because unreacted lactide monomer was not present. This agrees with the FTIR results. However, the difference effect of the initiators will be shown and discussed in other results.

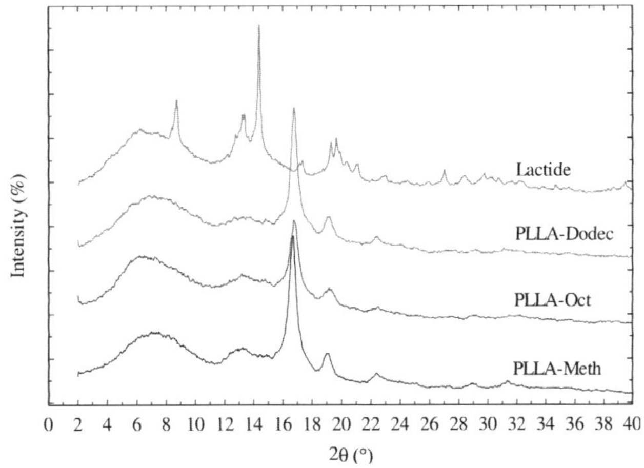

Figure 4. XRD patterns of lactide and PLLAs.

3.3. Molecular weight and yield percent

Number average molecular weights (M_n), weight average molecular weights (M_w) and polydispersity indices (PDI) of synthesized PLLAs were measured by gel permeation chromatography. The yield percent of PLLAs promoted with different initiators were calculated and shown in Table 1. It is notable that PLLA initiated by methanol (PLLA-Meth) displayed M_n (11,800 g/mol), M_w (14,400 g/mol) and yield percent (89%) higher than other PLLA samples and that the PLLA-Oct showed higher molecular weight and yield percent than PLLA-Dodec. The higher yield percent indicated a more complete reaction: we believe that it related to the effect of initiator on the polymerization initiationstep because all factors (ratio of monomer/initiator, catalyst type and amount, temperature and reaction time) were similar except for initiator type. The highest molecular weight of PLLA-Meth was observed even when the reaction time was equal. We attribute this to higher initiation rate. Based on the insertion-coordination mechanism, the propagation continues by repeatedly adding additional lactide molecules into the tin-oxygen bond [5], so all propagation rates are expected to be similar because all factors governing this step are similar. Therefore, the higher initiation rate is the only variable affecting the molecular weight. However, all PLLA samples had PDI values nearly 1 and it was observed that the molecular weight distribution peak was quite sharp (Figure 5), indicating that the number of polymer chains remained constant, equal to the number of the original initiating species and the polymer chains approach uniform length[8].These results indicate that methanol was the most efficient initiator for ring opening polymerization under our conditions.

Table 1. Molecular weight and yield percent of polylactide.

Sample	M_n (g/mol)	M_w(g/mol)	PDI	Yield (%)
PLLA-Dodec	5,460	6,200	1.14	65
PLLA-Oct	9,310	11,600	1.24	78
PLLA-Meth	11,800	14,400	1.22	89

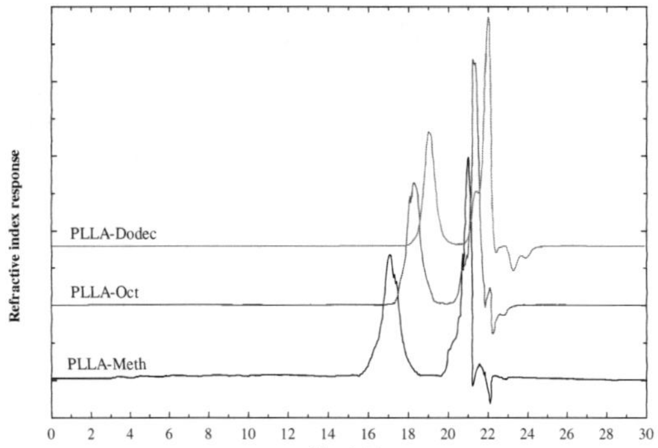

Figure 5. Molecular weight distribution graphs of PLLAs.

3.4. Thermal properties

Melting and crystallization behaviours were investigated using DSC. The temperatures obtained from the first heating, second heatingand cooling cycles at rates of 10 degC/min. The first heating scan data was discarded because it included the previous thermal history [10]. The results are shown in Figure 6 and Table 2. The data obtained from second heating scan (Figure 6(a)) indicates the effect of theinitiator on formation and crystallization during polymerization. PLLA synthesized in other laboratories showed melting temperatures (T_m) in the range 150-170°C [1, 7], while our experiments showed values in the range 130-160°C. It was observed that all PLLA samples have a single melting peak. It is notable that PLLA-Meth showed T_m at 167°C higher than other samples. Melting temperature differences correlated with their molecular weight (as shown in Table 1). For low molecular weight, mostly less than 10,000 g/mol, the polymer crystallinity will decrease and melting temperature also reduce, because the molecular surface is more irregular and individual polymer molecules do not pack together as well as longer strands leading to lower melting temperature. Saeidlou et al showed that the PLA melting temperature varies with number average molecular weight and reaches steady state about 100,000 g/mol [11]. Our experiments had number average molecular weights less than 10,000 g/mol and thus still in the rising part of Saeidlou et al's PLA melting point as a function of molecular weight curve (figure 5 in [11]). Polymer crystallization is made more difficult by long chains, which make exact chain alignment difficult. On the other hand, too low molecular weight polymers, eg. PLLA-Oct (5,460 g/mol), are difficult to crystallize because the short chains cannot fit into existing crystal locations, so the PLLA-Oct crystallinity was lowest.

The crystallization behaviour of all synthesized PLLAs were distinguished on cooling cycle, the thermograms shown in Figure 6(b). The crystallization temperature (T_c) of PLLA-Meth was 110°C and the PLLA-Oct was 108°C consistent with other results, eg. Zhanxin et al. found, for PLLA (Mw 200,000 g/mol, NatureWorks), T_c at 110°C [1].This is expected because crystal growth is achieved by further addition of folded polymer chain segments and only occurs for temperatures below the melting temperature (T_m of PLLA ~ 160°C) and above the glass transition temperature (T_g of PLLA~50°C). However, it is clear that T_c of PLLA-Dodec shows an unusual value (77°C), which correlates with its crystallization behavior. In summary, PLLA initiated by methanol has the best combination of thermal properties, including melting temperature, crystallization temperature and degree of crystallinity.

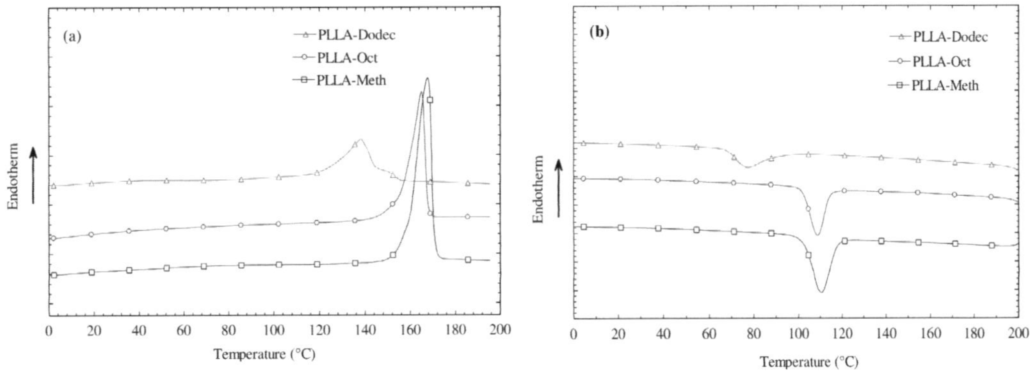

Figure 6. DSC thermograms of PLLAs: (a) second heating scan and (b) cooling scan.

Table 2. Thermal parameters of lactide and PLLAs obtained from DSC.

Sample	T_m (°C)	ΔH_m (J/g)	χ_c(%)	T_c (°C)	ΔH_c (J/g)
PLLA-Dodec	138	39	42	77	29
PLLA-Oct	165	50	54	108	39
PLLA-Meth	167	72	77	110	56

3.5. Thermal stability

Thermal decomposition was observed when the temperature exceeds 200°C. The TGA curves of temperature decomposition are shown in Figure 7 and the nominal values of the 10% weight loss temperature (T_{10}) are in Table 3. Lactide begins to decompose at 158°C. All PLLAs showed a single decomposition process and it was clear that PLLA-Meth had higher decomposition temperature (317°C). This indicates that PLLA-Meth had the highest thermal stability as well as the best values of other properties. PLLA-Meth showed the highest molecular weight which could undergo the high formation of chain entanglements [12]. The lowerer chain entangled PLA gets fragmented easily and subsequent volatilisation of the degradation products is easier. On the other hand, the higher entanglement, PLLA-Meth, makes the bond rupture difficult and increased the thermal stability [12-14].

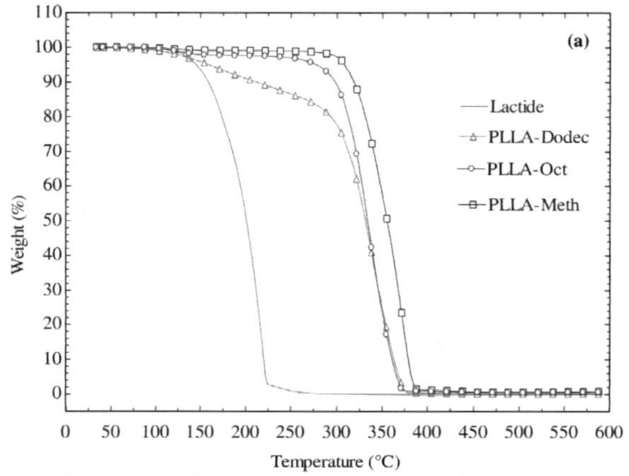

Figure 7. TGA curves of lactide and PLLAs.

Table 3. Degradation temperature of PLLAs.

Sample	T_{10}(°C)
Lactide	158
PLLA-Dodec	223
PLLA-Oct	299
PLLA-Meth	317

4. Discussion

Masutani and Kimura [6] Ryner et al [5] discussed coordination-insertion mechanism of ring opening and suggested that the first step of reaction is that the alcohol attacks the octoate ligand (catalyst). Our results showed that methanol was more effective. We attribute the strongest effect of methanol, the smallest alcohol, to lower steric hindrance when attacking the octoate ligand, resulting in a more complete polymerization, because it can attack the substrate more easily than longer primary alcohols (1-octanol and 1-dodecanol) which have more steric hindrance.

5. Conclusion

We concluded that methanol was the most effective of the alcohols studied as initiators for synthesis of L-polylactide (PLLA) through ring opening polymerization using stannous octoate as a catalyst: it led to the highest molecular weight, melting temperature, thermal stability and degree of crystallinity as well as the highest yield. Our study included 1-dodecanol, 1-octanol and methanol. Further, FTIR and XRD results verified that, when the reactions were initiated with those alcohols, it achieved a complete polymerization and perfect structure of PLLA. This implies that the methanol based PLLA would better qualify for a variety of applications.

Acknowledgement

We acknowledge financial support from the Research and Development Institute, Rajabhat Maha Sarakham University.

References

[1] Jing Z, Shi X, Zhang G, Li J, Li J, Zhou L and Zhang H 2016 *Polymer* **92** 210-21
[2] Lopes M S and Jardini A L 2014 *Chem. Eng. Trans.* **18** 331-6
[3] Lasprilla A J R, Martinez G A R and Hoss B 2011 *Chem. Eng.* **24** 985-90
[4] Swift G 1993 *Acc. Chem. Res.* **26** 105-10
[5] Ryner M, Stridsberg K, Albertsson A C, von Schenck H and Svensson M 2001 *Macromolecules* **34** 3877-81
[6] Masutani K and Kimura Y 2014 *PLA Synthesis. From the Monomer to the Polymer* (London: Royal Society of Chemistry)
[7] Qi F, Tang M, Chen X, Chen M, Guo G and Zhang Z 2015 *Eur. Polym. J.* **71** 314-24
[8] Xiao H, Lu W and Yeh J T 2009 *J. Appl. Polym. Sci.* **113** 112-21
[9] Erbetta C D A C, Alves R J, Resende J M, de Souza F R F and de Sousa R G 2012 *J. Biomater. Nanobiotechnol.* **3** 208
[10] Srithep Y, Pholharn D, Veang-in O and Yangyuen S 2017 *J. Polym. Eng.* **35** 335-344
[11] Saeidlou S, Huneault M A, Li H and Park C B 2012 *Prog. Polym. Sci.* **37** 1657-77
[12] Chalmers J M and Meier R J 2008 Comprehensive Analytical Chemistry *Molecular characterization and analysis of polymers (Vol. 53)* (Amsterdam: Elsevier)
[13] Carrasco F, Pagès P, Gámez-Pérez J, Santana O and Maspoch M L 2010 *Polym. Degrad. Stab.* **95** 116-25
[14] Mathew A P, Packirisamy S and Thomas S 2001 *Polym. Degrad. Stab.* **72** 423-39

IOP Conf. Series: Materials Science and Engineering **213** (2017) 012023 doi:10.1088/1757-899X/213/1/012023

Preparation of copper thin film mask by sputtering technique assisted by polymer mask photolithography

C Pakpum[1,2]

[1]Program in Applied Physics, Faculty of Science, Maejo University, Chiang Mai 50290, Thailand

E-mail: chupong@mju.ac.th

Abstract. This research aims to build cheaply homemade DC magnetron sputtering system for deposit metal thin films on substrate by assist of polymer mask photolithography. The metal thin film is able to use as a hard mask for patterning transfer via chemical wet etching or plasma dry etching process and it also can be used as a metal electrode in micrometre scale. Prepared copper thin films will be measured thickness by scanning electron microscope. Crystalline structure of the thin films will be examined by x-ray diffractometer. Elemental analysis contained in the films will be determined by energy dispersive x-ray spectrometer and morphology of the films will be revealed by scanning electron microscope.

1. Introduction

Many methods of metal thin film deposition processing techniques have been conducted for obtaining high quality of deposited film in conjunction with low cost of mass production. Investigated deposition techniques, up to now, composed of; direct current (DC) sputtering, radio frequency (RF) sputtering, thermal evaporation, electron beam physical vapour deposition, radio frequency (RF) ion plating, ultrasonic spray pyrolysis, wet chemical sol-gel process, chemical vapour deposition (CVD) and plasma enhanced chemical vapour deposition (PECVD). By comparison, DC sputtering has been considered as a process of choice providing good uniformity of large scale processing substrates and providing acceptable deposited thin films quality as well [1]. In fundamental, sputtering phenomena occurred from positive ions, exist in plasma that generated from collision process by energetic electrons gained energy from dense electric field between two electrodes supplied by DC high voltage in vacuum chamber, to bombard atoms on surface substrate called "target" (which generally placed on electrode that applied by negative voltage) resulted in removing atoms on target to deposit on another substrate surface called "substrate" placed on electrode that having higher electric potential eventually atoms from target accumulate on substrate to form the films [2]. The traditional process of polymer photolithography to create polymer patterns on silicon substrate were achieved by sequence of three steps process; coating light-sensitive polymer on substrate, irradiating ultraviolet (UV) light to polymer through the desired pattern (photo-mask) and chemical wet etching to UV exposed polymer. The cured portion by UV of polymer cause that part hardened and can withstand to chemical wet etching, eventually the polymer layer on substrate appeared the same pattern with photo-mask [3]. In this work, author aims to build cheaply homemade DC magnetron sputtering system for deposit copper (Cu) thin films on silicon (100) substrate by assist of polymer mask photolithography by means of the polymer mask will be used as a temporary mask to allow sputtered Cu atoms to deposit on silicon substrate area where dry film mask was unprotected. Prepared copper thin films will be

IOP Conf. Series: Materials Science and Engineering **213** (2017) 012023 doi:10.1088/1757-899X/213/1/012023

measured thickness by scanning electron microscope. Crystalline structure of the thin films will be examined by x-ray diffractometer. Elemental analysis contained in the films will be determined by energy dispersive x-ray spectrometer and morphology of the films will be revealed by scanning electron microscope.

2. Experimental methods

2.1. Fabricate homemade DC magnetron sputtering system

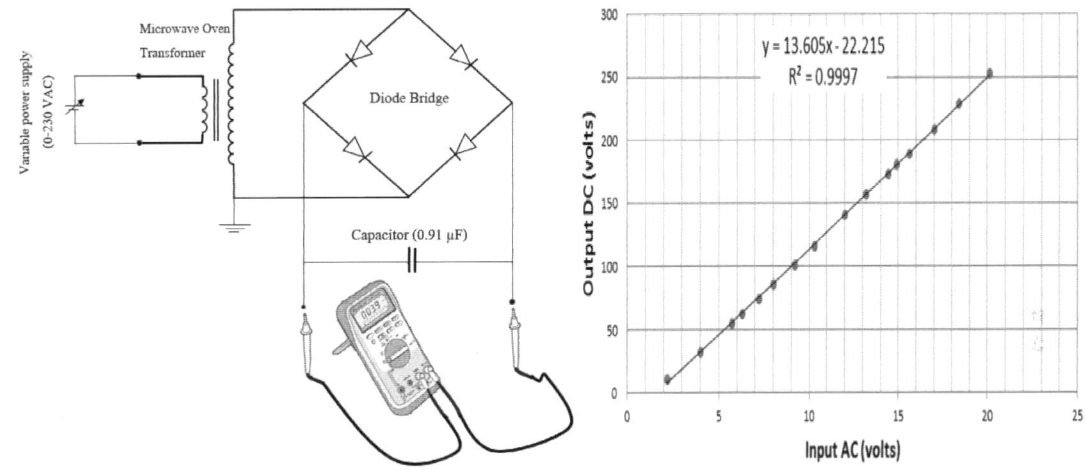

Figure 1. Illustrate AC to DC circuit (left) and performance to generate DC output (right).

In Figure 1 on left side, showed AC to DC circuit used to produce plasma in vacuum chamber while input voltage is AC (0-230 volts) and transformed to negative DC high voltage output at points where multimeter connected shown in the figure. After test the circuit, in Figure 1 on right side, with 20 volts (AC) input to the circuit results output 250 volts (DC) therefore this circuit should provide maximum 3,105 volts DC output while applying maximum AC input at 230 volts.

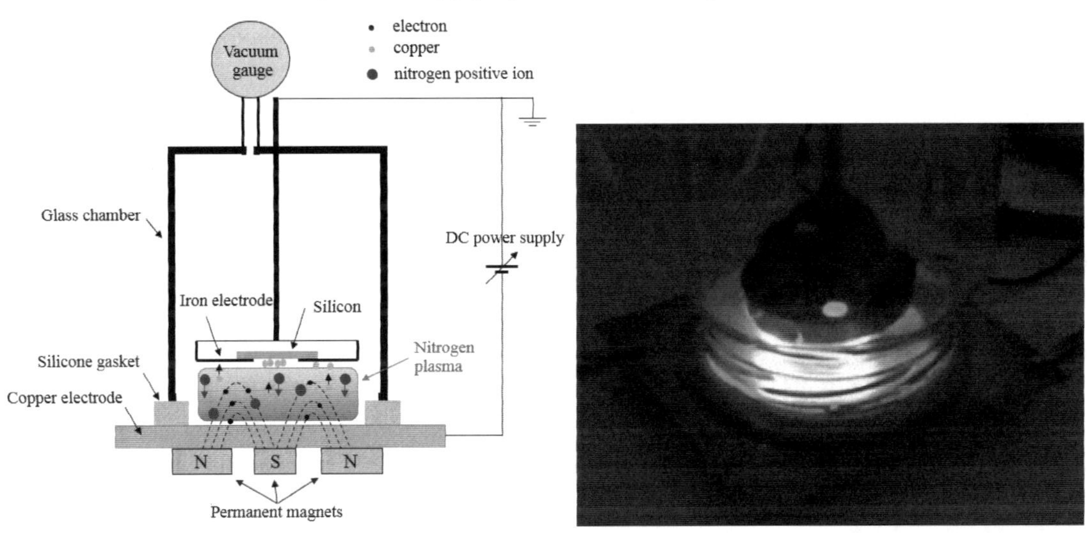

Figure 2. Illustrate schematic of a DC magnetron sputtering system (left) and the fabricated homemade DC magnetron sputtering system (right).

In Figure 2 on left side, showed the system was constructed by using a parallel of grounded iron electrode and cathode copper electrode used to ignite the plasma, plasma glow discharge is illustrated on the right side of Figure 2. An electric potential, generated from direct current, of a few hundred to several thousand volts is applied between the two electrodes causing strong electric field built-up between electrodes, a partial of neutral atoms in the process chamber is ionized by collision of energetic electrons (which move in a distance enough to gain kinetic energy from electric filed) that driven moving downward to electrode that bias with a negative voltage and in coupled with moving along magnetic field lines (that generated by the permanent magnets mounted on the back side of the copper electrode) in corkscrew pattern, while ions that generated from collision will move in opposite direction of electrons that are moved to grounded electrode in upward direction. The interaction of ions to electric fields increases the degree of ionization to the neutral gas (N_2 molecules and O_2 molecules residue in vacuum chamber) by collision process. It is not clear understanding of dissociation and ionization process of N_2 plasma yet, due to the participation complicated processes by the interaction of; molecular vibration process, electron vibration process, chemical process and surface kinetic process. Explanation of the reactivity of N_2 molecules contained in plasma may play a major role by molecules excitation by vibration process and charge-exchange processes (metastable states) [4]. To dissociate N_2 molecule generally required energy of about 9.756 eV [5] and such this amount of energy possible supply by electron collision process in low pressure circumstance comply in equation (1) and (2). In addition, vibrational dissociation and dissociative recombination according to equation (3) [6] are also possible;

$$N_2 + e^- \rightarrow N^+ + N + 2e^- \quad (1)$$

$$N_2 + e^- \rightarrow N_2^+ + e^- \quad (2)$$

$$e^- + N_2^+ \rightarrow N + N \quad (3)$$

2.2. *Sample preparation*

The (100) oriented p-type (1-40 Ω·cm) of 3-inch diameter single-side polished silicon (Si) wafer was cut to small pieces. To deposit patterned Cu thin film on Si surface, the Si pieces were processed with a conventional photolithography process; laminated with polymer dry film photoresist, UV exposure by using desired patterned mask and mask development by sodium carbonate developer. After that, the sample was put on the iron electrode at the position that on top of the hole at the center of iron electrode, by facing the patterned dry film down to the Cu electrode so that the unprotected Si surface from cured UV dry film will be deposited by sputtered Cu atoms from Cu electrode when sputtering process is performed. The sputtering process is carried out at vacuum pressure at 235 Pa at applied negative DC voltage at -400 volts and total sputtering time is 6 minutes (continuously). Next, the sample is dipped in concentrated NaOH solution in order to remove dry film in which this process step resulted in patterned of Cu thin film. Finally, the Cu thin film was determined the thickness, morphology, elemental analysis and crystalline structure by various characterization techniques, detailed in next section.

3. Results and discussion

3.1. *Thickness and morphology of copper thin film by scanning electron microscope (SEM)*

In Figure 3, illustrate the cross sectional SEM image by scanning electron microscope (SEM, model: JEOL JSM-5410LV). The deposition process was performed under vacuum pressure at 235 Pa with −400 volts bias on Cu electrode for 6 minutes depositing time, obtained Cu film with 250 nm in thickness on silicon substrate, resulted in 41.7 nm/min deposition rate. It is seen obviously that the Cu film was deposit uniformly over the area appeared in Figure 3. For the morphology of Cu film, it is observed that when magnified greater than ×10,000 appeared blurred images, therefore the morphology cannot be revealed.

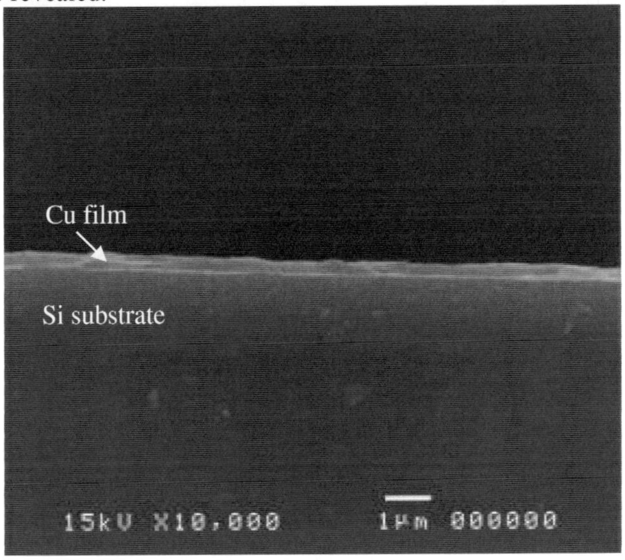

Figure 3. Cross sectional SEM image of Cu film on Si substrate.

3.2. Elemental analysis contained in copper thin film by energy dispersive x-ray spectrometer (EDX)
Elemental analysis of wide scan area of about 20μm×30μm (Figure 4 (left), by EDX measurement (equipped on SEM-JEOL JSM-5410LV) observed silicon (Si) atoms for 19.80 at% and copper (Cu) atoms for 80.20 at%. It is indicated that the deposited film is a thin copper film, the appearance of Si peak in the EDX spectra point out that the film is thin enough that energetic electron beam able to penetrate through the film reach the substrate surface and then collected signal from silicon substrate.

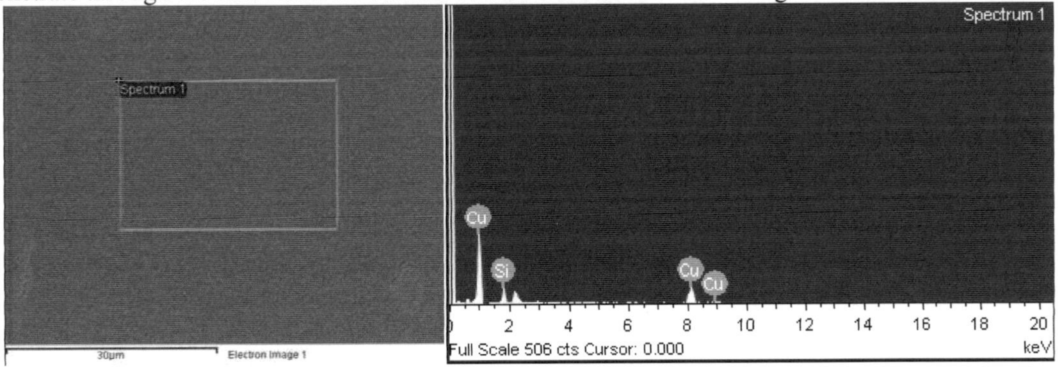

Figure 4. EDX scan area in pink rectangular (left) and EDX spectra of Cu film on Si substrate (right).

3.3. Crystalline structure of the copper thin film by x-ray diffractometer (XRD)
XRD spectra was recorded from X-ray diffractometer (model: Rigaku-MiniFlexII) at 2θ through a range of 2θ = 10° to 2θ = 80°, but 2θ = 60° – 76° was illustrated as shown in Figure 5. The dominant peaks were observed at 2θ = 69.3° and 2θ = 69.5° [7], along with very small peak at 2θ = 61.8° [8], were assigned for Si (100) of substrate material. In between 2θ = 20°– 50°, a broad peak [9] centered at 35° was detected and it is a characteristic of amorphous structure, as shown in Figure 6.

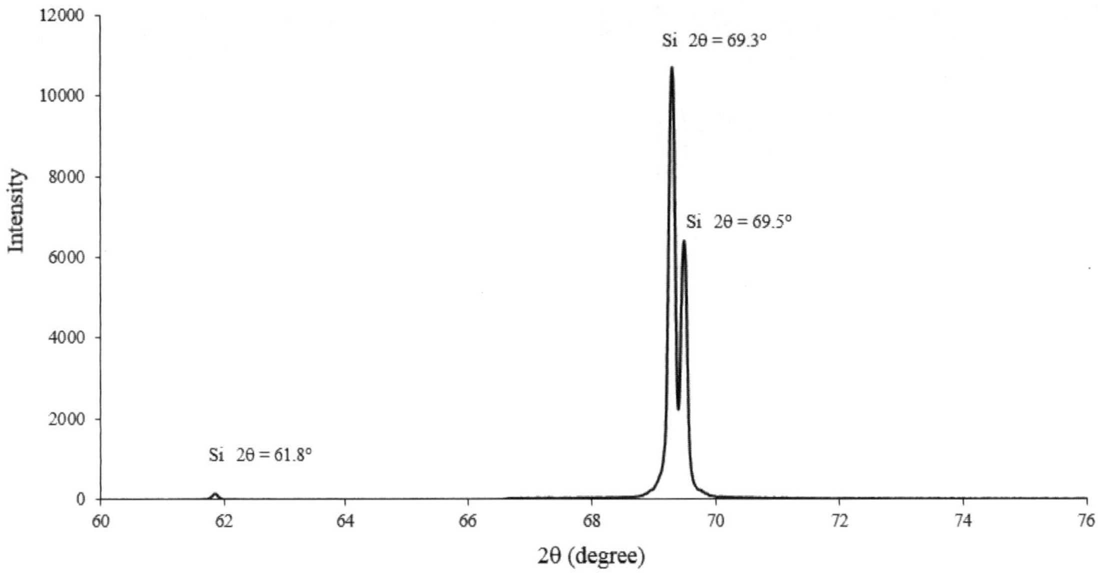

Figure 5. XRD pattern of 2θ = 60° – 76° for Si (100) substrate after deposition of copper thin film.

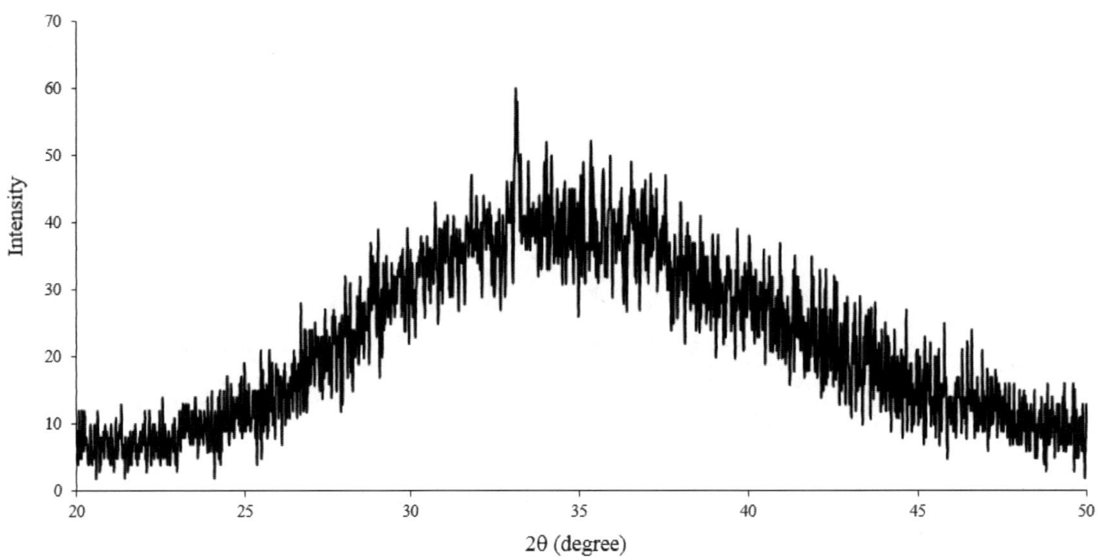

Figure 6. XRD pattern of 2θ = 20° – 50° for Si (100) substrate after deposition of copper thin film.

4. Conclusion

In this work, it is successfully to build a cheaply homemade DC magnetron sputtering system for deposit copper thin film on silicon substrate by assist of polymer mask photolithography. An EDX analysis revealed that the deposited layer on silicon substrate is copper film, as detected Cu atoms. The cross sectional SEM image showed that the Cu film has a thickness of about 250 nm that growth under deposit condition of maintaining voltage at –400 volts at vacuum pressure 235 Pa, in air plasma (which is N_2 based plasma) that the residue gas in vacuum chamber. Morphology of the Cu film cannot be observed by SEM because the limitation of magnification of SEM tool but XRD spectra indicated that the film is amorphous in nature.

Acknowledgement
The author would like to acknowledge the financial support (under grant number MJ.1-59-034) by the office of agricultural research and extension, Maejo University.

References

[1] Meng L J and Dos Santos M P 1996 *Thin Solid Films* **289** 65
[2] Limcharoen A, Pakpum C, Witit-anun N, Chaiyakun S and Limsuwan P 2012 *AMM.* **217-219** 712
[3] Pakpum C and Pussadee N 2016 *Surf. Coat. Technol.* **306** 299
[4] Guerra V, Sá P A and Loureiro J 2004 *Eur. Phys. J. Appl. Phys.* **28**(02) 125
[5] Frost D C and McDowell C A 1956 *Proc. of the Royal Society of London. Series A. Mathematical and Physical Sciences* **236**(1205) 278
[6] Tatarova E, Guerra V, Henriques J L and Ferreira 2007 *J. Phys.: Conf. Series* **71**(1) 012010
[7] Thakur M, Isaacson M, Sinsabaugh S L, Wong M S and Biswal S L 2012 *J. Power Sources* **205** 426
[8] Nezamdost J, Zolanvari A, Sadeghi H and Nezamdost J 2013 *Middle-East J. Sci. Res.* **18**(6) 845
[9] Lai T, Yin H and Lind M L 2015 *J. Membrane Sci.* **489** 264

IOP Conf. Series: Materials Science and Engineering **213** (2017) 012024 doi:10.1088/1757-899X/213/1/012024

Effect of annealing treatment on structure of Ag doped ZnO films deposited by thermal evaporation method

L N Wang[1,2,4], L Zhou[3], D Zhou[2], Y Yu[2], Z Q Luan[2], D L Tang[2], N Zhu[2], Z H Gao[2] N Liu[2] and L Z Hu[1]

[1]School of Physics and Optoelectronic Technology, Dalian University of Technology, Dalian, China
[2]Key Laboratory of Marine Biophysics of Liaoning Province, Dalian Ocean University, Dalian, China
[3]Department of Academics, Dalian Naval Academy, Dalian, China

E-mail: l.n.w@163.com

Abstract. The Ag doped ZnO films were prepared on sapphire substrates using simple thermal evaporation method. The effect of annealing treatment was investigated in detail. The results shows that the annealed Ag doped ZnO film has a prefer orientation along c-axis. The E_2 (high) mode in Raman spectra has a little blue shift which caused by crystal defects oxygen vacancies, meanwhile the compressive stress was exist in the film. After annealed, the intense of E_2 modes have great improvement which means the quality of the film was enhanced, meanwhile, the compressive stress released to some extent.

1. Introduction

Zinc oxide (ZnO) as a Ⅱ-Ⅵ semiconductor material, can be widely used in many fields, for instance nanogenerators [1-2], light-emitting diode (LED) [3-4], photodetectors [5], solar cells [6] and gas sensors etc [7-9]. As well known, because of the native defects such as interstitial zinc or oxygen vacancy ZnO expression n-type electrical conductivity, the application of ZnO based devices need both n-type and p-type ZnO. It is not easy to synthetise stable and reproduce p-type ZnO. Doping in ZnO with suitable elementary substance such as group Ⅴ(N,P,As and Sb), Group IB (Ag and Cu) can solve this problem efficiently [10]. Theoretical study has point out that based on its size and energy level, Ag can occupy the Zn site or interstitial to achieve p-type doping [11]. The studies also point out that Ag doping can adjust optical and electrical properties of ZnO thin films [12-14]. So it is necessary to synthesised stable and reproduceable Ag doped ZnO films [15-18].

Due to the different melting point, it is difficult to deposit Ag doped ZnO films (AZF), in this paper, we synthesised AZF using simple thermal evaporation method, and the effect of annealing treatment was studied by the image of scanning electron microscopy. Annealing treatment can improve the crystal quality to some extent; meanwhile it can make the dopant more effective.

2. Experiment

The AZF were deposited on sapphire substrates by thermal evaporation method. The substrate was cut into small pieces (1cm×1cm) then ultrasonically cleaned in acetone and ethanol for ten minutes, cleaning by deionized water repeatedly and subsequently dried in N_2 gas flow. The deposited process was divided into three steps: First, the metal Ag was used as vapour source deposited on sapphire substrate by using vacuum coating equipment (DM-300B). Secondly, the Zn film was deposited on Ag film using the same equipment. Finally, the samples were put into a horizontal tube furnace for oxidize. The furnace was heated into 400℃ for 60 minutes under constant flow of oxygen (100sccm). The diagram was shown in Figure 1. After oxidation, the AZF were annealed at 700℃ in oxygen ambient with flow rate of 100 sccm for 40 minutes, and then the AZF before and after annealed were obtained. The crystal structure properties of the samples were evaluated by X-ray diffraction (XRD). Surface morphology was observed by scanning electron microscope (SEM) and the optical properties of AZF were carried out using Raman spectra.

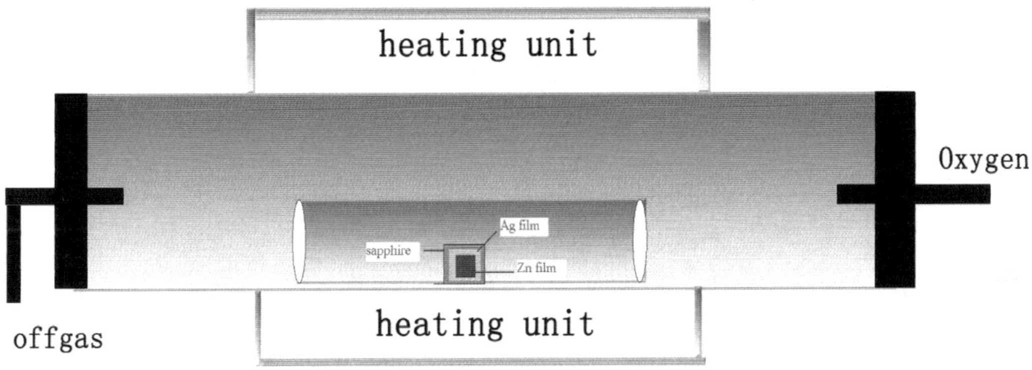

Figure 1. The illustration of experimental system.

3. Results and discussion

The surface morphology of Zn film marked as sample *a*, AZF heating at 400℃for 60 minutes marked as sample *b* and AZF annealed in oxygen for 40 minutes marked as sample *c* were shown in Figure 2a, b and c. The Zn film was composed of irregular pieces with many nanowires on the edge out of order. After oxidize sample *b* was composed of small grain arranged in order. There are also little small hole on the surface of the film, the amount of nanowires was reduced rapidly. As well known, the stand ZnO film was compact and normalizes, however due to Ag doping, there little changes in the surface ZnO film. When annealed in O_2, the film exhibit different feature: compact, regular grains mixed with little holes, only a few nanowires were found on the surface which indicates that after annealing treatment, the crystal quality have a great improvement.

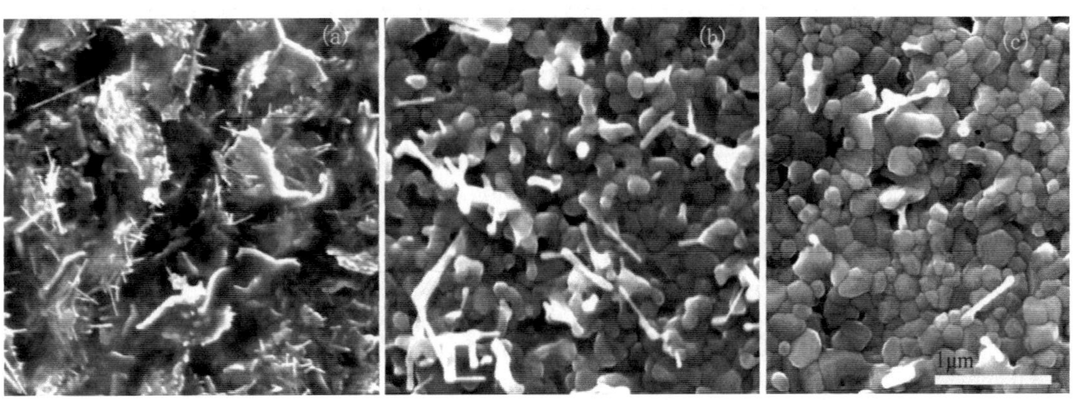

Figure 2. SEM image of Zn film(a); AZF heating at 400℃ for 60 minutes(b); AZF after annealed(c).

The XRD pattern of sample *b* and *c* were shown in Figure 3. There are seven peaks arrives at 31.63°, 34.33°, 36.18°, 62.75°, 65.08°, 72.57° and 78.18° respectively which corresponding to ZnO(100), (002), (101), (103), (200), (004) and (202) peak [19]. There are no phases caused by Ag oxides can be observed from the pattern. Among the seven peaks, ZnO (002) peak have a strong intense which means the sample have prefer orientation, the same orientation of c-axis which normal to the substrate. After annealing treatment, only ZnO (002) peak located at 33.95° with great intense could be found, this means the quality of the film have a great improvement. Compare with stand ZnO film (002) peak which located at 34.42°, both of the two samples have a slight shift to lower 2θ direction, this may be caused by the Ag doping, the exist of Ag atom affect the fabrication of ZnO film, this is consistent with SEM result. As we know the radius of Ag^{+} was 0.102nm bigger than the radius of Zn^{2+} which was 0.056nm[20], if Ag^{+}substitutes on Zn^{2+} sites, the 002 peak will shift to lower 2θ direction. The shift of sample *c* was bigger than sample *b*, which means after annealed more Ag^{+} was substitute Zn^{2+} sites[21], this may due to the annealing process the Ag layer permeate into ZnO films, so the 002 peak shift more.

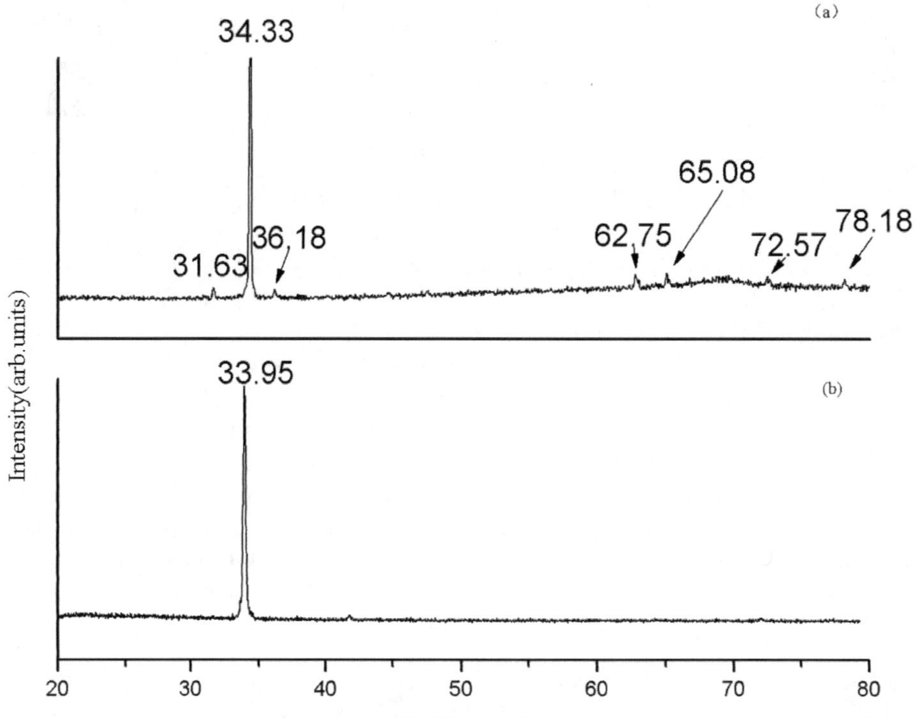

Figure 3. The XRD pattern of AZF before (a) and after (b) annealing.

Figure 4 shows the Raman spectrum of AZF before and after annealed. The six Raman active modes in ZnO crystal are $2E_1$, $2E_2$, $2A_1$. According to selection rules, only three of them can be observed from unpolarized Raman spectra which were E_2 (low), E_2 (high) and A_1 (LO). The Rman frequencies of the three modes were located at 101, 437 and 574 cm^{-1}, respective [22]. The E_2 (low) mode which associated with the vibration of the heavy Zn sublattice [23-25] and E_2(high) mode which associated with oxygen atoms appear at 96.03 cm^{-1} and 439.06 cm^{-1} could be clearly observed from Figure 4(a) respectively. The other six Raman peaks were from the sapphire substrate. Compare with stand ZnO, the E_2 (high) mode have slight shift to higher frequency direction, two reasons may be responsible. One is the crystal defect oxygen vacancies give rise to the lattice distortion, another one is the sample was under compressive stress. Because of the sensitivity on stress, E_2 (high) mode always used to explain the situation of stress in films, under compressive stress the Raman frequency would have shift to higher direction [26]. After annealing treatment, the intense of the E_2 (low) and E_2 (high) modes enhance a lot, the intensity of E_2 modes have strong relationship with crystal quality, the great intense of E_2 modes indicated the perfect crystal quality [27].Compare with as deposited Ag doped ZnO film, the E_2 (high) mode showed 1.75 cm^{-1} red shift, which means the compressive stress in the film release to some extent. From Figure 4 we can deduce that the annealing treatment improve the crystal quality of the film and make the compressive stress released.

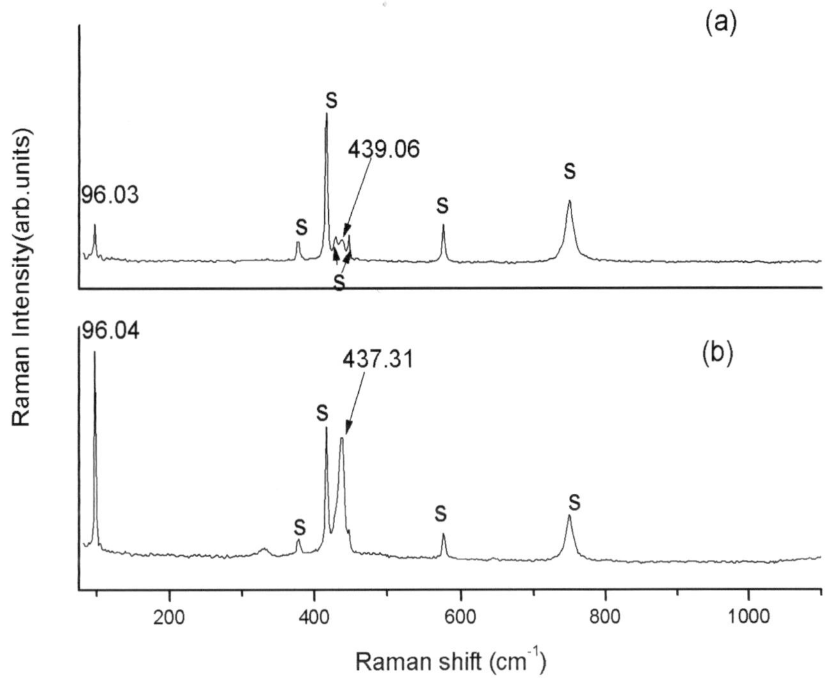

Figure 4. Raman spectra of AZF before(a) and after(b) annealing.

4. Conclusion

In summary, the Raman spectra of AZF before and after annealing treatment were deposited by simple thermal evaporation method on sapphire substrates. The effect of annealing treatment was investigated. The SEM image shows that after annealed the film was more compact, and the nanowires on the surface were reduced. The XRD results tell that the film was grown along c-axis and the Ag doping effect the fabrication of the film. Raman spectrum give information about the crystal quality, the film has perfect wurtzite structure and there was compressive stress in the film, after annealed, the stress was released to some extent.

Acknowledgement

This work was supported by the Nature Science Foundation of Liaoning Province (No. 201601285).

References

[1] Yin B, Qiu Y, Zhang H Q, Ji J Y and Hu L Z 2014 *Cryst. Eng. Comm.* **16** 6831
[2] Yin B, Qiu Y, Zhang H Q, Ji J Y, Lei J X, Luo Y M, Zhao Y and Hu L Z 2015 *J. Mater. Sci: Mater. Electron.* **26** 742

[3] Tsukazaki A *et al* 2005 *Nat. Mater.* **4** 42
[4] Tsukazaki A, Kubota M, Ohtomo A, Onuma T, Ohtani K, Ohno H, Chichibu SF and Kawasaki M 2005 *Japan J. Appl. Phys. Part 2* **44** 643
[5] Liang S, Sheng H, Liu Y, Huo Z, Lu Y and Shen H 2000 *J. Cryst. Growth* **225** 110
[6] Rau U and Schmidt M 2001 *Thin Solid Films* **387** 141
[7] Chen Y, Xu P C, Xu T, Zheng D and Li X X 2017 *Sens. and Actu. B* **240** 264
[8] Zou A L, Qiu Y, Yu J J, Yin B, Cao G Y, Zhang H Q and Hu L Z 2016 *Sens. and Actu. B* **227** 65
[9] Zou A L, Hu L Z, Qiu Y, Cao G Y, Yu J J ,Wang L N, Zhang H Q, Yin B and Xu L L 2015 *J. Mater. Sci.: Mater. Electron.* **26** 4908
[10] Paul R, Gayen R N, Biswas S , Bhat S V and Bhunia R 2016 *RSC Advances* **66** 61661
[11] Kim I S, Jeong E K, Kim DY, Kumar M and Choi S Y 2009 *Appl. Surf. Sci.* **255** 4011
[12] Ahn B D, Kang H S, Kim J H, Kim G H and Chang H W 2006 *J. Appl. Phys.* **100** 093701
[13] Gouvea C A K, Wypych F and Moraes S G 2000 *Chemosphere* **40** 433
[14] Abiyasa A P, Yu S F, Lau S P, Leong E S P and Yang H Y 2007 *Appl. Phys. Lett.* **90** 231106
[15] Mastour N, Ben H Z, Benchaabane A, Sanhoury M A and Kouki F 2016 *Phys. Lett. A* **380** 3866
[16] Mastour N and Bouchriha H 2013 *Org. Electron.* **14** 2093
[17] Ridene R, Mastrour N, Gamra D, Bouchriha H 2015 *Int. J. Mod. Phys. B* **29** 1550211
[18] Ridene S and Bouchriha H 2014 *Supe. Micr.* **75** 203
[19] Gayen R N and Paul R 2016 *Thin Solid Films* **605** 248
[20] Duan L, Gao W, Chen R Q and Fu Z X 2008 *Solid State Commun.* **145** 479
[21] Sahu D R 2007 *Microelectron J.* **38** 1252
[22] Arguello C A *et al* 1969 *Phys. Rev.* **181** 1351
[23] Yadav H K, Sreenivas K, Gupta V and Katiyar R S 2009 *J. Raman Spectrosc.* **40** 381
[24] Nusimovici M A and Birman J L 1967 *Phys. Rev.* **156** 925
[25] Pandiyarajan T, Udayabhaskar R and Karthikeyan B 2012 *Appl. Phys. A* **107** 411
[26] Yang L L, Yang J H, Wang D D, Zhang Y J and Wang Y X 2008 *Physica. E.* **40** 920
[27] Zhang R, Yin P G, Wang N and Guo L 2009 *Soli. Stat. Scie.* **11** 865

IOP Conf. Series: Materials Science and Engineering **213** (2017) 012025 doi:10.1088/1757-899X/213/1/012025

Morphology of polycaprolactone/needle-shaped hydroxyapatite (PCL/HAN) nanocomposite blends using ultrasound assisted melt blending

S Akhba[1,3], I Subuki[1], R W Sharudin[1] and M H Ismail[2]

[1]Faculty of Chemical Engineering, Universiti Teknologi Mara, 40170 Shah Alam, Malaysia
[2]Centre for Advanced Materials Research (CAMAR), Faculty of Mechanical Engineering, Universiti Teknologi Mara, 40170 Shah Alam, Malaysia

E-mail: suffiyana@salam.uitm.edu.my

Abstract. Polycaprolactone/Hydroxyapatite (PCL/HA) composite is an excellent material for fabricating tissue engineering scaffolds, especially as bone scaffolds. However, polarity differences between the hydrophilic HA and the hydrophobic PCL often leads to poor dispersion during the blending process. This is due to the tendency of the HA particles to agglomerate, and the difficulty to efficiently mix the polymer melt at high loading. The objective of this study is to introduce ultrasound waves during the conventional melt blending to overcome the problem. PCL and needle-shaped hydroxyapatite (HAN) were blended using an ultrasonically assisted extruder. The morphology of the PCL/HAN blend was investigated using field emission scanning electron microscopy (FESEM). This study found that the presence of ultrasound waves during melt blending was helpful in breaking up the HAN agglomerates. Thus, better dispersion of HAN was obtained, although some small agglomerates of HAN were still observed.

1. Introduction

Polycaprolactone (PCL) is a semi crystalline aliphatic polyester with a relatively low melting temperature that ranges between 50 and 60 °C, and a glass transition temperature of approximately -62.15°C [1]. It is employed in various biomedical applications, such as matrices for drug delivery, and tissue engineering scaffolds due to its biocompatibility, non-cytotoxicity, and low degradation rate [1]. Meanwhile, hydroxyapatite (HA, $Ca_{10}(PO_4)_6(OH)_2$) is a bioactive ceramic that is chemically similar to the inorganic matrix of natural bone. Thus, it has excellent osteoconductive and osteoinductive properties, which makes it commonly used for bone tissue applications. However, in the case of PCL-HA composites, a homogenization and intermixing problem was detected due to the hydrophilic and hydrophobic nature of HA and PCL, respectively. Therefore, this problem needs to be mitigated to utilize the beneficial effects of this composite.

Kim H-W (2007) applied oleic acid, an amphiphilic surfactant, to mediate the interaction between the hydrophilic HA and the hydrophobic PCL [2]. The mixing ratio of the HA/PCL was fixed at 0.3. The results showed that HA was uniformly distributed, which improved the mechanical properties of the composites. Raucci M.G *et al.* (2010) also utilized a similar approach, whereby an amphiphilic surfactant, Span85, had improved the interaction between PCL and HA, which was shown by the HA's homogenous dispersion [3]. The actual content of HA was at 34wt%. Meanwhile, Wang Y. *et al.*

(2010) suggested grafting PCL onto HA to improve the dispersibility of HA in the PCL [4]. Consequently, 20wt% of HA/PCL scaffold had shown better compressive strength (70Pa) and modulus (0.74MPa) compared to none grafted HA/PCL.

Other than that, Choi W.Y *et al.* (2010) and Rezaei A. and Mohammadi M.R (2013) were proposed another approach, which is *in situ* HA crystals were synthesized and precipitated with the PCL solution, instead of using a commercial HA [5-6]. However, the size, shape, and amount of HA could not be controlled precisely. Besides, other studies were introduced ultrafine-sized HA (with mean sizes of 0.5 μm, 5 μm, 10 μm, and 200 nm), and ultrasound wave during solution mixing in order to obtain HA well-dispersed [7-10]. However, this approach has one weakness: HA loading is limited due to the different polarity, which could lead to the agglomeration of HA in the PCL/HA composite [10].

All previously discussed approaches had solved the homogenization problem. However, the major drawback of these approaches is the use of large amounts of solvents, which require longer drying. This is the main concern for applications in bone tissue engineering where these solvents must be completely removed to avoid toxicity effect to the human body [11].

Due to this, some studies were introduced plasticized water soluble polymers, namely, polyethylene glycol (PEG) and polyethylene oxide (PEO) in melt blending to improve the intermixing process between PCL and HA at higher content [11-13]. However, PEO and PEG must be leached out prior to application.

This study was interested in introducing another approach, which was the ultrasonic-assisted melt blending process. Ultrasonic wave was expected to improve the homogenization of HA in the PCL blend. The main advantage of this approach is that it is a solvent-free blending method. This study has investigated the effect of ultrasound on the morphology of PCL/HA composite blends, and will be discussed in this paper.

2. Materials and method

2.1. Material

PCL (60,000 g/mol) in pellet form was purchased from Shenzhen Esun Industrial Co., Ltd. Needle-shaped hydroxyapatite powder was purchased from Berkeley. The density of the HAN powder was 2.97 g/cm^3 and has needle-shaped particles. The mean particle size distribution was 31.06 μm, which was analysed using a Malvern Instruments' Mastersizer 2000. Meanwhile, the molar ratio of calcium (Ca) and phosphorus (P), Ca/P was 1.64, determined by energy dispersive X-ray spectrometer (EDS). The FESEM micrograph in Figure 1 shows the detail of HAN's particle surface as received.

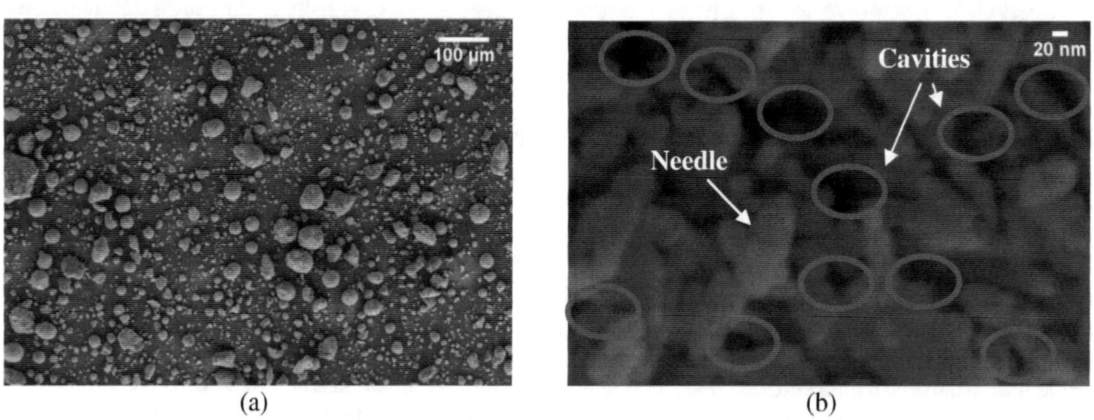

Figure 1. FESEM images for HAN, as received. (a) HAN particles, as received, tend to agglomerate, and have a mean particle size distribution of 31.06 μm, (b) HAN, as received, was presented in the form of agglomerates with many cavities or voids, and shaped like needles.

2.2. Methods

2.2.1. Sample preparations. All PCL-containing materials were dried under vacuum at 38 °C for a minimum of 10 hours, while HAN powder was dried at 120 °C for 10 hours to avoid moisture-induced degradation reactions [12]. HAN loading was varied from 10 to 30wt%. PCL and HAN were mixed in a shaker before undergoing melt blending in an extruder equipped with an ultrasonic placed at the end of the die. Melt blending was conducted at a temperature profile of 110, 100, 110, 110, and 100 °C from hopper to nozzle with a screw rotation speed of 9 rpm. Ultrasound was used at 100% cycle at 400 watt and 22 Hz. After the extrusion, the strands were cooled in a water bath, and consequently, pelletized. Then, the pelletized PCL/HAN nanocomposite blend was compressed using a hot press at temperature and pressure of 80 °C and 700 psi, respectively, to produce disc-shaped samples that were 20 mm in diameter and 1 mm thick. These discs were fractured using liquid nitrogen for the morphology study.

2.2.2. Morphology of PCL/HAN nanocomposite blends. The morphology of these PCL/HA composite samples were assessed using a field emission scanning electron microscope (FE-SEM). The samples were cross-sectioned, gold–coated, and analyzed using the FE-SEM (model GEMINI: ZEISS SUPRA 55VP) under secondary electron imaging.

3. Results and discussion

3.1. Morphology of PCL/HAN nanocomposite blends

The effectiveness of the ultrasonically assisted melt blending was investigated through the morphology of the PCL/HAN nanocomposite blends, as shown in Figure 2.

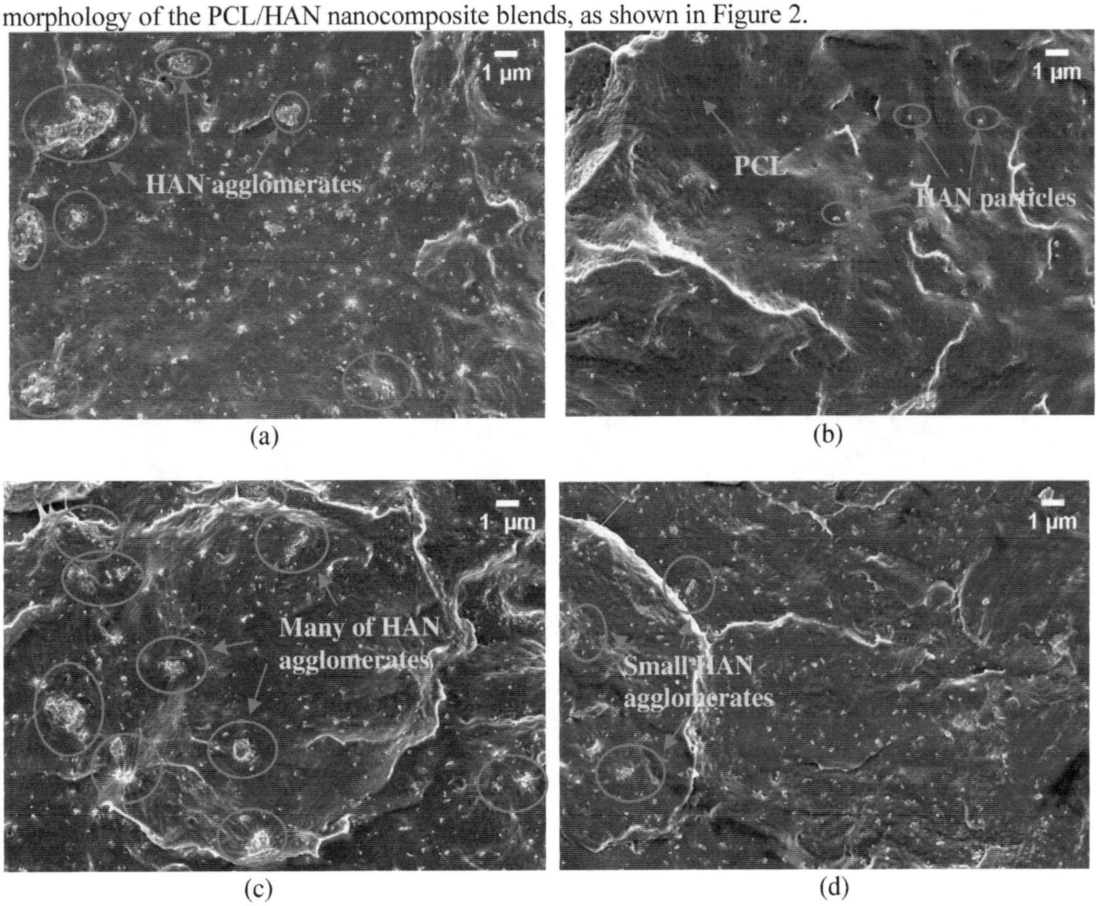

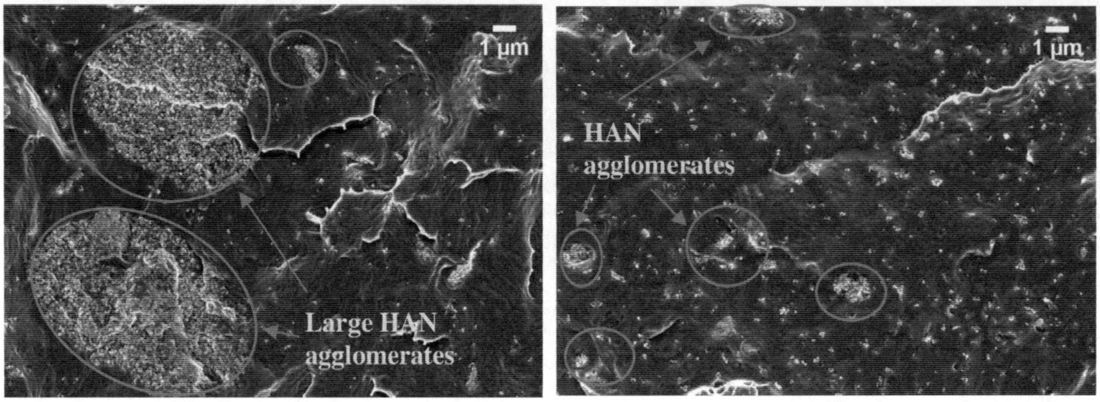

(e) (f)

Figure 2. (a), (c) and (e) are FESEM images for PCL/HAN nanocomposite blends without ultrasonic for 10, 20 and 30wt% HAN respectively. (b), (d) and (f) are FESEM images for PCL/HAN nanocomposite blends with the presence of ultrasonic for 10, 20 and 30wt% HAN, respectively.

From the micrographs shown in Figure 2, it was clear show the different morphology of PCL/HAN blends due to the influent of ultrasonic wave. In the presence of ultrasonic wave, HAN particles were evenly dispersed in the PCL matrix at 10wt% of HAN. However, some small HAN agglomerates were observed at 20wt% and 30wt% of HAN loadings. Meanwhile, in the absence of ultrasonic, small HAN agglomerates were found at 10wt% HAN and the agglomerates' size become larger at 20 and 30wt% of HAN.

These results showed that the as-received HAN's agglomerates could have experienced breakage, and were dispersed in the PCL matrix. According to the mechanism proposed by Zhong J *et al.* (2016) illustrated in Figure 3, the ultrasonic wave propagation in the cavities of HAN's agglomerates may have generated oscillatory pressure waves, and induced bubble expansion and contraction, which led to a breakage of the agglomerates [14]. This mechanism was helpful in reducing the size of HAN agglomerates during melt blending, thus leading to better dispersion of HAN particles.

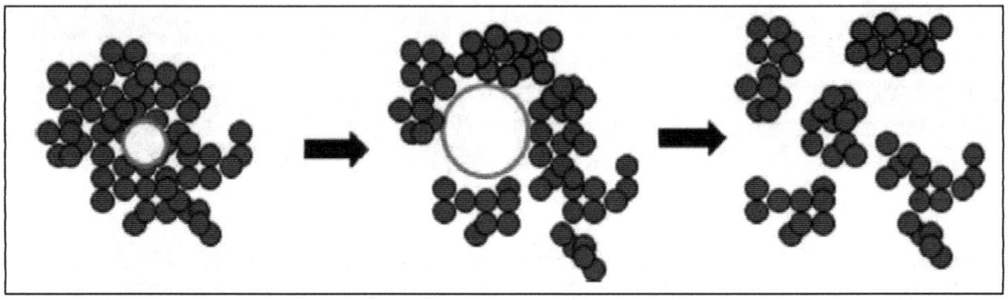

Figure 3. Inside cavitation mechanisms by ultrasonic wave in PCL/HA composite blends [14].

Nonetheless, even though the ultrasonic waves had reduced HAN agglomeration, some small agglomerates were still observed at higher HAN loading (20 and 30wt% HAN). This is because at higher loading, HAN particles have a stronger tendency to agglomerate [5]. Consequently, 400 Watt of ultrasonic waves was still ineffective to completely break HAN agglomeration. However, this method had improved the morphology of PCL/HAN nanocomposite blends compared to conventional melt blending.

4. Conclusion

Ultrasound waves were helpful in the breakage of HAN agglomerates and produced better morphology in PCL/HAN nanocomposites during the melt blending process compared to conventional met blending (without ultrasonic) which produced many and larger HAN agglomerates. This alternative method can be used to replace solution blending due to its solvent-free usage.

Acknowledgements

The researchers would like to acknowledge the financial support by the Fundamental Research Grant Scheme (FRGS, 600-RMI/FRGS 5/3 (89/2014), from Universiti Teknologi Mara (UiTM), Malaysia.

References

[1] Wang L, Zhou H, Wang X and Mi J 2015 Mechanism of bubble nucleation in poly (e-caprolactone) foaming at low temperature *Polymer* **79** 47-55
[2] Kim H W 2007 Biomedical nanocomposites of hydroxyapatite/polycaprolactone obtained by surfactant mediation *J. Biomed. Mater. Res. A* **83A** 169-77
[3] Raucci M G, D'Antò V, Guarino V, Sardella E, Zeppetelli S, Favia P and Ambrosio L 2010 Biomineralized porous composite scaffolds prepared by chemical synthesis for bone tissue regeneration *Acta Biomaterialia* **6(10)** 4090-9
[4] Wang Y, Dai J, Zhang Q, Xiao Y and Lang M 2010 Improved mechanical properties of hydroxyapatite/poly (ε-caprolactone) scaffolds by surface modification of hydroxyapatite *J. Appl. Surf. Sci.* **256** 6107-12
[5] Choi W Y, Kim H E, Oh S Y and Koh Y H 2010 Synthesis of poly (ε-caprolactone)/ hydroxyapatite nanocomposites using in-situ co-precipitation *J. Mater. Sci. Eng., C* **30** 777-80
[6] Rezaei A and Mohammadi M R 2013 In vitro study of hydroxyapatite/polycaprolactone (ha/pcl) nanocomposite synthesized by an in situ sol-gel process *J. Mater. Sci. Eng., C* **33** 390-6
[7] Kim M J and Koh Y H 2013 Synthesis of aligned porous poly (ε-caprolactone) (PCL)/hydroxyapatite (HA) composite microspheres *J. Mater. Sci. Eng., C* **33(4)** 2266–72
[8] Salerno A, Fanovich M A and Pascual C D 2014 The effect of ethyl-lactate and ethyl-acetate plasticizers on PCL and PCL–HA composites foamed with supercritical CO_2 *J. Supercrit. Fluids* **95** 394-406
[9] Ivanovic J, Knauer S, Fanovich A, Milovanovic S, Stamenic M, Jaeger P, Zizovic I and Eggers R 2016 Supercritical CO_2 sorption kinetics and thymol impregnation of PCL and PCL-HA *J. Supercrit. Fluids* **107** 486-98
[10] Salerno A and Domingo C 2015 Pore structure properties of scaffolds constituted by aggregated microparticles of PCL and PCL-HA processed by phase separation *J. Porous Mater.* **22** 425-35
[11] Liu L, Wang Y, Guo S, Wang Z and Wang W 2012 Porous polycaprolactone/ nanohydroxyapatite tissue engineering scaffolds fabricated by combining NaCl and PEG as co-porogens: Structure, property, and chondrocyte-scaffold interaction in vitro *J. Biomed. Mater. Res. B Appl. Biomater.* **100 B** 956-66
[12] Chui Z, Nelson B, Peng Y, Li K, Pilla S, Li W J and Shen C 2012 Fabrication and characterization of injection molded poly (ε-caprolactone) and poly (ε- caprolactone)/ hydroxyapatite scaffolds for tissue engineering *J. Mater. Sci. Eng., C* **32** 1674-81
[13] Salerno A, Zeppetelli S, Di Maio E, Iannace S and Netti P A 2012 Architecture and properties of bi-modal porous scaffolds for bone regeneration prepared via supercritical CO_2 foaming and porogen leaching combined process *J. Supercrit. Fluids* **67** 114-22
[14] Zhong J, Isayev A I and Zhang X 2016 Ultrasonic twin screw compounding of polypropylene with carbon nanotubes, graphene nanoplates and carbon black *Eur. Polym. J.* **80** 16-39

IOP Conf. Series: Materials Science and Engineering **213** (2017) 012026 doi:10.1088/1757-899X/213/1/012026

The effect of C_{60} on the optical absorption properties of HPPH photosensitizer

C H Zhou[1,2], Z H Sun[1], J C Zhang[1] and G Fan[1]

[1]Department of Chemistry and Chemical Engineering, Xianyang Normal University, Xiangyang 712000, China

E-mail: xbdxzch@126.com

Abstract. The optical absorption properties of the clinical photosensitizer (HPPH) and HPPH-C_{60} dyads (HPPH-C_{60} (1), HPPH-C_{60} (2) and HPPH-C_{60} (3)) have been calculated by density functional theory (DFT) and time dependent density functional theory (TD-DFT). Based on the optimized geometrical structures of three HPPH-C_{60} dyads, their binding energies, the charges populations and the excited energies are calculated specifically. The results show that the different binding sites between C_{60} and HPPH induce the different binding energy. The maximum binding energy and the significant charge transfer occur at C_{60} and nitrogen atoms of HPPH in HPPH-C_{60}(1). At the same time, the excited energies of three HPPH-C_{60} dyads are smaller than that of HPPH. The maximum absorption peak of the HPPH-C_{60}(1) reaches near infrared 961.69nm.

1. Introduction

In recent years, it is attracted great attention that the photosensitizer applying to photodynamic therapy. As the third generational photosensitizer, HPPH has remarkable curative effect in clinical treatment [1-3]. So it has been widely applying to cure skin cancer maculopathy adjuvant therapy. However, HPPH still belongs to porphyrin drugs, and porphyrin rings are modified by carbonyl groups and hydroxyl groups, it can reduce molecular symmetry and conjugation. The investigation shows that HPPH has hardly enough triplet state energy, which induces low singlet oxygen yields. So a low dosage of HPPH has slight therapeutic effect, it needs repeated treatment [4-5]. But too much oxygen radical will injure cytomembrane, which has great toxic and side effect. So that it is very urgent to improve the effect of treatment.

Fullerenes are a class promising bioactive materials [6-7], which can be apply to the anticancer and photodynamic therapy. For example, during the beginning of the twentieth century, the antitumor effect of C_{60}-PEG compound had been reported by Tabata[8]. It is found that C_{60}-PEG can stay at cancer cell for longer time and have better antitumor effect than clinical photosensitizer photofrin. In 2007, Witte reported that fullerene can interaction with cytochrome C. That is, the stable compound can be formed by fullerene photosensitizer and DNA, which can depress the reproduction of cancer cells when it injects into biological body [9]. In 2015, Francis D'Souza designed one kind of fullerene photosensitizer containing pyrrole and porphyrin groups, winch has great triplet efficiency and powerful infrared absorption ability [10]. He predicted that those fullerene derivatives can be as potential photosensitizer. At the same time, Bang ping et al. synthesized water-soluble carbon nano-materials and phthalocyanin compounds that can be used to photodynamic therapy [11]. In 2015, it is

IOP Conf. Series: Materials Science and Engineering **213** (2017) 012026 doi:10.1088/1757-899X/213/1/012026

proved that single-walled carbons nanotubes will effective promote the yields of singlet oxygen and enhance photodynamic effects [12].

The above extensive researches showed that complexed carbon nanomaterials such as fullerene combined with porphyrin, will enhance singlet oxygen yields of prophyrin photosensitizers and en-larged absorption peak of photosensitizers to vis-infrared regions. So it is so important to develop fullerene-porphyrin photosensitizer for promoting photodynamic effects. However, we still need further study the intrinsic properties of fullerene-porphyrin complexes. Such as, the combined sites, the modes of interaction and the change of absorption spectra for fullerene and porphyrin. So the structure and the optical absorption properties of the HPPH are calculated by DFT/TD-DFT in this paper. Based on above calculated results, according to possible interaction site of C_{60} and HPPH, it can get three kinds of stable composites, HPPH-C_{60}(1), HPPH-C_{60}(2) and HPPH-C_{60}(3). The binding energy, charge population and excited energy of three kinds of composites are also calculated using DFT/TD-DFT. The intrinsic reason for C_{60} promoting HPPH optical absorption properties is dis-cussed.

2. Computing method

Based on the density functional theory (DFT) with the M06-2X[13,14] exchange-correlations, geometry structures and vibration frequency of four complexes are calculated in this work. All examined complexes are no symmetry constraints. Using the adiabatic approximation of time dependent density functional theory TD-DFT/M06-2X/6-31+G(d, p), their excited energies and oscillator strengths are computed. For ground state structure optimizations, it is used that the split valence plus polarization basis sets 6-31+G(d, p) is for nonmetal (H, C, N, S, O) atoms. The counterpoise correction was used to correct for the basis set superposition error (BSSE) at the same level of theory as used for the geometrical optimizations [15-17]. BSSE is calculated with the counter poise procedure method advance by Boys and Bernardi[16]. The corrected interaction energy (ΔE) shown that $\Delta E = E(HPPH\text{-}C_{60})\text{-}EHPPH(C_{60})\text{-}EC_{60}(HPPH)$. In this formula, E(HPPH-C_{60}) is the energy of composites(HPPH-C_{60}(1), HPPH-C_{60}(2) and HPPH-C_{60}(3)). EHPPH (C_{60}) is the energy of HPPH based on all the atoms are set as puppet atoms with virtual orbital in C_{60}. EC_{60}(HPPH) is the energy of C_{60} based on all the atoms be set as puppet atoms with virtual orbital in HPPH. Based on calculated results from Gaussian09, the optical absorption spectra are simulated by SWizard program[18], and the peak half-width is set in 0.209eV(1685.83cm-1).

3. Results and discussion

3.1. The Ground-state structure of HPPH-C_{60} composites

Based on M06-2X/6-31+G(d, p), three structures of HPPH-C_{60} composites are completely optimized. The Figure 1 shows stable configurations of three composites without imaginary frequency. The weak interactions between C_{60} and HPPH are found in Figure 1. In composites HPPH-C_{60}(1), the weak interactions are from carbon atoms of C_{60} with nitrogen atoms of porphyrin ring, their shortest distance is 2.52Å. In another two composites, the weak interactions are between C_{60} and HPPH are 1.94 Å and 2.41 Å, respectively.

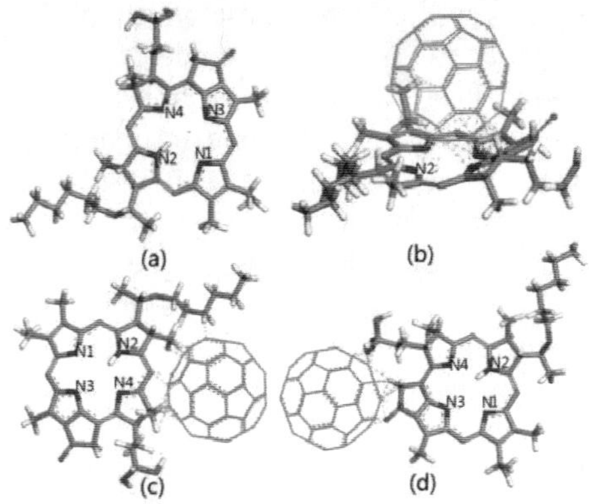

Figure 1. The structures of HPPH (a), HPPH-C_{60}(1) (b), HPPH-C_{60}(2) (c) and HPPH-C_{60}(3) (d).

It is found that the C=N and C=C bonds are elongated in three composites, comparing with those in HPPH. The biggest deviation is less than 3%, it means that the weak interaction slightly influence on geometrical structure of porpyrin ring. Based on BSSE correction, the binding energies of three composites (HPPH-C_{60}(1),HPPH-C_{60}(2) and HPPH-C_{60}(3)) are computed. Their binding energies are 0.4107eV, 0.2824eV and 0.3269eV, respectively. The computational formula of binding energy is as follow: Ebindingenergy=$|\Delta E|$=E(C_{60}-HPPH)-E(C_{60})-E(HPPH). From binding energies, it can find that the composites HPPH-C_{60}(1) has the biggest binding energy, the composites HPPH-C_{60}(2) has the least energy. So C_{60} trends to interact with porpyrin ring in HPPH. That is, the C_{60} can receive the charges from nitrogen atoms of porpyrin ring, and it is similar to many reports [18-22]. So the charge populations of three composites and HPPH are analyzed by natural bond orbital (NBO) theory. As Table 1 shown that the charge of N(1) is -0.2081e in HPPH, in three composites, the charge of N(1) is reduced, they are -0.2067e,-0.2065e,-0.2067e, respectively. It means that negative charges of nitrogen atoms are almost equal in three compounds. The charges of oxygen atoms in HPPH are affected by carbon atoms of C_{60}. For example, in HPPH-C_{60}(3), carbon atoms of C_{60} interact with the O(23) and O(24) of HPPH, and the charges of O(23) and O(24) is lower than in HPPH molecule. So from binding energies and charge populations of three composites, it can see that the C_{60} hardly influence geometrical structure of HPPH, and the charges transfer is from HPPH to C_{60} in all composites. Among three composites, the binding energy of HPPH-C_{60}(1) is the biggest, the charge transfer is also the most obvious for HPPH-C_{60}(1).

Table 1. The NBO charge populations of partial atoms for HPPH and three composites.

Charge /e	HPPH	HPPH-C_{60}(1)	HPPH-C_{60}(2)	HPPH -C_{60}(3)
N(1)	-0.2081	-0.2067	-0.2065	-0.2067
N(2)	-0.0243	-0.0269	-0.0242	-0.0245
N(3)	-0.0476	-0.0498	-0.0476	-0.0472
N(4)	-0.2033	-0.2002	-0.2032	-0.2032
O(17)	-0.1067	-0.1059	-0.1051	-0.1074
O(18)	-0.2550	-0.2533	-0.2545	-0.2548
O(19)	-0.1171	-0.1188	-0.1188	-0.1192
O(20))	-0.1595	-0.1578	-0.1599	-0.1627
O(21))	-0.2196	-0.2202	-0.2306	-0.2238
O(22)	-0.2661	-0.2698	-0.2691	-0.2699
O(23)	-0.1055	-0.1052	-0.1055	-0.0952
O(24)	-0.2559	-0.2562	-0.2556	-0.2530

3.2. The Absorption spectra simulation of HPPH and three composites

Absorption spectra of photosensitizers is an important character, so absorption spectra of three composites (HPPH--C_{60}(1), HPPH-C_{60}(2) and HPPH-C_{60}(3)) and HPPH are calculated and discussed in details.

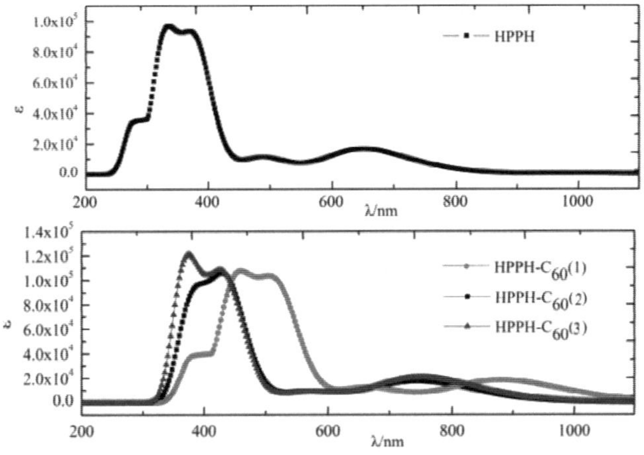

Figure 2. The absorption spectra of HPPH and its three composites.

Based on the calculated excited energies, absorption spectra of HPPH and three composites (HPPH--C_{60}(1), HPPH-C_{60}(2) and HPPH-C_{60}(3) are simulated in Figure 2. the first excited energy of HPPH is the biggest, the excited energies of HPPH-C_{60}(2) and HPPH-C_{60}(3) are almost similar. The longest absorption bands are 759.74nm and 765.40nm, and the extinction coefficient reach 10^5 in HPPH-C_{60}(2) and HPPH-C_{60}(3), so their longest absorption bands are the longer than that of HPPH. For the longest absorption wavelength of HPPH, the calculated value is 658.09nm, which is very close with the experimental value 660nm[23]. The first excited energy of HPPH-C_{60}(1) is the smallest among three composites, so the longest absorption band of HPPH-C_{60}(1) reaches 946.96nm. It is proved that C_{60} can extend the longest absorption band of HPPH. That is, C_{60} can improve the light injection efficiency of HPPH photosensitizer in photodynamic therapy.

Table 2. The frontier orbital's energies of HPPH and three compounds.

E /eV	HPPH	HPPH-C_{60}(1)	HPPH-C_{60}(2)	HPPH -C_{60}(3)
HOMO	-4.475	-4.445	-4.522	-4.593
LUMO	-3.091	-3.796	-3.827	-3.821
ΔE_{L-H}	1.384	0.649	0.695	0.772

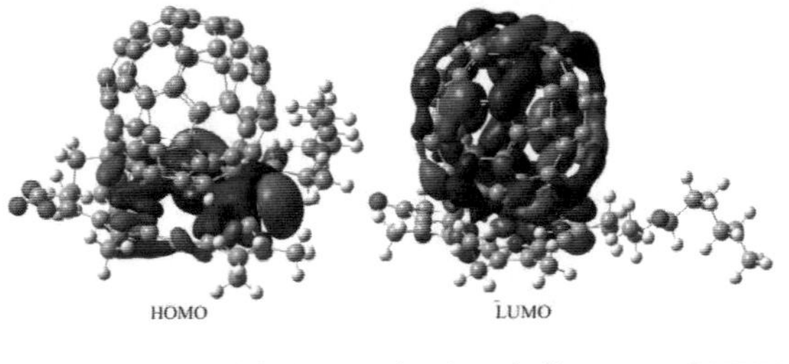

Figure 3. The frontier orbital constitutional diagram of HPPH-C_{60}(1).

To reveal the effects of electronic structures on absorption bands of composites, their frontier orbital energies and the atomic orbital components of HPPH-C_{60}(1) are calculated and analyzed. From Table 2, it can be seen that their highest occupied orbital energies are almost similar in three composites, but their lowest unoccupied orbital energies are reduce obviously, comparing with those orbitals in HPPH. It causes the energy gaps between the lowest unoccupied orbital and the highest occupied orbital of three composites are smaller than energy gap between the lowest unoccupied orbital and the highest occupied orbital of the HPPH. From Figure 3 shown, for HPPH-C_{60} (1), the highest occupied molecular orbital is still from porphyrin ring, and the lowest unoccupied molecular orbital is from π orbital of fullerene. The energy gap of HPPH-C_{60} (1) is the smallest among three composites. The frontier orbital populations of HPPH-C_{60}(2) and HPPH-C_{60}(3) are similar to HPPH-C_{60}(1), but their energy gaps are lower than that of HPPH-C_{60}(1). It is note that the energy gap of frontier orbitals (HOMO and LUMO) are not equally to the first excited energy, because the electronic excitation also considers the orbital symmetry besides the orbital energy.

4. Conclusion

In conclusion, based on optimized structures of HPPH photosensitizer and its three composites (HPPH-C_{60}(1), HPPH-C_{60}(2)and HPPH-C_{60}(3), their binding energies, charge populations and excited energies have been calculated by density functional theory. The calculated results show that the weak interactions between the donating electron atoms of HPPH and C_{60} resulting into three stable composites. Among three composites, HPPH-C_{60}(1) has the greatest binding energy, in which C_{60} interacts with nitrogen atoms of porphyrin ring and the obvious charge transfer can been found from nitrogen of HPPH to C_{60}. From the calculated excited energies, it can find that three composites have less excited energies than HPPH do, simulated absorption bands of three composites also are longer than that of HPPH. The above results due to the unoccupied orbitals of C_{60} are lower than the unoccupied orbitals of HPPH, as a result, occupied electrons can easily excited to unoccupied orbitals in composites. Especially, the absorption wavelength of HPPH-C_{60}(1) reaches 946.9nm, which belongs to infrared region of the solar light.

Acknowledgement

This work was supported by the Specialized Research Fund of Xian yang Normal University (No. 14XSYK013, 15XSYK040), the 56th China Postdoctoral Science Foundation (No. 2014M560758), and National Undergraduate Training Programs for Innovation and Entrepreneurship(NO. 201610722011).

References

[1] Zhao Z *et al* 2012 *Inorg. Chem.* **51** 812–21
[2] Ertan A, Zaheer I, Correa A M, Thosani N and Blackmon H S 2013 *World J. Gastroenterol* **19** 7106-13
[3] Komatsu T, Wang R M, Zunszain P A, Curry S and Tsuchida E 2006 *J. Am. Chem. Soc.* **128** 16297-301
[4] Perni S, Prokopovich P, Pratten J, Parkin I P and Wilson M 2011 *Photochem. Photobiol. Sci.* **10** 712-20
[5] Bakry R, Vallant R M, Najam-ul-Haq M, Rainer M, Szabo Z, Huck C W and Bonn G K 2007 *Int J Nanomedicine* **2** 639-49
[6] Stratton S P, Schaefer W H and Liebler D C 1993 *Chem. Res. Toxicol.* **6** 542-7
[7] Hamano T, Okuda K, Mashino T, Hirobe M, Arakane K, Ryu A, Mashiko S and Nagano T 1997 *Chem. Commun.* 1 21-2
[8] Tabata Y, Murakami Y and Ikada Y 1997 *Jpn. J. Cancer Res.* **88**(11) 1108-16
[9] Liu J, Ohta S I and Sonoda A J 2007 *J. Control. Release* **117** 104-10
[10] Obondi C O, Lim G N and D'Souza F 2015 *J. Phys. Chem. C.* **119** 176-85
[11] Wang L, Shi J, Liu R, Liu Y, Zhang J, Yu X, Gao J, Zhang C and Zhang Z 2014 *Nanoscale* **6** 4642-51
[12] Schmitt J, Heitz V, Sour A and Bolze F 2015 *Angew. Chem. Int. Ed.* **54** 169–73
[13] Zhao Y and Truhlar D G 2008 *Theor. Chem. Account* **120** 215-41
[14] Zhao Y and Truhlar D G 2006 *J. Chem. Phys.* **125** 194101
[15] Hay P J and Wadt W R 1985 *J. Chem. Phys.* **82** 270-83

[16] Tagmatarchis N, Kato H and Shinohara H 2001 *Phys. Chem. Chem. Phys.* **3** 3200-2
[17] Jeziorski B, Moszynski R and Szalewicz K 1994 *Chem. Rev.* **94** 1887
[18] Gorelsky S I 1998 *Swizard Program* (Toronto: York University)
[19] Schmitt J, Heitz V, Sour A and Bolze F *2015 Angew. Chem. Int. Ed.* **54** 169-73
[20] Cheng L, Wang C, Feng L Z, Yang K and Liu Z 2014 *Chem. Rev.* **114** 10869-939
[21] McCluskey D M, Smith T N, Madasu P K and Phillips J P 2009 *Appl. Mater. Interfaces* **1** 882-7
[22] Wang F, Cui X N, Lou Z R, Zhao J Z, Bao M and Li X W 2014 *Chem. Commun.* **50** 15627-30
[23] Goswami L N, White W H, Spernyak J A, Ethirajan M, Chen Y, Missert J R, Morgan J, Mazurchuk R and Pandey R K 2010 *Bioconjug Chem.* **21**(5) 816-27

The application of PA/CF in stab resistance body armor

M Q Yuan[1,2], Y Liu[1], Z Gong[1] and X M Qian[1]

[1]Beijing Institute of Technology, State Key Laboratory of Explosion Science & Technology, Beijing 100081, China

E-mail: myuan@bit.edu.cn

Abstract. Stab resistance body armor (SRBA) is an essential defensive equipment to protect human body against injuries from stabbing. The conventional SRBAs shared low wearing frequency since they are heavy and poor in flexibility. This paper designed a structured stab-resistance plate using the model of crocodile armor and manufactured using 3D printing technology-laser sintering (LS). CF(Carbon fiber) was applied to enhance the stab resistance properties of SRBA. The effects of the material and structure were analysed through the stab resistance property tests based on the national standard GA68-2008. It is found that the stab resistance property of flat plates sintered by PA powder and PA/CF are both weaker than that of the structured plate. The penetrating depth of PA/CF structured plate is significantly 2-mm-less than the pure PA structured plate. The SEM observations confirmed the conclusion that addition of the CF largely improved the plate stab resistance property. Moreover, using PA/CF structured plate to produce the stab resistance body armor would result in a weight reduction by about 30-40% as compared to the existing SRBA that was made up of metal plates, which could largely reduce the wearer physical burden and improve the wearing frequency.

Keywords: Laser sintering, Stab resistance body armor, Polyamide, carbon fiber, Bionic structure

1. Introduction

The number of terrorist attacks has increased rapidly in these years. In China, the use of gun is strictly restrained, thus the threat mainly comes from knives. According to a survey, 6,000 people were killed during the period 2011-2015 due to knife injuries. There were 500,000 cases reported, during 2009 to 2015, against illegal carriage of the restrained knives. Stab resistance body armor is essential equipment in protecting humans from knife injuries [1]. However the low wearbility results in low protection effect, which is mainly due to three reasons: firstly the conventional SRBA which used alloy materials is very heavy, of which the weight could reach 3.5-4.0 kg [2], causing heavy burden to the human body; secondly the SRBA shows low flexibility due to the integral plate design; thirdly the SRBA exhibits the low thermo-physical properties, of which the thermal insulation and vapour resistance are two times larger than that of a winter jacket [3]. The limited development of SRBA is mainly due to the limitation on material selection, structure design, and manufacturing methods: the conventional used materials include rigid panels (i.e. alumina, ceramic, etc.) or flexible panels (i.e. Kevlar fabric [2], UHMWPE fabric [3], etc.), which is either too hard or multi-layered due to the limitation on mechanical properties of the fabric. Therefore, an advanced type of SRBA is desired which would exhibit excellent stab resistance property while being light weight and more flexible.

The stab resistance property is achieved as a collaboration of the material mechanical property and the designed energy dissipation structure. For centuries, scientists and designers have studied natural bio-protection layers in animals trying to find inspirations to develop innovative body armors. The first biological armor system could be traced back to 540 million years ago to the Paleozoic era where such evolved to provide protections against penetrating predatory attacks [4]. Many modern-day animals also possess armors, including mammals (e.g., armadillos and pangolin), reptiles (e.g. alligators, crocodiles, lizards) and numerous fish [5], which provide new solution for SRBA design. Yang et al [4] examined scales from several fish and mammals and concluded that the flexibility has been increased without significantly sacrificing strength by owning a hierarchical structure with collagen fibers joining more rigid units. It is pointed out the understanding of such flexible dermal armors is essential in that it may provide a basis for new synthetic armor materials. John [6] studied five classifications of scales containing placoid, ganoid, osteoderms, elasmoid and the pangolin. The scale size was calculated and it is stated when arranging the individual scale elements in an imbricated assembly way, it is possible to create a hierarchical structure capable of providing effective and flexible levels of protection against localized threat while minimizing back face deformation.

However, the fact that the animal armor could provide excellent inspiration for the SRBA structure design, the complex structure is difficult to realize in conventional manufacturing methods. Therefore, one solution to be explored, in an attempt to address these issues of developing naturally inspired body armor featuring complicated interlinked assemblies, is that of additive manufacturing (AM) [7]. Laser sintering (LS) is an additive manufacturing technology that employs a laser to fuse polymer powder into a mass that has the desired three-dimensional shape [8]. In this process, laser scanning of one surface layer is followed by addition of a new layer of fresh powder to the material surface which is then laser scanned to provide a new layer. The process is repeated until the desired part is fabricated. Laser Sintering is a typical AM method, which enables the designers to freely fabricate a part without the restrictions imposed by conventional design, manufacturing, and assembling methods [9].

Additive manufactured stab resistant body armor has been studied in recent years using LS technology. Johnson [10] first investigated the use of polyamide to fabricate LS planar materials. All of these products were found to meet to the Body Armor Standards for UK Police (2007) level one, able to resist impact energy of 24 Joules and a maximum permissible knife penetration of 7 mm [11]. It was found that the armors manufactured using a 50:50 mixture of virgin and recycled powder performed much better than those fabricated from virgin powder. The mechanical properties of the composites improved remarkably due to the addition of carbon fiber [12]. The standard demanded by Stab Resistant Body Armor National Standard published by the Ministry of Public Security of the People's Republic of China in 2008 (GA 68-2008)[13] requires that an acceptable personal body armor exhibit zero penetration when the stab impact energy is 24 J. More work such as using new materials to produce lighter SRBAs is needed in the development of body armor.

Therefore, in this study, research was conducted on four types of laser sintered materials to evaluate whether polyamide/carbon fiber (PA/CF) can be used as a material for making SRBAs to meet the requirements of the GA 68-2008 standard. Experiments were initially conducted on flat samples of various thicknesses to identify the optimum thickness. In further work a structured plate was designed and fabricated to improve the performance of the flat plates. The angle and thickness were analysed and discussed. The advantages of CF application was compared and investigated. This study established use of laser sintered polyamide for fabrication of stab proof personal body armor, which identifies a new approach for design, manufacturing and assembly of personal protective clothing design.

2. Methodology

2.1. Test standard

All stab resistance tests of the novel armor materials conformed to the GA 68-2008 National Standard [13]. The test platform comprised of a drop hammer, a knife, the test specimen and the backing material (as shown in Figure 1). The total mass of the hammer and the knife was 2.4 kg, which was released vertically 1 m above the test material, in a free fall onto the test material which produced an impact energy of 24 J. The standard required that the body armor should resist any penetration from the test blade. The backing material was composed from top to the bottom of neoprene sponge, polyethylene plastic and natural rubber. All the tests were performed at ambient temperature of -20-55℃. The test apparatus used for the testing were purchased from Chengde Kecheng Testing Machine Co. LTD [14].

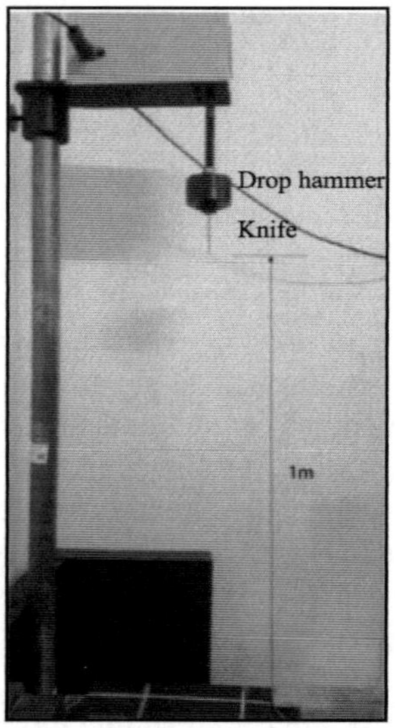

(a)

Figure 1. Stab resistance test platform.

2.2. Sample preparation

Four types of materials were employed for the LS fabrication: PA 3200, PA 4300, PA/GF and PA/CF. PA 3200 and PA 4300 were composed of polyamide 11 and 12 compositions. The PA/GF material was a composite comprising of 60.wt% polyamide and 40.wt% glass fiber. The PA/CF was a composite produced from70.wt% of polyamide and 30wt. % carbon fiber(CF). All the samples were fabricated using SolidWorks software. For each preparation, sample plates were fabricated with dimensions of 60×60 mm^2 as shown in Figure 2(a). For single layered specimens, the sample thicknesses were 9, 8, and 7 mm.

In order to test the excellent performance of PA/CF in different structures of the plates, the dimensions of the sample plates was set as 60×60 mm^2. In each direction (x and y), three pyramids were placed. The dimension of each pyramid was 20×20 mm^2, thus a total of 9 pyramids per test plate. As shown in Figure 2(b), the pyramid angle was measured from the back surface and the tilted pyramid ridge. The plate thickness was measured at the corner of the plate. The plate thickness was 6.5, 7, 7.5 mm separately, while the pyramid angle were 20, 25 and 30° which was accomplished via Solid works software.

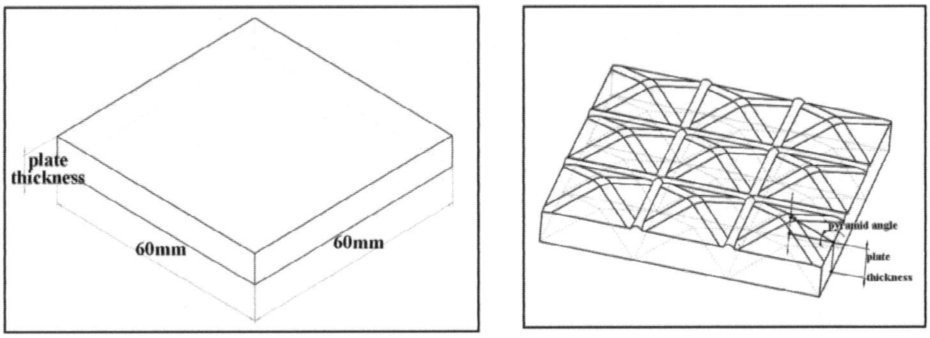

(a) (b)

Figure 2. Overview of specimen structure (a) planar plate, (b) pyramid structured plate: a- pyramid angle, h-plate thickness, s-pyramid element length.

3. Results and discussion

3.1. Planar plates

Experiments were performed using the LS fabricated samples. Figure 3 shows an overview of the four types of specimens that were tested. The detailed experimental results are shown in Table 1. All the flat specimens failed in the experiment when the plate thickness is 7mm.

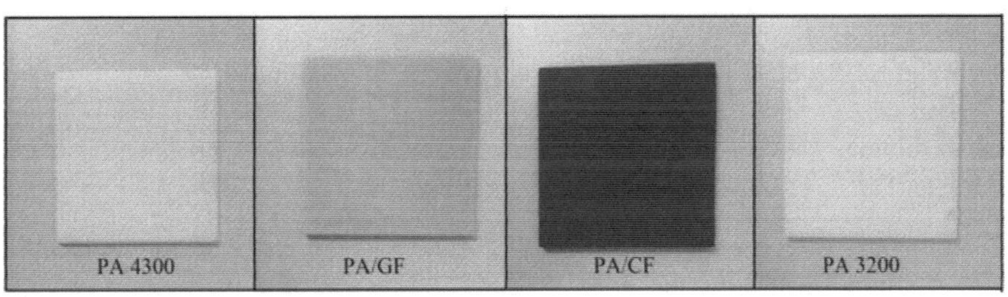

Figure 3. Overview of the four tested specimens before the stab resistance experiment, from left to right, laser sintered P4300, PA/GF, PA/CF, PA3200.

Table 1. Experimental results from the stab resistance experiments for four planar laser sintered materials. (√: success; ×: failure.)

Experiment arrangement		Material			
		PA 3200	PA4300	PA/GF	PA/CF
Plate	Single layered	9(×)	9(×)	9(√)	9(√)
thickness		8(×)	8(×)	8(√)	8(√)
(mm)		7(×)	7(×)	7(×)	7(×)

The specimens that passed the single layer experiment included the 8-mm-thick laser sintered PA/GF, 9-mm-thick laser sintered PA/GF, 8-mm-thick laser sintered PA/CF, and 9-mm-thick laser sintered PA/CF. Figure 4 shows a photo of the 9-mm-thick laser sintered PA 3200 specimen after the test, which was broken into two pieces. The highlighted black square shows the notch area where the blade entered. The 24 J-impact energy generated by the falling knife resulted in a brittle fracture in the specimen and the crack divided the specimen into 2 pieces. It was concluded that PA/CF had potential for penetration resistant applications.

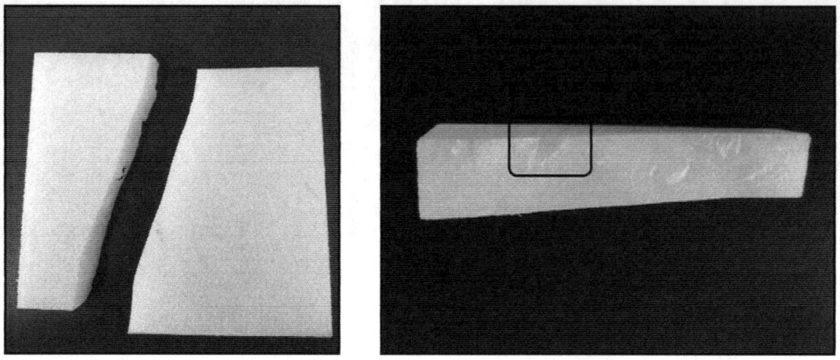

(a) (b)

Figure 4. 8-mm-thick laser sintered PA 3200 after the stab resistance experiments (a) front view of the broken sample, (b) cross section view of the broken sample.

3.2. Structured plates

The structured plates were used to testify the stab resistant performance difference between PA 3200 and PA/CF. Experiments were performed with pyramid angles of 20°, 25°and 30°. The resulting plate thickness of the fabricated samples ranged between 6.5 to 7.5 mm. The detailed experimental results are shown in Table 2. As shown, the laser sintered PA 3200 exhibited a 50% rate of success in all the experiments, which included specimens that were 7-mm- thick with tilt angles of 25°and 30°and 7.5-mm-thick samples with tilt angles of 20°, 25°and 30°. The 6.5 mm thick samples failed in every test, because they were too thin. Test plates with thickness and greater tilt angles produced a higher possibility of success. For the laser sintered PA/CF composite, all the experiments were successful when the plate thickness ranged from 6.5 to 7.5 mm and the tilt angles ranged from 20° to 30°. The area densities of the laser sintered plates were measured before the tests, as shown in Table 2. The area densities were in the range of 7.27-8.14 kg/m^2, resulting in a total weight of 2.2-2.4 kg. This represented a 31- 37% less weight than the conventional SRBA.

The thickness of the structured plate played a fundamental role during the impact energy dissipation, longer sliding distance within the plate lead to less impact energy on the bottom of the plate. The pyramid angle helped to deflect the force of the knife tip; however, an acute angle produced a thinner pyramid and a shorter route to dissipate the kinetic, therefore the impact energy was sufficient to penetrate the plate. To achieve the optimal performance, there was a balance to be achieved between the plate thickness and pyramid angle: thicker plate leads to longer energy dissipation route as well as higher area density; larger pyramid angle leads to more improved energy dispersion efficiency as well as a week support from the bottom and shorter energy dissipation route.

Table 2. Area density of the structured specimens (Note: ×: failure.)

Material	Thickness (mm)	Area density (kg/m^2) α=20°	α=25°	α=30°
PA 3200	6.5	6.53(×)	6.76(×)	6.88(×)
	7	7.51(×)	7.27(√)	7.56(√)
	7.5	7.90 (√)	7.68(√)	7.60(√)
PA/CF	6.5	7.41 (√)	7.04(√)	6.58(√)
	7	7.96 (√)	7.77(√)	7.29(√)
	7.5	8.14 (√)	8.02(√)	7.92(√)

Figure 5 shows the front and back views of the structure of the laser sintered PA 3200 and PA/CF following the penetration tests. Each specimen had a pyramid angle of 20° with plate thicknesses of

6.5, 7, and 7.5 mm from left to the right. It was found that the 6.5-mm-thick and 7-mm-thick laser sintered PA 3200 specimens were broken into pieces, while all of the laser sintered PA/CF remained intact with three small holes in each plate. No holes were observed at the back of the laser sintered PA/CF plates. The results indicated that the addition of the CF increased the penetration resistance of the sample.

Considering the experimental results and the total weight (area density), the optimum parameters needed to produce a successful stab resistant body armor fabricated using laser sintered PA/CF are: thickness of 6.5 mm with tilt angle of 30°. If the body armor was fabricated with 0.3 m^2 of protection, which is required by the GA 68 -2008 standard, the total weight would be 2.0 kg, which represent 57% of the weight of a conventional SRBA.

(a) (b)

Figure 5. Experimental test results from the structured laser sintered PA 3200 and PA/CF, with tilt angle of 20°, plate thicknesses of 6,5 7, 7.5 mm from left to the right, first row: front view, second row: back view, (a) laser sintered PA 3200, (b) laser sintered PA/CF.

The specimen structures were examined using scanning electron microscopy (SEM). Figure 6 shows the cross section of specimens that were laser sintered using PA 3200 and PA/CF. The plates were 7 mm thick with a pyramid angle of 20°. The laser sintered pure PA (top row) exhibited a smooth interface under 500×and 1000×magnification. The shear marks indicated the knife stabbing imprint and the track of the crack. The laser sintered PA/CF exhibited a very different interface. No agglomeration occurred and the carbon fiber were found to be homogeneously dispersed. The addition of the CF appeared to have prevented the knife from penetrating the bulk of the material by dissipating the impact energy. Thus the CFs appeared to contribute to the plate integrity based on the experimental results. In addition, it was demonstrated that the addition of CF increased sample's mechanical properties as a result of the high surface to volume ratio, and the high aspect ratio of the CFs [15].

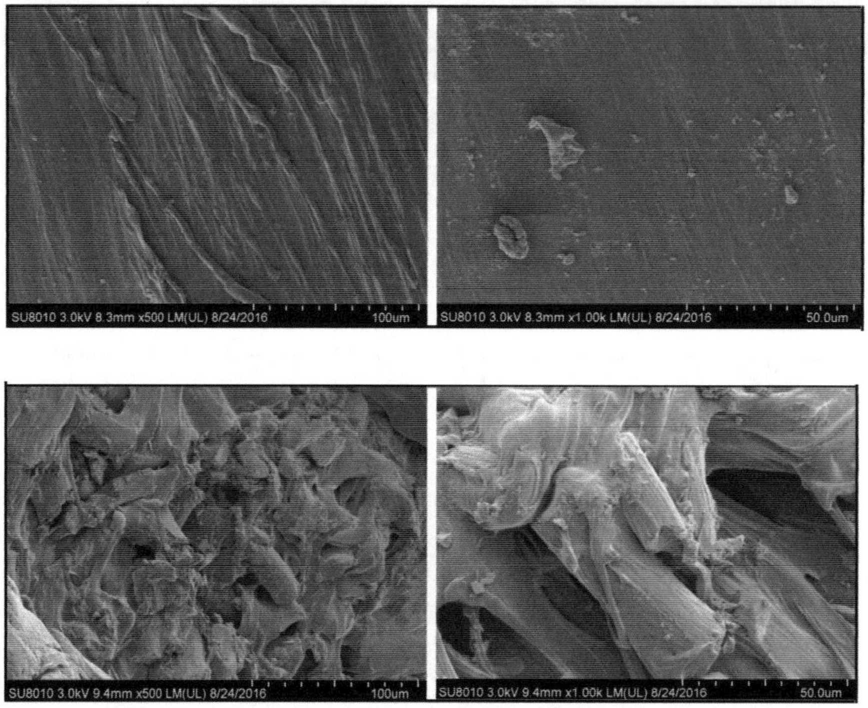

Figure 6. SEM images of the cross section of the structured specimen with a thickness of 7 mm and tilt angle of 20°, laser sintered PA 3200 (top row), laser sintered PA/CF (bottom row).

4. Conclusion

In summary, natural animal armor inspired the structural design of stab resistance body armor (SRBA), and Laser Sintering (LS) technology was used to rapidly prototype the characteristics and the complexity of the desired structure. The pyramid structured plate was designed and optimum parameters were tested, discussed and achieved. The mechanism of the impact energy dissipation by the pyramidal structures of the novel SRBA was studied and it was found that this structure dispersed the deformation energy of the knife and minimized the damage to the plate. Four kinds of LS materials were prepared and tested based on the GA 68-2008 standard.

It was concluded that PA/CF was an optimum material for preparing a new armor for the desired stab resistant application. In the planar experiment, a minimum thickness of 8 mm of planar laser sintered PA/CF was required to achieve a suitable penetration resistance to an impact energy of 24 J based on the GA 68-2008 standard. In the structured experiment, the stab resistance property increased with the plate thickness and the angle. The addition of the CF to the materials mix improved the product stab resistance due to the improved mechanical properties of the product and a high aspect ratio. Laser sintered PA/CF with a plate thickness of 6.5 mm and a pyramid angle of 30°exhibited appeared to be the optimum composition for the desired application with area density of 6.58 kg/m^2. The total weight of the proposed SRBA would represent a 43% reduction of the weight of the conventional SRBA.

Acknowledgement

We would like to thank the National 135 Program 'National Key Research Program' (grant number: 2016YFC0802800) the National 125 Program 'National Science Support Program' (grant 2015BAK10B03) and the Independent Research Project of State Key Laboratory of Explosion Science and Technology (grant number YBKT16-13) for their supports.

References

[1] Chen X G and Yang D 2010 Use of 3D angle-interlock woven fabric for seamless female body armor: Part 1: Ballistic evaluation *Text Res. J.* **80**(15) 1581–8

[2] Paddy C D, Phil J H and Nancy J R 2014 Body armor: the effect of load, exercise and distraction on landing forces *J. Sport. Sci.* **32**(4) 301-6

[3] Yuan M, Jiang J, Yang J and Qian X 2016 Investigation on Thermal Insulation and Vapor Resistances of Stab-Resistance Body Armor Using Thermal Manikin Method *Transactions of Beijing Institute of Technology* **36**(6) 560-73

[4] Caron J 2008 Ancient worms in armour *Nature* **451** 133–4

[5] Wen Y, Irene H C, Bernd G, Elizabeth A Z, Robert O R and Marc A M 2013 Natural Flexible Dermal Armor *Adv. Mater.* **25** 31–48

[6] Andrew A J 2014 Establishing design characteristics for the development of stab resistant Laser Sintered body armour (UK: Loughborough University) (*Doctoral thesis*)

[7] Tuck C J 2007 Rapid manufactured textiles *Int. J. Comp. Integ. M.* **20**(1) 96-105

[8] Johnson A 2014 Establishing design characteristics for the development of stab resistant Laser Sintered body armour (UK: Loughborough University) (*Doctoral thesis*)

[9] Song J, Ortiz C and Boyce M C 2011 Threat-protection mechanics of an armored fish *J. Mech. Behav. Biomed.* **4**(5) 699-712

[10] Johnson A, Bingham G A and Wimpenny D I 2013 Additive manufactured textiles for high-performance stab resistant applications *Rapid Prototyping J.* **3**(19) 199-207

[11] Home Office Scientific Development Branch 2007 *HOSDB Body Armour Standards for UK Police, Part 3: Knife and Spike Resistance* (39-07-C)

[12] Zhao Y H, Zhang Y F, Bai S L and Yuan X W 2016 Carbon fibre/graphene foam/polymer composites with enhanced mechanical and thermal properties *Composites Part B* **94** 102-8

[13] Ministry of Public Security of the People's Republic of China 2008 *Stab resistnace body armor for police, People's Republic of China Public Safety National Standard* (GA 68-2008)

[14] CHENGDE KECHENG TESTING MACHINE CO., LTD, *http://www.kcsyj.com/main/index.asp*

[15] Winey K I and Vaia R A 2007 Polymer nanocomposites *Mrs Bull* **32** 314–9

IOP Conf. Series: Materials Science and Engineering **213** (2017) 012028 doi:10.1088/1757-899X/213/1/012028

Structural parameter effect of porous material on sound absorption performance of double-resonance material

C Fan[1,3], Y Tian[1], Z Q Wang[2],J K Nie[1],G K Wang[1]and X S Liu[1]

[1]Global Energy Interconnection Research Institute, Beijing, 102211, China
[2]State Grid Henan Electric Power Company, Zhengzhou, 450018, China

E-mail: fanchao006@126.com

Abstract. In view of the noise feature and service environment of urban power substations, this paper explores the idea of compound impedance, fills some porous sound-absorption material in the first resonance cavity of the double-resonance sound-absorption material, and designs a new-type of composite acoustic board. We conduct some acoustic characterizations according to the standard test of impedance tube, and research on the influence of assembly order, the thickness and area density of the filling material, and back cavity on material sound-absorption performance. The results show that the new-type of acoustic board consisting of aluminum fibrous material as inner structure, micro-porous board as outer structure, and polyester-filled space between them, has good sound-absorption performance for low frequency and full frequency noise. When the thickness, area density of filling material and thickness of back cavity increase, the sound absorption coefficient curve peak will move toward low frequency.

1. Introduction

Recently, with the rapid development of economy and the increase of urban scale, many power transmission projects are under way[1]. Due to shortage of land source, many substations are located in urban densely-populated areas and disturb the lives and work of the nearby residents, being the urban substations noise a special cause of considerable public concern. Urban substations noise is featured by medium and low frequency, long wavelength, slow attenuation, and strong penetration to the building, with its frequency spectrum mainly concentrated on 50Hz and its octave[2-5]. Common noise reduction materials, for example mineral wool, cannot meet the requirements of absorbing low frequency and full frequency noise.

In view of the noise feature and service environment of urban power substations, our research team used the idea of compound impedance[6-7], using the advantages of resonance sound-absorption materials for low-frequency noise[8-11]and porous sound-absorption materials for high-frequency noise[12-16]. We finally developed a new-type of composite acoustic board with double-resonance structure, being the first resonance cavity filled with porous sound-absorption materials. Compared with commonly used materials, the absorbing band of our acoustic board is broadened and its low-frequency absorption ability is enhanced. The double-resonance structure consists of micro-porous fiber composite acoustic board, which is made with aluminum fibrous sound-absorption materials with an area density of 500 g/m^2 and thickness of 1mm[17-19], and a micro-porous acoustic board with the hole spacing of 6mm and thickness of 1mm[20-21]. The filling material is a kind of weather-resistant, economic and environmental friendly polyester[22-24]. The sound-absorption properties of these materials are shown in Figure 1. Through the combination of polyester and double-resonance materials with different structural parameters, the authors analyze the influence of structural parameters of the new-type of composite acoustic board on its sound-absorption performance, which is significant to the structure optimization and sound-absorption material development.

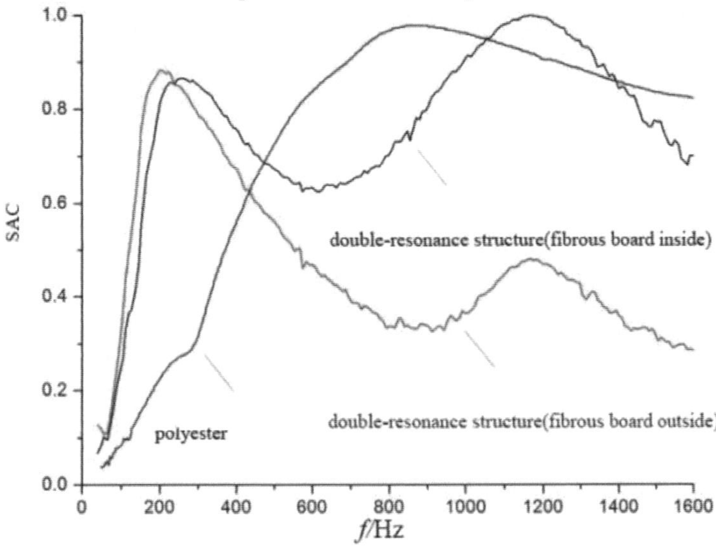

Figure 1. Sound absorption coefficient curves of different materials.

2. Experiment

The sound-absorption performance of the new-type of composite acoustic board mainly depends on the property of compound impedance, which is related to the assembly order, the area density and thickness of the filling material, and the thickness of the back cavity. Thus, in order to acquire good acoustic property, this research conducted several acoustic tests and evaluations according to GB/T18696.2-2002 Acoustics-Determination of Sound absorption coefficient and impedance in impedance tubes-Part 2: Transfer function method[25]. The test equipment is B&K impedance tube, as shown in Figure 2. The samples include aluminum fibrous acoustic board, micro-porous board and polyester, as shown in Figure 3. The experimental plan is shown in Table 1.While testing, the test samples were composited according to the compound mode of experimental plan. The composite was put into test equipment for testing, and the test results were evaluated.

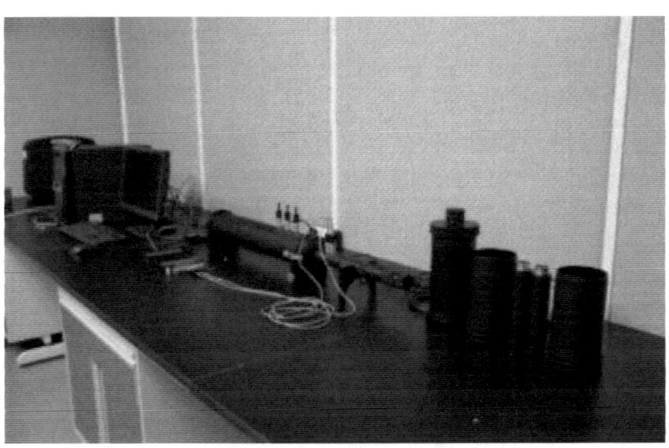

Figure 2. B&K impedance testing system.

(a) aluminum fibrous acoustic board (b) polyester (c) micro-porous board

Figure 3. Test samples.

Table 1. Experimental plan of the new-type composite acoustic board.

No.	Outer structure	Inner structure	Filling materials	Thickness of filling materials	Area density of filling materials	Thickness of back cavity
1#	micro-porous board	aluminum fibrous acoustic board	polyester	20cm	1500g/m^2	0cm
2#	aluminum fibrous acoustic board	micro-porous board	polyester	20cm	1500g/m^2	0cm
3#	aluminum fibrous acoustic board	micro-porous board	polyester	20cm	1500g/m^2	10cm
4#	aluminum fibrous acoustic board	micro-porous board	polyester	20cm	1500g/m^2	20cm
5#	aluminum fibrous acoustic board	micro-porous board	polyester	10cm	1500g/m^2	0cm
6#	aluminum fibrous acoustic board	micro-porous board	polyester	25cm	1500g/m^2	0cm
7#	aluminum fibrous acoustic board	micro-porous board	polyester	20cm	1000g/m^2	0cm
8#	aluminum fibrous acoustic board	micro-porous board	polyester	20cm	2000g/m^2	0cm

3. Result and discussion

3.1. Influence of assembly order

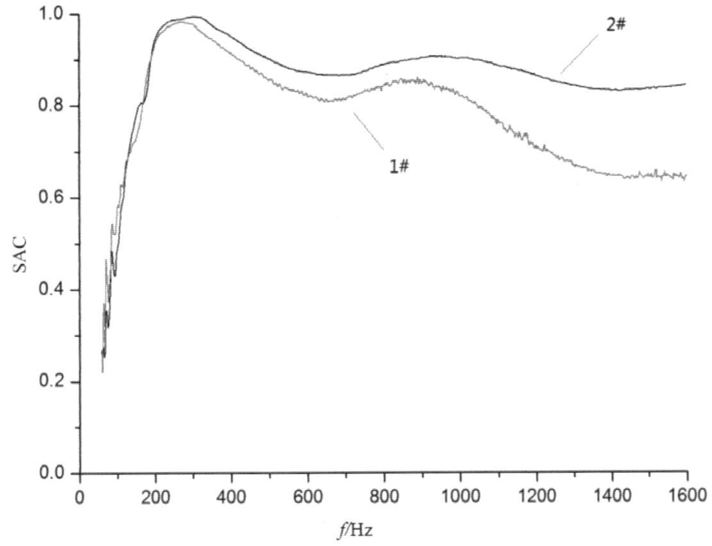

Figure 4. Comparison of sound absorption coefficient (SAC) curves for the new-type of composite acoustic boards with different assembly orders.

Figure 4 shows the comparison of sound absorption coefficient (SAC) curves for the new-type of composite acoustic boards, with the micro-porous board and the aluminum fibrous acoustic board in different positions, and filled by polyester with a thickness of 20cm and area density of 1500g/m2, being the thickness of back cavity of0cm. The results show that, when the other parameters are the same, the acoustic board containing aluminum fibrous acoustic board outside and micro-porous board inside exhibits the following behavior: the SAC remains near 0.8 after it reaches the peak, and it does not decrease much at higher frequencies. In comparison, the new-type of acoustic board with aluminum fibrous acoustic board inside and micro-porous board outside exhibits the following behavior: the SAC is nearly the same as the one with different assembly order up to300 Hz, while after 300Hz, a difference shows up. Especially after 1000Hz, SAC decreases rapidly, and reaches the value of 0.6 at1600Hz. The reason of this behavior is that the micro-porous board mainly absorbs noise through the effect of Helmholtz resonance, which means that it needs the resonance of the back cavity to dissipate sound energy. When the cavity is filled with porous sound-absorption material, the thickness of the cavity is reduced and the resonance effect is suppressed. While the aluminum fibrous acoustic board has the property of acoustic resistance due to its inclusion of aluminum fiber felt, the filling of porous sound-absorption materials can optimize the impedance matching and acquire perfect sound absorption. Thus, the structure with aluminum fibrous acoustic board outside and micro-porous board inside is more beneficial to the absorption of urban substation noise.

3.2. Influence of the back cavity thickness

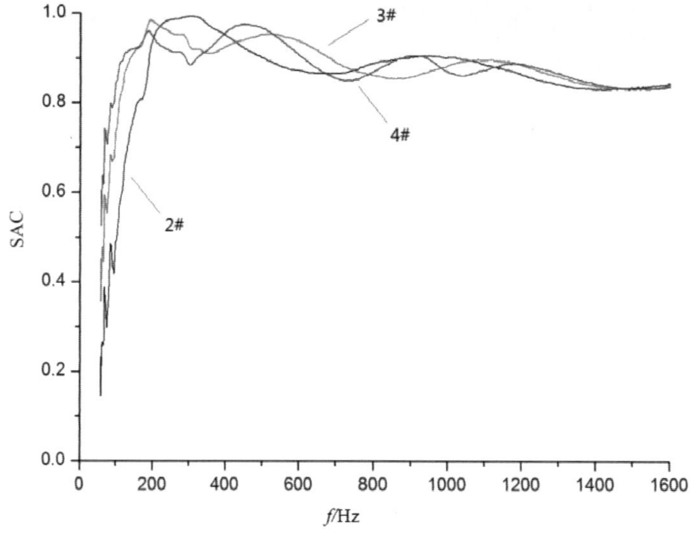

Figure 5. Comparison of sound absorption coefficient (SAC) curves for the new-type of composite acoustic boards with different thicknesses of back cavity.

Figure 5 shows the comparison of SAC curves for the new-type of composite acoustic boards with different thicknesses of back cavity, being filled by polyester with a thickness of 20cm and area density of 1500g/m^2, where the outer structure is aluminum fibrous acoustic board and the inner structure is micro-porous acoustic board. The test results show that when the thickness of the back cavity increases, the peak of SAC curve moves toward low frequency. The reason for this behavior is that the low-frequency wavelength is long, and the back cavity can resonate easily and dissipate the noise energy when the thickness is large. It's shown in Figure 5 that, when the thickness of back cavity is 20cm, its SAC reaches 0.85 at100Hz, and it's not lower than 0.84 in the range of 100~1600Hz, which guarantees that the SAC curve for the new-type of composite acoustic board covers the low, medium and high frequency of substation noise.

3.3. Influence of the filling material thickness

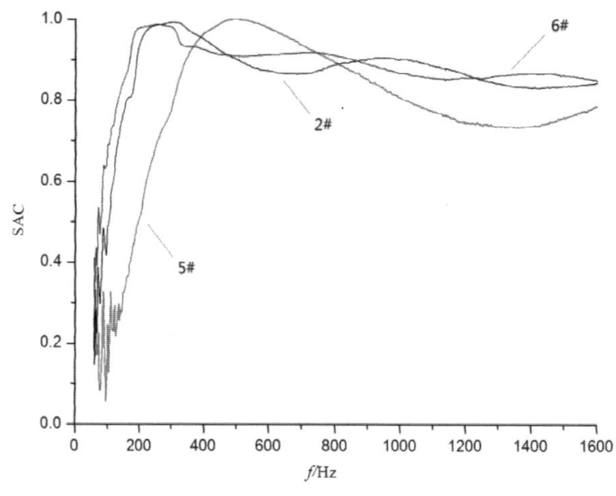

Figure 6. Comparison of sound absorption coefficient (SAC) curves for new-type of composite acoustic boards with different thicknesses of filling material.

Figure 6 shows the comparison of SAC curves for the new-type of composite acoustic boards filled by polyester with the area density of 1500g/m^2 and different thicknesses, where the outer structure is aluminum fibrous acoustic board and the inner structure is micro-porous acoustic board. The test results show that, when the other parameters are the same, the increase of thickness of the filling material moves the peak of SAC curve towards low frequency, and the full-frequency absorption ability is enhanced, while the enhancing effect is weakened when the thickness increases. The reason is that the filling material will increase the thickness of the first resonance cavity, and enhance the sound-absorption performance for medium and low frequency noise. Meanwhile, the increase of thickness can increase the material acoustic resistance, which is helpful to the absorption of medium and high frequency noise. Thus, the full-range sound-absorption performance is strengthened. Furthermore, the dissipation of sound energy increases with the thickness of filling material, while the increase rate becomes lower. It is also reflected in the SAC curves that the moving trend of curve peak and the enhancing effect of sound-absorption performance become weakened. Figure 6 also shows that, when the thickness increases from 20cm to 25cm, the SAC curve does not change significantly.

3.4. *Influence of the filling material area density*

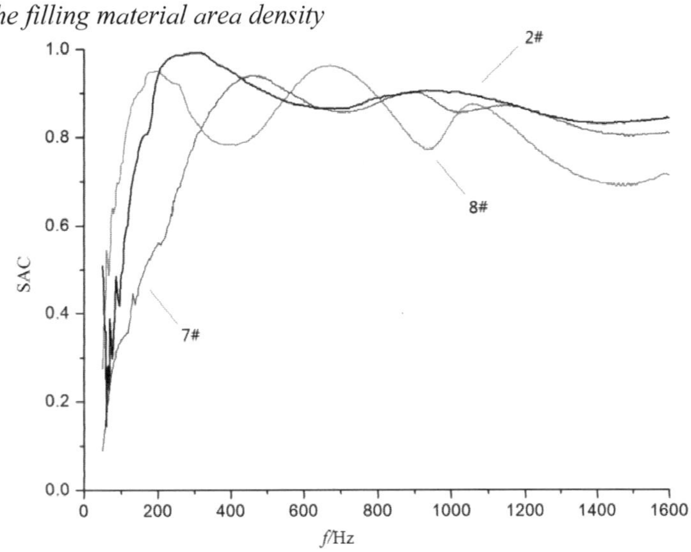

Figure 7. Comparison of sound absorption coefficient (SAC) curves for the new-type of composite acoustic boards with different area densities of filling material.

Figure 7 shows the comparison of SAC curves for the new-type of composite acoustic boards filled by polyester with a thickness of 20cm and different area densities, where the outer structure is aluminum fibrous acoustic board and the inner structure is micro-porous acoustic board. The test results show that, when the other parameters are the same and the area density increases, the SAC curve peak of the new-type of composite acoustic board moves toward low frequency, and the full-frequency acoustic performance is enhanced to some degree, while the moving trend and the overall sound absorption performance become weakened as the area density continuously increases. The reason is that the amount of fiber increases with the area density and, as a result, the polyester becomes much denser and the pore channels inside becomes tiny and twisted, which consumes more sound energy. It is shown in the SAC curves that the full-frequency sound absorption performance is enhanced at the beginning stage of the area density increase, while after the threshold, the increasing area density means that most of the space is occupied being harder to the sound wave to enter the sound-absorption material, and the overall sound absorption performance is weakened. As shown in Figure 7, when the area density increases from 1500 g/m^2 to 2000 g/m^2, the enhancing effect is not obvious and the fluctuation of sound absorption performance appears in medium and high frequency.

4. Conclusions

Adopting the test method of impedance tube, this paper researched on the influence of structural parameters, including assembly order, thickness and area density of filling material, and back cavity thickness and on the sound-absorption performance of the double-resonance material. The results provide a guide for the development of noise reduction material for urban substation noise:

1) Influence of assembly order: To make full use of the acoustic impedance property of aluminum fibrous board, this research designs a new-type of acoustic board with aluminum fiber as outer structure, micro-porous board as inner structure, and polyester filling inside the cavity. This assembly order can optimize the impedance matching and broaden the sound-absorption frequency.

2) Influence of the back cavity thickness: As the low frequency wavelength is long, when the thickness of back cavity is large, it is easy for the new-type of composite acoustic board to resonate, and the sound absorption curve peak will move toward low frequency.

3) Influence of the filling material thickness: The thickness increase of the filling material has the same effect as the thickness increase of first resonance cavity on sound-absorption performance. Besides, due to the increase of flow resistance caused by the filling material, the full frequency sound-absorption performance of the composite acoustic board is enhanced. After the threshold, the increase rate becomes lower and lower, and the enhancing effect is weakened.

4) Influence of the filling material area density: When the area density of the filling material increases, the pore channels inside the material will become tiny and twisted, this is helpful to the sound energy consumption. After the threshold, the increasing area density means that most of the space is occupied, being harder to the sound wave to enter the sound-absorption material, and the overall sound absorption performance is weakened.

References

[1] Liu Z Y U 2005 *ltra-high Voltage Power Grid* (Beijing: China Economic Publishing House) pp 4-7

[2] Chen C M, Gao Y and Liu S T 2014 Study on noise prediction model and control schemes for substation *The Scientific World J.* **2014** 696429

[3] Wang Z H, Zhou J G, Su L and Jiang J H 2008 Source and characteristic analysis of audible noise for converter substation *East China Electric Power* **36**(11) 16-8

[4] Ye J B and Chen Y 2005 Analysis and discuss of substation environment noise test *Guangdong Electric Power* **18**(10) 53-6

[5] Wang G K, Fan C, Nie J K, Tian Y and Xiao W M 2014 Research on noise feature of urban substations *Noise and Vibration Control* **12**(34) 196-9

[6] Lee F C and Chen W H 2001 Acoustic transmission analysis of multi-layer absorbers *J. Sound and Vibration* **248**(4) 621-34

[7] Xu Z Y, Huang Q B and Yao B Y A 1999 Study of the Acoustical Properties of Compound Impedance in Sound Absorbing Structure *J. Huazhong U. Sci.* **1** 53-5

[8] Tian Y, Chen J S, Liu Z G, Lu W J, Chen X and Han Y 2016 Application of Microporous Fiber Composite Acoustic Board in the Noise Control Engineering of Urban Substations *Smart Grid* **10** 988-92

[9] Rostand T, Thomas D and Philippe L 2011 Experimental investigation of holes interaction effect on the sound absorption coefficient of micro-perforated panels under high and medium sound levels *Appl. Acoust* **72**(10) 777-84

[10] Lee Y Y, Lee E W M and Ng C F 2005 Sound absorption of a finite flexible micro-perforated panel backed by an air cavity *J. Sound Vib.* **287**(1-2) 227-43

[11] Fan C, Nie J K, Xiao W M, Chen X, Han Y and Geng H J 2014 Research on sound-absorption materials for substation noise reduction *Electric Power* **47**(4) 158-61

[12] Delany M E and Bazley E N 1970 Acoustical characteristics of fibrous absorbent material *J. Acoust Soc. Am.* **3**(2) 105-16
[13] Jesus A, Romina del R, Jaime R and Jorge A 2011 An inverse method to obtain porosity, fibre diameter and density of fibrous sound absorbing materials *Arch. Acoust* **36**(3) 561-74
[14] Lu T J, Audrey H and Ashby MF 1999 Sound absorption in metallic foams *J. Appl. Phys.* **85**(11) 7528-39
[15] Zhong X Z 2012 *Building sound absorption materials and sound insulation materials* (Beijing: Chemical Industry Press)
[16] He D L 2012 Research progress and development trend of porous absorption materials *Mater. Rev.* **26** 303-6
[17] Wang Z M and Yu W Z 2005 Theoretical analysis on acoustical characteristics aluminum fiber board *Technical Acoustics* **24**(3) 183-5
[18] Nie J K, Kong X F, Wang B, Lu Ch Y, Jin X F and Xiao W M 2015 The research on serviceability and preparation process of aluminum fiber sound-absorbing panel in transformer station noise control *Noise and Vibration Control* **12**(35) 340-3
[19] Sun F G, Chen H L, Wu J H and Feng K 2009 Sound absorbing characteristics of fibrous metal materials at high temperatures *Appl. Acoust* **71** 221-35
[20] Maa D Y 1997 General theory and design of microperforated-panel absorbers *Chinese Journal of Acoustics* **3** 193-202
[21] Pei C M, Zhou B, Li D K and Chang D Q 2015 Study on the Composite Sound Absorber Made up of Porous Materials and MPP *Noise and Vibration Control* **35**(5) 35-8
[22] Zhong X Z 2005 Experimental study on sound absorbing properties of polyester fiber board *Audio Engineering* **10** 10-4
[23] Massimo G and Francesco P 2005 A simple empirical model of polyester fibre materials for acoustical applications *Appl. Acoust* **66**(12) 1383-98
[24] Peng L, Wang J, Fu F, Wang D and Zhu G 2015 Experimental Study on Sound Absorption of Wood Fiber/Polyester Fiber Composite Materials *Journal of Building Materials* **18**(1) 172-6
[25] Acoustics-Determination of sound absorption coefficient and impedance tubes-Part 2: Transfer function method (*GB/T 18696.2-2002*)

IOP Conf. Series: Materials Science and Engineering **213** (2017) 012029 doi:10.1088/1757-899X/213/1/012029

Synthesis, curing of hydroxyl-terminated liquid fluoroelastomer: thermal, chemical resistant and mechanical properties

N Li[1], J J Wu[2], Y Q Wang[2], J H Wu[2], J J Zhang[3], J Z Huo[3] and Y P Bai[1,4]

[1]School of Chemistry and Chemical Engineering, Harbin Institute of Technology, Harbin 150001, China
[2]College of Materials Science and Engineering, Key Laboratory of Carbon Fiber and Functional Polymers of Ministry of Education, Beijing University of Chemical Technology, Beijing 100029, China
[3]FoShan NanHai Gotop Packaging Material Co., Ltd, Foshan 528225, China

E-mail: 413957845@qq.com

Abstract. A novel hydroxyl-terminated liquid fluoroelastomer (HTLF) was synthesized; FT-IR, GPC and ^{1}H-NMR result show that HTLF was synthesized successfully. Then HTLF was cured with HDI trimer, the structure of cured product was characterized by FT-IR. Compared with traditional solid fluorine elastomer, Cured HTLF has higher char yield and faster thermal decomposition rate. It ascribed to the introduction of ring structure the cross-link structure formed in the curing process. The study of chemical resistant properties and tensile strength of cured fluoroelastomer film suggested that Cured HTLF was provided with similar mmechanical properties, chemical resistant property than solid fluorine elastomer.

1. Introduction

Liquid fluoroelastomers (LFs) are a class of polymer material with some special functions, such as curable, chemical resistance, low surface energy, thermostability, easy processing processing and molding, good mobility and so on; they are wildly used in the field of special coating, adhesive, sealant, and medicine industry. Since the early 1960s, Dupont and 3M Company successfully developed the LFs on a large scale [1-4], greatly improve its application in industry. According to the present research, fluoroelastomer synthesis via fluorinated monomers including glycol[5] polyether glycol [6-9] side group [10] and end-capper [11]. Nowadays, liquid fluoroelastomer has not only been used as a processing aid, but also be used as raw materials of coating, sealant, and adhesive. However, compared with traditional solid fluorine elastomer, mechanical properties and thermal resistance of LFs are poor. So far, there are none of LFs and their cured products are provided with similar mechanical properties and thermal resistance with solid fluorine elastomer [12-17].

In this work, a novel hydroxyl-terminated liquid fluoroelastomer(HTLF) which is a liquid fluoropolymer with reactive end group was prepared. What is more, HTLF can be cured with the polyiscocyanate. Compared with traditional solid fluorine elastomer, it was provided with similar mechanical properties, chemical resistant property and more excellent thermal resistance. According to the present research, this kind of LFs has not been reported now.

IOP Conf. Series: Materials Science and Engineering **213** (2017) 012029 doi:10.1088/1757-899X/213/1/012029

The structure of HTLF was characterized and the cure reaction and properties of HTLF with HDI trimer were studied. The study of chemical resistant property, tensile strength and TGA of cured fluoroelastomer film(Cured HTLF) indicate that above properties of the product is still excellent compared with traditional solid fluorine elastomer. What is more, Cured HTLF has better machinability than traditional solid fluorine elastomer. This novel liquid fluoroelastomer is suitable for sealing the fuel tanks of aircraft and coating the external surfaces of building owing to overcoming the practical limitations in size and shape of high molecular weight fluoroelastomer which is mostly formed by compression molding.

2. Experimental

2.1. Materials and chemical agents

Carboxyl-terminated liquid fluoroelastomer(CTLF) were synthesised by our laboratory[18]. Idium borohydride ($NaBH_4$), iodine(I_2), dibutyltin dilaurate, anhydrous sodium sulfite (Na_2SO_3), hydrochloric acid (HCl), sulfuric acid (H_2SO_4), tetrahydrofuran(THF), acetone were commercially obtained. HDI (HT-100, the mass percent of isocyanate=21.8±0.3) was purchased from YANTAI WANHUA POLYURETHANE CO., LTD.

2.2. The synthesis of hydroxyl-terminated liquid fluoroelastomer(HTLF)

HTLF was prepared *via* reduction reaction of CTLF. A typical preparation method is as follow. To a stirred solution of $NaBH_4$(10 mmol) in dry THF(20 mL) a solution of iodine (5 mmol) in dry THF(10 mL) was added dropwise under argon at 0°C over 45 min. Next, carboxyl-terminated liquid fluoroelastomer s (10g, 3.85mmol carboxyl group) in dry THF (100mL) was added and then the temperature was raised to 70°C, the reduction reaction was allowed to proceed for 8h. The mixture was then cooled to 0°C and the excess hydride was carefully destroyed by dropwise addition of 2mol/L hydrochloric acid (10 mL). Filtered the mixture, separating, Then the product was added deionized water. The resultant product was concentrated by a rotary evaporator. A viscous yellow liquid was obtained. (Yield is 91%)

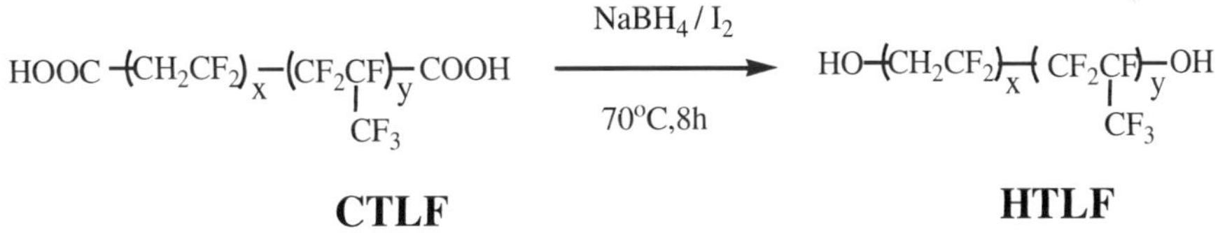

Scheme 1. The synthesis route of hydroxyl terminated liquid fluorine elastomer (b).

2.3. Curing of hydroxyl-terminated liquid fluoroelastomers

HTLF (5 g) was mixed thoroughly with 0.5 g HDI trimer. The mixture was shaped in mold and cured at 60℃ for 3h .

IOP Conf. Series: Materials Science and Engineering **213** (2017) 012029 doi:10.1088/1757-899X/213/1/012029

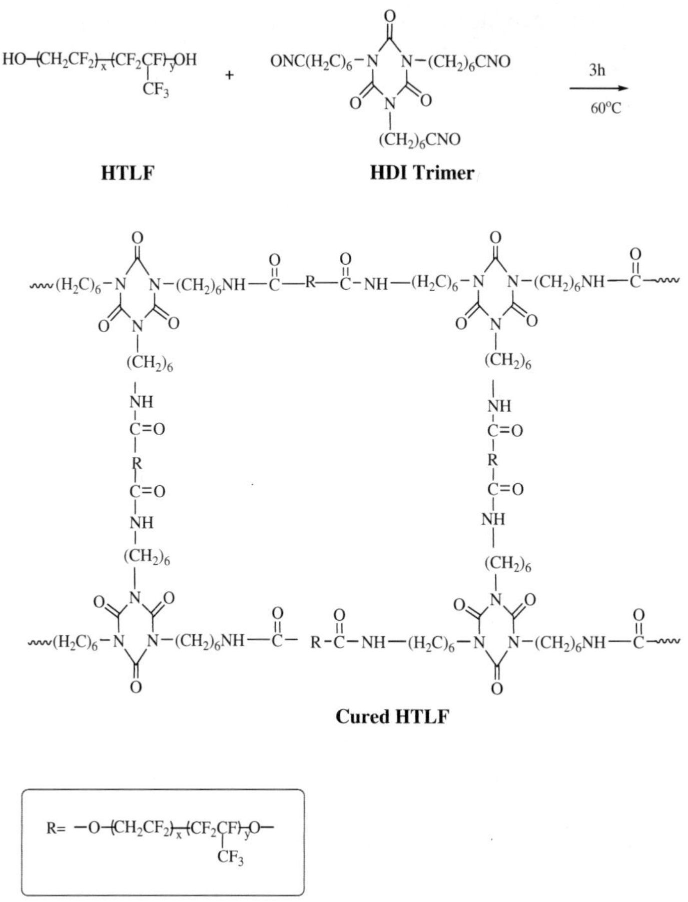

Scheme 2. Cure mechanism of hydroxyl-terminated liquid fluoroelastomer (**b**) with HDI Trimer.

3. Results and discussion

*3.1. Characterization of hydroxyl-terminated liquid fluoroelastomers(**HTLF**)*

3.1.1. FT-IT. FT-IR spectra of HTLF and CTLF are shown in Figure1. These two compounds exhibit absorption peaks at 1396,1204,and 886 cm^{-1} ascribed to stretching vibration of $-FCH_2-$, $-CF_2-$, and $-CF_3$, respectively, which clearly shows that they have the same backbone structure. The new peak at 3500cm^{-1} in the spectrum of HTLF is ascribed to the stretching vibration of hydroxyl groups, indicating the formation of $-CH_2OH$. While, IR spectra at 1730 cm^{-1} which characteristic of the carbonyl group disappeard, showing that the reduction of the carboxylic end groups was complete. However, after this quantitative reduction evidenced by means of IR spectroscopy, it appears that the amount of CH=CF was higher.

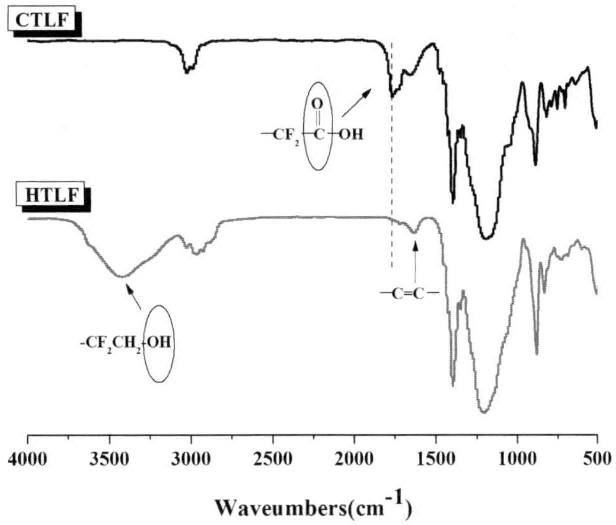

Figure 1. FT-IR spectra of hydroxyl terminated liquid fluoroelastomer (**HTLF**) and its precursor carboxyl terminated fluoroelastomer (**CTLF**).

3.2. ^{1}H-NMR

To future characterize the liquid fluoroelastomers, NMR spectroscopy was utilized. ^{1}H-NMR spectra of hydroxyl terminated liquid fluoroelastomer (**HTLF**) and its precursor carboxyl terminated liquid fluoroelastomer (**CTLF**) are shown in Figure 2. Compared with the precursor, the fluoroelastomer exhibits two new peaks, chemical shift of 1.6 ppm ascribed to the hydroxyl groups and 1.6 ppm ascribed to the -CH_2 group, they confirms the formation of hydroxyl-terminated liquid fluoroelastomer .

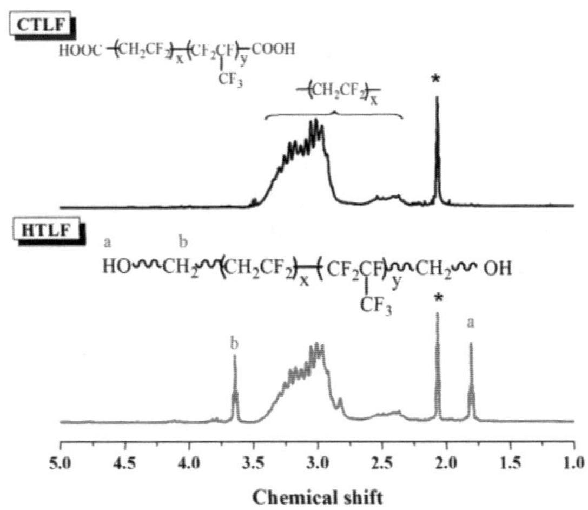

Figure 2 1H-NMR spectra of hydroxyl terminated liquid fluoroelastomer (**HTLF**) and its precursor carboxyl terminated liquid fluoroelastomer (**CTLF**) in acetone-d6.The solvent peaks are marked with asterisks.

3.3. Characterization of Cured HTLF

IOP Conf. Series: Materials Science and Engineering **213** (2017) 012029 doi:10.1088/1757-899X/213/1/012029

3.3.1. FT-IR. FTIR spectra of the cured fluoroelastomer are given in Figure 3. The spectra of their precursor, HDI trimer is also given in the figure for comparison. Compared with the spectra of b, the cured fluoroelastomer exhibits absorption peaks at 1396, 1204, and 886 cm^{-1} ascribed to stretching vibration of -FCH-, $-CF_2-$, and $-CF_3$, respectively. Like its precursor, the cured fluoroelastomer exhibits new peaks at 765cm^{-1}, 1433cm-1, 1466cm-1, 1532 cm-1, assigned to benzene ring of HDI trimer. The peak at 1689 cm-1 in the spectrum of the cured fluoroelastomer is ascribed to the stretching vibration of carbonyl. Furthermore, the cured fluoroelastomer exhibits a new peak at 3372 cm^{-1}, which confirms the formation of -NH-.The result of FTIR spectra also shows that the liquid fluoroelastomer s can be cured by HDI trimer.

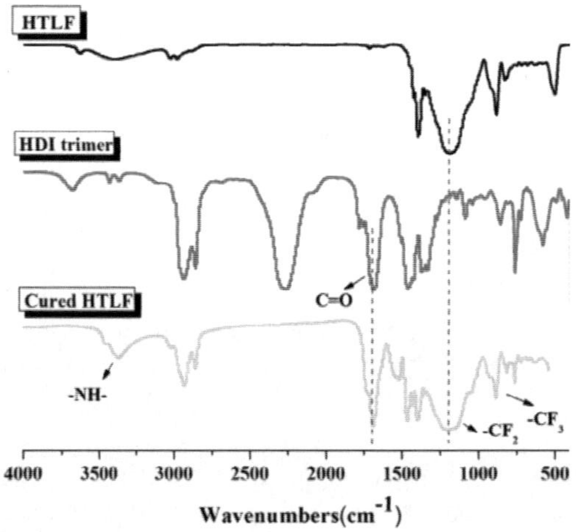

Figure 3. FT-IR spectra of hydroxyl terminated liquid fluoroelastomer (**HTLF**), HDI trimer and the **Cured HTLF**.

3.4. Thermal property of the cured fluoroelastomer(Cured HTLF)

TGA analysis of hydroxyl terminated liquid fluoroelastomer (**HTLF**) and **Cured HTLF** were shown in Figure 4, solid fluoroelastomer (Poly(VDF-HFP) is given for comparison. In order to evaluate the thermal decomposition rate, T_5 is defined as the temperature at which the polymers lose 5% of their original weight. It is shown that T_5 of **Cured HTLF, HTLF** and Poly(VDF-HFP) is 335°C, 240°C and 223°C respectively. Obviously, after cured, the thermal decomposition rate of hydroxyl terminated liquid fluoroelastomer is slower than solid fluoroelastomer. Besides, the char yield of (Poly(VDF-HFP) at 750°C is 11%, while the char yield of **Cured HTLF** remains at 29%. Based on the study of the thermal degradation rate and the char yield of these two kinds of fluorine elastomers, it is found that the cured product based on hydroxyl terminated liquid fluoroelastomer has better thermal resistance than the solid fluorine elastomer. The result is ascribed to the introduction of ring structure (HDI Trimer) improve the thermal resistance of the cured product, what is more, cross-link structure formed in the curing process, it make the thermal decomposition rate of liquid fluoroelastomer further reduced.

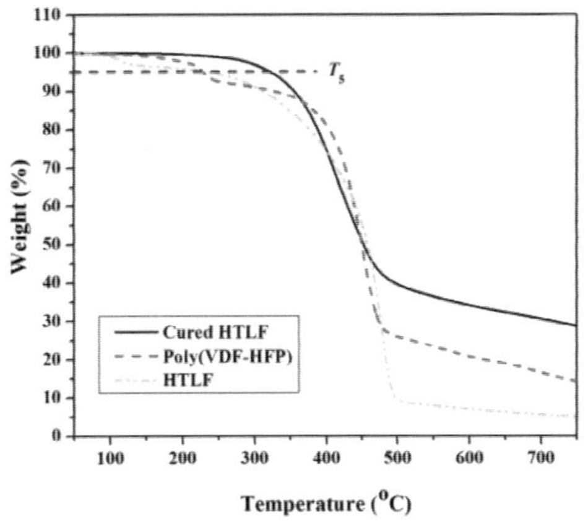

Figure 4. TGA curve of hydroxyl terminated liquid fluoroelastomer (**HTLF,** green dashed line), **Cured HTLF** (black solid line) and solid fluoroelastomer (Poly(VDF-HFP), red dashed line) under N_2 (10°C/min).

3.5. Chemical resistant properties of cured fluoroelastomer film(Cured HTLF)

In order to verify the chemical stability of cured fluoroelastomer in different media, we made the cured fluoroelastomer film immerse in 37% HCl, 45% NaOH, jet aircraft oil and cyclohexane for 7 days respectively. The common solid fluoroelastomer poly(VDF-HFP) is for comparison. To evaluate the chemical stability behavior of polymers, "change in mass" was be used. Change in mass was calculated as follows:

$$\text{Change in mass } (\%)=[(M_2-M_1) / M_1] \times 100\%$$

The result is shown in Figure 5, and Table1 is detail. It clear that the cured liquid fluoroelastomer exhibits almost the same chemical resistance as poly(VDF-HFP). No matter in the organic solvent or in the acid-base reagent, the change in massof Cured HTLF has no more than 5%, it suggested that the chemical resistance of the product cured by liquid fluorine elastomer is still excellent. This attributes to low surface energy of "C-F" group in main chain and side chain. Property of "surface enrichment" make the fluorine content in the surface of the material is enriched, so little fluorine could make fluoropolymer has good hydrophobicity and lipophobicity. It unrelated to product structure.

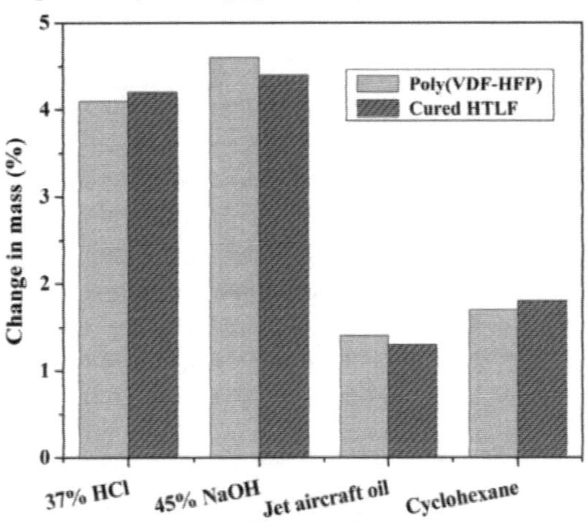

Figure 5. Change in mass(%)of solid fluoroelastomer (Poly(VDF-HFP) and Cured liquid fluoroelastomer (Cured HTLF).

Table 1. Change in mass(%) of Poly(VDF-HFP) and Cured HTLF.

Media solvent	Weight change (%)	
	Poly(VDF-HFP)	Cured HTLF
37% HCl	4.1	4.2
45% NaOH	4.6	4.4
Jet aircraft oil	1.4	1.3
Cyclohexane	1.7	1.8

3.6. Mechanical properties of cured fluoroelastomer film(Cured HTLF)

In order to verify the mechanical properties of cured fluoroelastomer, we research the tensile strength and elongation at break of Cured HTLF, traditional solid fluorinelastomer it was shown in shown in Table2. While, the date of common solid fluorine elastomer (poly(VDF-HFP)) for compare. Five measurements were made and the data were averaged. It is shown that Cured HTLF has the similar tensile strength and elongation at break with Poly(VDF-HFP), attribute to bond energy of "C=O" is 733kJ/mol and "C=N" is 533kJ/mol, althrough the precursor of Cured HTLF is liquid, the high bond energy of cross-linked structure make Cured HTLF have excellent mechanical property compared with solid fluorine elastomer.

Table 2. Change in mass(%) of Poly(VDF-HFP) and Cured HTLF.

	Poly(VDF-HFP)	Cured HTLF
Tensile strength (MPa)[a]	20	21
Elongation at break (%)[a]	352	349

[a] Five measurements were made and the data were averaged.

4. Conclusion

In summary, a novel hydroxyl-terminated liquid fluoroelastomer (HTLF) was synthesized and cured with HDI trimer. The structure of product was characterized by FT-IR and ^{1}H-NMR. Through the study of the product, Cured HTLF show similar chemical resistant and mechanical property with traditional solid fluorine elastomer. It is attributing to "surface enrichment" and high bond energy of cross-linked structure. Amazingly, Cured HTLF shows more excellent thermal resistant than traditional solid fluorine elastomer. It is ascribed to the introduction of ring structure (HDI Trimer) improve the thermal resistance of the cured product, what is more, cross-link structure formed in the curing process, it make the thermal decomposition rate of liquid fluoroelastomer further reduced. This kind of LFs has not been reported now. This novel liquid fluoroelastomer is suitable for sealing the fuel tanks of aircraft and coating the external surfaces of building owing to overcoming the practical limitations in size and shape of high molecular weight fluoroelastomer which is mostly formed by compression molding.

Acknowledgments

This work was financially supported by International Cooperative Project (Harbin Institute of Technology 2014DFR40370) and International Cooperative Project (Wuxi HIT Limited Corporation & Research Institute of New Materials BZ2015024).

IOP Conf. Series: Materials Science and Engineering **213** (2017) 012029 doi:10.1088/1757-899X/213/1/012029

References

[1] Close D 1983 *U.S. Patent* 4,423,18312-27
[2] Hochgesang P J and Jacks J W 2006 *Fluoroelastomer composition: U.S. Patent* **7,098,270** 8-29
[3] Close D1982 *U.S. Patent* 4,347,268 8-31
[4] Sandberg C L, Mullins J M and Paul S 1963 *U.S. Patent* **3080347A** 4-31
[5] Ho T and Wynne K 1992 *Macromolecules* **25** 3521
[6] Tonelli C, Ajroldi G, Turturro A and Marigo A 2001 *Polymer* **42** 5589
[7] Tonelli C, Ajroldi G, Turturro A and Marigo A 2001 *Polymer* **42** 9705
[8] Tonelli C and Ajroldi G 2003 *J. Appl. Polym. Sci.* **87** 2279
[9] Bassi M, Tonelli C and Meo A 2003 *Macromolecules* **36** 8015
[10] Kim Y, Lee J, Ji Q and McGrath J 2002 *Polymer* **43** 7161
[11] Tang Y, Santerre J, LaboW R and Taylor D 1996 *J. Appl. Polym. Sci.* **62** 1133
[12] Gallagher G A 1962 *U.S. Patent* **3069401** 8-11
[13] Ameduri B and Boutevin B 2005 *Prog. Polym. Sci* 644-87
[14] Maclachlan J D 1978 *Polym-Plast. Tech. Eng.* **11** 41-53
[15] Vijayakumar G, Karthick S N, Paramasivam R and Ilamaran C 2012 *Polym-Plast. Tech. Eng.* **51** 1427-31
[16] Abdelhamida M I, Aboelwafaa A M, Elhadidya H and Habiba A 2012 *Inter. J. Polym. Mater. Polym. Biomater* **61** 505-19
[17] Ameduri B, Boutevin B and Kostov G 2001 *Prog. Polym. Sci.* **26** 105-87
[18] Li J, Lu Y, Liu Y, Li Y, Zhang X and Qi S 2014 *Polym-Plast. Tech. Eng.* **53**(1) 46-53

IOP Conf. Series: Materials Science and Engineering **213** (2017) 012030 doi:10.1088/1757-899X/213/1/012030

Prediction of elastic properties of pyrolytic carbon based on orientation angle

Y S Song[1,2], L H Qi[1] and Y XLi[1]

[1]School of Mechatronic Engineering, Northwestern Polytechnical University, Xi'an, China

E-mail: songyongshan@mail.nwpu.edu.cn

Abstract. In the present work, a finite element computation micromechanics model was established to evaluate mechanical properties of pyrolytic carbon (PyC). The proposed computation of PyC considering the texture, i.e. the orientation distribution of the coherent domains, was modeled by a normal distribution that depends on orientation angle (OA). With the proposed model, different textures of PyC, were tested under six different loading modes. The calculated elastic modulus were coincided with the reported experimental results approximately. In particular, the influence of OA of PyC on Young's modulus was investigated. The results show that the Poisson's ratio and Young's module of PyC tend to be isotropic when OA is higher than 110° andthe shear modulus becomes isotopic with OA higher than 50°. The results demonstrate that the proposed model would provide a valuable numerical tool to predicate the elastic properties of PyC.

1. Introduction

C/C composites have been widely used for aircraft, aerospace, car and nuclear industry due to their excellent mechanical properties at high temperature [1-2]. Pyrolytic carbon is used as matrix of C/C composites produced by chemical vapor infiltration. In order to predict the mechanical properties of C/C composites, one has to fully investigate the mechanical properties of PyC. Several experimental methods have been used to measure the elastic properties of PyC, such as ultrasonic pulse-echo experiments [3] and sharp indentation tests [4]. But due to the significant anisotropy of PyC, accuracy of the experiment is not satisfied. In order to predict the mechanical properties of PyC, a numerical approach was required.

On the submicron scale, the microstructure of PyC can be described as a set of coherent domains with different orientations. This can be performed by textures, which are classified as: isotropic, low-textured (LT), medium-textured (MT) and high-textured (HT). OA <50°, from 50°to 80°, 80°to 180°, and correspond to HT-, MT- and LT-carbon, respectively[5].

Pyrolytic carbon can be described by a small number of texture components. A texture component is a domain orientation for which the orientation distribution function (ODF) reaches a maximum value. In the neighborhood, the ODF is decreasing in an anisotropic way. A lot of ODF of domains have been reported by different experimental characterization methods [6-8]. However, only Piat [6] utilized the real physical 2D spatial distribution of carbon structural units and established the correlation of ODF and OA.

Micromechanical methods can establish relationship between microstructure and elastic properties and provide all the properties of materials. Micromechanical models for predicting the mechanical properties of pyrolytic carbon have been reported in the literatures [7-9].

It is suggested that the studies for pyrolytic carbon are still insufficient. The investigations of pyrolytic carbon based on orientation angle were rarely studied. The objective of this work is to establish a detailed mechanical model for predicting the elastic properties of pyrolytic carbon with different orientation angles.

The ODF of domains is modeled by a Gauss distribution based on OA. Then, the finite element model is established and the texture is characterized by monte carlo method. In the last part of the paper, the elastic properties with different OA are determined numerically.

2. Model development

The RVE can be regarded as a set of cuboids and shown in Figure 1. Each cuboid represents a coherent domain with same size but different orientations. The unit normal c on each domain can be represented by its endpoint on the sphere around the center O and thus be described by spherical coordinates {σ, φ}. The angle φ is distributed in the interval [0, 2π) uniformly, and the angle σ corresponds to the Gauss density function [6]:

$$f(\alpha)=\left(\sigma\sqrt{2\pi}\right)^{-1}\exp\left(\alpha^2/2\sigma\right)\quad \sigma\in\left(-\frac{\pi}{2},\frac{\pi}{2}\right) \tag{1}$$

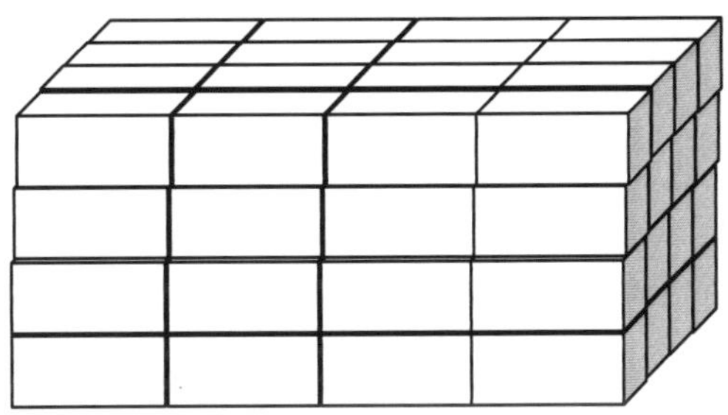

Figure 1. Microstructure of PyC.

The RVE and finite element model is showed in Figure 2 where brick solid element SOLID95 with twenty nodes is used. The finite element approach based on strain energy is finally used to predict effective elastic properties of PyC.

Figure 2. Unit cell and FE mesh.

In this study, special boundary [10-12] conditions are applied on the RVE to evaluate effective elastic properties. The macroscopic behaviors of the RVE can be characterized by the effective stress tensor σ and strain tensor ε in the elastic regime. They are interrelated by the stiffness matrix C^H.

$$\sigma=C^H\varepsilon \tag{2}$$

where $\sigma = \frac{1}{V}\int_{\Omega}\sigma d\Omega$, $\varepsilon = \frac{1}{V}\int_{\Omega}\varepsilon d\Omega$. Consideration the case of 3D orthotropic materials, equation corresponds to

$$\begin{Bmatrix} \sigma_x \\ \sigma_y \\ \sigma_z \\ \tau_{yz} \\ \tau_{zx} \\ \tau_{xy} \end{Bmatrix} = \begin{bmatrix} C_{11} & C_{12} & C_{13} & & & \\ C_{21} & C_{22} & C_{23} & & & \\ C_{31} & C_{32} & C_{33} & & & \\ & & & C_{44} & & \\ & & & & C_{55} & \\ & & & & & C_{66} \end{bmatrix} \begin{Bmatrix} \varepsilon_x \\ \varepsilon_y \\ \varepsilon_z \\ \gamma_{yz} \\ \gamma_{zx} \\ \gamma_{xy} \end{Bmatrix} \tag{3}$$

The strain energy of the RVE is equal to:

$$\begin{aligned} E = &\int_{\Omega}\tfrac{1}{2}\left(\sigma_{11}\varepsilon_{11} + \sigma_{22}\varepsilon_{22} + \sigma_{33}\varepsilon_{33} + \sigma_{12}\varepsilon_{12} + \sigma_{23}\varepsilon_{23} + \sigma_{31}\varepsilon_{31}\right)d\Omega \\ &-\tfrac{1}{2}\left(\overline{\sigma_{11}\varepsilon_{11}} + \overline{\sigma_{22}\varepsilon_{22}} + \overline{\sigma_{33}\varepsilon_{33}} + \overline{\sigma_{12}\varepsilon_{12}} + \overline{\sigma_{13}\varepsilon_{13}} + \overline{\sigma_{31}\varepsilon_{31}}\right)V \end{aligned} \tag{4}$$

Suppose a unit initial strain is imposed in direction 1, i.e. $\varepsilon^{(1)} = (1\ \ 0\ \ 0\ \ 0\ \ 0\ \ 0)^T$. The corresponding average stress is then obtained by Eq. (4):

$$\overline{\sigma}^{(1)} = \left(C^H_{1111}\ \ C^H_{1122}\ \ C^H_{1133}\ \ 0\ \ 0\ \ 0\right)^T \tag{5}$$

By replacing $\varepsilon^{(1)}$ and $\overline{\sigma}^{(1)}$ in Eq. (5), the strain energy can be obtained as following expression:

$$E^{(1)} = \tfrac{1}{2}\overline{\sigma}^{(1)}\overline{\varepsilon}^{(1)}V - \tfrac{1}{2}C^H_{1111}V \tag{6}$$

The matrix coefficient C^H_{1111} can be derived as:

$$C^H_{1111} = 2E^{(1)}/V \tag{7}$$

3. Results and discussion

The elastic constants C1111 = 40.016, C3333 = 18.185, C1122 = 20.021, C1133 = 12.779, and C2323 = 1.776 [GPa] are used for the domains with transversely isotropic material symmetry [4]. In high texture PyC (OA less than 50°), Taylor et al.[13] performed nanoindentation tests (Berkovich indenter) perpendicular to deposition direction and the measured values of Young's modulus were between 16.1 and 26.2 GPa. These values were in good agreement to the calculated elastic modulus (between 18.4 and 26.5 GPa).

The predicted extensional (EX and EZ) modulus of the pyrolytic carbon with various values of OA are presented in Figure 3. It is observed that with the increase of OA, EX is decreased while EZ is slight decreased firstly, and then increases. However, when OA>110°the influence of OA on EX and EZ seems negligible. It can be found that When OA>110°, PyC tends to be an isotropic material. The effect of the OA of PyC on its microstructure is quite significantly.

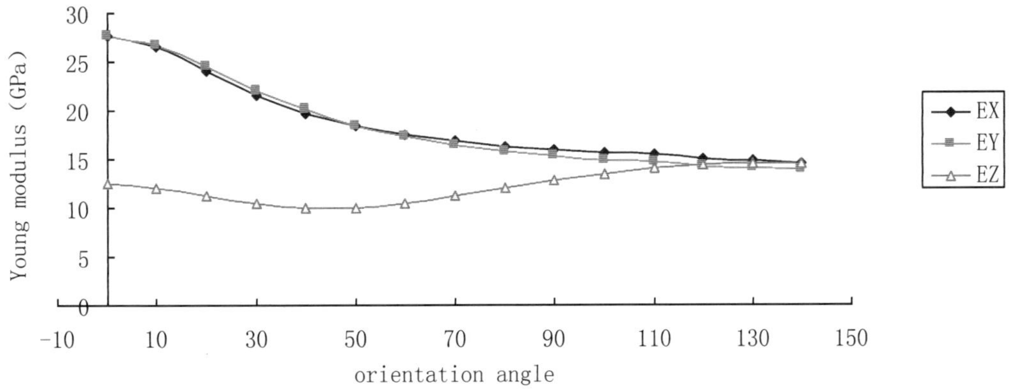

Figure 3. Young's moduli with different orientation angles.

The predicted shear (GXY and GYZ) moduli of the PyCs are presented in Figure 4. It is observed that when OA>50°, PyC tends to an isotropic material. With the increase of OA, GXY is decreased from 10 to 4.5 Gpa while GYZ is increased from 2 to 4 Gpa.

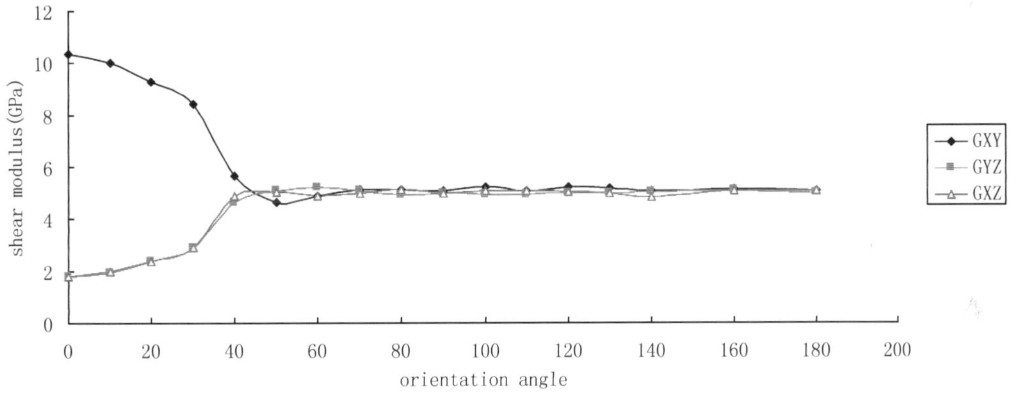

Figure 4. Shear moduli with different orientation angles.

Values of the Poisson's ratio with different OA are presented in Figure 5. When OA<50°, Vxy decreases firstly, and then increase with the increase of OA, Vxz and Vyz have an opposite tendency. PyC tends to be isotropic in poisson ratio with OA>110°.

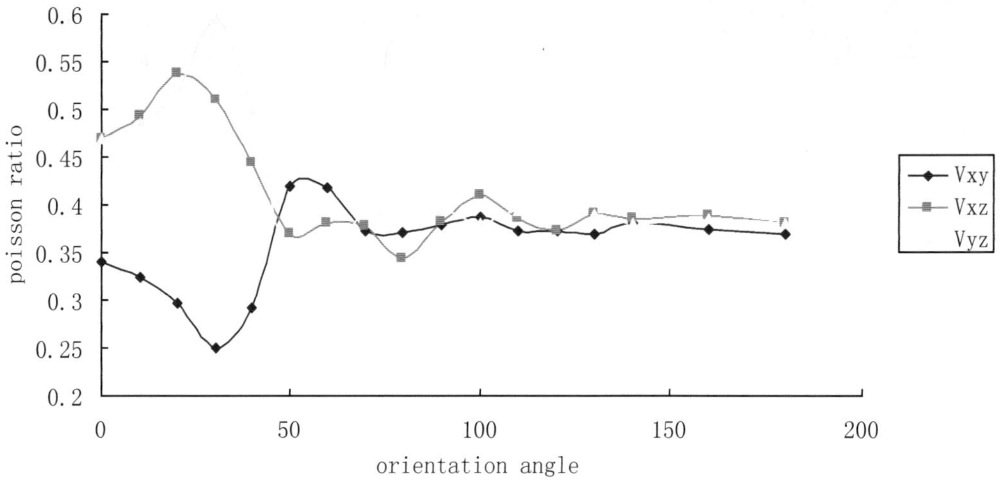

Figure 5. Poisson ratio with different orientation angles.

4. Conclusions

The elastic properties of PyC on the micrometer scale have been determined based on the elastic properties of the domains and the orientation distribution of the domains. A Gauss distribution function based on OA has been used as a texture component model.

A numerical study of PyC using ANSYS was performed. The numerical results indicate that the strong elastic anisotropy of PyC induces a large gap, and it is found that the HT pyrolytic carbon films and measured values of Young's modulus between 16.1 and 26.2 GPa. These values are in good agreement to the calculated elastic moduli(between 18.4 and 26.5 GPa). The bounds and the approximations of the effective elastic properties depend only on one concentration parameter which completely specifies OA.

There are significant gradients in the elastic properties of the material throughout the whole domain of the values for OA and thus throughout the different texture degrees of HT, MT and LT. PyC tends to be the poisson ratio and Young's module of an isotropic material with OA>.110°, shear moduli of which with OA>50°.

Acknowledgements

This work has been supported by the National Natural Science Foundation of China under Grant No.50472203 and the Doctorate Foundation of Northwestern Polytechnical University.

References

[1] Reznik B and Huttinger K J 2002 On the terminology for pyrolytic carbon *Carbon* **40**(4) 621-4
[2] Reznik B, Gerthsen D and Huttinger K J 2001 Micro- and nanostructure of the carbon matrix of infiltrated carbon fiber felts *Carbon* **39**(2) 215-9
[3] Gebert J M *et al* 2010 Elastic constants of high-texture pyrolytic carbon measured by ultrasound phase spectroscopy *Carbon* **48**(12) 3647-50
[4] Diss P, Lamon J, Carpentier L, Loubet J and Kapsa P 2002 Sharp indentation behavior of carbon/carbon composites and varieties of carbon *Carbon* **40**(14) 2567-79
[5] Papadakis E P and Bernstein H 1963 Elastic moduli of pyrolytic graphite *J. Acoustical Sci. America* **35**(4) 521-4
[6] Piat R, Reznik B, Schnack E and Gerthsen D 2004 Modeling of effective material properties of pyrolytic carbon with different texture degrees by homogenization method *Compos. Sci. Technol.* **64** 2015-20

[7] Sauder C and Lamon J 2005 Prediction of elastic properties of carbon fibers and CVI matrices *Carbon* **43**(10) 2044-53
[8] Sauder C, Lamon J and Pailler R 2005 The tensile properties of carbon matrices at temperatures uo to 2200◦ *Carbon* **43**(10) 2054-65
[9] Böhlke T, Jöchen K, Piat R, Langhoff T A, Tsukrov I and Reznik B 2010 Elastic properties of pyrolytic carbon with axisymmetric textures *Tech. Mech.* **30**(4) 343-53
[10] Rao M V, Mahajan P and Mittal R K 2008 Effect of architecture on mechanical properties of carbon/carbon composites *Compos. Struct.* **83** 131-42
[11] Swain M V and Field J S 1996 Investigation of the mechanical properties of two glassy carbon materials using pointed indetners *Philos. Mag. A.* **74** 1085-96
[12] Richter A, Ries R, Smith R, Henkel M and Wolf B 2000 Nanoindentation of diamond, graphite and fullerene films *Diamond Relat. Mater.* **9** 170-84
[13] Taylor C, Wayne W and Chiu W 2003 Residual stress measurement in thin carbon films by raman spectroscopy and nanoindentation *Thin Solid Films* **429**(1-2) 190-200

IOP Conf. Series: Materials Science and Engineering **213** (2017) 012031 doi:10.1088/1757-899X/213/1/012031

Experimental study on relationship between processing parameters and stress wave propagation during automated fiber placement process

H Y Fu[1], W Q Li[1], S Z Sun[1] and Z Y Han[1,2]

[1]School of Mechatronics Engineering, Harbin Institute of Technology, No. 92, Xidazhi Street, Harbin 150001, China

Email: hanzy@hit.edu.cn

Abstract. Automated fiber placement (AFP) is an important manufacturing method of composites, which has been widely used in the field of aerospace. Unreasonable processing parameters could lead to some manufacturing defects including pores, bubbles and cracks. In this paper, the propagation characteristics of stress waves are believed to be closely related to the defects during AFP process. Experiments are conducted to collect stress wave signal under different processing parameters (pressure, velocity and temperature) during manufacturing process. And the relationship between the processing parameters and the characteristics of stress waves is explored by Control Variate Method (CVM). Finally, the effects of laying parameters on stress amplitude, response speed and duration are summarized.

1. Introduction

Due to the advantages of light weight, high strength and flexible forming methods, carbon fibre/epoxy laminates are widely applied to manufacture aircraft components and their prospects become more and more popular in the field of aerospace [1-2]. AFP, which is rapidly developed in the past 30 years, is an important manufacturing method for carbon fiber composite materials [3-4]. It is suitable for forming structures with complex surfaces and smaller structures than general ones because of its flexible and controllable width. AFP could utilize single or multiple narrows to realize a given total prepreg band width [5-6]. If forming laying process is unreasonable, the laminates could be forced to cause various defects such as pores, bubbles, cracks, which will affect their mechanical properties.

The formations of defects in composite material are closely associated with the characteristics of stress waves. For instance, excessive shear stress between the adjacent layers could cause delamination, and the crack tip could result in stress concentration phenomenon leading to further extension of micro cracks [7-9]. The effect of the structure of mediums on stress waves could provide some references to evaluate the quality of products made by different materials. At present, the internal defect detection of wood structure have maturely developed based on stress wave theory [10-12]. The relationship among stress waves velocity, the moisture content of wood and tree species has been found. And there is statistically linear relationship between wave resistance and bending modulus of elasticity [13]. It is more complicated that stress waves propagate in the anisotropy laminates because of the interaction among incident waves, reflected waves and scattered waves. Therefore, study of the characteristics of stress waves during laying process does not only make contribution to optimize laying process, but also provide the reference of manufacturing defects detection.

The porosity of composite laminates is proved to affect stress wave characteristics. Ju Y [14-15] prepared different porosity samples by imitating natural structure of gravel and sand, and conducted SHPB experiments for cylindrical samples. By analyzing the experiment results, they found that the amplitudes and the amount of wave peaks could grow up with the increase of porosity, and stress waves could affect the form of detects when they propagate in the composite laminates. Yan G [16] used high-order plate theory to analyze the scattering in composite laminates and established a damage detection system through calculating the speed of stress wave propagation. Guo Z [17] built a relational model between detects and stress waves for thin plate under dynamic loads, and revealed that the formation of detects results from the interaction of stress waves.

In this paper, the effect of processing parameters on the characteristic of stress waves is discussed by variable parameter experiments. The variable-controlling approach is used to compare the effect of laying pressure, laying temperature and laying velocity on stress waves by analyzing the real-time signal. Finally, the effects of laying parameters on stress amplitude, response speed and duration are summarized.

2. Experiment setup for collection of stress wave signal

In an attempt to reveal the relationship between laying parameters and the characteristics of stress wave propagation, three factors and three levels of normal experiments are designed. Different parameters have different impacts on the quality of products. Laying velocity not only decides productivity, but also influences the time of laying pressure and pre-heating. Appropriate laying velocity could ensure the stickiness and smoothness of laminates as well as an acceptable level of efficiency. The force situation of tows is determined by the laying pressure. Overloading could lead to excessive deformation and under loading could cause incomplete join, bubbles and even warps. Viscidity/overlay ability is affected by laying temperature. The viscidity of tows increases with temperature rising. Stronger viscidity makes it difficult to separate the tow from the wrap, and weaker viscidity could lead to slip between the tow and the mold. Therefore, it is important to ensure a group of appropriate laying parameters.

Table 1. Experimental parameters.

Parameter	Factor1	Factor2	Factor3
F(mm/s)	10	15	20
P(N)	350	600	850
T(℃)	40	50	60

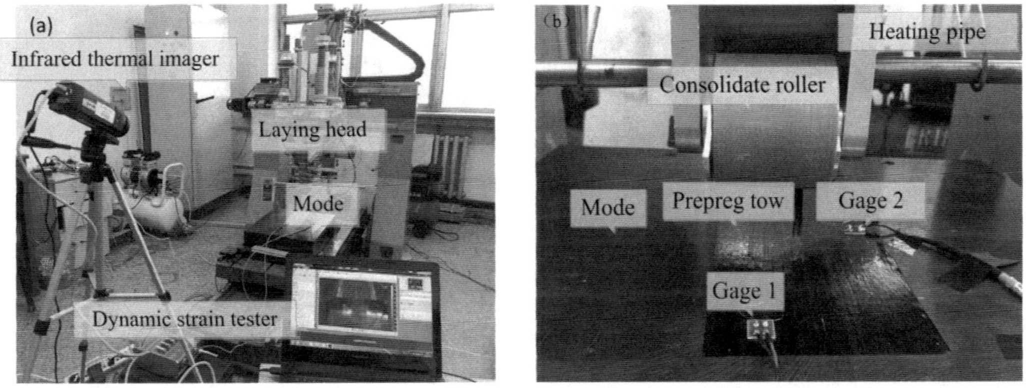

Figure 1. Laying stress wave test experiment platform: (a) Laboratory equipment; (b) Laying roller and foundation.

IOP Conf. Series: Materials Science and Engineering **213** (2017) 012031 doi:10.1088/1757-899X/213/1/012031

Experimental facilities are shown in Figure 1, the experimental platform is based on the five-axes automated fiber placement machine produced by HIT numerical control laboratory, which can adjust laying pressure by air pump, laying velocity by servo motor and laying temperature by variable power heater. Experimental material is No.6511 carbon fiber prepreg with the width of 5mm produced by Guangwei company, China. The properties of the prepreg are shown in Table 2. Dynamic strain tester with strain gages is used to collect the real-time stress signal from test point, which its resolution can achieve to 0.1με and the strain detection error can be no more than 0.02%. The strain gauge resistances are 120Ω±2% and the sensitivity coefficient can reach to 2.0±1%. The temperature is observed by infrared thermal imager, which can detect the temperature range of 0-300℃. Because of the differences (such as elasticity modulus and density) between fiber prepregs and the rectangle metal mold that is made by No.45 steel with the thickness of 5cm. If the strain gauges are pasted to the surface of mold directly, the excessive stiffness of mold will limit the deformation of the stress gauge which can not reflect the actual deformation of the fiber prepreg. Before collecting stress wave signal, five layers (100mm×300mm) of prepregs have been laid on the mold as a foundation in order to avoid the impact of slight deformation of the mold. Two strain gages are attached to the specify locations on the fifth layer and kept distance from the laying path to prevent the roller press the strain gages directly. The stress signal is collected when the roller moves, and there are five tows distributed in a layer, which results in five cycles in a group of the collected signal.

Table 2. Material parameters.

Matrix type	Prepreg resin content	Cured ply thickness	Prepreg width	Normalised tensile strength	Normalised tensile modulus
6511	33%	0.14mm	5mm	1846Mpa	120Gpa

3. Analysis of experiment results

3.1. Laying pressure

Laying a tow is divided into four processes as shown in Figure 2. Firstly, the consolidation roller drops down, and loads to the foundation. Secondly, when the roller reaches the predetermined pressure, it would roll toward the strain gage for attaching the tow onto the surface of the upper layer. Then the roller continues to move forward to the end position, and unloads at the end position. The stress waveforms collected at test point show a regular change, which is closely related to the relative position of roller and strain gages.

When laminates are energized, stress spreads out in the form of waves and reflects when propagating to the boundary of the laminates. The stress wave propagates in the form of transverse and longitudinal waves in the laminates with boundary; the direction of transverse wave propagation is perpendicular to the direction of vibration, while the direction of longitudinal wave propagation is the same as the direction of vibration. The incident stress wave interacts with the reflected stress wave to form a complex stress state.

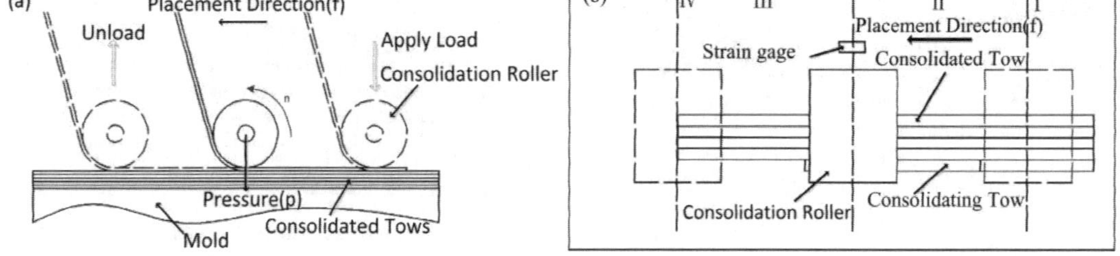

Figure 2. Laying process: (a) Laying process front view; (b) Laying process top view.

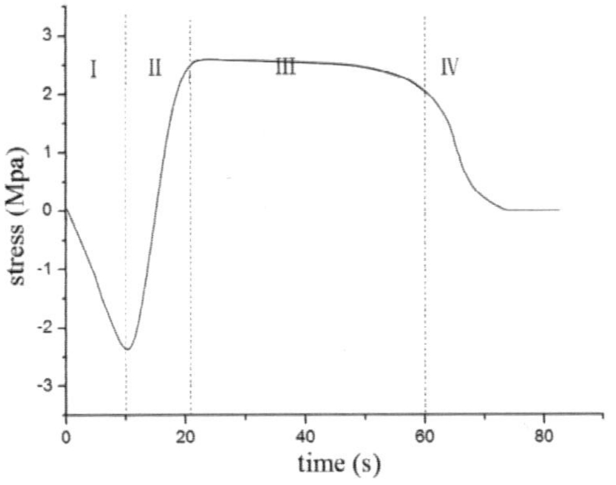

Figure 3. The stress wave waveform during laying process.

The stress wave signal of a certain laying process is shown in figure 3. In the first stage of this experiment, the laying head press vertically and the distance between the laying head and the test point is not changed, the stress state of the test point mainly dependents on the interaction of the incident stress wave with the reflected stress wave. In the following stage, as the laying head rolling forward, friction is created in the fibre direction. The stress state of the test point is not only affected by the stress wave but also by the small tension in the fiber direction. This complex stress state varies with the distance between the excitation and the test point. As the distance between the roller and the test point is shortened, the effect of friction is becoming more obvious which reach to the maximum in the vicinity of the test point and then remain stable relatively. At the end of laying path, the stress gradually reduces to zero in unloading process. It is worth noting that the magnitude of the stress wave is positively correlated with the magnitude of the pressure according to Figure 4.

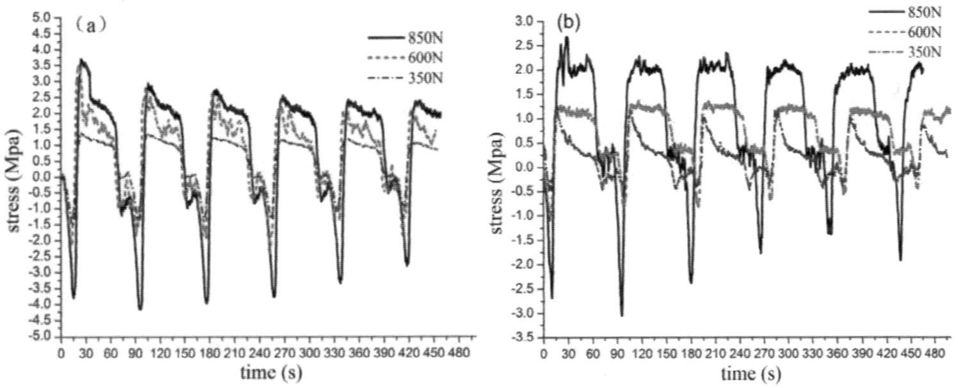

Figure 4. Real-time stress wave curves of laying process under different pressures: (a) f=15mm/s, t=60℃; (b) f=15mm/s, t=40℃.

3.2. Laying temperature

Figure 5 shows that the real-time stress wave signal is collected at different temperatures under the same speed and pressure. In order to find the effect of temperature parameters on the stress wave during the laying process, three different parameters of 40℃, 50℃and 60℃ are selected. It could be seen from Figure 4 that the higher laying temperature is, the greater the maximum of the stress wave is, especially at the speed of 20mm/s and the pressure of 350N. The maximum of stress wave at the temperature of 60℃ reaches 1.7Mpa. The maximum of stress wave at 50℃ is about 0.8Mpa, but it is only 0.6Mpa at 40℃.

IOP Conf. Series: Materials Science and Engineering **213** (2017) 012031 doi:10.1088/1757-899X/213/1/012031

From the waveform point of view, the waveform of high temperature stress is generally steeper than that of lower temperature. On the one hand, it is due to the temperature drift of the strain gauge that is affected by different temperature effects. On the other hand, the degrees of melting and fluidity of the epoxy resin in the prepreg increase with temperature, which makes stress status more sensitive.

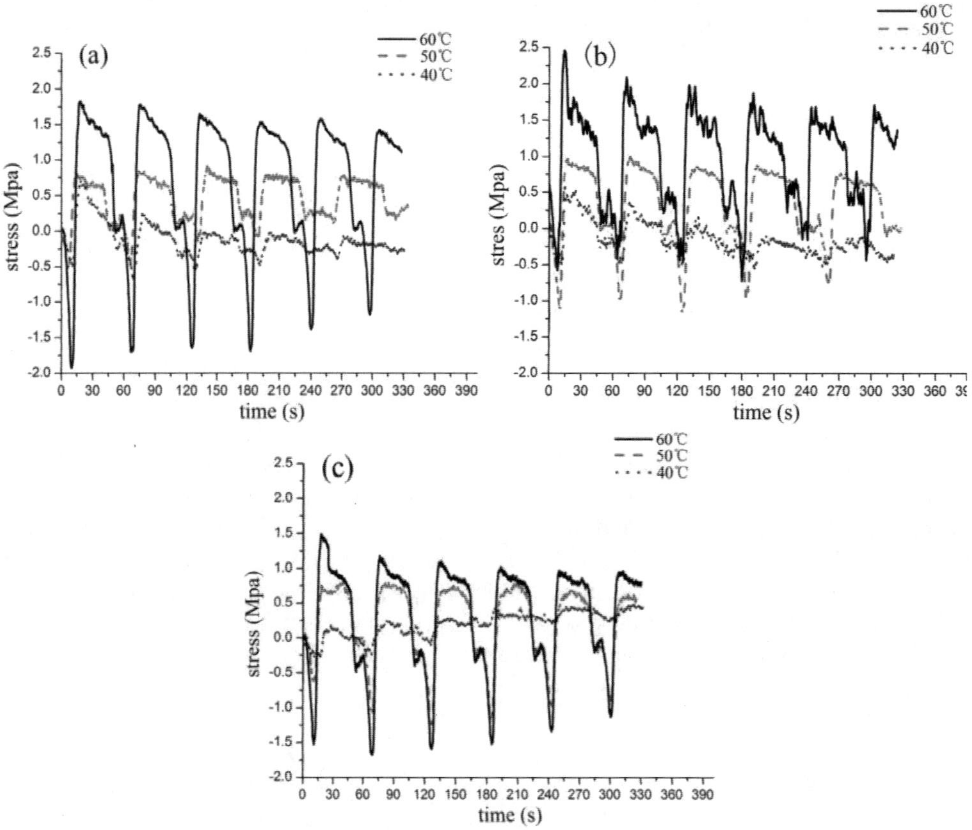

Figure 5. Real-time stress wave curves of laying process under different temperatures: (a) f=20mm/s, p=850N; (b) f=20mm/s, p=600N; (c) f=20mm/s, p=350N.

3.3. Laying velocity

The laying head presses the tow onto the foundation with a certain pressure, which is equivalent to applying a continuous excitation to the foundation. The stress fluctuation generated by the excitation propagates outside in the form of the shear wave and the longitudinal wave through the material. When stress wave is propagated to boundary, it would reflect. Also the pores and the cracks in the material will scatter the stress waves. The reflection and scattering of stress waves are accompanied by the propagation of energy. To a certain extent, the magnitude and the speed of the stress waves can indicate the amount of energy absorbed by the foundation.

Since the data of this experiment is collected while roller moves, the continuous excitation of the laying roller makes its response signal very complicated. Taking account of this situation, the sampling frequency is set to 10Hz because the collected stress wave signal is the accumulation effect of complex high frequency stress wave signal in a short time at this frequency. It can be seen from the above figures, the faster the laying speeds is, the shorter the stress waves are, and the higher the frequency is. The effect of speed on the stress waves is not noticeable. In addition, the stress waveforms under the same laying temperature and pressure are similar.

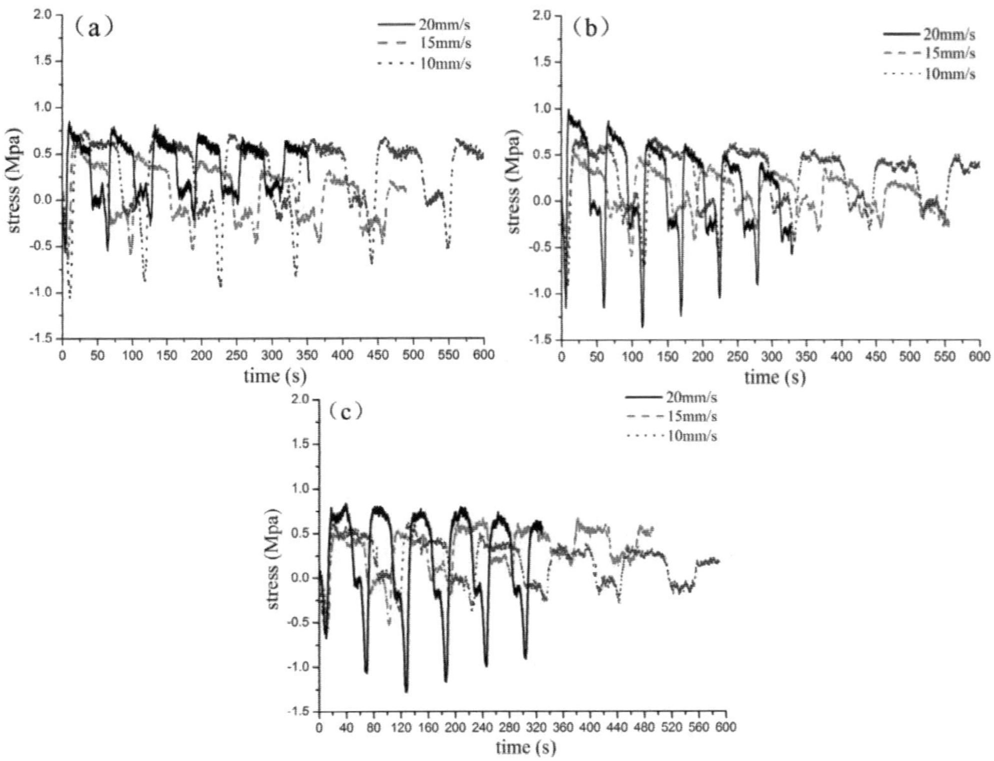

Figure 6. Real-time stress wave curves of laying process under different velocities: (a) t=50℃, p=850N; (b) t=50℃, p=600N; (c) t=50℃, p=350N.

In addition, with the purpose of studying the effect of different parameters on the magnitude of the stress wave an analysis is conducted. The correlation between the amplitude of stress wave data and parameters in 27 groups of experiments is analysed by means of gray relational analysis theory. By calculating the relative correlation between the data and the different parameters of each group, it can be found that the maximum correlation with the stress amplitude is the laying pressure (correlation coefficient 0.56), followed by temperature (0.53), the smallest is speed (0.51).

4. Conclusion

Laying pressure mainly determines the curves amplitude of the stress waves. With the different loading statuses and the different distances between the pressure roller and the strain gage, the curves of stress waves also present different trends. The fluidity of resin is affected by the laying temperature. The response speed of the stress wave is quick under the condition of strong fluidity of resin. And the increase of the temperature causes the thermal stress of the material to increase, where the stress curve amplitude is enlarged. The response speed of the stress waves collected at the test point has positive correlation to the laying speed, which is the most obvious in the latter two stages. The decay rate of the stress wave decreases with laying speed, and laying speed has no significant effect on the amplitude of the stress wave. Future work is going to explore the relationship between stress waves and defects.

Acknowledgments

This study is financially supported by National Science and Technology Major Projects of China (2014ZX04001091).

References

[1] Xiao J, Li Y and Wen L W 2009 *Materials China* **28** 28
[2] Han Z Y, Sun S Z, Shao Z X and Fu H Y 2016 *Adv. Mater. Sci. Eng.* ID 5480352 14
[3] Puck A and Schurmann H 2002 *Compos. Sci. Technol.* **62** 1633
[4] Elsherbini Y and Hoa S V 2017 *Compos. Sci. Technol.* **146** 49
[5] Han Z Y, Wang Z Y and Bian D Y 2014 *Fiber Composites* **04** 3
[6] Han Z Y, Sun S Z, Shao Z X and Fu H Y 2016 *Adv. Compos. Mater.* **2016** 1
[7] Chen H R and Yang Q S 1991 *Fiber. Reinforced. Pla.* **06** 1
[8] Ridene S and Bouchriha H 2014 *J. Phys. Chem. Solids.* **75** 203
[9] Melek M, Saidi H, Ben Ahmed A, Ridene S and Bouchriha H 2016 *J. Int. Light. Elect. Op.* **127** 4903
[10] Feng H L, Li G H and Fang Y M 2010 *Sys. Simul.* **6** 1490
[11] Guntekin E, Emiroglu Z G and Yilmaz T 2013 *J. Bio. Resour.* **1** 231
[12] Xu H D and Wang L H 2011 *Sci. Sil. V. Sin.* **47** 123
[13] Olson S E, Desimio M P and Swenson E D 2010 *Struct. Health. Monit.* **24** 1347
[14] Ju Y, Yang Y M, Mao Y Z, Liu H Band Wang H J 2009 *Science China Press.* **5** 904.
[15] Ju Y, Wang H J, Yang Y M, Hu Q Ang and Peng R D 2010 *SCI. China. Press.* **6** 711
[16] Zhou L and Yan G 2006 *Smart. Struct. Syst.* **2** 141
[17] Guo Z L and Ren G W 2014 *Explosion. Shock. Wave.* **34** 52

IOP Conf. Series: Materials Science and Engineering **213** (2017) 012032 doi:10.1088/1757-899X/213/1/012032

High performance UV and thermal cure hybrid epoxy adhesive

C F Chen[1,3], S Iwasaki[1], M Kanari[1], B Li[2], C Wang[2] and D Q Lu[2]

[1]AG Product Development, Henkel Japan Ltd., Yokohama, Japan
[2]AG Product Development, Henkel (China) Co., Ltd., Shanghai, P. R. China

E-mail: chunfu.chen@jp.henkel.com

Abstract. New type one component UV and thermal curable hybrid epoxy adhesive was successfully developed. The hybrid epoxy adhesive is complete initiator free composition. Neither photo-initiator nor thermal initiator is contained. The hybrid adhesive is mainly composed of special designed liquid bismaleimide, partially acrylated epoxy resin, acrylic monomer, epoxy resin and latent curing agent. Its UV light and thermal cure behavior was studied by FT-IR spectroscopy and FT-Raman spectroscopy. Adhesive samples cured at UV only, thermal only and UV + thermal cure conditions were investigated. By calculated conversion rate of double bond in both acrylic component and maleimide compound, satisfactory light curability of the hybrid epoxy adhesive was confirmed quantitatively. The investigation results also showed that its UV cure components, acrylic and bismalimide, possess good thermal curability too. The initiator free hybrid epoxy adhesive showed satisfactory UV curability, good thermal curability and high adhesion performance.

1. Introduction

Radiation curing, especially ultraviolet (UV) light curing adhesive has been widespread used in various industrial applications. It is typically one component packed room temperature stable, solvent free and can cure very fast, suitable for high speed production applications [1]. Most common UV curing adhesives are free radical curable acrylic adhesive that is mainly composed of acrylic monomer, oligomer and photo-initiator. Photo-initiator can absorb light energy to generate free radical that will immediately initiate polymerization of acrylic compositions. In this curing process, however, several small molecules are generated as byproducts. In addition, photoinitiator cannot consume completely. In the case of using benzyl dimethyl ketal (BDMK) as photoinitiator, for example, E. Sitzmann [2] described that there are at least three small molecules generated, such as benzaldehyde, methyl benzoate. These small molecular byproducts and remained photo-initiator cannot be chemically bonded to cured adhesive and thus will potentially contaminate substrates or damage adhesion performance. UV curing adhesive without photoinitiator containing will have no such problem. Maleimide compounds have been studied years for use in photoinitiator free UV curing systems [3-8]. UV light cure performance has been identified. Till today, however, there is still almost no commercial success in this approach.

Epoxy resin adhesive is one of most widely used reactive adhesives, especially for structural bonding applications where high performance is required [9]. Epoxy resin adhesive normally needs relatively long time cure that will not meet fast curing requirements for some high production efficiency applications. By combination with UV curing acrylic components, UV thermal hybrid epoxy resin adhesive has been recently commercialized and used in actual applications such as special electronics parts assembly.

The UV thermal hybrid epoxy adhesive meets well with both high speed production and high performance requirements. For some sensitive high precise substrate bonding applications such as fine semiconductor packaging or display assembly, there are big concerns on contaminants from low molecule chemicals such as by products, remained photointiator during UV curing process on uncured adhesive compositions.

In this study, a new type one component UV thermal curable hybrid epoxy adhesive is designed. The hybrid epoxy resin adhesive is mainly composed of liquid bismaleimide compound, partially acrylated epoxy resin, acrylic monomer, epoxy resin and latent curing agent. UV cure and thermal cure behavior was investigated by FT-IR and FT-Raman spectroscopy. Glass adhesion performance was confirmed.

2. Experimental

2.1. Materials and Chemicals

Epikote 828, standard grade bisphenol A glycidyl ether epoxy resin (epoxy equivalent weight: 184 ~ 194, viscosity: 12,000 ~ 15,000 mPas/25℃), was supplied by Mitsubishi Chemical Corporation. Uvacure 1561, partially acrylated bisphenol A epoxy resin (epoxy equivalent weight: 450, viscosity: 1,000 mPas/60℃), was supplied by Daicel Allnex Ltd. Trimethylolpropane triacrylate (TMPTA) was supplied by Arkema Incorporation. Liquid bismaleimide compound was supplied by Henkel Corporation. KBM-403, 3-Glycidoxypropyltrimethoxysilane, was supplied by Shin-Etsu Chemical Co., Ltd. VDH-J (fine powder latent curing agent), 4-Isopropyl-2, 5-dioxoimidazolidine-1, 3-di(propionodihydrazide), was supplied by Ajinomoto Fine-Techno Co., Incorporation. Talc was supplied by Nippon Talc Co., Ltd. Fumed silica was supplied by Nippon Aerosil Co., Ltd. All raw materials are used directly without any further treatments. Chemical structure of main reactive materials was shown in Figure 1.

2.2. Adhesive sample preparation

The adhesive sample was composed of 28.0% Epikote 828, 22.0% Uvacure 1561, 10.0% TMPTA, 5.0% bismaleimide, 15.0% VDH-J, 1.0% KBM-403, 16.0% talc and 3.0% fumed silica. All components were added and premixed at first in a planentary centrifugal mixer. The premixed sample was further mixed on a triple roller mixer and then thoroughly degassed via full vacuum condition before use. Viscosity was measured by E-type viscometer. Pot life was determined by measuring viscosity increase during storage at 25℃ to 2 times of the initial one.

2.3. Cured sample preparation, glass transition temperature determination and adhesion strength measurement

Batch type metal halide lamp was used for UV cure. Light intensity was measured by Ushio UIT-101 UV meter. Conventional oven was used for thermal cure. Typical UV cure condition was 30 seconds at $100mW/cm^2$ and thermal cure condition was 60 minutes at 120℃. UV fixture time was determined by Henkel standard method with use of slide glass as substrate. Glass transition temperature was measured by Seiko DMS 6100 dynamic mechanical analysis (DMA). Glass adhesion strength was measured according to test method ASTM D 2095. Normal slide glass was used as substrate.

2.4. Analysis and Characterization

Varian 610-IR Fourier Transform infrared spectroscopy (FT-IR) was used to measure infrared spectrum. LabRAM HR-800 Raman Spectroscopy was used to measure FT-Raman spectrum. Conversion rate of both acrylic and maleimide double bond as well as epoxy group was calculated by peak area decrease of each specific absorption with use of absorption peak at $1505cm^{-1}$ attributed to aromatic ring absorption, as internal reference. The $1405cm^{-1}$ absorption peak was monitored for acrylic double bond [10]. The 690cm^{-1} peak area was monitored for maleimide double bond [11]while the 915cm^{-1} peak was monitored for epoxy group [12].

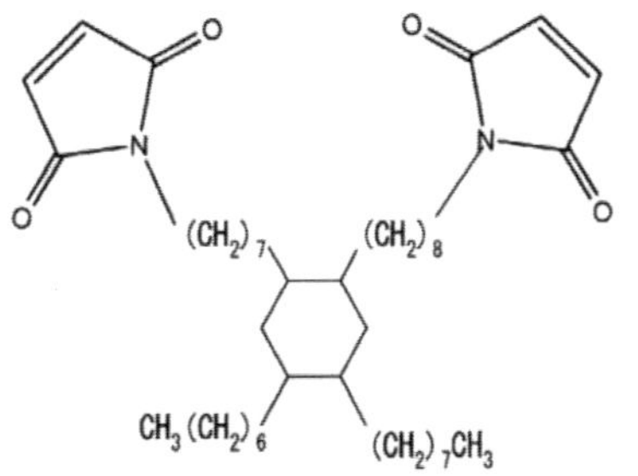

Bismaleimide compound

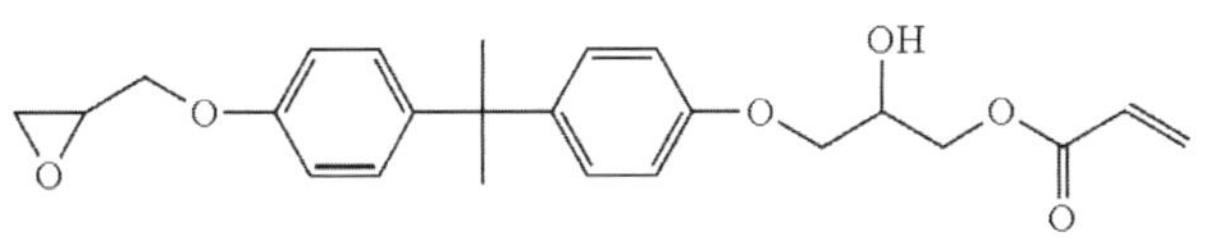

Partially acrylated bisphenol A epoxy resin

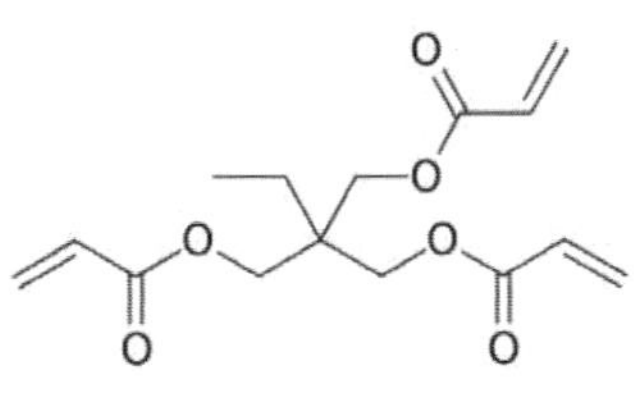

TMPTA

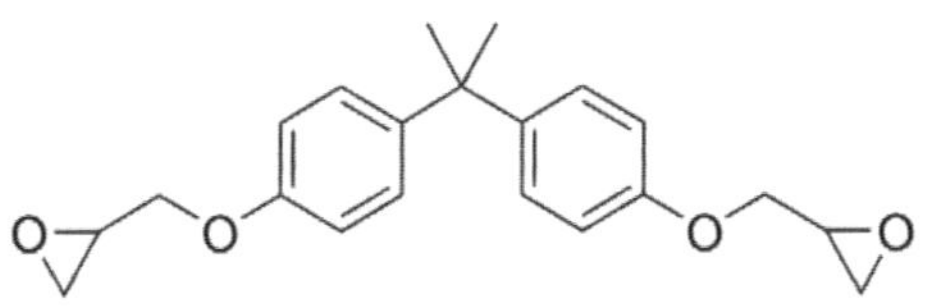

Bisphenol A epoxy resin

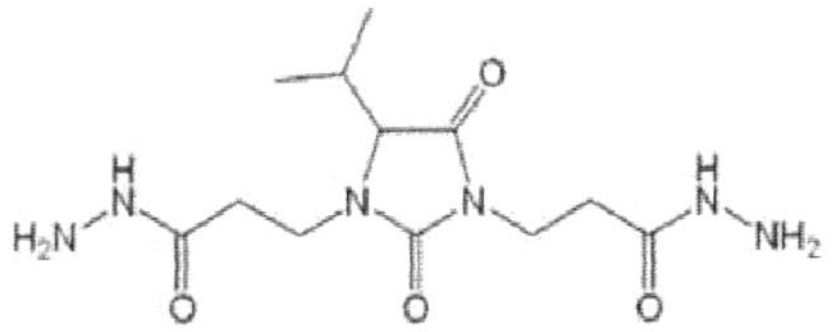

VDH-J

Figure 1. Chemical structure of key reactive materials used.

3. Results and discussion

3.1. Adhesive property and glass adhesion performance

The prepared adhesives were a high viscous milky white paste. Its viscosity measured at 25℃ was 29,6000 mPas. Pot life was 5 days at 25℃ storage. Cured sample possessed good thermal property. Glass transition temperature was 112℃.

As described previously, the adhesive sample prepared does not contain any conventional photo-initiator or thermal initiator. It is complete initiator free composition. Its UV fixture time was measured as 5 seconds at 100 mW/cm^2. This result showed that the adhesive sample possessed fast UV fixing property that can well meet high speed production requirement in various applications.

Glass adhesion strength was measured and compared at three cure conditions, i.e, UV cure only, thermal cure only and UV + thermal cure conditions. Glass adhesion strength results were summarized in Table 1. At UV cure stage, quite strong adhesion strength was obtained already. It means that this adhesive sample, even with no photoinitiator contained, showed quite good UV cure behavior. As expected, adhesion strength was much improved with post thermal cure process where epoxy part could be cured.

The sample showed very strong adhesion on glass substrate with cohesion failure. Adhesion strength at thermal cure only condition was some lower than that cured at UV and thermal cure but higher than that cured with UV only condition.

Table 1. Glass adhesion strength results.

Cure condition	Glass adhesion strength (N/mm^2)	Failure mode
UV cure only 100 mW/cm^2 x 30 sec	0.89	Interface failure
UV+ thermal cure 100 mW/cm^2 x 30 sec + 120℃ x 60 min	3.10	Cohesion failure
Thermal cure only 120℃ x 60 min	2.21	Cohesion failure

3.2. FT-IR measurements and analysis

FT-IR was performed to measure and analyse quantitatively cure behaviour of the adhesive sample. IR Spectrum of adhesive samples cured at all three conditions as well as non-cure sample was measured. Figure 2 showed and compared the measured FT-IR spectrum of adhesive sample cured with UV only condition at 100mW/cm^2 for 30 seconds, thermal cure only at 120℃ for 60 minutes and UV + thermal cure condition and that of non-cure sample. Changes of the 1405cm^{-1} absorption peak attributed to acrylic double bond, the 690cm^{-1} peak attributed to maleimide double bond and the 915cm^{-1} peak attributed to epoxy group were especially compared. As can be seen clearly that both C=C double bonds, the acrylic peak at 1405cm^{-1} and maleimide peak at 690cm^{-1}, disappeared completely after UV and thermal curing.

The conversion rate was further quantitatively calculated following the decrease of the 1405cm^{-1} absorption peak attributed to acrylic double bond, the 690cm^{-1} peak attributed to maleimide double bond and the 915cm^{-1} peak attributed to epoxy group. The calculation results were summarized in Table 2. As can be seen, conversion rate 62% of acrylic and 95% of maleimide double bonds had been achieved at this UV cure condition. This result confirmed again that major part of acrylic and almost all maleimide double bonds had been reacted during this UV cur condition. As expected, epoxy group cured only at thermal cure condition.

Very interestingly, it was found that remained uncured acrylic double bonds at UV cure process continued to react and the conversion rate increased eventually to 100% at post thermal cure condition. At thermal cure condition without UV cure, it was also found that conversion rate of acrylic double bond and maleimide double bond achieved 67% and 95%. As described previously, the adhesive sample does not contain any thermal initiator component such as peroxide as usually used in UV thermal hybrid epoxy resin adhesive to assure complete curing of the remained acrylic composition for better performance. Nevertheless, UV cure components, i.e. acrylic and maleimide components, of the adhesive sample showed very good thermal curability. From epoxy resin part, conversion rate of epoxy group of adhesive sample cured at thermal cure only condition was some lower than that cured at UV + thermal cure condition. Based on this result, we supposed that acrylic and maleimide double bonds reacted with the epoxy curing agent. As well know, acrylic double bond can react with amines via Michael addition mechanism is shown in Scheme 1 [13-14]. Most probably, acrylic and maleimide double bond reacted with N-H in hydrazide group via Michael addition reaction.

Table 2. Conversion rate of C=C group and epoxy group measured by FT-IR.

Cure condition	C=C conversion rate (%)		Epoxy conversion rate (%)
	acrylic	bismaleimide	
UV cure only 100 mW/cm^2 x 30 sec	62	95	0
UV+ thermal cure 100 mW/cm^2 x 30 sec + 120℃ x 60 min	100	96	85
Thermal cure only 120℃ x 60 min	67	95	69

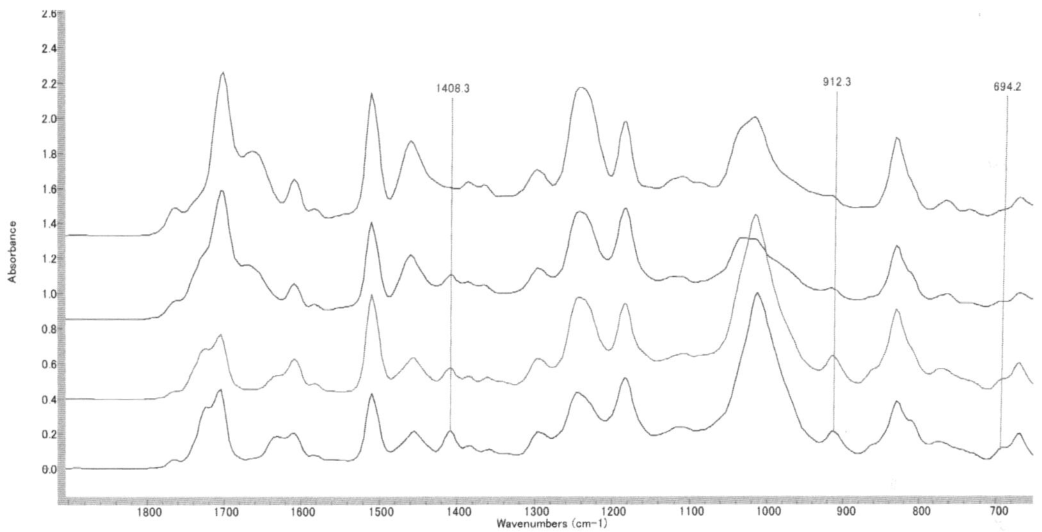

Figure 2. FT-IR spectrum of uncure sample (the bottom curve in blue colour), sample cured with UV only (the 2nd curve from bottom in light blue colour), sample cured with thermal only (the 3rd curve from bottom in purple colour) and sample cured with both UV and thermal (upper curve in red colour).

Scheme 1. Michael addition reaction between acrylic compound and amines.

3.3. FT-Raman measurements and analysis

FT-Raman analysis is a typical non-destructive analysis method with high sensitivity suitable for very trace sample measurements. The adhesive samples bonded with glass substrates cured at the same three conditions were directly used to measure FT-Raman spectrum. The conversion rate was calculated following the decrease of the 1405cm^{-1} absorption peak area attributed to acrylic double bond. Conversion rate result was summarized in Table 3. As can be seen, FT-Raman results coincided very well with FT-IR results shown in Table 2. Although there was slight difference in conversion rate values, UV and thermal cure degree trend was the same. 78% conversion rate of acrylic double bond was achieved at UV cure condition and it further increased to 97% during post thermal cure process. 88% conversion rate of acrylic double bond was obtained at thermal cure only condition. Acrylic double bond was confirmed again to possess both good UV curability and further thermal curability.

IOP Conf. Series: Materials Science and Engineering **213** (2017) 012032 doi:10.1088/1757-899X/213/1/012032

Table 3. Conversion rate of acrylic double bond measured by FT-Raman spectroscopy.

Cure condition	Acrylic C=C conversion rate (%)
UV cure only 100 mW/cm^2 x 30 sec	78
UV+ thermal cure 100 mW/cm^2 x 30 sec + 120℃ x 60 min	97
Thermal cure only 120℃ x 60 min	88

4. Conclusion

New type one component UV and thermal curable hybrid epoxy adhesive was successfully developed. The hybrid epoxy adhesive is complete initiator free composition. Neither photoinitiator nor thermal initiator is contained. The hybrid adhesive showed fast UV fixture property and very strong adhesion performance. Both FT-IR and FT-Raman results confirmed that its UV cure components possess not only good UV curability but also good thermal curability.

References

[1] Herold J and Kluke M 2012 *Radtech Report* **3** 27
[2] Sitzmann E, Fuchs A and Worstatzky D 2004 *Handbook of Coatings Additives* **2** 61
[3] Hoyle C E, Jonsson S, Shimose M and Sundell P 1997 *ACS Symp. Ser.* **673** 133
[4] Andersson H and Hult A 1997 *J. Coat. Technol.* **69** (865) 91
[5] Kuang W, Sabahi M and Nguyen C 2004 *RadTech Technical Proc.* **e/5** 86
[6] Vazquez C P, Joly-Duhamel C and Boutevin B 2009 *Macromol. Chem. Phys.* **210** 269
[7] Fan S L, Boey F Y C and Abadie M J M 2007 *Express Polym. Lett.* **1**(6) 397
[8] Bongiovanni R, Sangermano M, Malucelli G and Priola A 2005 *Prog. Org. Coat.* **53** 46
[9] Petrie E M 2006 *Epoxy Adhesive Formulations* **1** 6
[10] Scherzer T and Decker U 1999 *Vibr. Spectr.* **19** 365
[11] Cholake S T, Mada M R, Raman R K S, Bai Y and Zhao X 2014 *Defense Science Journal* **64**(3) 314
[12] Ohtsuka K, Kimura H, Ikeshita S, Nakao H and Tsubota S 2015 *High Perfromance Polymers* **28**(5) 591
[13] Wang D, Zheng Z, Hong C, Liu Y and Pan C 2006 *J. Polym. Sci., Part A: Polym. Chem.* **44** 6226
[14] Cheng W, Wu D and Liu Y 2016 *Biomacromolecules* **17** 3115

IOP Conf. Series: Materials Science and Engineering **213** (2017) 012033 doi:10.1088/1757-899X/213/1/012033

Effect of gas inlet angle on the gas-assisted extrusion forming of polymer melt

Z Ren[1,2,3] and X Y Huang[2]

[1]Key Laboratory of Optic-Electronic and Communication, Jiangxi Science and Technology Normal University, Nanchang, 330038, China
[2]School of Mechanical and Electrical Engineering, Nanchang University, Nanchang, 330031, China

E-mail: renzhong0921@163.com

Abstract. In this paper, the effect of gas inlet angle on the gas-assisted extrusion (GAE) forming of polymer melt was studied by means of numerical simulation method. The geometric models and the corresponding finite element meshes of four different gas inlet angles (0°, 30°, 60°, and 90°) were established. The computed fluid dynamic software package Polyflow was used. The shear stress, normal stress, and first normal stress difference of melt at the gas/melt interface were obtained. The results show that the influence of gas inlet angle at 30 on the gas-assisted extrusion forming of melt is lest, which can provide the technique guidance for the optimal designing of the gas-assisted die for the polymer melt.

1. Introduction

In the polymer process, the extrusion forming [1] is one of important method in the plastic manufacturing field, such as plastic bar, tube, profiled bar, etc. Since the polymer melt has the characteristic of high viscoelasticity, some serious extrusion problems, e.g., extrudate swell [2], melt fracture [3], extrudate distortion [4], will be generated under the large shear rate or shear stress. The mechanisms of generating the above mentioned problems were the elastic energy storage recovery mechanism, large first normal stress, and the shear stress focuses at outlet of metal die. In order to overcome these problems, some methods have been studied by many scholars, such as, material modification [5], fluoride coating on the surface of die [6], vibration extrusion [7], GAE [8], etc. Among these methods, the GAE method is a well-established, most-promising, and greens environmental method of extrusion forming. The GAE forming have been proved to greatly decrease the shear stress of melt on the channel of die by using stable gas thin-layers formed between melts and die's wall. Many studies showed that the gas process parameters (e.g., gas pressure [9], gas flow [10], gas temperature, etc) have the great influences on the extrusion effect of polymer melt. However, the study about the effect of gas layer's inlet angle on extrusion forming of polymer melt is very few. In this paper, the influence of gas layer's inlet angle on extrusion forming of polymer melt was numerically investigated. Studies results show that the influence of gas layer's inlet angle on the extrusion forming of polymer melt is significant. The study results can provide the good guidance for the GAE forming of polymer melt in practice.

2. Numerical simulations

IOP Conf. Series: Materials Science and Engineering **213** (2017) 012033 doi:10.1088/1757-899X/213/1/012033

2.1. Numerical model

The geometric models of GAE forming for melt with different gas layer's inlet angles are shown in Figure 1.

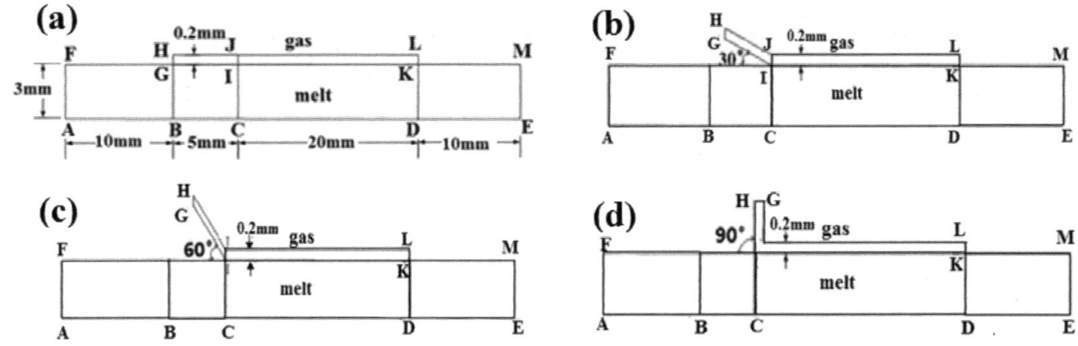

Figure 1. Geometric model of GAE forming for melt with different gas layer's inlet angles. (a) 0 angle, (b)30 angle, (c) 60 angle, (d) 90 angle.

In Figure 1, AEMF is the region of the melt, where ADKF is the melt in the inner part of the die, DEMK is the melt outer part of the die. GKLH is the region of the gas channel in the inner die. Since the symmetric characteristic of geometric model, the half of geometric model was used in the numerical simulation. The width of AF is 3mm, the width of gas layer is 0.2mm, the length of AB, BC and CD is 10mm, 5mm, and 20mm, respectively. The length of DE is 10mm. The length of gas layer is also 20mm. In this paper, four different gas inlet angles were considered, which are 0°, 30°, 60°, and 90°, which are given in Figure 1(a-d), respectively.

2.2. Controlling equations

In the finite element numerical computing, the following basic hypotheses should be satisfied: 1) the melt's flow behavior is looked as steady, isothermal, and laminar flow. 2) The melt is looked as the incompressible and viscoelastic fluid. 3) The flow of gas layer is also laminar, and the turbulence effect was neglected. 4) The inertia, thermal dissipation, and the gravity effects of melt and gas are all neglected.

According to the above mentioned hypotheses, the controlling equations are given as follows,

$$\textit{Continuity equation: } \nabla \cdot v_k = 0 \qquad \mathrm{k} = I, II \tag{1}$$

$$\textit{Momentum equation: } \nabla p_k - \nabla \cdot \tau_k = 0 \tag{2}$$

where $k = I, II$ denote melt and assisted gas, respectively. v denotes flow velocity vector. ∇ denotes Hamilton operator. p denotes melt's and gas's pressure. τ denotes extra stress tension.

PTT constitutive equation [11] was used to describe the viscoelasticity and rheology of melt, i.e.,

$$\begin{aligned} &\exp\left[\frac{\varepsilon\lambda}{(1-\eta_r)\eta} tr(\tau_1)\right]\tau_1 + \lambda\left[\left(1-\frac{\xi}{2}\right)\overset{\nabla}{\tau_1} + \frac{\xi}{2}\overset{\Delta}{\tau_1}\right] \\ &= 2(1-\eta_r)\eta D \end{aligned} \tag{3}$$

where η denotes melt's total viscosity. η_1 denotes Non-Newtonian component viscosity, η_2 denotes the Newtonian component viscosity, $\eta_r = \eta_2 / \eta$ denotes the viscosity ratio. λ denotes the melt's relaxing time. The parameters of ε and ξ correlate with melt's tensile and shear. D denotes strain-rate of tensor. $\overset{\Delta}{\tau_1}$ and $\overset{\nabla}{\tau_1}$ are the lower andupper convected derivative of the extra stress tensor τ_1, respectively.

PTT model's parameters are given in Table 1.

Table 1. PTT model's parameters.

Parameter	η (Pa.s)	λ (s)	ε	ξ	η_r
Value	2700	0.2	0.18	0.23	0.12

2.3. Boundaries setting condition

1) inlet: Supposing that the flow behaviors of melt and gas are full developed, which satisfies the following relationship,

$$\partial v_y / \partial y = 0, v_x = 0 \tag{4}$$

where v_x and v_y are the melt's and gas's flow velocities at radial and axial direction, respectively.

2) wall: No-slip condition is imposed on the die's wall, which is satisfied as follows,

$$v_n = v_s = 0 \tag{5}$$

where *n*, and s are the normal and tangential direction, respectively.

3) gas/melt interface: The flow and thermal transmitting of melt and gas at the melt/gas interface is continuous, and the relative slip between the melt and the gas is neglected. The following conditions should be satisfied,

$$v_n = 0, f_n = 0, f_s = 0 \tag{6}$$

4) symmetric boundary: On the symmetric boundary, the flow and thermal transmitting of melt and gas at the melt/gas interface is also continuous. The relationship is similar to the gas/melt interface boundary.

5) free boundary: The relationships should be obeyed as follows,

$$f_n = 0, f_s = 0 \text{ and } v_n = 0 \tag{7}$$

6) outlet: Supposing there are no any trances and entangle velocities on the outlet of melt, i.e.,

$$f_n = 0, \ v_s = 0 \tag{8}$$

3. Results and analyses

3.1. Shear stress distribution

To analysis the effect of gas layer's inlet angle on GAE of melt, the melt's shearing stress distribution at four gas layer's inlet angles were obtained, which are given in Figure 2(a-d), respectively.

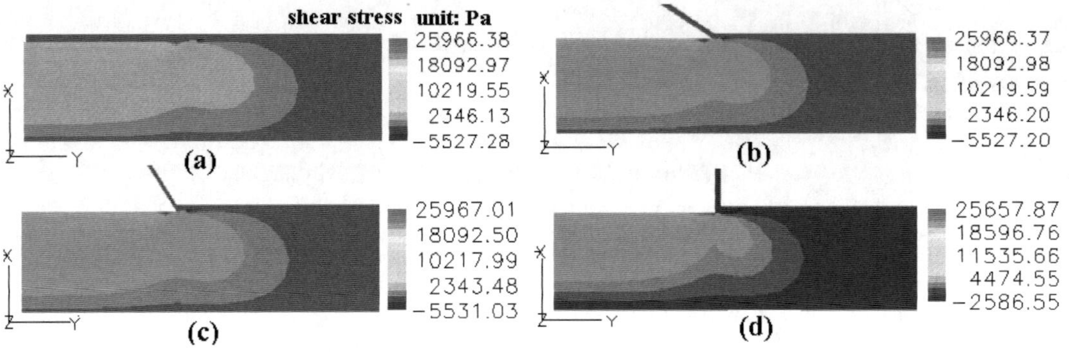

Figure 2. Shear stress distribution of melt at four kinds of different gas layer's inlet angles. (a) 0 angle, (b)30 angle, (c)60 angle, (d)90 angle.

In Figure 2, the melt's shear stress with the gas layer's inlet angle of 90° is less than others. The melt's shearing stress at the gas layer's inlet angle of 60° is largest.

3.2. Normal stress distribution

To further investigate the effect of gas layer's inlet angle on the extrusion forming of melt, normal stress of melt distributions were obtained, which are shown in Figure 3(a-d), respectively.

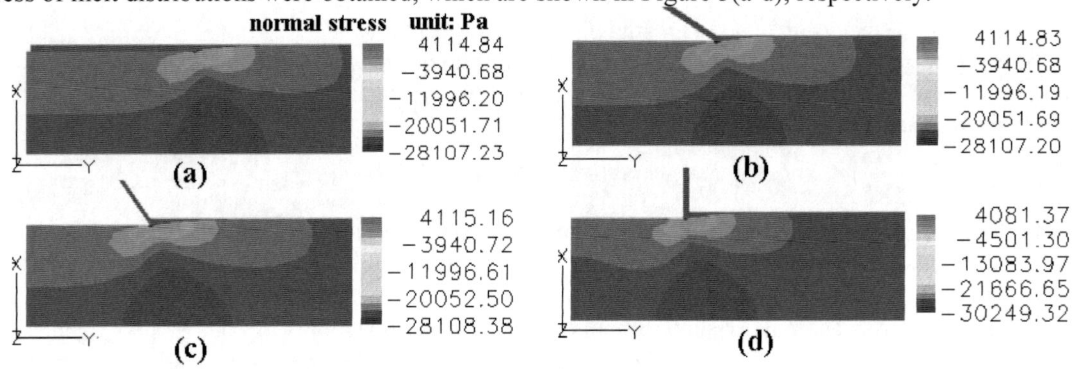

Figure 3. Normal stress distribution of melt at four kinds of different gas layer's inlet angles. (a)0 angle, (b)30 angle, (c)60 angle, (d)90 angle.

In Figure 3, the melt's normal stress with gas layer's inlet angle of 90° is largest, but the normal stress of melt at the angle of 0° and 30° are lest.

3.3. First normal stress difference distribution

Then, the melt's first normal-stress different at four kinds of gas layer's inlet angles were gotten, which are presented in Figure 4 (a)-(d), respectively.

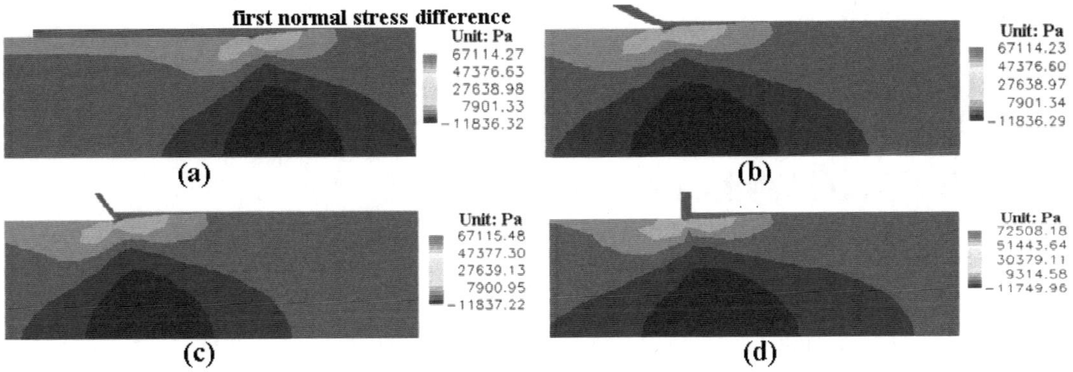

Figure 4. Melt's first normal stress differences of at four gas inlet angles. (a)0 angle; (b)30 angle; (c)60 angle; (d)90 angle.

In Figure 4, the melt's first normal-stress differences at angle of 30° are less than that of others. Moreover, the first normal-stress difference value at the angle of 90° is largest, which indicates the influence of gas layer's inlet angle at 90° is largest but the influence of gas inlet angle at 30° is lest. Based on the above mentioned analyses and results, it can be seen that the effect of gas layer's inlet angle of 30° is best, which can be used to design the gas inlet angle for the die of GAE forming.

4. Conclusion

In the GAE forming of melt, optimal design of GAE die is very key work. To optimally design the GAE die, the effect of gas layer's inlet angle on the extrusion forming of melt was investigated in this work, the finite element simulations based on Polyflow software was used. Shear stress, normal stress, and first normal-stress differences of melt based on four kinds of gas layer's inlet angles (0°, 30°, 60°, and 90°) were obtained. The results show that the effect of gas inlet angles at 30° on the melt's GAE forming is best.

Acknowledgements

This work was financially supported by the National Natural Science Fund (61650402), the Scientific Education Department Science and Technology project of Jiangxi province (No. GJJ150814), Young top-notch talents project of school level (2014QNBJRC004).

References

[1] Ren Z, Huang X Y and Liu H S 2016 *IOP Conf. Series: Mater. Sci. Eng.* **137** 012007
[2] White J L and Huang D 2010 *Polym. Eng. Sci.* **21** 1101
[3] Vlachopoulos J and Alam M 2010 *Polym. Eng. Sci.* **12** 184
[4] Jallouli I, Kallel T, Ayadi A and Halouani F 2010 *Phys. Chem. News* **53** 22
[5] Rungruangsuparat S, Patcharaphun S and Sombatsompop N 2016 *Polym. Testing* **57** 184
[6] Zagrebaev S A, Orlova E A, Alekseev V V, Zhmurin V G, Orlov M A, Shirshov Y N, Torbenkova I Y and Tychinskii P I 2015 *Russ. Phys. J.* **57** 1283
[7] Chen J, Chen Y, Li H, Lai S Y and Jow J 2010 *Ultrason. Sonochem.* **17** 66
[8] Ren Z, Huang X Y, Liu H S, Deng X Z and He J T 2015 *J. Appl. Polym. Sci.* **132** 12365
[9] Ren Z, Huang X Y and Liu H S J 2016 *Sichuan Univ. (Eng.Sci. Ed.)* **48** 200
[10] Yu Z, Huang Y B, Zhang K, Dong T W and Li T 2016 *Polym. Mater. Sci. Eng.* **32** 113
[11] Thien N P and Tanner R I 1997 *J. Non-Newtonian Fluid Mech.* **2** 353

IOP Conf. Series: Materials Science and Engineering **213** (2017) 012034 doi:10.1088/1757-899X/213/1/012034

The crystal morphology and mechanical properties based on poly(l-lactic acid)/silica nanocomposites

Y H Cai[1,2]

[1]Chongqing Key Laboratory of Environmental Materials & Remediation Technologies, Chongqing University of Arts and Sciences, Chongqing-402160, P.R. China

E-mail: caiyh651@163.com

Abstract. Nano silica (SiO_2) was introduced into Poly(L-lactic acid) (PLLA) matrix to prepare PLLA/ SiO_2 nanocomposites, and the crystal morphology, crystallization behavior and mechanical performance were investigated. The XRD experimental data indicated that nano SiO_2 could improve the crystallization of PLLA, and PLLA/ SiO_2 nanocomposites exhibited sharp diffraction peak after isothermal crystallization. In addition, through the POM analysis, PLLA/ SiO_2 sample had the typical spherulite structure, and the size of the spherulite became larger with the increase of crystallization temperature. The tensile testing showed that a small amount of SiO_2 could improve and retain the mechanical performance of PLLA

1. Introduction

Silica (SiO_2) widely exists in the earth's crust, and also was widely employed in many fields such as ceramics materials[1], glass materials[2], polymer filler[3], etc. Pei et al.[4] reported that a core-shell SiO_2@MgO with flower structure was synthesized, and the synthesized core-shell SiO_2@MgO sample showed excellent performance for the removal of crystal violet, and the maximum adsorption capacity was 2244.85 mg/g, exhibiting a promising environmental remediation material. Meng and coauthor[5] synthesized a magnetic sonophotocatalyst Fe_3O_4@SiO_2@TiO_2 nanocomposites to enhance biodegradability of organophosphate pesticide. The experimental results showed that the Fe_3O_4@SiO_2@TiO_2 nanocomposites had good photocatalytic performance and stability. That is, the Fe_3O_4@SiO_2@TiO_2 nanocomposites were used for the fifth time, the CODcr removal efficiency was still very high (about 62.38%).

Poly(L-lactic acid) (PLLA), as very promising biodegradable materials[6], has achieved excellent development for application[7]. The hydroxyapatite and the bone morphogenetic protein-2 were introduced into PLLA to prepare the bioactive scaffold material, the animal experiments indicated that PLLA/hydroxyapatite/bone morphogenetic protein-2 can significantly induce and promote bone tissue formation [8]. The poly (L-lactic acid)/ZrO_2 nanofibrous membranes were fabricated using electrospinning technique, and the antimicrobial activity and release behavior showed that poly(L-lactic acid)/ZrO_2 nanofibrous exhibted well controlled release and better antimicrobial activity against Staphylococcus aureus[9]. In addition, PLLA was used in nanocomposites[10-11], biomaterials[12], etc. However, there exist some critical defects including slow crystallization rate and low heat resistance to reflect the application of PLLA. Adding SiO_2 into PLLA matrix is a value technology method for improving the performance of PLLA. Wu *et al*[13] reported that modified SiO_2 through ring-opening of L-lactide or nucleophilic addition reaction was added into PLA matrix to solve the low melt strength of PLLA. Results showed that long grafted chains modified on the surface of SiO2 through nucleophilic addition reaction were contributed to a significant enhancement in the melt strength of PLA.

In this work, we fabricated PLLA/ SiO_2 nanocomposites using traditional melt blending technology, and their crystal morphology; crystallization and mechanical properties of PLLA/SiO_2 nanocomposites were investigated using polarized optical microscopy (POM), X-ray diffraction (XRD), and tensile testing.

2. PLLA/SiO_2 nanocomposites

Dry nano-SiO_2 and PLLA was blended in a counter-rotating mixer, and the detailed operating procedure is similar to that used for other PLLA based materials[14].

3. Results and discussion

3.1. Crystallization and morphology

The crystallization behavior is very important research topics for semi-crystalline polymer. Figure 1 shows the XRD curves of SiO_2, PLLA and PLLA with different SiO_2 content without heat treatment. It is very clear that the diffraction peaks of PLLA and PLLA/ SiO_2 are very wide, showing that the crystallization degree is very low. However, the diffraction peaks of PLLA/ SiO_2 nanocomposites after isothermal crystallization at 105°C for 20 min become very sharp (See Figure 2), but the diffraction peak of the neat PLLA still is very wide, these discrepancy indicated that SiO_2 play an important role in improving the crystallization of PLLA, this also further confirmed by literature reports about the nucleating effect of SiO2 for PLLA [15]. In addition, the strongest diffraction peak of PLLA at 2θ =16.7° belong to diffraction from (110) planes, and the others peaks at 2θ=14.9° , 19.0° , 22.3° occur from the (010) planes, (203) planes and (205) planes, respectively [16].

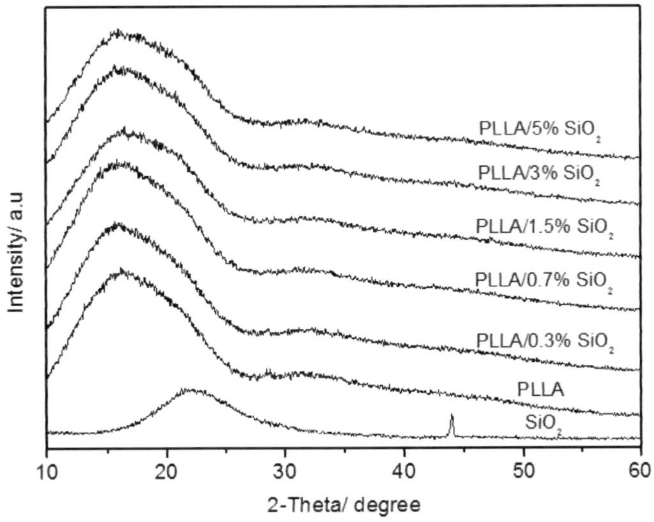

Figure 1. XRD curves of SiO_2, PLLA and PLLA/ SiO_2 with heat treatment.

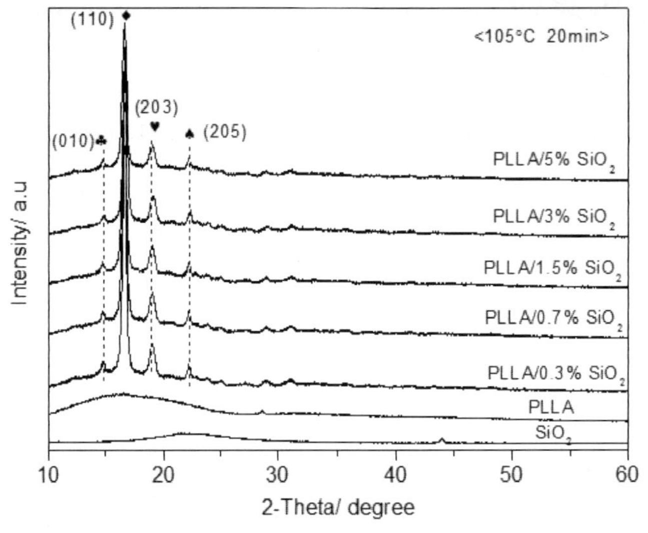

Figure 2. XRD curves of SiO_2, PLLA and PLLA/ SiO_2 after isothermal crystallization at 105°C for 20 min.

The crystal morphology is also very key factor for application, because the morphology of PLLA is related to its performance. Figure 3 shows the POM of PLLA/3% SiO_2 at different crystallization temperature. It can be seen that the PLLA/3% SiO_2 exhibits typical spherulite, and the spherulites become bigger with an increase of crystallization temperature, the reason is that the PLLA crystals become more perfect at higher crystallization temperature through the ordered arrangement of molecular chain with stronger motility.

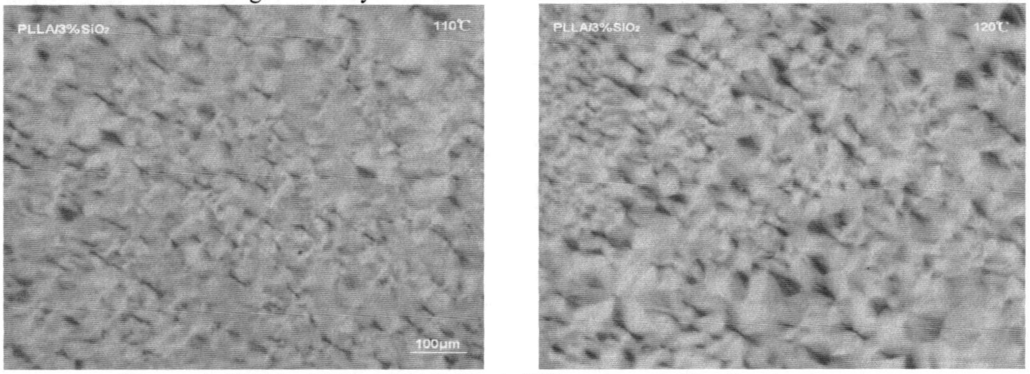

Figure 3. POM of PLLA/3%SiO2 at different crystallization temperature.

3.2. Mechanical Properties

Mechanical performance directly affects the application of PLLA. The mechanical performance of PLLA/SiO_2 nanocomposites was investigated using the electronic tensile tester. Table.1 shows the tensile strength and elongation at break of PLLA and PLLA/SiO_2 nanocomposites. Apart from 1.5% SiO_2, the tensile strength and elongation at break of PLLA decreases with the increasing of SiO_2 content, and the different result of PLLA/1.5%SiO_2 may result from the dispersibility of SiO_2 in PLLA matrix and testing error. Upon the addition of 0.3% SiO_2, the tensile strength and elongation at break have the maximum value, they are 49.7MPa and 7.58%, respectively. This result indicates that a small amount of SiO2 can improve and retain the mechanical performance. In contrast, a large amount of SiO_2 reduces the mechanical properties.

Table 1. Mechanical properties of PLLA and PLLA/SiO_2 nanocomposites.

Sample	tensile strength (MPa)	elongation at break (%)
PLLA	47.6±2.4	5.2±0.6
PLLA/0.3%SiO_2	49.7±1.7	7.6±0.5
PLLA/0.7%SiO_2	46.9±2.1	5.9±0.3
PLLA/1.5%SiO_2	48.1±1.3	6.3±0.4
PLLA/3%SiO_2	45.7±1.7	5.5±0.4
PLLA/5%SiO_2	44.5±0.9	2.7±0.3

4. Conclusion

The crystallization behavior, morphology and mechanical properties of PLLA/SiO_2 nanocomposites were studied, and the XRD and POM measure indicated that SiO_2 could accelerate the crystallization of PLLA. The addition of SiO_2 could also increase the nuclear density of PLLA matrix, but the crystal morphology still was typical spherulite structure. In addition, 0.3% SiO_2 made the tensile strength and elongation at break of PLLA exhibits the maximum value.

Acknowledgements

This work was supported by National Natural Science Foundation of China (project number 51403027), Natural Science Foundation of Chongqing Municipal Science and Technology Commission (project number cstc2015jcyjBX0123), Scientific and Technological Research Program of Chongqing Municipal Education Commission (project number KJ1601101), and Innovation Team Project of Chongqing Municipal Education Commission (project number CXTDX201601037).

References

[1] Zhang L X, He F, Xie J L and Liu J 2015 Structure and Luminescent Properties of Sm^{3+} Doped SrO-MgO-SiO_2 Glass Ceramics *Journal of Wuhan University of Technology-Materials Sciecne Edition* **30**(2) 282-7

[2] Guo X G, Zhai C H, Kang R K and Jin Z J 2015 The mechanical properties of the scratched surface for silica glass by molecular dynamics simulation *J. Non-Cryst. Solids* **420** 1-6

[3] Li N, Huang Y J, Xia T, Yang Q and Li G X 2013 Phase Behavior of PS/PVME/SiO_2 Blends Under Oscillatory Shear Flow Field *Chemical Journal of Chinese Universities-Chinese* **34**(12) 2896-902

[4] Pei Y Y, Wang M, Tian D, Xu X F and Yuan L J 2015 Synthesis of core–shell SiO_2@MgO with flower like morphology for removal ofcrystal violet in water *J. Colloid Interface Sci.* **453** 194-201

[5] Meng L R, Shi J J, Zhao M and He J 2014 Synthesis of Magnetic Sonophotocatalyst and its Enhanced Biodegradability of Organophosphate Pesticide *Bulletin of the Konrean Chemical Society* **35**(12) 3521-6

[6] Blanco I 2014 End-life prediction of commercial PLA used for food packaging through short term TGA experiments: Real chance or low reliability? *Chin. J. Polym. Sci.* **32**(4) 681-9

[7] Blanco I and Siracusa V 2013 Kinetic study of the thermal and thermo-oxidative degradations of polylactide-modified films for food packaging *J. Therm. Anal. Calorim.* **112**(3) 1171-7
[8] Yang X Z and Li Z S 2015 Influence of hydroxyapatite and BMP-2 on bioactivity and bone tissue formation ability of electrospun PLLA nanofibers *J. Appl. Polym. Sci.* **132** 42249.1-9.7
[9] Wang H L, Ma X K, Li Y N, Jiang S W, Zhai L F, Jiang S T and Li X J 2013 Synthesis, antimicrobial and release of chloroamphenicol loadedpoly(l-lactic acid)/ZrO_2nanofibrous membranes *Int. J. Biol. Macromol.* **62** 494-9
[10] Amirian M, Chakoli A N, Sui J H and Cai W 2013 Thermo-mechanical properties of MWCNT-g-poly (l-lactide)/poly (l-lactide) nanocomposites *Polym. Bull.* **70**(10) 2741-54
[11] Gardella L, Basso A, Prato M and Monticelli O 2013 PLA/POSS Nanofibers: A Novel System for the Immobilization of Metal Nanoparticles *ACS Applied Materilas & Interfaces* **5**(16) 7688-92
[12] Boccafoschi F, Mosca C, Ramella M, Carmagnola I, Chiono V, Ciardelli G and Cannas M 2013 Biological evaluation of materials for cardiovascular application: The role of the short-term inflammatory response in endothelial regeneration *Journal of Biomedical Materilas Research Part A* **101**(11) 3131-40
[13] Wu F, Zhang B, Yang W, Liu Z Y and Yang M B 2014 Inorganic silica functionalized with PLLA chains via grafting methods to enhance the melt strength of PLLA/silica nanocomposites *Polymer* **55** 5760-72
[14] Fan Y Q, Zhu J, Yan S F, Chen X S and Yin J B 2015 Nucleating effect and crystal morphology controlling based on binary phase behavior between organic nucleating agent and poly(L-lactic acid) *Polymer* **67** 63-71
[15] Zhang Y, Deng B Y and Liu Q S 2014 Rheology and Crystallization of PLA containing PLA-grafted nanosilica *Plast. Rubber Compos.* **43**(9) 309-14
[16] Hergeth W D, Lebek W, Stettin E, Witkowski K and Schmutzler K 1992 Particle formation in emulsion polymerization, 2. Aggregation of primary particles *Die Makromolekulare Chemie* **193**(7) 1607-21

IOP Conf. Series: Materials Science and Engineering **213** (2017) 012035 doi:10.1088/1757-899X/213/1/012035

Moisture absorption of starch based biocomposites reinforced with water hyacinth fibers

H Abral[1,2] and J Hartono[1]

[1]Mechanical Engineering, Andalas University, Padang, Indonesia

E-mail: abral@ft.unand.ac.id

Abstract. Bioplastic based on tapioca starch (TSB) is very sensitive on moisture; meanwhile this substance may be used to replace synthetic plastic. This paper reports effect of Water Hyacinth Fibers (WHF) content on performance moisture absorption of starch based biocomposites. WHF content in the TSB matrix was varied in 1, 3, 5, and 10% respectively. The samples were placed in closed room with high relative humidity (RH) of 99% at 25^{0}C with different duration for 30 and 960 min respectively. The result showed that moisture absorption in the beginning was increased rapidly, and then achieved a level steady state. After that, significant swelling of the sample occurred for further duration in 960 min. Gradient of the swelling was decreased as increasing the fibers content in the TSB matrix.

1. Introduction

Starch based bioplastic has advantages such as low cost, wide availability, and total composability without toxic residues [1]. For this reason, the bioplastic provides a great attention to replace synthetic polymers such as for packaging applications [2]. General procedure for processing a bioplastic involves the granular disruption by the combination of temperature, shear, and a plasticizer, which is usually water and/or glycerol [3][4]. However, weakness of the bioplastic is easy to absorb moisture [5]. Therefore, several techniques have been investigated to minimize the weakness, for example, the bioplastic mixed with fillers [6][7][1]. After adding cellulose as filler in bioplastic, moisture resistance of the biocomposites improved as reported by last work of K. Kaewtatip and J. Thongmee [8].

Cellulose of plants is considered to be one of the most promising renewable resources [9]. As the aquatic weed, water hyacinth plant demonstrated the feasibility of producing cellulose nanofibers, which can have potential applications for reinforcement in the near future [10]. In the present work, WHF in form of pulping has been incorporated in TSB. Moisture absorption characterization of the biocomposite in various WHF content was measured. The originality of this study is to observe the characterisation of moisture absorption in region of steady state achieved.

2. Methods

2.1. Measuring chemical compositions

Chemical composition of WHF tested was respectively hemicellulose, cellulose and lignin. Measuring methods used were ASTM 1104-56, and testing by Technical Association of the Pulp and Paper Industry (TAPPI) standard T9M-54 and TAPPI T13M-54 respectively.

2.2. Manufacture of biocomposites
Extracted WHF was of freshwater hyacinth plant growing in local swamp in Kabupaten 50 Kota, Indonesia. It was cut for 1 cm long, and cleaned with fresh water. Then, it was placed in a close room isolated by plastic under sunlight. The studied samples were dried in the room after 5 days. Next steps was that the dried WHF samples were entered in 25% NaOH solution to be assimilated by digester for 6h at about T=130^{0}C, 2 bar. Finally the pulp was neutralized by water until pH 7, and then blended by mixer for 10 min. The pulp was screened by using screen printing with T61 mesh. The next step was to dry the wet pulp using sunlight.

Tapioca powder was used as material for making bioplastic matrix for this research. The powder was from a commercial product with name "Cap Pak Tani", bought in local market. Content of water, amylose, and amylopectin in the powder was 19, 15, and 85%, respectively. TSB was processed as following (i) Aqua of 140 ml was mixed WHF based on volume fraction of 1, 3, 5, and 10%, respectively. The mixing was homogenized for 5 min, then, added by tapioca powder of 10 g. Further steps were homogenizing of the solution while 5 min. Glycerol of 3 ml was added to the solution, then, homogenized for 5 min; (ii) Magnetic stirrer set up at temperature of 150^{0}C, 500 rpm, was used for mixing the solution average 18 min in order to get gelatinization; (iii) The gel was poured into a glass mould making biocomposite film. Drying process of all tested samples was in universal oven (Memmert UN-55) for 20 h at 50^{0}C. The studied samples were released of the mould after 24 h in room temperature.

2.3. Moisture absorption
All tested samples were dried in oven (Universal Oven Memmert UN-55) until achieving constant weight. Then, they were stored in closed chamber in which both relative humidity (RH) and temperature were 99% and 25^{0}C, respectively. Duration of the sample in the chamber was varied, there was 30 min, and even there was 960 min. The samples were taken out from the chamber to be weighted by using precision balance of 0.1mg (Kenko). Percentage of moisture absorption in the sample was calculated by difference between dried weight and wet weight divided by dried weight in % unit [11].

3. Results and analysis

3.1. Chemical composition
Of this research, after digesting and blending for 10 min., chemical compositions content of the studied WHF was measured that hemicellulose, lignin, and cellulose was 3.5, 3.8, and 67.0%, respectively. As it can be seen, content of hemicelluloses and lignin was very low due to the chemical and mechanical treatment. Meanwhile, cellulose content of 67.0% was high enough, and more dominant in the WHF. This result was in good agreement with last studies [12][13].

3.2. Moisture absorption
Figure 1 shows performance of moisture absorption for all studied samples. As control, both pure starch based bioplastic, and 25%NaOH treated digested pulp of WHF were also measured of own moisture absorption behaviour. All samples were stored in closed chamber with very high RH of 99% at temperature of 25^{0}C. Duration of the sample in the closed chamber was varied, both 30 and 960 min. In the beginning, the studied samples exhibited high rate of moisture absorption due to a high different concentration of relative humidity between the inner dry samples and the humid environment of the chamber. For the initial 30 min. in the chamber, it was gauged that moisture absorption of sample TSB composites (1, 3, 5, and 10% WHF), and 100%WHF was 51, 49, 51, 54, 53, and 24 % per h, respectively. It can be seen from the data that the lowest moisture absorption was found in 100%WHF, meanwhile TSB as well as the biocomposites film displayed similar high capacity in absorbing moisture. Hence, existing of the fibers provides an increase resistance of water uptake. Diffusion of water molecules was hindered by the fibers in the matrix. Among 960 min placed in the humid

chamber, the biocomposites show swelling due to nature hydrophilic samples. Highest swelling was observed in pure WHF, and the lowest one was on 10% WHF in TSB.

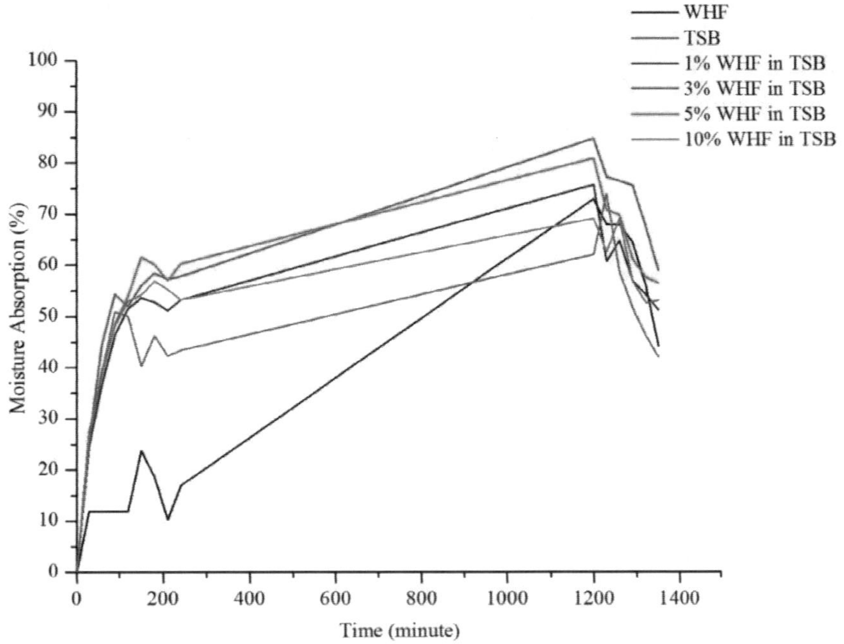

Figure 1. Performance of moisture absorption for all studied samples.

For further 30 min observation, the rate was decreased continuously to achieve equilibrium level. As it can be seen in Figure 1, that TSB, started from 90 up to 240 min, displayed the fluctuation of moisture absorption. In this case, moisture content in the tested sample was never in steady state condition. The sample cannot keep an equilibrium point of water content. Moisture was probably diffused and evaporated easily into and from the tested sample. While the sample was stored during 16 h in closed chamber, the result was very surprised that the moisture absorption was still increased continuously, measured additionally of 2%. Then, TSB sample was stored again in closed chamber, and for each 30 min long it was taken out to measure the moisture absorption of the sample. However, the result shows that the moisture absorption was decreased to tend an equilibrium level again. This is due to the fact that amount of moisture was evaporated from the sample. Surprisingly, all studied samples also have exhibited same behaviour that they have very high sensitivity on the moist environment, and cannot keep a constant moisture content in environment with varied relative humidity.

3.3. SEM photograph

Figure 2 shows SEM photograph of fracture surface for tensile testing. As shown in Figure 2a, beach marking may due to low crack propagation of starch based thermoplastic. Meanwhile, agglomeration as well as clearance between fibers and matrix may provide possibility to water molecule to enter, thus increasing the moisture absorption of the biocomposite. It is noted that frature surface of the fiber may inform good adhesion bonding between matrix and fiber.

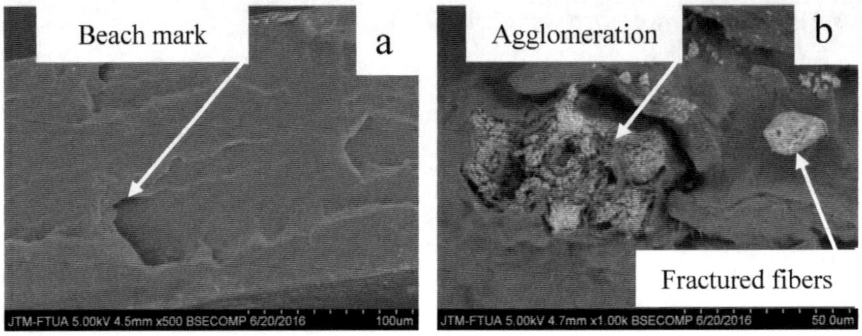

Figure 2. SEM photograph of fracture surface of, a) TSB matrix, b) 3% WHF in TSB matrix.

4. Conclusion

Variation of WHF in the TSB was 1, 3, 5, and 10%, respectively. Increase in volume fraction of WHF in the TSB matrix improved generally moisture absorption resistance of the biocomposites. In the beginning, all materials absorb water rapidly until achieving a steady state for short duration in humid chamber. After that, swelling of the biocomposites was significantly increased in increasing the duration in humid chamber. It was then slightly lower as further WHF content in the TSB.

Acknowledgments

Acknowledgement is addressed from the authors to Directorate General of Higher Education, Indonesia for supporting research finance of KLN 2017.

References

[1] Xie F, Pollet E, Halley P J and Averous L 2013 *Prog. Polym. Sci.* **38** 1590-628
[2] Djoković V *et al* 2009 *Biointerfaces* **73** 30-5
[3] Edhirej A, Sapuan S M, Jawaid M and Zahari N I 2016 *Starch/Stärke* **68** 1–11
[4] Salaberria A M, Labidi J and Fernandes S C M 2014 *Chem. Eng. J.* **256** 356–64
[5] Cotonieto-Morales M B *et al* 2015 *Starch/Staerke* **67** 355–64
[6] Müller C M O, Laurindo J B and Yamashita F 2011 *Industrial Crops and Products* **33** 605–10
[7] Lomelí-Ramíreza M G, Kesturb S G, Manríquez-Gonzálezc R, Iwakiria S, de Muniza G B and Flores-Sahagun T S 2014 *Carbohydr. Polym.* **102** 576–83
[8] Kaewtatip K and Thongmee J 2012 *Mater. Des.* **40** 314–8
[9] Wan Y Z, Luo H, He F, Liang H, Huang Y and Li X L **2009** *Compos. Sci. Technol.* **69** 1212–7
[10] Thiripura Sundari M and Ramesh A 2012 *Carbohydr. Polym.* **87** 1701–5
[11] Abral H *et al* 2014 *Mater. Des.* **58** 125–9
[12] Shibata S, Senaha I and Abral H 2014 *Bioresources* **9** 4430–40
[13] Abral H, Andriyanto H, Samera R, Sapuan S M and Ishak M R 2012 *Polym. Plast. Technol. Eng.* **51** 37–41

IOP Conf. Series: Materials Science and Engineering **213** (2017) 012036 doi:10.1088/1757-899X/213/1/012036

A study on PVDF-HFP gel polymer electrolyte for lithium-ion batteries

W Liu[1,2], X K Zhang[1], F Wu[1] and Y Xiang[1,2]

[1]School of Energy Science and Engineering, University of Electronic Science and Technology of China, No.2006 Xiyuan Avenue, West High-Tech Zone, Chengdu, Sichuan, China

E-mail: liuwen_work@163.com; xyg@uestc.edu.cn

Abstract. In this paper, poly (vinylidene fluoride-co-hexafluoropropylene) (PVDF-HFP) gel polymer electrolyte was fabricated via solvent casting method in order to improve the performance of the lithium-ion batteries. By comparing the physical and electrochemical properties of PVDF-HFP gel polymer electrolyte with three different proportions, the optimization of the PVDF-HFP gel polymer electrolyte was obtained: 10wt% PVDF-HFP - 80wt% tetrahydrofuran (THF) / acetone - 10wt% 1mol L^{-1} lithium perchlorate ($LiClO_4$) in elthylene carbonate (EC) and diethyl carbonate (DEC). The optimized PVDF-HFP gel polymer electrolyte displayed a high conductivity of 1.06×10^{-3} S cm^{-1} at room temperature, a high lithium transference number of 0.36 and a good thermal stability within 100℃. Moreover, the discharge specific capacity was 135.1 mAh g^{-1}, and the charge / discharge efficiency was 99.1% at 0.1C rate. Therefore, the fabricated PVDF-HFP gel polymer electrolyte was an effective gel polymer electrolyte to be applied on lithium-ion batteries.

1. Introduction

Lithium-ion batteries have been widely investigated in order to fulfill the ever-increasing demand in portable devices and electric vehicles[1]. According to the physical form of electrolyte in lithium-ion batteries, the electrolyte can be divided into three categories: liquid electrolyte, gel electrolyte and solid electrolyte. Gel polymer electrolyte is an electrolyte formed by polymer network structure filled with liquid electrolyte, and it contains solid phase and liquid phase. In the gel polymer electrolyte, the liquid electrolyte is well retained in the solid polymer matrix, and this structure avoids leakage of the liquid electrolyte. Compared with solid electrolyte and liquid electrolyte, gel polymer electrolyte has many advantages, such as high voltage, high specific capacity, long cycle life, rapid charge and discharge, small size, thin thickness, etc. Therefore, gel polymer electrolyte is the main focus of recent development, because it has high conductivity of the liquid electrolyte and safety of the solid electrolyte[2-3].

The gel polymer electrolyte essentially includes three parts, i. e. polymer matrix, plasticizer and lithium salt. The choice of polymer matrix determines whether the gel polymer electrolyte has excellent properties. In general, the polymer matrix must have the characteristics of good electrochemical performance, excellent thermal stability, low crystallinity, large dielectric constant and strong mechanical property. Various polymers, i. e. polyacrylonitrile (PAN)[4], polyoxyethylene (PEO)[5], polymethylmethacrylate (PMMA)[6], polyvinylidene fluoride (PVDF) and poly (vinylidene fluoride-co-hexafluoropropylene) (PVDF-HFP)[7], have been developed as polymer matrix for gel electrolytes. Especially, PVDF with strong electro-withdrawing functional groups (-C-F) shows a high dielectric constant (ε =8.4) ,which is beneficial for the dissolution of lithium salts to keep a high concentration of charge carriers[8]. Furthermore, since the copolymerization effect of VDF and HFP, PVDF-HFP has a relatively lower crystallinity compared to PVDF[9-10]. Therefore, PVDF-HFP contains more amorphous domains to capable of trapping large amounts of liquid electrolytes. Hence, PVDF-HFP is regarded as the most promising matrix as gel polymer electrolyte material.

The ionic conduction in gel polymer electrolyte is resulted from the diffusion of ions through their free volumes, so it is important that lithium salts be dispersed in polymers at molecular level[11]. The influence of plasticizers on the gel polymer electrolyte has a strong interaction with the specific nature of the plasticizer, such as dielectric constant, viscosity, polymer-plasticizer interaction and ion-plasticizer coordination[12]. Generally, the cyclic carbonate (such as ethylene carbonate or propylene carbonate) has a high dielectric constant, and the chain-like carbonate (such as diethyl carbonate or dimethyl carbonate) has a low viscosity. A mixture plasticizer consists of the cyclic carbonate and the chain carbonate is selected to support the demands of the working temperature and conductivity. As we all know, $LiClO_4$ has small ionic radius and low dissociation energy, and it can be dissolved in most of the organic solvents[13]. So $LiClO_4$ is selected as a lithium salt in this study. In this work, PVDF-HFP gel polymer electrolytes with three different proportions were fabricated. We obtained an effective gel polymer electrolyte to be applied on lithium-ion batteries.

2. Experimental

2.1. Materials

PVDF-HFP (kynar LBG, M_w~500000) polymer was obtained from Arkema, France. Reagent grade anhydrous $LiClO_4$, elthylene carbonate (EC) and diethyl carbonate (DEC) were purchased from Aladdin bio-chem Technology Co., shanghai, China. The solvent tetrahydrofuran (THF) and acetone were bought from Chengdu Kelong Chemical Reagent Company, China. The $LiCoO_2$ cathode was purchased from Kejing Star Technology Co., Shenzhen, China.

2.2. Gel polymer electrolyte preparation

PVDF-HFP gel polymer electrolyte was synthesised using a solvent casting method. The solvent casting method is the polymer dissolved in a co-solvent system composed of two or more organic solvents to form gel, then the gel was coated on a glass plate and the volatile solvent was evaporated to obtain gel polymer electrolyte membrane. First, $LiClO_4$ was dried under vacuum at 100℃ for 10 h, and then dissolved in the solution composed of EC and DEC (volume ratio of EC/DEC=1:1) to obtain 1mol L^{-1} $LiClO_4$ solution after vigorous stirring at 50℃ for 2h. Second, PVDF-HFP polymer was dissolved in solution of THF and acetone (volume ratio of THE/acetone=1:1) under persistent stirring at 50℃ for 6h. Third, the $LiClO_4$ solution was added to PVDF-HFP polymer solution and stirred for another 6h at 50℃ to obtain the gel precursor. At last, the gel precursor was casted on a spotless glass plate at room temperature. With the evaporation of THF and acetone solvents, a white gel polymer electrolyte membrane was formed, followed by stripped off from glass substrate and punched into a circle shape with a radius of 9 mm.

2.3. Characterization techniques

Scan electron microscopy (Bruker S3400N) was employed for morphology observation. Thermo gravimetric analysis (TGA) was carried out for thermal stability measurement using a Q50 TGA instrument under N_2 atmosphere from 20-500 ℃ with a temperature ramp of 5℃/min. The ionic conductivity was investigated by Electrochemical Impedance Spectroscopy (EIS) method (Parstat 2263 advanced electrochemical system, Princeton Applied Research) in a 2032 coin cell with a GPE sandwiched structure between two stainless (SS) in the frequency range of 0.1 Hz to 10^5 Hz at an alternative signal of 10 mV. The linear sweep voltammetry (LSV) technique was tested on a

Li/GPE/Li cell with the SS as the working electrode and lithium foil as both the reference and counter electrodes. The scanning rate was 5 mV/s over a range of 0-6V.

The chronoamperometry method was carried out to measure the lithium transference numbers, and the step voltage of Li/GPE/Li cells was 0.3V. In short, the initial current I_0 and the steady-state current I_{SS} that spreading through the cell were recorded. Moreover, the initial resistance R_0 and the steady-state resistance R_{SS} were obtained through the impedance spectroscopy monitored by the same cell. Charge-discharge cycling tests of the Li/GPE/$LiCoO_2$ cell was performed at 0.1C rate within the voltage range of 3.0-4.2V *vs.* Li/Li^+ at room temperature.

3. Results and discussion

3.1. Morphology structure

Gel polymer electrolyte membrane not only in its surface has a micro-porous structure, but also its internal is micro-porous structure. This micro-porous structure forms a 3D network structure in gel polymer electrolyte[14]. Table 1 lists the components, pore size and conductivity values of PVDF-HFP gel polymer electrolyte. Figure 1 shows the surface morphology of GPE1, GPE2, GPE3 and the cross-sectional morphology of GPE3. With the increasing of PVDF-HFP ratio and the proportion of solvent decreasing, the concentration of PVDF-HFP become larger and the tendency of polymer swelling to network structure is weaker. So, the pore size of the gel polymer electrolyte membranes is smaller. Nevertheless, the larger pore size means that the network is sparse, and the mechanical strength of gel polymer electrolyte is inoperable. Thus, we think that the pore size is inversely proportional to the mechanical strength [15]. By adjusting the concentration of PVDF-HFP, we can achieve the balance of the mechanical strength and pore size of gel polymer electrolyte.

Table 1. The components, pore size and conductivity values of PVDF-HFP gel polymer electrolytes.

sample	PVDF-HFP	THF/acetone	1mol/L $LiClO_4$ -EC/DEC	Pore size (μm)	Conductivity ($S\ cm^{-1}$)
GPE1	2wt%	90wt%	8wt%	6.88	2.23×10^{-3}
GPE2	5wt%	90wt%	5wt%	3.87	1.27×10^{-3}
GPE3	10wt%	80wt%	10wt%	3.15	1.06×10^{-3}

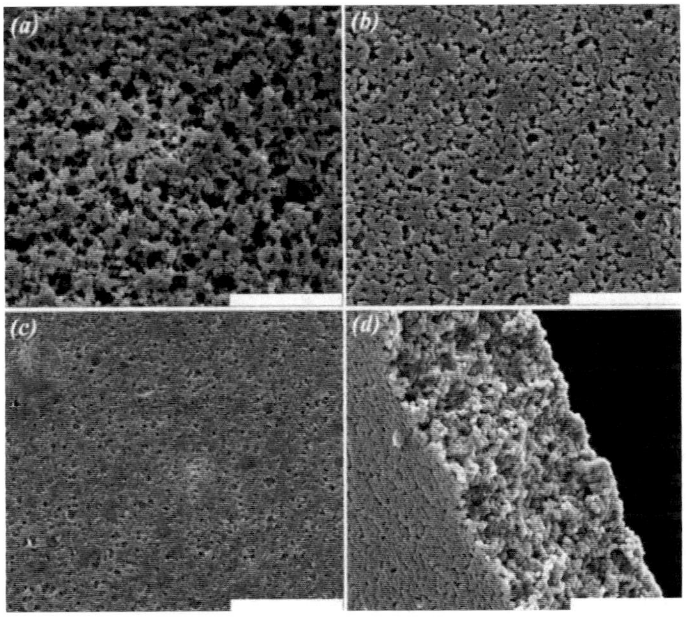

Figure 1. The surface morphology of (a) GPE1, (b) GPE2, (c) GPE3 and (d) the cross-sectional morphology of GPE 3. (Scale=50 μm).

3.2. Ionic conductivity

With the concentration of PVDF-HFP increasing, the impendence increases and the conductivity decreases. Gradual increased curves of the impedance values are revealed in Figure 2a. The conductivity of the gel polymer electrolyte is inversely proportional to the impendence. The conductivity is related to the liquid absorption, and the rate of liquid absorption depends on the gel process. This micro-porous network allows the liquid electrolyte to be embedded in the polymer network to gel the polymer electrolyte. Then, the liquid electrolyte is stored in the pores of the polymer network. The ability of the gel polymer electrolyte to adsorb the liquid electrolyte affects the conductivity. GPE1 has the largest pore size, so the impendence is minimal and the conductivity is maximum. The conductivity of the GPE3 is 1.06×10^{-3} S cm^{-1} , and it also satisfies the demand of gel polymer electrolyte for lithium-ion batteries.

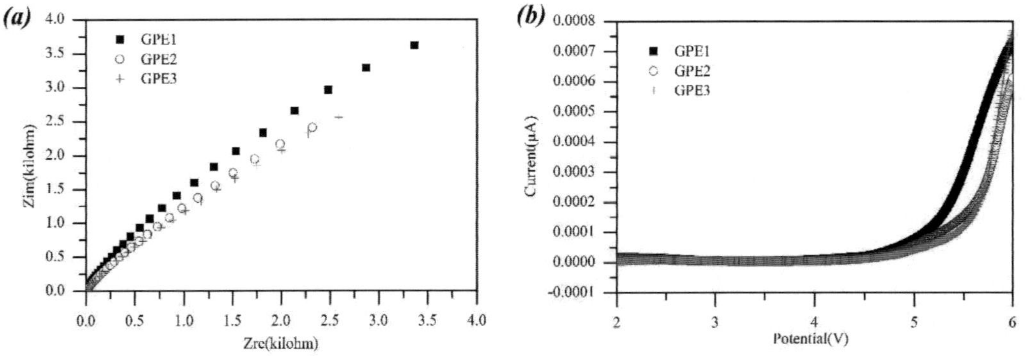

Figure 2. (a) The impendences of GPE1, GPE2, GPE3; (b) the electrochemical stability window values of GPE1, GPE2 and GPE3.

3.3. Electrochemical stability

The curve of current response to the voltage expresses the decomposition voltage of the gel polymer electrolyte, that is, the corresponding electrochemical stability window value. As shown in the Figure 2b, the currents of the GPE1, GPE2 and GPE3 start to increase sharply around 4.5V. That means the oxidation and decomposition voltages of the three kinds of gel polymer electrolytes are about 4.5V. This shows that prepared PVDF-HFP gel polymer electrolyte can be used with $LiCoO_2$ cathode material for preparing high performance lithium-ion batteries.

3.4. Thermal stability

With the temperature rising, the mass of the gel polymer electrolyte gradually decline shown in Figure 3. There are three significant mass losses between 30℃ and 500℃. The gel polymer electrolyte is basically stable within 100 ℃. A little mass loss may be due to the evaporation of residual solvents such as THF and acetone; There are 20% reduction in mass at 100-200℃, and it maybe because of PVDF-HFP preliminary decomposition and DEC evaporation[16-17]; It has a 10% mass loss at 300℃, and this situation probably due to the evaporation of EC and $LiClO_4$; The mass of the gel polymer electrolyte begin to decrease significantly at 460℃, and it probably because of the decomposition of the great mass of PVDF-HFP. Usually, the operating range of lithium-ion battery is 40-70 ℃. The prepared PVDF-HFP gel polymer electrolytes have good thermal stability within 100℃.

3.5. Lithium transference number

In Table 1, the maximum conductivity (2.23×10^{-3} S cm^{-1}) is exhibited by the sample GPE1, but the mechanical strength of the GPE1 is too terrible to operate. Considering the conductivity, mechanical strength, electrochemical stability and thermal stability of the three gel polymer electrolytes, the various properties of the GPE3 are excellent. Therefore, we selected GPE3 as the preferred sample. Further, we measured the lithium transference number of the GPE3 and obtained a value of 0.36. The impendences and currents before and after polarization are shown in the Figure 4a. High lithium transference number can reduce the polarization of the lithium-ion battery during charging and discharging. Generally, the lithium transference number of the gel polymer electrolyte is about 0.4, so that the prepared PVDF-HFP gel polymer electrolyte shows excellent battery performance.

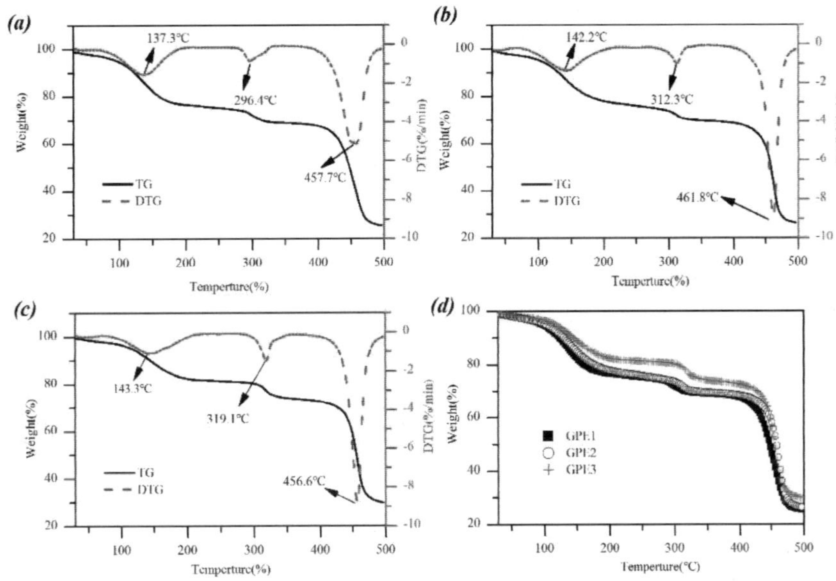

Figure 3. TG and DTG curves of (a) GPE1, (b) GPE2 and (c) GPE3, and (d) TG curves of GPE1, GPE2 and GPE3.

3.6. *Cycle stability*

The charge and discharge specific capacity and Coulombic efficiency of the GPE3 are revealed in Figure 4b. The actual specific capacity of the $LiCoO_2$ cathode material is 140mAh g^{-1}. At 0.1 C rate, the discharge specific capacity of GPE3 up to 135.1mAh g^{-1}, and the Coulombic efficiency of GPE3 reach to 99.1%. It indicates that the compatibility between prepared PVDF-HFP gel polymer electrolyte and $LiCoO_2$ cathode material is good.

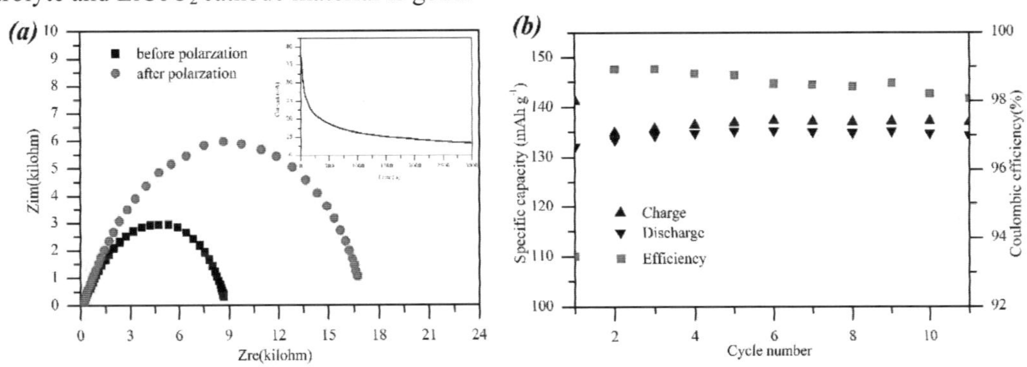

Figure 4. (a) The impendences and currents before and after polarization of GPE3; (b) the charge and

discharge specific capacity and Coulombic efficiency of the GPE3.

4. Conclusions
In summary, the prepared PVDF-HFP gel polymer electrolyte has been demonstrated to be an effective gel polymer electrolyte to be applied on lithium-ion batteries. The optimized PVDF-HFP gel polymer electrolyte expressed a high conductivity of 1.06×10^{-3} S cm^{-1} at room temperature, the electrochemical window value of 4.5 V, a high lithium transference number of 0.36, a good thermal stability within 100℃, the discharge specific capacity of 135.1 mAh g^{-1} and the charge / discharge efficiency of 99.1% at 0.1C rate. Our work supplies a basic PVDF-HFP gel polymer electrolyte. Later, we can optimize and improve the performance of this PVDF-HFP gel polymer electrolyte by adding nano-fillers, such as SiO_2, TiO_2, Al_2O_3, etc.

References
[1] Wu B, Wang L P, Li Z L, Zhao M J, Chen K H, Liu S H, Pu Y Q and Li J Z 2016 *J. Electrochem. Soc.* **163** 2248–52
[2] Tsao C H and Kuo P L 2015 *J. Membr. Sci.* **489** 36-42
[3] Kim S U, Yu D M, Kim T H, Hong Y T, Nam S Y and Choi J H 2015 *J. Ind. Eng. Chem.* **23** 316-20
[4] Gopalan A L, Santhosh P, Manesh K M, Nho J H, Kim S H, Hwang C G and Lee K P 2008 *J. Membr. Sci.* **325** 683-90
[5] Kim J S, Seo J I and Bae J Y 2009 *Polym-Korea* **33** 544-50
[6] Liao Y H, Rao M M, Li W S, Tan C L, Yin J and Chen L 2009 *Electrochim. Acta* **54** 6396-402
[7] Manuel S A 2006 *Eur. Polym. J.* **42** 21-42
[8] Subadevi R, Sivakumar M, Rajendran S, Wu H C and Wu N L 2011 *Ionics* **18** 283-9
[9] Karuppasamy K, Reddy P A, Srinivas G, Tewari A, Sharma R, Shajan X S and Gupta D 2016 *J. Membr. Sc.i.* **514** 350-7
[10] Croce F, Focarete M L, Hassoun J, Meschini I and Scrosati B 2011 *RSC* **4** 921-7
[11] Tsuchida E, Ohno H and Tsunemi K 1983 *Electrochim. Acta* **28** 591-5
[12] Rajendran S and Sivakumar P 2008 *Physica B* **403** 509-16
[13] Rajendran S, Sivakumar P and Babu R S 2007 *J. Power Sources* **164** 815-21
[14] Idris N H, Rahman M M, Wang J Z and Liu H K 2012 *J. Power Sources* **201** 294-300
[15] Cao J H, Zhu B K and Xu Y Y 2006 *J. Membr. Sci.* **281** 446-53
[16] Li H, Chen Y M, Ma X T, Shi J L, Zhu B K and Zhu L P 2011 *J. Membr. Sci.* **379** 397-402
[17] Saikia D and Kumar A 2004 *Electrochim. Acta* **49** 2581-9

IOP Conf. Series: Materials Science and Engineering **213** (2017) 012037 doi:10.1088/1757-899X/213/1/012037

Study on epoxy resin modified by polyether ionic liquid

X C Jin[1], L Y Guo[1,2], L L Deng[1] and H Wu[1]

[1]School of Petrochemical Engineering, Shenyang University of Technology, 111003, Liaoyang, China

E-amil: lyguo1981@163.com

Abstract. Chloride 1-carboxyl polyether-3-methyl imidazole ionic liquid (PIIL) was synthesized. Then blended with epoxy resin(EP) to prepare the composite materials of PIIL/EP, which cured with aniline curing agent. The structure and curing performance of PIIL/EP were determined by FT-IR and DSC. The effects of the content of PIIL on strength of EP were studied. The results show that the PIIL was the target product. The strength was improved significantly with increase of the PIIL content. The obvious rubber elasticity of PIIL/EP after cured was showed when the content of PIIL accounts for 40% and the impact strength was up to 15.95kJ/ m^2.

1. Introduction

Epoxy resin (EP) is one of the most widely used thermosetting resins. The toughening modification has been the focus of research at home and abroad. The most traditional methods of toughening epoxy resins are the so-called "soft toughening" and "hard toughening"[1]. New modification methods appear continuously. Such as National Taiwan University Hsieh et al[2] ,who studied the graft interpenetrating polymer networks (EP) and polyurethane elastomers (PUR) The tensile strength of EP was improved, and the epoxy polyurethane acrylic acid (IPN) coating was synthesized by Shah et al[3]. The tensile strength and bond strength of the coating were improved by the excellent corrosion resistance of the coating. Yang Weijiang et al[4] optimized the synthesis of epoxy acrylate resin emulsion by changing the glass transition temperature (Tg) and thermal decomposition temperature (Td).

Ionic liquids have been widely utilized in the fields of organic synthesis, separation, and extraction, as well as in material preparation, because of their excellent performance. The structure of green solvent ionic liquid has a unique "tailored" design[5]' if we can design a rational structure, it can react with epoxy groups with long chain ionic liquids, prepared with epoxy resin modified with long chain structure, its impact toughness will be greatly improved in theory, at the same time the thermal properties and processing performance can be improved. Therefore, this article on the existing research and synthesis of novel functional ionic liquids - carboxyl polyether ionic liquid, the E-44 type epoxy resins curing under the action of a small amount of aniline. Preparation of polyether ionic liquid / epoxy resin composite material (PIIL/EP), and the performance of heat shock. Study on toughness of polyether ionic liquid for epoxy resins, and expounds its toughening mechanism.

2. Experimental

2.1. Reagents and instruments

IOP Conf. Series: Materials Science and Engineering **213** (2017) 012037 doi:10.1088/1757-899X/213/1/012037

Materials and equipments: Epichlorohydrin (AR), N- methyl imidazole (AR), epoxy resin, E-44 (6101), aniline (content above 99.5%), ethylene glycol (AR), three boron fluoride ether (content 47.5%). MAGNA-IR750 infrared spectrometer (FTIR), Mettler 821e/400 DSC differential thermal scanner, 2-XZ-4 type rotary vane vacuum pump, XJJ-50 Charpy impact test machine, SFX-2L type DF-101S instrument, rotary evaporation magnetic stirrer, DZF-6050 type vacuum drying box.

2.2. Preparation of polyether ionic liquid

Hydroxyl-terminated polyepichlorohydrin was prepared by the cationic ring-opening polymerization of epichlorohydrin[6], and chloroacetic acid was added into a three-neck flask with a condenser pipe (molar ratio of 1:2.2), with acetonitrile as the solvent. After mixing at 80 °C for 1 h, a small amount of Na_2CO_3 was placed into the system in batches until insolubility, with continuous reaction for 6 h to 8 h. Hydrochloric acid was dripped into the reaction system to regulate the system for acerbity. The solvent was removed by rotary evaporation under the conditions of 70 °C, 0.09 MPa. The excess chloroacetic acid was removed by washing with deionized water for 3 to 5 times. A small amount of water was removed under the conditions of 70 °C and 0.09 MPa by rotary evaporation again. The carboxyl-terminated polyepichlorohydrin intermediate was then obtained.

The above intermediate and N-methylimidazole were added to the three-neck flask at a molar ratio of 1:1 and mixed in a 80 °C thermostatic waterbath for 24 h. The reaction was then stopped. The product was washed with ethyl acetate for 3 times and vacuum dried for 12 h. Finally, the [HOOC-PECH-MIM]Cl catalyst, which is a yellow-brown sticky liquid, was obtained. The yield was 74%. Its chemical structure is shown in Figure 1.

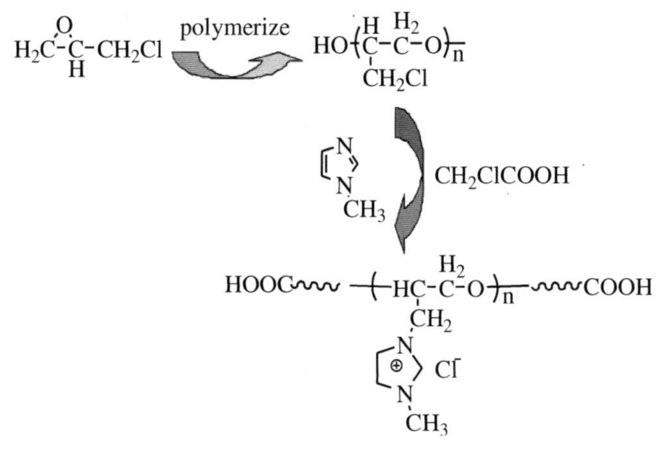

Figure 1. Chemical structure of [HOOC-PECH-MIM]Cl.

2.3. Preparation and curing of PIIL/EP

The 50g epoxy resin was mixed with a certain amount of polyether ionic liquid (10%, 20%, 30%, and 50%, respectively) for the epoxy resin, and then was mixed into the round bottom flask while stirring at 2h for 80℃. The reaction system into brown viscous, adding 5g aniline curing agent, its quality for the added weight of the epoxy resin 10% with stirring to cast. For the stage of curing it was 100-150 ℃, the curing process below 100℃ for 1h, 120 ℃ for 2h and 150 ℃ for 1h.

3. Results and discussions

3.1. Chemical structure and characterization of PIIL/EP

Fourier transform infrared spectroscopy was used to characterize the chemical structure of the epoxy resin before and after modification. The results are shown in Figure 2.

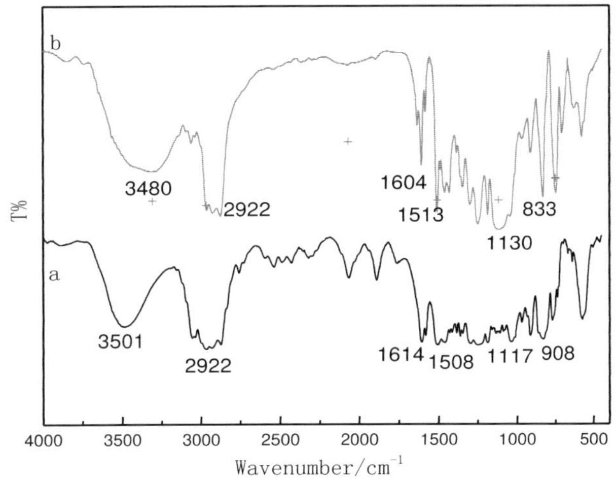

Figure 2. IR spectrum of EP (a)and PIIL/ EP (b) (a-EP; b-40%PIIL/EP).

Figure 2 is the EP and polyether ionic liquid modified PIIL/EP composite system (PIIL content of 40%) infrared spectrum. Figure indicates that the polyether ionic liquid has a strong and broad hydroxyl absorption peak near 3501cm^{-1}. In the vicinity of the 1115cm^{-1}, the absorption peak of the in-plane symmetric stretching vibration of the 2922cm^{-1} is observed. The in-plane shear vibration absorption peak of 1429cm^{-1} appears in the vicinity of C-O. You can see from curve B did not change the basic peak and absorption peak of 2822cm^{-1}, 1429cm^{-1}, 1115cm^{-1}. Absorption peak and absorption peak of 3480 cm^{-1} decreased significantly and widely, and moving towards the lower wavenumber towards 908cm^{-1} characteristic absorption peaks of epoxy group which indicated that the polyether ionic liquid has been transformed into the most epoxy group.

3.2. Determination of thermal properties of PIIL/EP

The curing properties of epoxy resin were tested by DSC before and after the modification. DSC data was taken on a Mettler 821e/400 DSC instrument over a temperature range 25-350 °C at a heating rate of 10 °C·min-1. The results are shown in Figure 3.

IOP Conf. Series: Materials Science and Engineering **213** (2017) 012037 doi:10.1088/1757-899X/213/1/012037

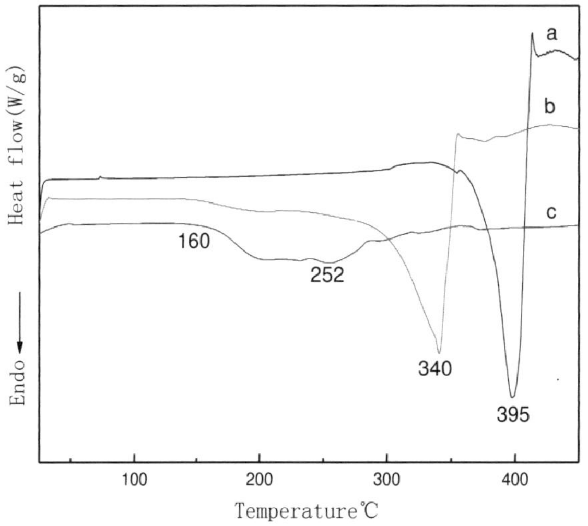

Figure 3. DSC spectrum of EP and PIIL/ EP (a-EP; b-20%PIIL/EP; c-40%PIIL/EP).

DSC spectra of EP and PPIL/EP are blended in Figure 3. The peak temperature of the decomposition of pure epoxy resin is 395 DEG C, and the endothermic peak temperature of PIIL/EP blend system (b) with a small amount of ionic liquid. The results showed that in the blend system, all of the polyether ionic liquids were received on the epoxy resin the ionic liquid and epoxy resin became a single system. With the increase of the ionic liquids, PIIL/EP blends (c) endothermic peak temperature is reduced to 252 DEG C, which may be the introduction of polyether ionic liquid. Its hydroxyl terminated epoxy resin has formed ring opening reaction of crosslinking and copolymerization. Hydroxyl has a certain role in the curing epoxy resin, so the endothermic peak temperature of the decomposition decreases. When the ionic liquid dosage is too large, the ionic liquid and the epoxy resin cannot be excessive reaction, which leading to the endothermic reaction of blends complexly and no obviously endothermic peak, the curve of 160 DEG C should be excess decomposition temperature of ionic liquid.

3.3. Determination of impact properties of PIIL/EP

The impact test machine was used to test the impact bar, and the effect of the addition of polyether ionic liquid on the impact strength of epoxy resin was studied. The results are shown in table 1.

Table 1. Effect of PIIL content on impact strength of epoxy resin.

Sample	1	2	3	4	5	6
PIIL content（%）	0	10	20	30	40	50
Impact strength（kJ/m^2）	2.13	5.23	9.52	14.22	15.95	15.98

According to the data in Table 1, the impact strength of the unmodified epoxy resin is only 2.13 kJ/m^2. After ionic liquid was modified, the impact strength increased significantly. With the increasing content of ionic liquid, the impact strength increased. When the percentage of ionic liquid for epoxy resin quality reached 40%, the impact of strength was up to 14.21kJ/m^2 while the content of ionic liquid was 40%, the increasing impact modified epoxy resin strength of the amplitude tends to be stable. The results show that the bigger ionic liquid content is the longer the curing time of epoxy resin. Therefore, it can be concluded that the imidazole type polyether ionic liquid is a very effective toughening agent for epoxy resin, and its content is no more than 40%.

The polyether ionic liquid has active hydroxyl terminated reaction with epoxy resin and a flexible long chain structure. It is liquid at room temperature, with high boiling point and low volatility, its structure of hydrophilic lipophilic group indicates that ionic liquids have very good solubility and compatibility of polymers with the coupling effect during the curing process. It can be uniformly dispersed in the polymer resin; the improvement of impacted strength is conducive. The ionic liquid simultaneously was connected to long chain with the structure of the epoxy resin. It will make epoxy resin glass transition temperature decreased and the active groups of ionic liquids will make certain cross-linked epoxy resin. The improvement of the impact toughness of polymers was conducive. However, when the ionic liquid reaches a certain use level, the crosslinking degree of epoxy resin reaches the maximum value, and the toughness of epoxy resin is no longer obviously.

Therefore, the compatibility and curing properties of epoxy resin toughened by imidazole type polyether ionic liquid are much better than that of traditional toughening agent. This study provides a good theoretical basis and broad prospects for the development of functionalized ionic liquids in the preparation of high performance environmental friendly polymer materials.

4. Conclusions

In this article, the synthesis of imidazole based ionic liquids and the modification of epoxy resin were studied. The conclusions are as follows:

(1) Results of IR and TG analysis show that the chemical reactions of imidazole based ionic liquid polyether and epoxy resin or the polyether ionic liquid modified EP has good toughness, with the increase of the amount of ionic liquid, the decomposition temperature after heat modified by epoxy resin decreased gradually.

(2) Modification of epoxy resin with the use of imidazole type ionic liquid, the impact strength increased significantly. With the increasing amount of ionic liquid, the impact strength increases, but the amount of ionic liquid is 40%, the impact strength of the modified epoxy resin can be as high as 15.95kJ/m^2.

References

[1] Ong S, Ismail J, Bakar M A, Rahman I A, Sipaut C S and Chee C K 2009 Polyurethane-modified epoxy resin: Solventless preparation and properties *J. Appl . Polym. Sci.* **111**(6) 3094-103

[2] Hsieh K H 1990 Activiated adhesive for aluminum bonded pepairs Polym *Sci. Part B: Polym. Phy.* **28**(5) 623-7

[3] Shah A and Ryntz R A 1990 Epoxy resinsmodifiedwith polycarbonate polyurethane *J. Coat. Technol.* **62**(785) 63-7

[4] Yang W J and Wang Y X 1998 Polymerization and characterization of epoxy-polyacrylate IPN material *Polym. Mater. Sci. Eng.* **14**(1) 24-7

[5] Zhang H, Wu J, Zhang J and He J S 2005 1-Allyl-3-methylimidazolium chloride room temperature ionic liquid: A new and powerful nonderivatizing solvent for cellulose *Macromolecules* **38**(20) 8272-7

[6] Guo L Y, Zhang B, Ma X Y, Chang X T and Li C Y 2015 Synthesis of carboxyl-terminated polyether ionic liquids and reinforced and toughened phenolic resins *Appl. Chem. Ind.* **44**(10) 1825-8

IOP Conf. Series: Materials Science and Engineering **213** (2017) 012038 doi:10.1088/1757-899X/213/1/012038

Dissolution behaviour of polyethersulfone in diglycidyl ether of bisphenol-A epoxy resins

J L Zhou[1], C Cheng[1], H Zhang[1], Z Y Sun[2], S Zhu[2,3] and M H Yu[1,2,3]

[1]College of Materials Science and Engineering, Donghua University, Shanghai 201620, China
[2]Center for Civil Aviation Composites, Shanghai 201620, China

E-mail: pigsisterz@sina.com; yumuhuo@dhu.edu.cn

Abstract. Polyethersulfone (PES) is considered as an important and useful toughener for epoxy resins (EPs). Generally, PES was first dissolved in EP oligomer to form a homogeneous solution, which considerably affected the properties of cured EPs. Turbidity measurement was carried out to characterize the dissolution behaviour. Different PES/EP systems exhibited the same dissolution behaviour, albeit different swelling rate (R_S) and dissolution rate (R_D). Compared to molecular weight, the particle size and hydroxyl content of PES exerted considerable effects on dissolution. The smaller epoxy equivalent these diglycidyl ether of bisphenol-A (DGEBA) oligomers possessed, the better dissolubility they had; the more small molecules they contained, the more apparent the swelling observed. Moisture prevented the diffusion of the DGEBA oligomer molecules into PES. With increasing temperature, R_S and R_D first rapidly increased and then slowly increased, and the optimal dissolution temperature was 120°C. At the final steady, the turbidity increased linearly with increasing PES concentration; in addition, these values can be used as reference for homogeneous solution. Only one glass transition temperature (T_g) was observed for PES/EP systems, indicative of the good compatibility between PES and EP.

1. Introduction

Cured EPs have been widely used in coatings, adhesive materials, insulating materials, matrix resins for composites and many other fields [1-6]. Unfortunately, because of the high crosslink density, the cured resins tend to have a low resistance to brittle fracture [7]. The lack of toughness limits their further applications which demand better fracture resistance [8-9]. Therefore, toughening EPs has been an important issue for a long time.

PES is one of the most suitable commercially thermoplastic resins to toughen EPs, based on its good toughness, proper modulus, chemical resistance and thermal properties, and so on [10-12]. PES can improve the toughness without significantly decreasing the modulus and T_g of cured EPs [6]. The academic researches on PES toughening of EPs have long been focused on the morphological structures, mechanical and thermal properties, and toughening mechanism [7, 9, 6, 13-15].

Most of thermoplastic toughened systems are prepared starting from a homogeneous solution of thermoplastic polymers in epoxy oligomers. During the epoxy-hardener curing reaction, the miscible between thermoplastic and epoxy resin decreased, and the phase separation occurred. These phase separations lead to toughness of cured epoxy resins improved [16]. Some researchers have mentioned the preparation of thermoplastic and EP systems, including solvent method and melt method [17]. In Alessi S's research [18], PES was first dissolved in dichloromethane and then added to EP anhydride system, the solvent was removed at 120ºC for 1.5h. When solvents are used in EPs, they must be completely evaporated prior to cure, or otherwise bubbles could be formed during curing process, and causing weak joints [1]. Therefore, it's necessary to minimize the use of solvents. In Mimure K's study [13], PES, EP and curing agent compound was stirred at 135ºC, until PES dissolved homogeneously in the mixture. In Yu Y's research [14], PES was added to the stirring epoxy monomer at 150ºC under nitrogen gas to form homogeneous mixture, then the curing agent was added to the mixture. Obviously, it is important to reach total dissolution of PES in order to obtain a homogeneous solution.

To date, few researches concerning the dissolving behavior and critical influences of PES in EP oligomer have been reported, to the author's knowledge. L. Bonnaud [19] used cloud-point curves to research the influence of different factors on the miscibility of the DGEBA/ thermoplastic systems, and a phenomenological model was proposed to predict the epoxy precursor and thermoplastic blend miscibility. Unfortunately, cloud-point curves had a low automation degree and a large impact on human factors.

In this study, we adopted turbidimetry [20-21] as a novel and simple technique to study the dissolution behavior of PES/EP oligomer systems for the first time. The dissolution behavior of different systems was explained, and the influence of solubility temperature, moisture, the properties of PES and DGEBA were studied. These results had important implications for determining the dissolving process conditions, distinguishing the final state of solution, and improving the raw material. At the same time, this study can provide a scientific basis for doing further research on toughening mechanisms.

2. Materials and methods

2.1. Materials

Two types of commercial PES and three trades of DGEBA were used. Their properties were shown in Table.1 and Table.2, and formulas were shown in Figure 1.

Table 1. Properties of the PES particles.

	PES1	PES2
Commercial code	Ultrason E2020P	Virantage VW 10200RFP
Company	BASF	Solvay Specialty Polymers
Country	Germany	Germany
Average molecular weight (g/mol)	55000	46500
-OH end groups (%)	>50 (typically>75)	32
Glass transition temperature (ºC)	225	220
Particle size (μm)	20	63

Table 2. Properties of the DGEBA according to the suppliers.

	EP1	EP2	EP3
Commercial code	DER331	CYD-128	CYD-144
Supplier	Dow Chemical Co.	YueYang Petrochemical Ltd	YueYang Petrochemical Ltd
Country	USA	China	China
Epoxy equivalent (g/eq)	182-192	184-200	210-240

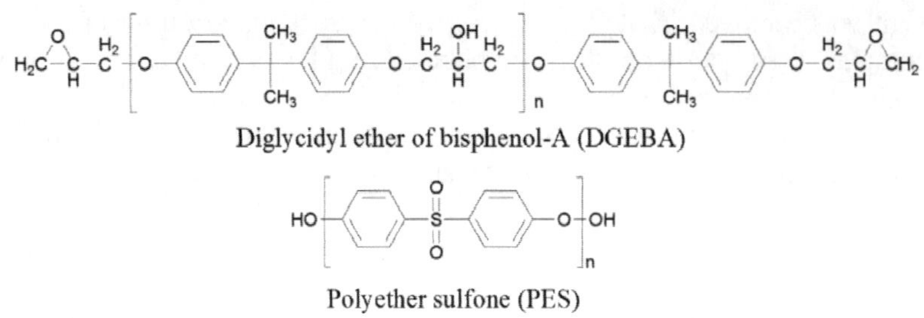

Figure 1. Molecular structure of DGEBA and PES.

2.2. Preparation of raw material

In order to avoid the influence of absorbed moisture, the raw materials should be dried before using. PES1 powder was dried 4h at 140°C in a vacuum drier. PES2 powder should be dried in vacuum oven at 130°C for 4h. Three DGEBA oligomers were dried at 70°C for 10h in vacuum drier.

2.3. Turbidimetric measurements

Turbidity was measured by WGZ-3B (Xinrui, China) turbidimeter with a scattered light detector perpendicular to the light beam. With the help of turbidimeter, the intensity of light scattered from these insoluble particles in transparent solution can be tested, what's more, the content of these insoluble particles can be characterized quantitatively. The units of turbiduty are called Nephelometric Turbidity Units (NTU). The range of the nephelometer is between 1 and 1000NTU. Formazine calibration solutions are used for correction.

2.4. Dissolution of PES in EP oligomers

A precise quantity of PES and EP oligomers were added to a beaker, after rapid stirring, the PES/EP systems were defoamed in vacuum oven at 70°C. Then, the mixture was transferred to a clear glass vial slowly until reach the calibration line. After that, the glass vial was defoamed in vacuum oven at 70°C again. Following the above steps, these samples with different systems and different concentrations were prepared. The turbidity of these samples was measured every 3 minutes until a stable NTU value was reached under designed condition. Unfortunately, the turbidity of systems with high PES contents at the initial dissolution could not be reached because the solution is completely opaque.

2.5. Characterization

2.5.1. X-ray diffraction. The crystal structure of PES powder was characterized by X-ray diffraction (XRD). XRD patterns of PES were obtained with X-ray diffractometer (D/max-2550 PC, Rigaku Corporation, Japan) using 18 kW Cu-Kα (λ=1.5406 Å) radiation. The scanning range was set from 5° to 80°.

2.5.2. Thermal platform microscope. High speed mixer (DFY-1000C, BILON, China) was used to mix PES powder and EP with different weight ratios. The dissolving process of PES/EP blends was observed using thermal platform microscope at 120°C. The photos were shot every 30 seconds.

2.5.3. GC-MS. The three trades of EPs were analyzed by gas chromatography mass spectrometry (GC-MS, QP2010Ultra, Shimadzu, Japan) system at 500°C for 30min.

2.5.4. Differential scanning calorimetry (DSC). Differential scanning calorimetry scans were performed on a Mettler 822 DSC (Mettler, Switzerland) in a heating-cooling-heating cycle at a heating rate of 10°C/min under a nitrogen flux of 50mL/min. The weight of each simple was approximately 5-

8mg. To check the reproducibility and uniformity, the procedure of every solution was repeated two more times.

3. Results and discussion

3.1. Dissolving process of PES in EP oligomers

Turbidimetry was chosen to characterize the dissolution behavior of different PES/EP systems for the first time, as shown in Figure 2-5. Like the other polymer dissolution behavior, the dissolution of PES powder in EP oligomers did not instantaneous, but involved two transport processes, namely solvent diffusion (swell) and chain disentanglement (dissolve). It can be further described that PES powder in EP oligomers at first swelled (turbidity values increased), then started to dissolve (turbidity values decreased), at the end of the dissolution process, the clear and transparent solution was obtained (turbidity values became stable). The dissolution behavior can be explained as follows: the molecular structure of PES and EPs had some same functional groups (Figure 1); PES is the amorphous character (shown as Figure 2b, only one prominent peak at $2\theta=18.1°$ [22-23]), the molecular interactions were weak, and the EP oligomers can diffuse into PES easily [24]. Due to plasticization of PES by EP oligmomers, the volumetric expansion of PES powder occurred. After time has passed, EP molecular entered into the intermolecular chains of PES, which led chain disentanglement, in other words, PES powder dissolved. The amorphous structure was widely considered to be one of the main causes of the good compatibility between PES and EP. On the other hand, PES can dissolve in NMP, DMAc, DMF, while, EP oligomers can dissolve in acetone. Based on the theory of "similarity and intermiscibility", the PES and EP is both polar molecule, and the polarity of PES is stronger.

The other information can be obtained from this turbidity versus time curves: the slope of upward section (from the start point to the peak point) can be used to characterize R_S; the slope of downward section (from peak point to equilibrium point) can be used to characterize R_D; different systems have same dissolution behavior but different R_S and R_D.

3.2. The influence of solution composition

The influences of different PES on dissolution behavior were shown in Figure 2. At the beginning of dissolution process, the turbidity values of PES2/EP1 were smaller than that of PES1/EP1, that maybe because particle size of PES1 (20μm) was smaller than PES2 (63μm). On the other hand, R_S and R_D of PES1/EP1 were bigger than PES2/EP1, which can be attributed to the difference of particle size and hydroxyl content (Table 1.). But, the dissolution behavior did not follow the rule that the dissolution rate decreased with increased polymer molecular weight. We can indicate that, compared to molecular weight, particle size and hydroxyl content of PES had more important influence on dissolution process. In order to observe the dissolving process visually, the macroscopic and microscope photographs were supplied in Figure 2a at the corresponding position.

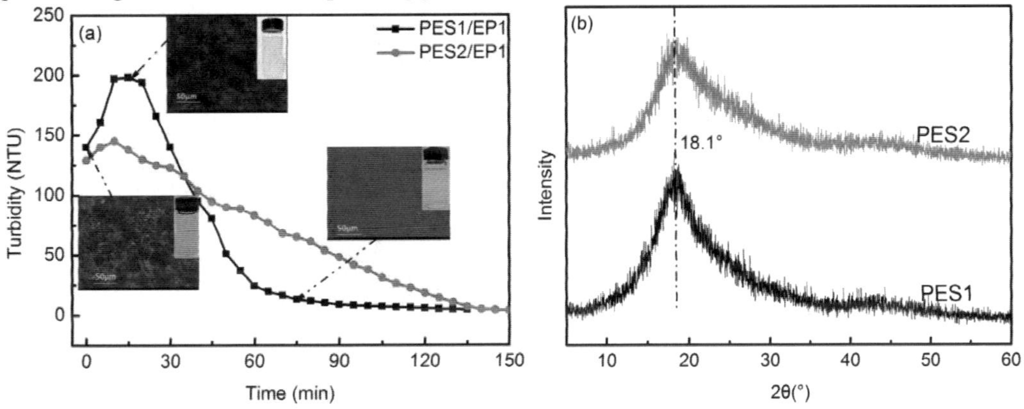

Figure 2. Turbidity versus time for PES/EP systems (PES 0.1 wt%) at 120 °C (a) The influence of PES on dissolution behaviour; (b) XRD patterns of PES powder).

The influences of different EPs on dissolution were shown in Figure 3. The turbidity peak value of PES1/EP2 and PES1/EP3 was larger than that of PES1/EP1. This phenomenon maybe because there were more small molecules in EP2 and EP3, as shown in the upper-right corner of Figure 3b, these small molecules can enter into PES easier, which caused the powder swelled. R_D and dissolution time of PES1/EP1 and PES1/EP2 were almost the same, and larger than that of PES1/EP3, which can be explained that the molecular weight (for DGEBA, the molecular weight is twice of epoxy equivalent) of EP1 and EP2 was close to each other and smaller than EP3 (Figure 3c). In conclusion, the small molecules in EP mainly affected the swelling process, while, the epoxy equivalent mainly affected R_D and dissolution time.

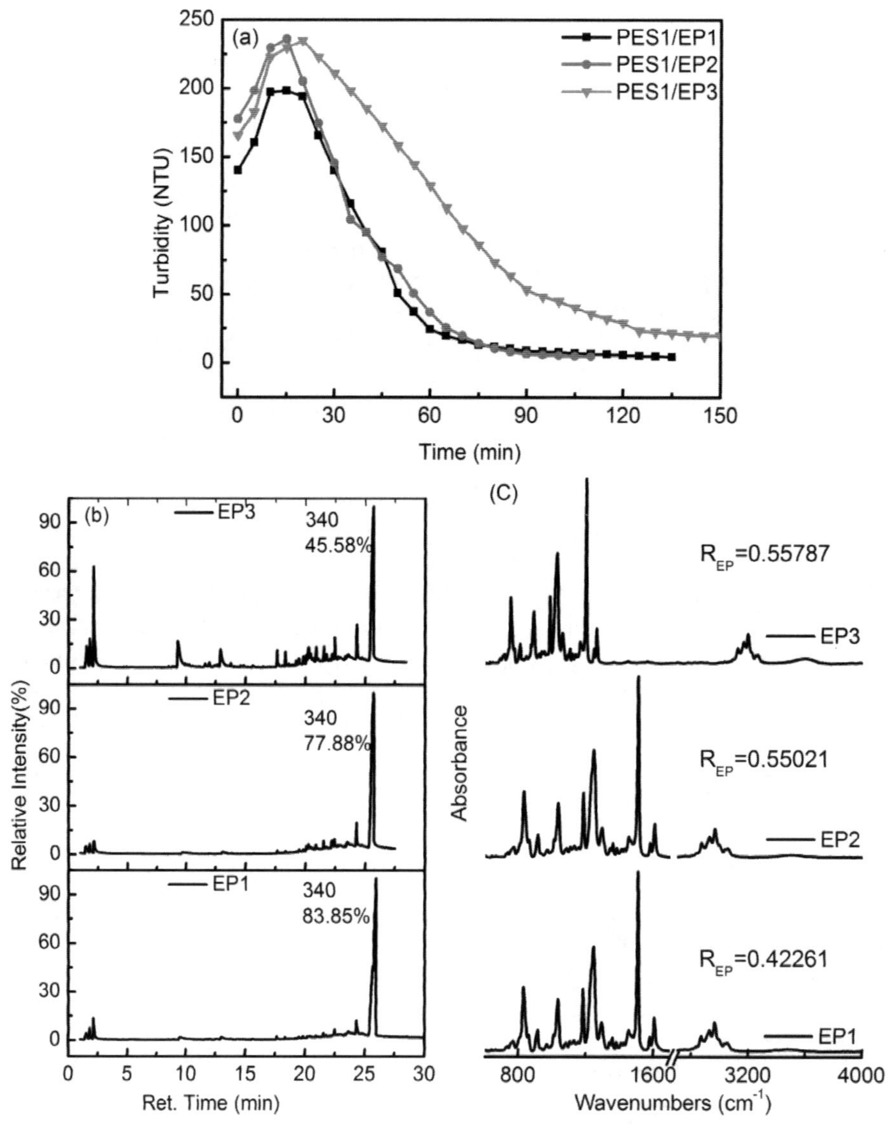

Figure 3. Turbidity versus time for PES/EP systems (PES 0.1 wt%) at 120 °C (a)The influence of EPs on dissolution behaviour; (b) thermal volatiles detected as the TICs at 500 °C; (c) FTIR transmission spectra of three types of EPs and the relative content of epoxy group (R_{EP}).

3.3. The influence of moisture

In order to study the influence of moisture absorption of PES on dissolution, PES1/EP1 and undried PES1/EP1 systems were chosen, as shown in Figure 4.

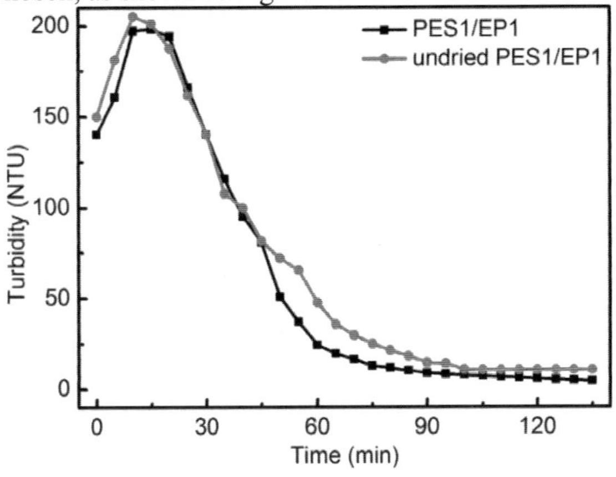

Figure 4. Turbidity versus time for PES and undried PES/EP systems (PES 0.1 wt%) at 120 °C.

The turbidity values of PES1 were bigger than that of undried PES. Based on ISO62 (equilibrium 23°C/50% r.h.), the moisture absorption of PES1 was 1.00%. This moisture occupied the free volume of PES, and prevented EP oligomer molecules diffusing into PES. Therefore, PES should be dried before use.

3.4. The influence of dissolution temperature

Temperature has an important effect on dissolution process. In order to observe the dissolution process of PES in DGEBA oligomer at different temperature clearly, PES1/EP1 with wt0.1% PES was prepared. With the increasing of temperature, R_S and R_D both increased, and the time of swelling and dissolution reduced. In detail, from 100°C to 120°C, the above values changed significantly, and from 100°C to 120°C, and the above values changed slightly, as shown in Figure 5. These results showed that the swelling and dissolution process of PES speeded with temperature increased, this may be due to an increase in temperature corresponded to faster motion of the molecules, and produce a certain volume expansion, while, when temperature up to 120°C, there were almost no variation of molecule motion and volume. Therefore, the dissolution temperature was chosen 120°C.

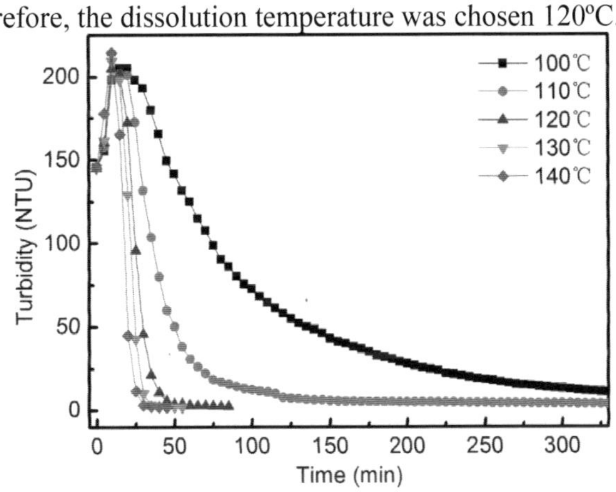

Figure 5. Turbidity versus time for five PES1/EP1 system at different temperature (PES 0.1wt%).

3.5. The influence of solution concentration

The relationship between turbidity and concentration (shown in Figure 6), can be used to characterize the dissolving capacity of PES in EP oligomers. The turbidity values of final steady state (turbidity values became stable) increased with the concentration increasing. There was no an abrupt change of turbidity in every system except PES1/EP3, therefore, the maximum solubility of PES was not reached. Sample with higher PES content could not be reached because the solution became too viscous to stir and measure. For PES1/EP3 system, when PES content was larger than 4%, the turbidity increased rapidly. This indicated that the solvents may be need for EP3 to dissolve PES. These values of turbidity of different systems can be used to determine the final dissolution state.

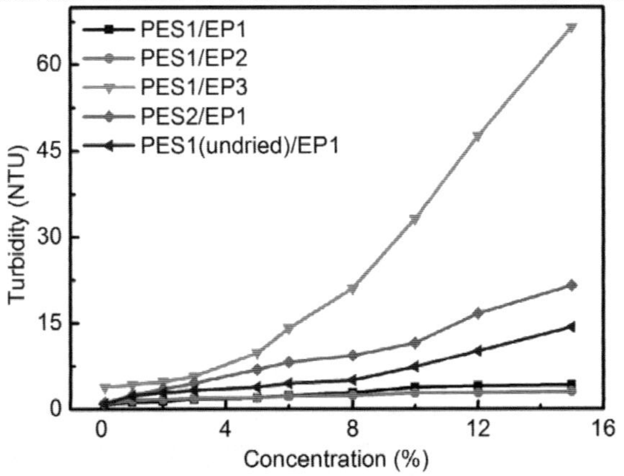

Figure 6. Turbidity versus PES concentration for five PES/ EP systems at the final steady state.

3.6. Thermodynamic compatibility

In fact, DSC has been proposed to be a simple, rapid and effective method to evaluate the compatibility between PES and DGEBA epoxy oligomer[25]. As shown in Figure 7, only one T_g was found in different systems, what's more, T_g of mixture located between T_g of PES and T_g of EPs. These results showed that the system composed by PES and EP was thermodynamic compatibility.

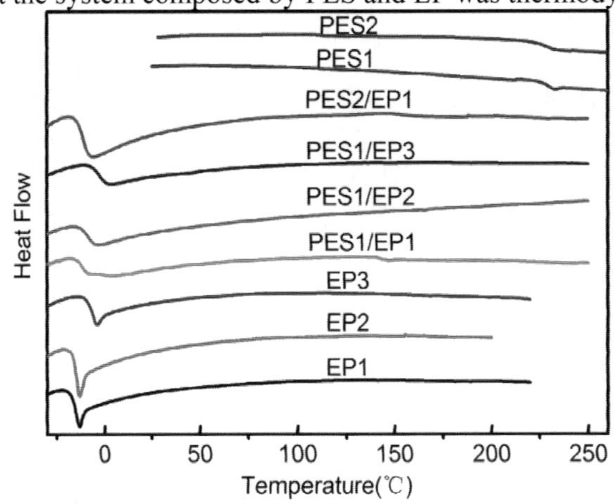

Figure 7. DSC curves of different DGEBA and system (PES 15 wt%).

4. Conclusions

It was demonstrated that turbidity can be used to characterize the dissolution behavior of PES in DGEBA epoxy oligomer. Different systems had same dissolution behavior but different R_S and R_D; what's more, R_S and R_D can be used to compare the dissolubility of different systems. For PES, the influence of particle size and hydroxyl content was more important than that of molecular weight. Moisture could occupy the free volume of PES and effect dissolving process. In order to improve the dissolubility of PES, it's necessary to decrease the particle size, increase the hydroxyl content, and dry PES before dissolution. EPs with higher molecular weight had smaller R_S and R_D. Melt method is suitable for EPs with lower epoxy equivalent. The turbidity values at the final steady state increased linearly, with the increasing of PES concentration. DSC curves of different system had only one T_g, the result showed that the mixture of PES and DGEBA epoxy oligomer was thermodynamic compatibility.

Acknowledgements
This work is supported by Shanghai Science and Technology Committee (No. 15JC1400302).

References

[1] Petrie E 2005 *Epoxy adhesive formulations (*New York: McGraw-Hill Education)
[2] Karger-Kocsis J 2010 *Macromol. Chem. Phys.* **211** 1836-42
[3] Fan M, Liu J, Li X, Cheng J and Zhang J 2013 *Thermochim. Acta.* **554** 39-47
[4] Burger, N., Laachachi A, Mortazavi B, Ferriol M, Lutz M, Toniazzo V and Ruch D 2015 *Int. J. Heat Mass Transfer* **89** 505-13
[5] Lee J, Bhattacharyya D, Zhang M and Yuan Y 2015 *Composites Part B* **78** 515-9
[6] Lee S E, Jeong E, Lee M Y, Lee M K and Lee Y S 2016 *J. Ind. Eng. Chem.* **33** 73-9
[7] Yu Y, Zhang Z, Gan W, Wang M, Li S 2003 *Ind. Eng. Chem. Res.* **42**(14) 3250-6
[8] Hodgkin J, Simon G and Varley R 1998 *Polym. Adv. Technol* **9**(1) 3-10
[9] Yang G, Zheng B, Yang J P, Xu G S and Fu SY 2008 *J. Polym. Sci., Part A: Polym. Chem* **46**(2) 612-24
[10] Paul D R and Yampol'skii Y P 1993 *Polymeric gas separation membranes* (Florida: CRC press)
[11] Zhou X M and Jiang Z H 2005 *J. Polym. Sci., Part B: Polym. Phys* **43**(13) 1624-30
[12] Linares A and R Benavente 2009 *Polym. J.* **41**(5) 407
[13] Mimura K, Ito H and Fujioka H 2000 *Polymer* **41**(12) 4451-9
[14] Yu Y, Wang M, Gan W, Tao Q and Li S 2004 *J. Phys. Chem. B* **108**(20) 6208-15
[15] Del Saz-Orozco B, Ray D, Kervennic A, McGrail P and Stanley W F 2016 *Mater. Des.* **93** 297-303
[16] Hameed N 2007 *Composites Part A* **38**(12) 2422-32
[17] Xu Y J and Liu S B 2010 *Thermosetting Resin* **6** 016
[18] Alessi S, Conduruta D, Pitarresi G, Dispenza C and Spadaro G 2011 *Polym. Degrad. Stab.* **96**(4) 642-8
[19] Bonnaud L, Bonnet A, Pascault J, Sautereau H and Riccardi C 2002 *J. Appl. Polym. Sci.* **83**(6) 1385-96
[20] Li X D, Lan A, Tang J W, Zhang Y, Gibril M E and Yu M H 2015 *Trans. Tech. Publ.* **1073** 282-5
[21] Mazza M, Catana D A, Vaca-Garcia C and Cecutti C 2009 *Cellulose* **16**(2) 207-15
[22] Khayet M and García-Payo M C 2009 *Desalination* **245**(1) 494-500
[23] Li J F 2009 *Appl. Surf. Sci.* **255**(9) 4725-32
[24] Sharma S, Dhiman N, Pathak D and Kumar R 2016 *Ionics* **22** 1865-72
[25] Kumar R, Sharma S, Pathak D, Dhiman N and Arora N 2017 *Solid State Ionics* **305** 57-62

Structural ATR-IR analysis of cellulose fibers prepared from a NaOH complex aqueous solution

Y P Yang[1], Y Zhang[1,2], Y X Lang[1] and M H Yu[1,2]

[1]State Key Laboratory for Modification of Chemical Fibers and Polymer Materials, College of Materials Science and Engineering, Donghua University, Shanghai 201620, People's Republic of China

E-mail: zyue021@126.com; yumuhuo@dhu.edu.cn

Abstract. In this study, regenerated cellulose fibers were successfully prepared from cellulose/NaOH/thiourea/urea aqueous solution through an efficient extrusion dissolution method. The microstructure of the regenerated cellulose fibers were analysed by using attenuated total reflection infrared (ATR-IR) spectroscopy and the corresponding two-dimensional (2D) correlation spectroscopy. The results showed that the fibers demonstrate a typical crystalline cellulose II and a relatively high crystallinity. Improved crystal structure and tensile strength were obtained in the regenerated cellulose fibers due to improved multi-roller drawing process. The total crystalline index (TCI) and tensile strength of the fibers at different spinning stages were determined in the range of 0.46-0.54 and 1.06-2.30 cN/dtex, respectively. The described cellulose regeneration pathway provided an environmentally friendly and simple method, which could have a significant influence on current cellulose regeneration industries.

1. Introduction

In the past 100 years, the viscose rayon [1] and cuprammonium rayon [2], which was an alternative to the viscose route, were extensively done. As well as known, wood or cotton linter pulp were using to produce these fibers, which could regard as an ideal substitution of the natural fibers. However, both the viscose and cuprammonium process resulted in the generation of hazardous byproducts (H_2S, CS_2 and heavy metals) that needed an additional treatment to meet the aqueous and gaseous emission standard. Therefore, the traditional approaches were not used to produce regenerated cellulose fibers in developed countries [3]. Increasing concerns over developing much more environmentally friendly spinning systems as well as using renewable raw materials were thus highly valuable.

Throughout the years, NaOH-based cellulose dissolution systems are regard as one of the alternatives for regenerating cellulose fibers, as the viscose fiber production machine can almost be used to prepare the fiber. For example, cellulose/NaOH [4-5], cellulose/NaOH/ZnO [6-7], cellulose/NaOH/thiourea [8-9], cellulose/NaOH/urea [10-11], and cellulose/NaOH/thiourea/urea [12-13] solutions have already been processed to regenerate fibers during wet-spun process. The advantages can be represented by the low cost and environmentally friendly solvent systems and production essentially non-toxic byproducts. Moreover, the solubility of cellulose can be significantly enhanced by adding urea or thiourea to NaOH aqueous solution. In addition, the physical of properties of regenerated cellulose fibers prepared by NaOH-based method are similar to NMMO-based fibers [8].

Previously, we presented an efficient extrusion dissolution method to prepare cellulose spinning dope in NaOH/thiourea/urea aqueous solution [14]. The shortening of the dissolution time to about 10 s and the increment of the dissolution temperature from -10 ℃ [13] to -2 ℃ were of great importance in this method. Hence, regenerated cellulose fibers can achieve a relatively better physical properties by the more stable cellulose spinning dope prepared by this new higher efficiency extrusion dissolution approach. In this work, we focus on regenerated cellulose fibers prepared from 7 wt% cellulose/NaOH/thiourea/urea aqueous solution which dissolved by twin-screw extruder at -2 ℃. Thereafter, we present attenuated total reflection infrared (ATR-IR) spectroscopy and two-dimensional (2D) correlation spectroscopy (2DCOS) analysis to investigate the effects of the staged spinning process on the structure of regenerated cellulose fibers.

2. Materials and methods

2.1. Materials

Cotton linter, with degree of polymerization (DP) of 350, was supplied by Longma Corporation (Jiangsu Province, China). All the chemical reagents were of analytic grade and used without further purification.

2.2. Preparation of cellulose fibers

Dried cotton linter pulp was dispersed into NaOH/thiourea/urea (9:6.5:6) aqueous solution at room temperature. This mixture was vigorously stirred for 5 min to form a cellulose suspension. Then the suspension was immersed immediately into a co-rotating twin-screw extruder (L/D = 40) precooled at -2 ℃. Subsequently, the complete dissolution occurred within about 10 s. Finally, the transparent cellulose extrudate was filtered and further deaerated at room temperature before wet spinning.

Wet-spinning equipment was used to produce regenerated cellulose fibers at room temperature (Figure 1). The cellulose spinning dope was extruded through a spinneret (the diameter is 0.08 mm) under a pressure of 0.1 MPa. Thereafter regenerated cellulose fibers was obtained by a multi-roller drawing bath: the coagulation bath (7.6 wt% H_2SO_4/15.2 wt% Na_2SO_4/1.25 wt% $ZnSO_4$ aqueous solution, 15 ℃), second bath (water, 45 ℃) and third bath (water, 98 ℃), respectively. All samples were dried to constant weight before tests under vacuum at 60℃.

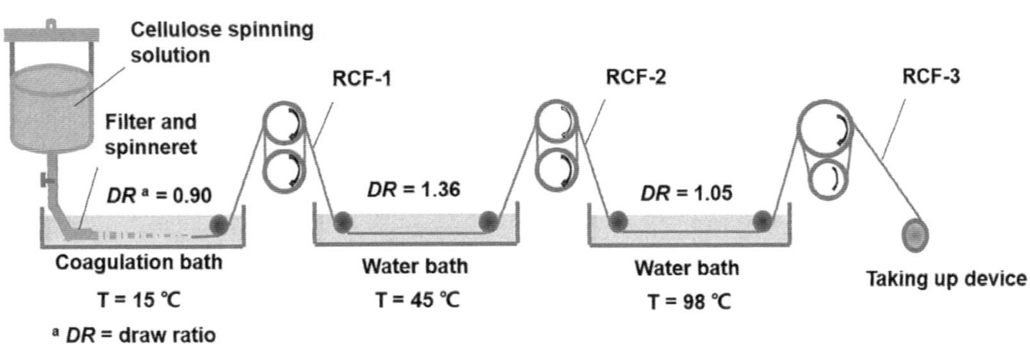

Figure 1. Schematic diagram of the wet-spun process of regenerated cellulose fibers.

2.3. Characterization

A universal tensile tester (XQ-1C, Donghua University, China) was used to calculate the tensile strength of the dried fibers with a gauge length of 20 mm according to GBT14337. At least 15 samples were tested for each sample and the tensile strength was calculated averagely. ATR-IR spectra were recorded by using a Nicolet 6700 FT-IR spectrometer (Thermo Fisher Scientific, America). Sixteen scans were averaged to obtain each spectrum at a resolution of 4 cm^{-1}. The spectra were subject to ATR correction, smoothing and baseline correction and being normalized. The two-dimensional correlation spectroscopy (2DCOS) was performed using the SpectraCorr software (Thermo Electron Corporation, 2004).

3. Results and discussion

3.1. Spectral behaviour of regenerated cellulose fibers

It is well known that the Fourier transform infrared spectroscopy technique was one of the well-defined tools to study the structure of polymers[15-16]. For cellulose-based polymers, a comparative study was applied to characterise the structural changes of cellulose in the 1500-850 cm^{-1} region by O'Connor and co-worker[17]. Figure 2 shows the infrared spectra characteristic of the cellulose pulp and regenerated cellulose fibers at different spinning stages in the region of 4000-600 cm^{-1}. The structural changes can be identified by the main infrared spectral differences of the cellulose pulp and regenerated cellulose fibers. The 896, 1107, 1161 and 1430 cm^{-1} absorption bands, which are assigned to group C_1 frequency, ring asymmetric stretching, C-O-C asymmetric stretching, and CH_2 symmetric bending vibration in cellulose I respectively, can be study the type of crystallinity changes[18]. These bands in the crystalline cellulose I spectrum disappear from the spectrum of the regenerated cellulose fibers, which indicates crystalline cellulose II, with amorphous cellulose, occupies the main content in the regenerated cellulose fibers. Furthermore, the 1261, 1313 and 1366 cm^{-1} absorption bands, which are assigned to CH bending vibration in cellulose II, CH_2 wagging vibration in cellulose II, and CH bending vibration in cellulose I and cellulose II respectively, show a predominance of crystalline cellulose II in the regenerated cellulose fibers[19]. Moreover, the 893 cm^{-1} absorption band also corroborate crystalline cellulose I can be almost negligible in the regenerated cellulose fibers. Owing to the 893 cm^{-1} is assigned to group C_1 frequency in cellulose II[20].

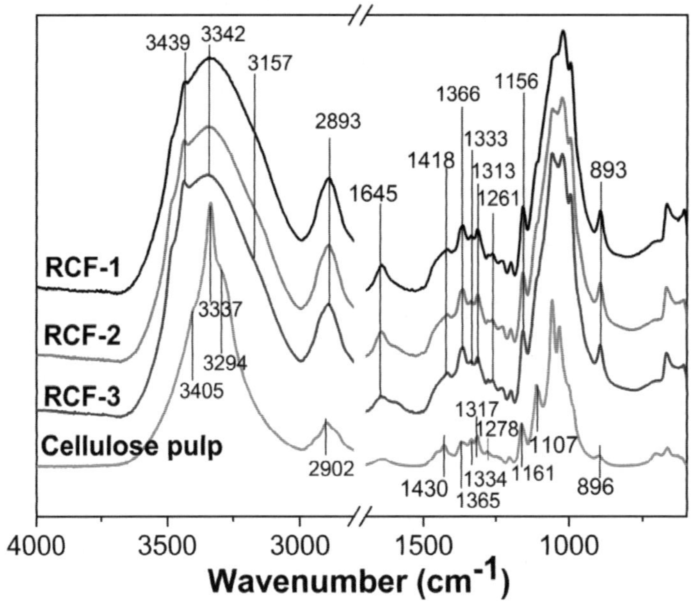

Figure 2. Comparative ATR-IR spectra of cellulose pulp and regenerated cellulose fibers at different spinning stages.

3.2. 2D correlation spectroscopy analysis of regenerated cellulose fibers

The generalized two-dimensional correlation spectroscopy (2DCOS), which is applied to represent spectroscopic intensity fluctuations, proposed by Noda[21]. The information of the spectral intensity changes, as well as the improvement of the apparent spectral resolution can be analysed by the 2DCOS. As shown in figure 2, the 3439, 3342, and 3175 cm^{-1} absorption band (–OH stretching hydrogen bonds) in the spectra of the regenerated cellulose fibers are typically presented in the crystalline cellulose II but not observed in the crystalline I spectra[20]. Synchronous 2DCOS in the 3600 - 3100 cm^{-1} (the O-H stretching vibration regions) of four samples are shown in figure 3. The intensities of bands in the region 3600 – 3100 cm^{-1} of the regenerated cellulose fibers decrease, while the peak area variations increase as compared to the cellulose pulp. This results indicate that the hydrogen bond interaction reduce and the broad peaks occur, leading to the lower crystallinity in the regenerated cellulose fibers than the cellulose pulp. Moreover, the changes in the peak shape and position of intermolecular and intramolecular hydrogen bonds demonstrate the regenerated cellulose fibers are changed into crystalline cellulose II during the regeneration process. Furthermore, the intensities of the whole broad O-H stretching vibration regions in the regenerated cellulose fibers increase, resulting in the crystallinity increase along the spinning line. Figure 4 shows synchronous 2DCOS in the 910 - 800 cm^{-1} of four samples. The group C_1 frequency vibration, 896 cm^{-1} in crystalline cellulose I, shift to 893 cm^{-1} in cellulose II. Meanwhile, the peak area and intensity variations in the regenerated cellulose fibers increase along the spinning line. This peak behavior is in well agreement with the above conclusions mentioned in the 3600 – 3100 cm^{-1} absorption band.

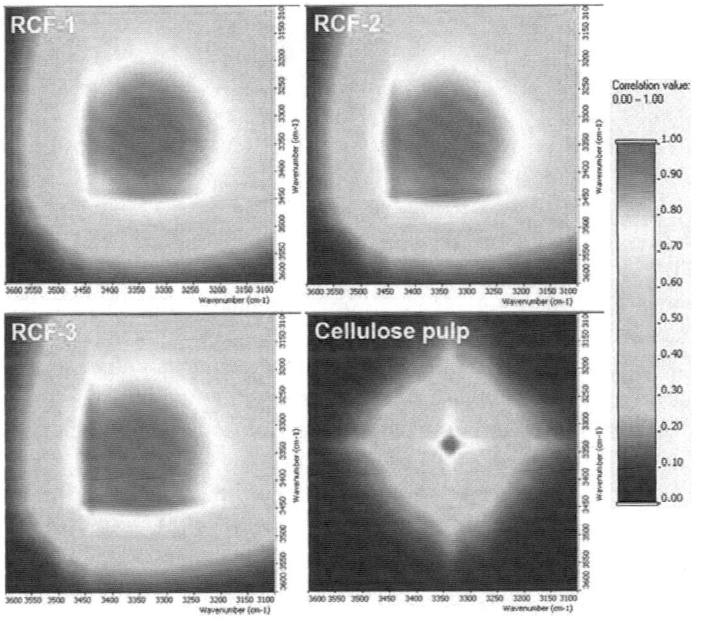

Figure 3. Synchronous 2DCOS in the range 3600 – 3100 cm $^{-1}$ of cellulose pulp and regenerated cellulose fibers at different spinning stages.

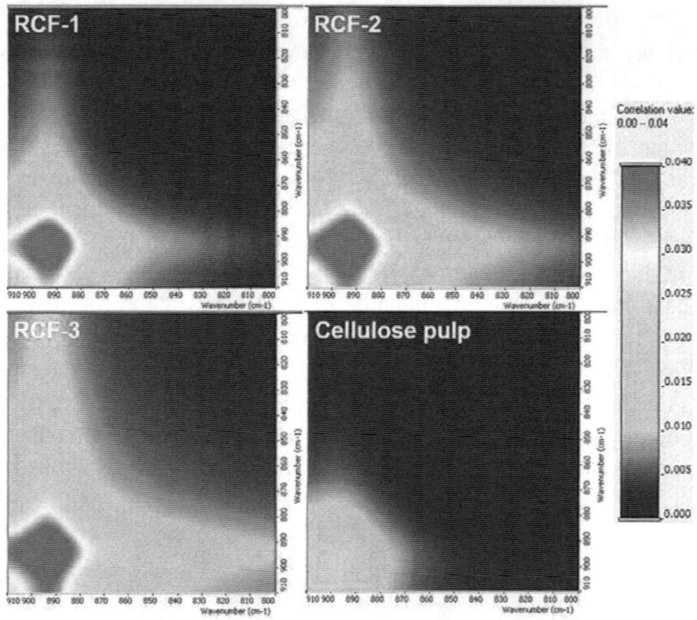

Figure 4. Synchronous 2DCOS in the range 910 – 800 cm^{-1} of cellulose pulp and regenerated cellulose fibers at different spinning stages.

3.3. Crystallinity analysis

The crystallinity FTIR index is a significant and well defined tool to analyse the crystallinity of cellulose-based polymers, because the index is proportional to the degree of crystallization[20]. Consequently, the absorbance ratios A_{1366}/A_{2902} and A_{1366}/A_{2893} focus on determining the total crystalline index (TCI) of the cellulose pulp and regenerated cellulose pulp respectively. The results of the cellulose pulp and regenerated cellulose fibers at different spinning stages are shown in Table 1. Meanwhile, Figure 5 shows the TCI of the regenerated cellulose fibers, and the tensile strength is also depicted for comparison. The tensile strength increases as the TCI increases. Furthermore, the sample RCF-3 has the highest TCI and tensile strength, where the values are 0.54 and 2.30 cN/dtex, respectively. This phenomenon corroborates that the multi-roller drawing spinning process plays an important role in crystal structure and tensile strength of the regenerated cellulose fibers, and the crystal structure appears to have a significant impact on the tensile strength.

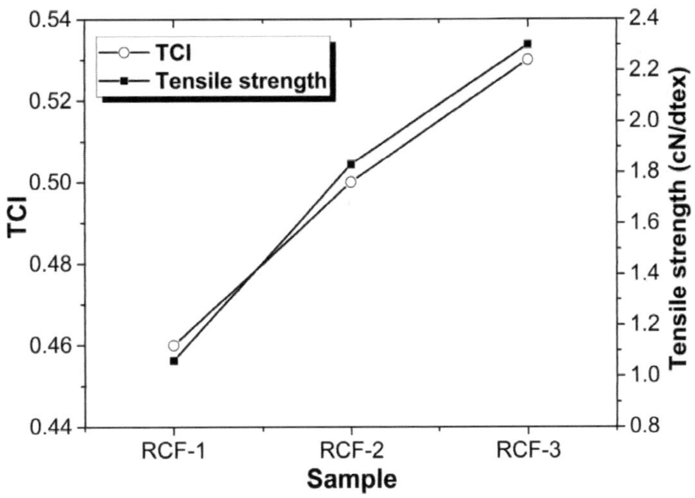

Figure 5. TCI of regenerated cellulose fibers at different spinning stages.

Table 1. Total crystalline index (TCI) and tensile strength of cellulose pulp and regenerated cellulose fibers.

Samples	A_{1366}	A_{2893}[a]/A_{2902}[b]	TCI	Tensile strength (cN/dtex)
RCF-1	0.243	0.532	0.46	1.06
RCF-2	0.267	0.530	0.50	1.83
RCF-3	0.268	0.500	054	2.30
Cellulose pulp	0.096	0.166	0.58	

[a] A_{2893} is the absorbance of 2893 cm^{-1} band from regenerated cellulose fibers.
[b] A_{2902} is the absorbacne of 2902 cm^{-1} band from cellulose pulp.

4. Conclusions

Attenuated total reflection infrared (ATR-IR) spectroscopy and the two-dimensional correlation spectroscopy (2DCOS) were applied to characterize the regenerated cellulose fibers. The qualitative analysis of the ATR-IR spectra showed that the regenerated cellulose fibers was mainly composed of crystalline cellulose II and amorphous cellulose, as the characteristic spectral bands at 1430, 1161, 1107 cm^{-1} were not found for the regenerated cellulose fibers. Moreover, the 893 cm^{-1} absorption band, assigned to group C_1 frequency in cellulose II, also corroborate crystalline cellulose I can be negligible in the regenerated cellulose fibers.

The total crystalline index (TCI) results indicated the crystallinity decreased when cellulose pulp was changed into regenerated cellulose fibers. However, the increase in TCI was accompanied by an increase in the tensile strength of the regenerated cellulose fibers along the spinning line. Comparative analysis of the 2DCOS in the range 3600 – 3100 cm^{-1} and 910 – 800 cm^{-1} showed a significant change in the peak area, shape, position and intensity between the cellulose pulp and regenerated cellulose fibers. This behaviour and the trend of TCI values were consistent with the ATR-IR analysis results. Therefore these results corroborated the staged spinning process were in favor of obtaining improved crystal structure, as well as the tensile strength of the regenerated cellulose fibers prepared by the described extrusion dissolution method.

Acknowledgments
This work was supported by National Natural Science Foundation of China (51503032) and Shanghai Sailing Program (14YF1405200)

References

[1] Klemm D, Heublein B, Fink H P and Bohn A 2005 *Angew. Chem. Int. Ed.* **44** 3358-93
[2] Fink H P, Weigel P and Purz H J 2001 *Prog. Polym. Sci.* **26** 1473-524
[3] Reddy N and Yang Y Q 2015 *Innovative Biofibers from Renewable Resources* (German: Springer Berlin Heidelberg)
[4] Yamane C, Mori M, Saito M and Okajima K 1996 *Polym. J.* **28** 1039-47
[5] Budtova T and Navard P 2016 *Cellulose* **23** 5-55
[6] Fu F Y, Zhou J P, Zhou X M, Zhang L N, Li D X and Kondo T 2014 *ACS Sustainable Chemistry& Engineering* **2** 2363-70
[7] Vehviläinen M, Kamppuri T, Grönqvist S, Rissanen M, Maloney T, Honkanen M and Nousiainen P 2015 *Cellulose* **22** 1653-74
[8] Chen X M, Burger C, Fang D F, Ruan D, Zhang L N, Hsiao B S and Chu B 2006 *Polymer* **47** 2839-48
[9] Ruan D, Zhang L N, Lue A, Zhou J P, Chen H, Chen X M, Chu B and Kondo T 2006 *Macromol. Rapid Commun.* **27** 1495-500
[10] Cai J, Liu Y T and Zhang L N 2007 *Adv. Mater.* **19** 821–5
[11] Luo X G and Zhang L N 2013 *Food Research International* **52** 387-400
[12] Zhang S, Li F X, Yu J Y and Gu L X 2009 *Fibers and Polymers* **10** 34-9
[13] Wang W C, Li F X, Yu J Y, Navard P and Budtova T 2015 *Cellulose* **22** 1333-45
[14] Yang Y P, Zhang Y, Dawelbeit A M and Yu M H 2016 *Proc. Int. Conf. on Polymer and Composite Materials (Hangzhou, China)* vol **137** (UK: IOP Publishing)
[15] Yang H, Yan R, Chen H, Dong H L and Zheng C 2007 *Fuel* **86** 1781-8
[16] Kumar R and Sekhon S S 2013 *Ionics* **19** 1627-35
[17] Nelson M L and O'Connor R T 1964 *J. Appl. Polym. Sci.* **8** 1311-24
[18] Nelson M L and O'Connor R T 1964 *J. Appl. Polym. Sci.* **8** 1325-41
[19] Marchessault R H and Liang C Y 1960 *J. Polym. Sci., Part A: Polym. Chem.* **43** 71-84
[20] Carrillo F, Colom X, Suñol J J and Saurina J 2004 *Eur. Polym. J.* **0** 2229-34
[21] Noda I 1989 *J. Amer. Chem. Soc.* **111** 8116-18

IOP Conf. Series: Materials Science and Engineering **213** (2017) 012040 doi:10.1088/1757-899X/213/1/012040

Preparation and characterizations of flame retardant polyamide 66 fiber

Y Y Li[1], K Liu[1] and R Xiao[1,2]

[1]State Key Laboratory for Modification of Chemical Fibers and Polymer Materials, College of Materials Science and Engineering, Donghua University, Shanghai 201620, People's Republic of China

Email: xiaoru@dhu.edu.cn

Abstract. The polyamide 66 (PA66) is one of the most important thermoplastic materials, but it has the drawback of flammability. So the flame retardant PA66 was prepared by condensation polymerization using nylon salt and DOPO-based flame retardant in this paper. Then the flame retardant PA66 fiber was manufactured via melt spinning. The properties of flame retardant PA66 and flame retardant PA66 fiber were investigated by relative viscosity, differential scanning calorimetry (DSC), tensile test, vertical burning test (UL94) and limiting oxygen index (LOI) test. Although the loading of the DOPO-based flame retardant decreased the molecular weight, the melting temperature, the crystallinity and the mechanical properties of flame retardant PA66, the flame retardancy properties improved. The flame retardant PA66 loaded with 5.5 wt% of DOPO-based flame retardant can achieve a UL94 V-0 rating with a LOI value of 32.9%. The tenacity at break decreased from 4.51 $cN \cdot dtex^{-1}$ for PA66 fiber to 2.82 $cN \cdot dtex^{-1}$ for flame retardant PA66 fiber which still satisfied the requirements for fabrics. The flame retardant PA66 fiber expanded the application of PA66 materials which had a broad developing prospect.

1. Introduction

Polyamide 66 (PA66) is one of the thermoplastic materials widely used in manufacture of plastics, films, fibers and textiles due to its excellent mechanical properties and chemical resistance[1-5]. However, the limiting oxygen index (LOI) is 24.2% which limits its applications. Thus the flame retardancy has become a necessary requirement for PA66[6-8]. DOPO-based flame retardant is a kind of environmentally friendly flame retardant containing the phosphorus linking pendent group. There have several literatures focused on the DOPO-based flame retardant polyesters and epoxy because of the efficient flame retardancy and low toxic gas evolution during combustion[9-11]. Wang et al[9] prepared 9,10-dihydro-10[2,3-di(hydroxycarbonyl)propyl]-10-phosphaphenanthrene-10-oxide (DDP) and nano α-ZrP flame retardant polyester. The LOI improved from 21.2% to 32.6%, which indicated that the fire retardant performance was significantly improved. However, there hardly any literatures focused on the DOPO-based flame retardant PA66 and PA66 fibers.

In this paper, flame retardant PA66 was prepared by condensation polymerization of nylon salt and DOPO-based flame retardant. Then the flame retardant PA66 fiber was prepared by melt spinning. The properties and characterizations of flame retardant PA66 and flame retardant PA66 fiber were investigated by relative viscosity, differential scanning calorimetry (DSC), tensile test, vertical burning test (UL94) and LOI test.

2. Experimental details

2.1. Preparation of flame retardant PA66 and flame retardant PA66 fiber

Flame retardant PA66 samples were prepared in a GSH-5L polymerization autoclave. At the start of the polymerization, 1500 g nylon salt, 82.5 g DOPO-based flame retardant, 27.5 g hexamethylene diamine (HMDA) and 1000 g deionized water were charged into the autoclave. Then the mixture was heated to 210°C and the pressure was kept at 1.75 MPa. Then the reactor was gradually decompressed to atmosphere pressure and the temperature was heated to 270°C. Then the polymerization was kept at -0.1 MPa at 275°C. Finally, the product was removed from the autoclave and cut into chips.

The PA66 and flame retardant PA66 pellets were dried at 120°C for 48 h under vacuum before melt spinning process. The fibers were prepared using a spinner instrument equipped with a spinneret containing 36 orifices, whose diameter was 0.3 mm. The fibers were collected at a take-up velocity of 600 $m\cdot min^{-1}$. Finally, the fibers were drawn at a draw ratio of 3.5.

2.2. Characterizations

The relative viscosity was measured at a concentration of 0.01 $g\cdot mL^{-1}$ in H_2SO_4 with an Ubbelohde viscometer at 20°C. The DSC measurements were carried out on a TA Q20 instrument. The samples were heated to 320°C at 40 $°C\cdot min^{-1}$ and remaining at this temperature for 5 min to eliminate the thermal prehistory. Then the samples were cooled to 100°C with a rate of 10 $°C\cdot min^{-1}$ and heated again to 320°C at the same rate. The vertical burning test was conducted according to ASTM D3801-2010 and the LOI test was measured according to ASTM D-2863. The mechanical properties of flame retardant PA66 fiber were carried out at room temperature with a drawing speed of 40 $mm\cdot min^{-1}$ and a clamping distance of 20 mm. Orientation tests were taken using a sound velocity measuring instrument under a tension of 0.1 $g\cdot dtex^{-1}$ at the speed of sound propagation of 20 cm and 40 cm, respectively.

3. Results and discussion

3.1. Molecular weight

It can be seen from Table 1 that the addition of DOPO-based flame retardant had an effect of lowering the molecular weight. The molecular weight decreased from 1.87×10^4 $g\cdot mol^{-1}$ for PA66 to 0.92×10^4 $g\cdot mol^{-1}$ for flame retardant PA66. The reason was the incorporation of DOPO-based flame retardant held back the polymerization of PA66 chains and the relative viscosity decreased.

Table 1. The molecular weight of PA66 and flame retardant PA66.

Samples	$\overline{M_\eta}$ ($g\cdot mol^{-1}$)	Relative viscosity
PA66	1.87×10^4	2.80
flame retardant PA66	0.92×10^4	2.12

3.2. Thermal transitions and crystallinity

In order to study the thermal transitions and crystallinity of PA66 and flame retardant PA66, DSC measurement was carried out. The DSC heating curves and cooling curves of PA66 and flame retardant PA66 are shown in Figure 1. The related DSC results are summarized in Table 2. According to the DSC curves, the melting temperature (T_m) of flame retardant PA66 was lower than that of PA66. The crystallization temperature (T_c) of flame retardant PA66 was detected at 208.59°C which was also lower than that of PA66. Besides, the crystallinity decreased from 33.70% for PA66 to 23.73% for flame retardant PA66. These phenomena were due to the incorporation of the DOPO-based flame retardant to the polymerization system of nylon salt. The DOPO-based flame retardant destroyed the regularity of PA66 main chain, decreased the molecular weight of PA66 and disturbed the

IOP Conf. Series: Materials Science and Engineering **213** (2017) 012040 doi:10.1088/1757-899X/213/1/012040

intermolecular interaction of PA66, which caused the reduction of the intermolecular interaction between PA66 molecules and resulting in a lowering of T_m, T_c and the crystallinity[12].

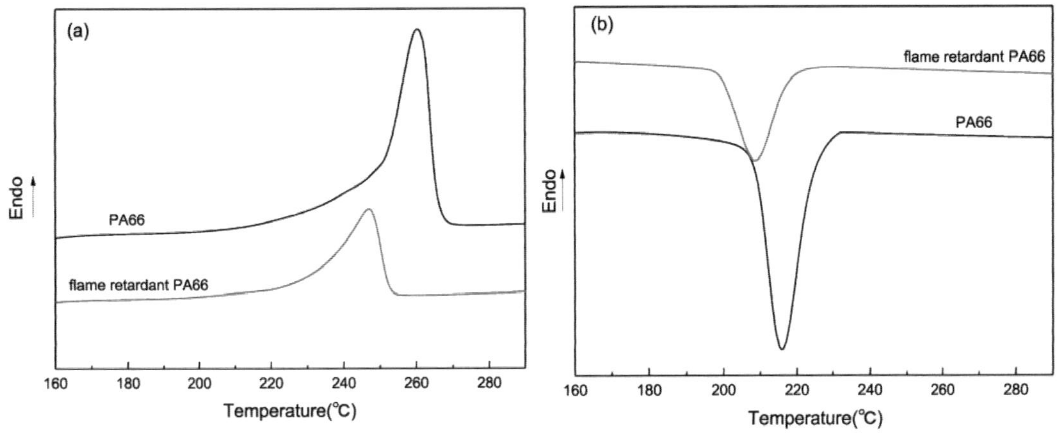

Figure 1. DSC heating curves (a) and cooling curves (b) of PA66 and flame retardant PA66.

Table 2. The DSC results of PA66 and flame retardant PA66.

Samples	T_m (°C)	ΔH_m (J·g^{-1})	T_c (°C)	Crystallinity (%)
PA66	260.43	65.71	215.87	33.70
flame retardant PA66	247.01	46.27	208.59	23.73

3.3. Flame retardancy properties

The vertical burning test and LOI test are widely accepted as important tests for determining flammability. It can be observed that PA66 showed bad flame retardancy properties. There existed many melt drippings during combustion which can burn the cotton and achieved a V-2 rating in the vertical burning test. As the incorporation of the DOPO-based flame retardant, the self-extinguishing time after ignition and the melt drippings of flame retardant PA66 decreased obviously. The vertical burning test result of flame retardant PA66 with 5.5 wt% of additives achieved a V-0 rating because it showed hardly any melt drippings and the cotton couldn't be ignited. The improvement of the flame retardancy of flame retardant PA66 can also be observed from the LOI test. The LOI value of flame retardant PA66 increased from 24.2% to 32.9%, which was higher than 28% and reached the standard "hard to burn". It was worth nothing that DOPO-based flame retardant made a great contribution to improve the flame retardancy properties of PA66.

Table 3. The vertical burning test and LOI test results of PA66 and flame retardant PA66.

Samples	Vertical burning test		LOI value (%)
	Whether the cotton be ignited	Rating	
PA66	Yes	V-2	24.2
flame retardant PA66	No	V-0	32.9

3.4. Mechanical properties of flame retardant PA66 fiber

As can be seen in Table 4, both the tenacity at break and elongation at break of flame retardant PA66 fiber decreased compared with PA66 fiber, which was inevitable in obtaining flame retardant fibers.

When the content of flame retardant was 5.5 wt%, the strength of the flame retardant PA66 fiber still attained 2.82 $cN\cdot dtex^{-1}$, which satisfied the requirements for fabrics. The decrease of the mechanical properties of flame retardant PA66 fiber was due to the decrease of the molecular weight and the crystallinity[13].

Table 4. The mechanical properties of PA66 fiber and flame retardant PA66 fiber.

Samples	Tenacity at break ($cN\cdot dtex^{-1}$)	Elongation at break (%)
PA66 fiber	4.51	51.37
flame retardant PA66 fiber	2.82	33.04

3.5. Orientation of flame retardant PA66 fiber
The degree of orientation was represented by orientation factor, which reflects the level of molecular chains arranged parallel to their direction of the preferred orientation unit for reference. Compared with the PA66 fiber, the sound velocity, orientation factor and modulus of flame retardant PA66 fiber decreased. The reason is mainly due to the incorporation of DOPO-based flame retardant, which lowered the molecular weight and the crystallinity, resulting in an attenuation of the sound waves. In addition, the regularity of the PA66 chains was broken up by the presence of the flame retardant, which went against to the axial alignment of the macromolecules in the fibers and affected the propagation velocity of sound waves between PA66 chains, resulting in the decrease of sound velocity, orientation factor and modulus of flame retardant PA66 fiber.

Table 5. The orientation test results of PA66 fiber and flame retardant PA66 fiber.

Samples	Sound velocity ($km\cdot s^{-1}$)	Orientation factor	Modulus ($gf\cdot d^{-1}$)
PA66 fiber	2.29	0.66	59.26
flame retardant PA66 fiber	2.12	0.60	50.79

4. Conclusions
The flame retardant PA66 and flame retardant PA66 fiber was prepared by condensation polymerization using nylon salt. Although the presence of DOPO-based flame retardant decreased the molecular weight, the melt temperature, the crystallinity and the mechanical properties of PA66, the flame retardancy properties improved. The flame retardant PA66 can achieve a V-0 rating according to the UL94 criterion with a LOI value of 32.9%, which showed that the flame retardant PA66 had relatively satisfactory flame retardancy. The breaking strength, elongation break, sound velocity, orientation factor and modulus of flame retardant PA66 fiber decreased compared to PA66 fiber. The tenacity at break of flame retardant PA66 fiber still attained 2.82 $cN\cdot dtex^{-1}$, which satisfied the requirements for fabrics.

Acknowledgements
This study was supported by Natural Science Foundation of China (No.21374015) and the Program of Introducing Talents of Discipline to Universities (No.111-2-04).

References
[1] Huang X, Li B, Shi B and Li L 2008 *Polymer* **49** 1049-55
[2] Bai Y, Huang L, Huang T, Long J and Zhou Y 2013 *Polymer* **54** 4171-6
[3] Peng J, Walsh P J, Sabo R C, Turng L S and Clemons C M 2016 *Polymer* **84** 158-66
[4] Xiao J, Tan Y, Song Y and Zheng Q 2016 *RSC Adv.* **6** 41392-403
[5] Sha K, Hu Y L, Wang Y H and Xiao R 2014 *Mater. Res. Innovations* **18** S4-843-S4-7
[6] Jimenez M, Duquesne S and Bourbigot S 2013 *Polym. Degrad. Stabil.* **98** 1378-88
[7] Naik A D, Fontaine G, Samyn F, Delva X, Louisy J, Bellayer S, Bourgeois Y and Bourbigot S

2014 *RSC Adv.* **4** 18406-18
[8] Zhan Z, Xu M and Li B 2015 *Polym. Degrad. Stabil.* **117** 66-74
[9] Wang D Y, Liu X Q, Wang J S, Wang Y Z, Stec A A and Hull T R 2009 *Polym. Degrad. Stabil.* **94** 544-9
[10] Hamciuc C, Serbezeanu D, Carja I D, Vlad-Bubulac T, Musteata V E, Forrat Pérez V, Guillem López C and López Buendia A M 2013 *J. Mater. Sci.* **48** 8520-9
[11] Liu Y L, Wu C S, Chiu Y S and Ho W H 2003 *J. Polym. Sci.: Polym. Chem.* **41** 2354-67
[12] Kumar R and Sekhon S S 2013 *Ionics.* **19** 1627-35
[13] Li Y, Lin Y, Sha K and Xiao R 2017 *Text. Res. J.* **87** 561-9

IOP Conf. Series: Materials Science and Engineering **213** (2017) 012041 doi:10.1088/1757-899X/213/1/012041

Improved adhesion performances of aramid fibers with vinyl epoxy via supercritical carbon dioxide modification

M L Qin[1], H J Kong[2], M H Yu[1,3] and C Q Teng[1,3]

[1]State Key Laboratory for Modification of Chemical Fibers and Polymer Materials, College of Materials Science and Engineering, Donghua University, Shanghai 201620, People's Republic of China
[2]School of Materials Engineer, Shanghai University of Engineering Science, Shanghai 201620, People's Republic of China

E-mail: Yumuhuo@dhu.edu.cn ; cqteng@dhu.edu.cn.

Abstract. In this paper, aramid fibers were treated under supercritical carbon dioxide ($SCCO_2$) with isocyanate terminated liquid nitrile rubber to improve the adhesion performances of vinyl epoxy composites. The interfacial shear strength (IFSS) of vinyl epoxy composites was investigated by micro-bond test. The results indicate that the surface modification of aramid fibers in $SCCO_2$ was an efficient method to increase the adhesion performances between fibers and vinyl epoxy. Fourier transform infrared spectroscopy (FTIR), X-ray photoelectron spectroscopy (XPS), and scanning electron microscopy (SEM) were adopted to investigate the surface structure and composition of aramid fibers. The flexural strength and interlaminar shear strength (ILSS) of treated aramid fibers/vinyl epoxy composites was improved by 18.1% and 28.9% compared with untreated aramid fibers, respectively. Furthermore, the fractured surfaces of the composites were observed by SEM, which showed that the interfacial adhesion of composites has been remarkably changed.

1. Introduction

Aramid fiber (AF) has triggered tremendous attention due to its unique characteristics, such as high tensile and modulus strength, low density and high thermal resistance[1-3], and it can be used as reinforced materials of advanced composites in a wide range of industrial fields such as aerospace, sports, automobiles and aviation [4-8]. However, due to the poor adhesion performances between aramid fibers and matrix, the application of the composites materials was limited[9]. Numerous efforts have been devoted to improve the interfacial characteristics of the aramid fiber/matrix composites, such as chemical etching and grafting[10-14], plasma treatment[15-17], ultrasound[18] and polymer coating[19-22]. However, these methods require harsh reaction conditions and may destroy the tensile strength of the aramid fibers.

Supercritical carbon dioxide ($SCCO_2$) has been widely used in many fields arising from its low viscosity and negligible surface tension. In this context, examples include polymer processing, food, cosmetics and pharmaceuticals industries[23-26]. $SCCO_2$ is an effective method for surface modification of fibers, owing to its liquid and gas properties and liquid-like density[27]. For example, Zhao et al. employed the mixture of $SCCO_2$ and organometallic complex for impregnation of the aramid fibers to prepare electro conductive composite fabrics[28]. Zheng and co-workers systematically investigated the effects of the $SCCO_2$ fluid on the structures and properties of the PMIA fiber [29-30]. Our previous work also demonstrated the improved interfacial shear strength of the PPTA fiber-based epoxy composites by modification of the fibers in $SCCO_2$ with Hexamethylene Diisocyanate[7]. However, this method can only improve the interfacial adhesion toward epoxy. In order to extend the scope of this strategy to aramid fiber for other resins composites, it requires one to introduce a new type of isocyanate in $SCCO_2$ condition, which can enhance the interfacial adhesion of composites based on various resins such as epoxy, rubber and vinyl epoxy resin.

In this study, we introduced a long flexible chain segment with isocyanate terminated liquid nitrile rubber in $SCCO_2$, which can improve interfacial adhesion of the vinyl epoxy resin. Isocyanate terminated liquid nitrile rubber, on the one hand, contains CH=CH group that can react with the vinyl group of the vinyl epoxy resin under curing agents to improve interfacial adhesion of vinyl epoxy composites, on the other hand, can improve interfacial adhesion of the epoxy composites in $SCCO_2$. Based on these points, isocyanate terminated liquid nitrile rubber was firstly adopted to treat the surface of aramid fibers in $SCCO_2$ for improving the interfacial adhesion of the vinyl epoxy resin. The micro-debond tests was utilized to characterize the IFSS of the aramid fibers with vinyl epoxy. The effects of treatment pressure, time, temperature and concentration on the IFSS in $SCCO_2$ with isocyanate terminated liquid nitrile rubber was discussed, and the results indicated that the IFSS was improved by 44.4% under the optimum treatment conditions. The chemical microstructures and compositions of the aramid fiber were investigated by FT-IR and XPS. The morphology of the aramid fibers was investigated by scanning electron microscopy (SEM). To illustrate the effect of the mechanical properties of composites, the ILSS were also discussed. In additional, SEM was carried out to illustrate the tensile fractured surfaces of composites.

2. Experiment

2.1. Materials

Isocyanate terminated liquid nitrile rubber was synthesized in our lab, carbon dioxide (CO_2) gas (99.99% pure) was supplied by Shanghai Chenggong Gases Co., Ltd. Acetone with analytical regent grade was purchased from Sinopharm Chemicl Reagent Co., Ltd. The aramid fibers (K49) with a diameter of 15 μm were supplied by DuPont Co., Ltd. The fibers were extracted with acetone overnight and dried at 60 °C for 6 h before use. Vinyl epoxy resin (MFE-2) was purchased from Sino Polymer Co., Ltd, China. Cobalt (II) naphthenate was bought from Sinopharm Chemicl Reagent Co., Ltd. 2-Butanone peroxide was purchased from Shanghai Aladdin biological Co., Ltd. The ratio of the vinyl epoxy resin and curing agent is100:1.5:1 (resin: 2-Butanone: cobalt (II) naphthenate).

2.2. Treatment of aramid fibers in $SCCO_2$

The treatment of aramid fibers in $SCCO_2$ was demonstrated in previous studies [7]. Briefly, the aramid fibers were purified with acetone and washed with distilled water at room temperature, and then dried in a vacuum oven at 100 °C for 12 h to remove residues. The pretreated aramid fibers were added in isocyanate terminated liquid nitrile rubber-acetone solution by a stainless steel filter. Then CO_2 was pressurized through a high-pressure pump at a desired temperature for several hours. Finally, the treated aramid fibers were obtained after washing by acetone for several times and dried at 80 °C in a vacuum oven for overnight.

2.3. Preparation of epoxy resin composites

The vinyl epoxy resin and curing agent were added prior to hand mixing for approximately 20 min. The vinyl epoxy resin and curing agent were mixed and degassed by a vacuum pump. Next, the resin mixture was poured into the mold, followed by adding pretreated aramid fibers. Then aramid fabric was stacked in the mold cavity. The samples were precured at 80 °C for 2 h and then postcured at 130 °C for 3 h under a certain pressure. Then the cured specimens were slowly cooled to room temperature.

2.4. Characterization

An Infrared spectrometer (Nicolet 8700) was used to measure the change of the surface structure of the aramid fibers at room temperature ranging from 4000 to 500 cm^{-1}.The surface morphology of the aramid fibers and tensile fractured surfaces of the composites were analyzed by scanning electron microscopy (FE-SEM, Su8010, Hitachi, Japan). The surface chemical composition of the aramid fibers was performed by X-ray photoelectron spectroscopy (XPS) on a RBD upgrade PHI-5000C ESCA system (USA). The XPS spectra was provided by Mg Kα radiation (h=1253.6 eV) source with power of 250 w, high voltage of 14.0 kV at a detection angle of 54 °. The IFSS of the fiber/epoxy composites was evaluated by micro-bond technique, the epoxy resin mixture was dropped on the aramid fibers, and the microdroplet resin were formed on the monofilament fiber then the specimens were cured at 120 °C in a vacuum oven for 6 h. The micro-bond tests were measured on an XQ-1 fiber tensile testing machine with an upper clamp displacement rate of 0.1 mm/min. The IFSS was calculated by the following equation:

$$IFSS = F/\Pi dL \quad (1)$$

where F is the maximum pullout force, d is the fiber diameter, and L is the microdroplet embedded length of the aramid fibers in vinyl epoxy resin. The average value of the sample showed was an average of 30 tests.The flexural strength and ILSS of the aramid fiber/vinyl epoxy composites were measured at room temperature using a LD24 microcomputer control electronic testing machine (Shenzhen, China). The flexural test was performed on a three point bending using ASTM D790 system, while ILSS was measured using ASTM D2344 system. More than five parallel samples were measured for each test.

3. Result and discussion

3.1. Interfacial adhesion of vinyl epoxy resin to aramid fiber

Figure 1 shows the IFSS of vinyl epoxy resin to aramid fibers under different treatment conditions. Figure 1a demonstrates that the IFSS increases firstly and then decreases with the increase of treatment concentration under the treatment conditions at 100 °C, 13 MPa and 60 min. It can be observed that the optimum concentration was 0.15%. The best value of the IFSS was acquired when the treatment time was 60 min when the treatment temperature was 100 °C, pressure was 13 MPa and concentration was 0.15% (Figure 1b). As shown in Figure 1c, the IFSS of the epoxy resin to aramid fiber increases to maximum at 100 °C, and then decreases when temperature higher than 100 °C. Meanwhile, it can be seen from Figure 1d that the IFSS displays the same tendency with Figure 1c. From the above results, one can conclude that the optimum processing conditions were achieved with treatment temperature of 100 °C, treatment time of 60 min, pressure of 13 MPa and concentration of 0.15%. The IFSS was improved by 44.4% under the optimum treatment conditions. In brief, the adhesion properties of vinyl epoxy resin to aramid fibers were obviously improved after $SCCO_2$ treatment with isocyanate terminated liquid nitrile rubber. This reveals that $SCCO_2$ treatment with isocyanate terminated liquid nitrile rubber was an effective approach to improve the interfacial properties. To further illustrate this reason, we propose a possible reaction mechanism, as shown in Figure 2. The reason was that CH=CH group would undergo a cross-linking reaction with CH=CH of vinyl epoxy resin in the presence of curing agents. In addition, the as-generated $-NH_2$ groups can react with epoxy to improve the interfacial adhesion of the epoxy resin. The related results were reported in our previous work[7].

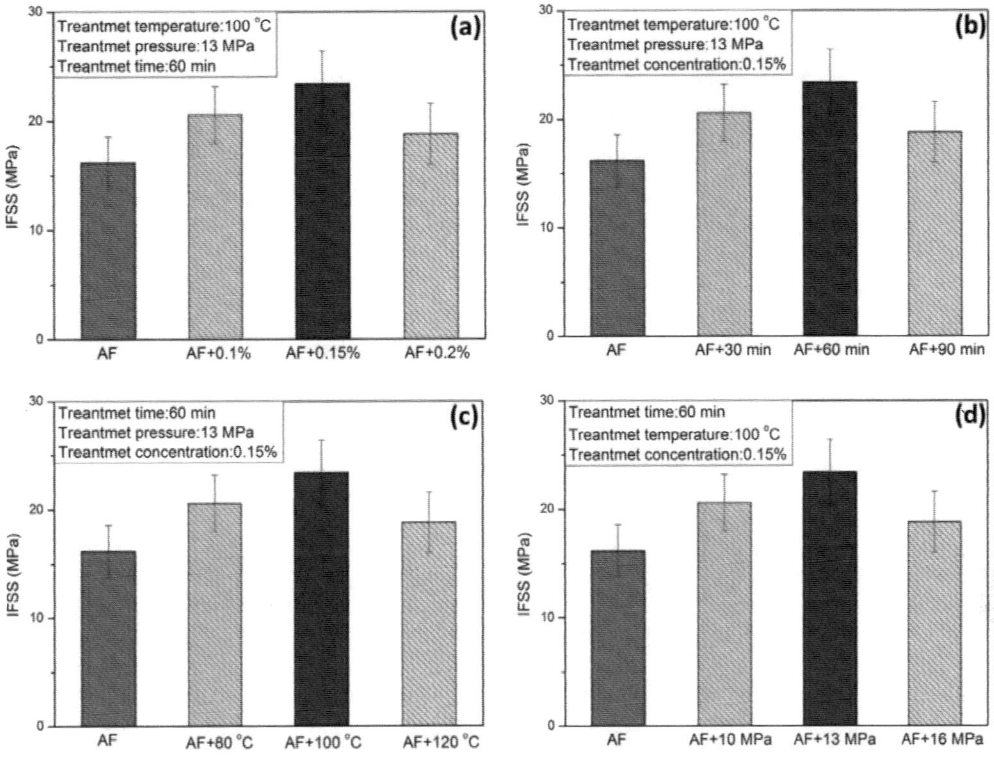

Figure 1. IFSS of aramid fibers/vinyl epoxy composites at different treatment conditions: concentration (a), time (b), temperature (c) and pressure (d).

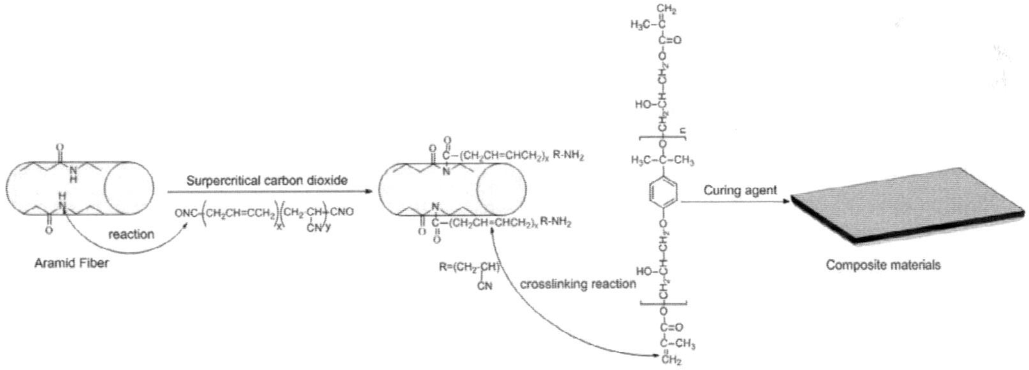

Figure 2. The proposed mechanism of the reaction between aramid fibers and vinyl epoxy resin.

3.2. Surface chemical structure of the modified aramid fibers

FTIR was employed to investigate the chemical structures of the aramid fibers before and after treatment in $SCCO_2$ condition with isocyanate terminated liquid nitrile rubber. As shown in Figure 3a, the characteristic absorption bands at 3320 cm^{-1} were assigned to hydrogen bonded N-H band, 1645 cm^{-1} was originated from amide I, and 1542 cm^{-1} was assigned to amide II, respectively. Compared to those of untreated aramid fiber, new absorption bands at 2919, 2847 and 1450 cm^{-1} were attributed to CH_2 asymmetric stretch vibration, CH_2 symmetric stretch vibration and CH stretching vibration plane, respectively. Furthermore, the absorption bands at 2284, 2238 and 967 cm^{-1} can be assigned to CNO, CN and CH=CH stretch vibration. After treatment with isocyanate terminated liquid nitrile rubber in

$SCCO_2$, the intensity of the characteristic bands at 1511 cm^{-1} became stronger (Figure 3b), this shift may be attributed to the combination of CO-NHR. The results showed that new groups (NH_2) were formed during this treatment, which is favorable to produce reaction with epoxy resin. It is possibly that–CNO groups can react with $-NH_2$, -NHCO and –COOH groups in aramid fibers to form a certain degree of grafting and cross-linking[31].

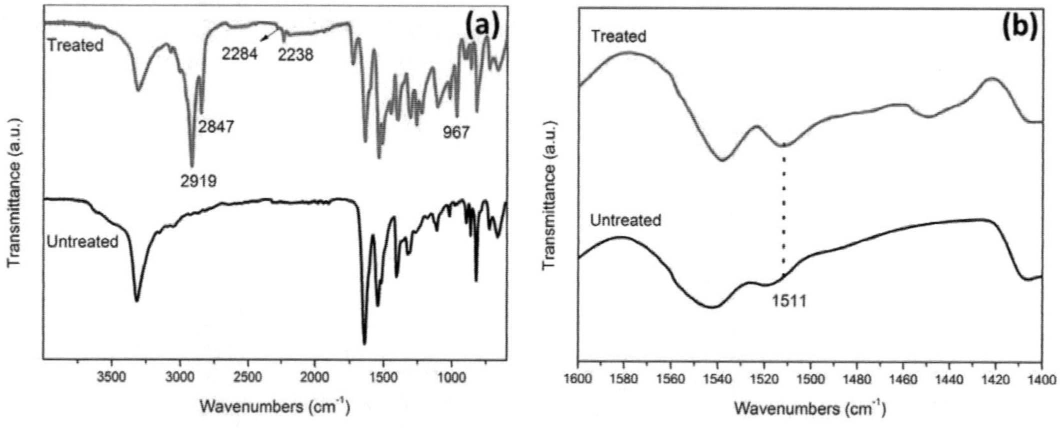

Figure 3. FTIR spectrum of aramid fibers before and after treatment in $SCCO_2$ with isocyanate terminated liquid nitrile rubber with wavenumber ranging from 600-4000 cm^{-1} (a) and 1400-1600 cm^{-1} (b).

3.3. Surface chemical composition of the treated aramid fibers

The surface chemical composition of the aramid fibers was observed by XPS measurements. It can be seen from Table 1 that after treated in $SCCO_2$, the atomic concentration of oxygen (O) and the ratio of oxygen to carbon (O/C) were slightly increased from 20.4 to 20.6 and 0.26 to 0.27, respectively. It is worth nothing that the atomic concentration of nitrogen (N) increased from 1.2% to 2.1% and the ratio of nitrogen to carbon (N/C) increased from 0.015 to 0.023. To investigate the change of the chemical interactions, as shown in Figure 4, the C 1s spectrum was deconvolved in four components, i.e., C-C (284.2 eV), C-N/C=O (285.5 eV), C=O (287.3 eV) and O-C=O (288.5 eV) groups. As a comparison, the same components in the original sample after the $SCCO_2$ treatment were also observed, with some slight changes on the fiber surface. As shown in Table 2, the content of C-C group decreased from 89.8% to 89.1%, while the C-O/C-N and C=O groups respectively increased from 1.4% to 2.1, and 3.3% to 4.4%. In addition, the O-C=O group decreased slightly after $SCCO_2$ treatment. These results revealed the presence of some functional groups formed on aramid fiber surface. The reason was that there were some polar groups could be introduced through reactions between CNO and the fiber surface after $SCCO_2$ treatment. It also can well explain the improved adhesive properties between aramid fiber and epoxy resin matrix.

Table 1. Surface elements of aramid fibers before and after treatment in $SCCO_2$ with isocyanate terminated liquid nitrile rubber.

	Atomic concentration (%)			Atomic ratio	
Samples	C	N	O	O/C	N/C
Untreated AF	78.4	1.2	20.4	0.26	0.015
Untreated AF	77.3	2.1	20.6	0.27	0.027

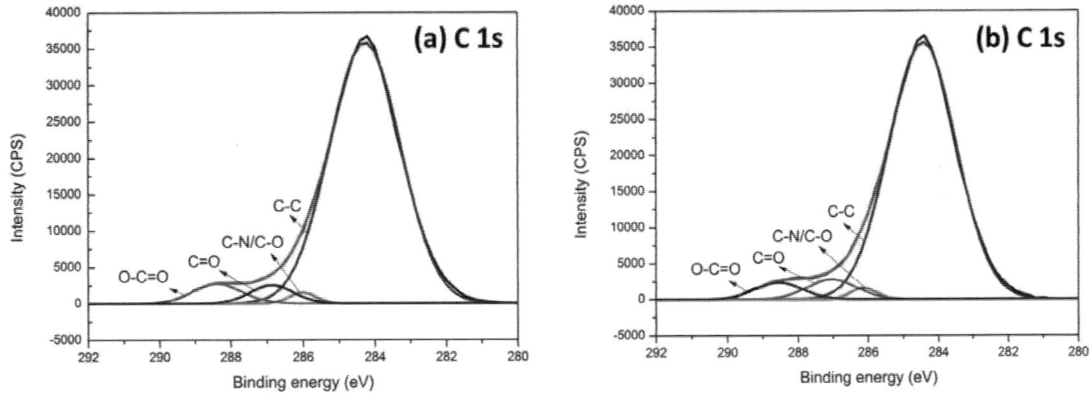

Figure 4. C 1s spectra of the untreated aramid fibers (a) and treated ones in $SCCO_2$ with isocyanate terminated liquid nitrile rubber (b).

Table 2. Deconvolution analysis of C 1s peaks of the untreated and treated aramid fibers in $SCCO_2$ with isocyanate terminated liquid nitrile rubber.

	Content of functional group (%)			
Sample	C-C	C-N/C-O	C=O	O-C=O
Untreated	89.8	1.4	3.3	4.8
Treated	89.1	2.1	4.4	4.4

3.4. Surface morphology of the aramid fibers

The surface morphology of the aramid fibers before and after $SCCO_2$ treatment with isocyanate terminated liquid nitrile rubber was estimated by SEM. As shown in Figure 5 a, the surface of the pristine fibers is extremely smooth and clean, which is similar to our previous report[7]. Compared to Figure 5a, the aramid fiber after treated in $SCCO_2$ with isocyanate terminated liquid nitrile rubber, a much rough surface was observed due to the formation of a thin layer of the polymer coating (Figure 5b). The structure transformation enables the compatibility of the aramid fibers and the vinyl epoxy resin, thus, can enhance the adhesion properties of the aramid fibers/vinyl epoxy composites.

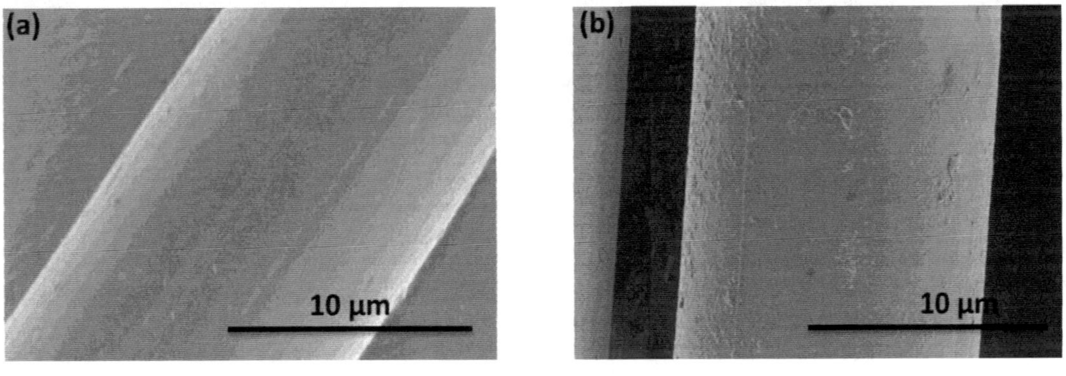

Figure 5. Surface morphology of the untreated aramid fibers (a) and those treated in SCCO2 with isocyanate terminated liquid nitrile rubber (b).

3.5. Mechanical properties of the aramid fibers reinforced vinyl epoxy resin composites

The flexural properties of the aramid fibers/vinyl epoxy resin composites were initially measured by three-point flexural method. As shown in Figure 6a, it is apparent that the flexural strength of the aramid fibers treated by $SCCO_2$ of vinyl epoxy composites was 18.1% higher than that of the untreated

IOP Conf. Series: Materials Science and Engineering **213** (2017) 012041 doi:10.1088/1757-899X/213/1/012041

aramid fibers . This indicates that $SCCO_2$ could improve interfacial adhesion between the fibers and the matrix. The ILSS was used to further investigate the interfacial adhesion between fibers and matrix. It can be seen from figure 6b that after treated in $SCCO_2$ with isocyanate terminated liquid nitrile rubber, the ILSS was increased from 42.5 MPa to 54.8 MPa. The improvement of flexural strength and ILSS of the aramid fibers/vinyl epoxy composites was given rise to the cross-linking reaction between CH=CH bonds and vinyl groups of vinyl epoxy resin.

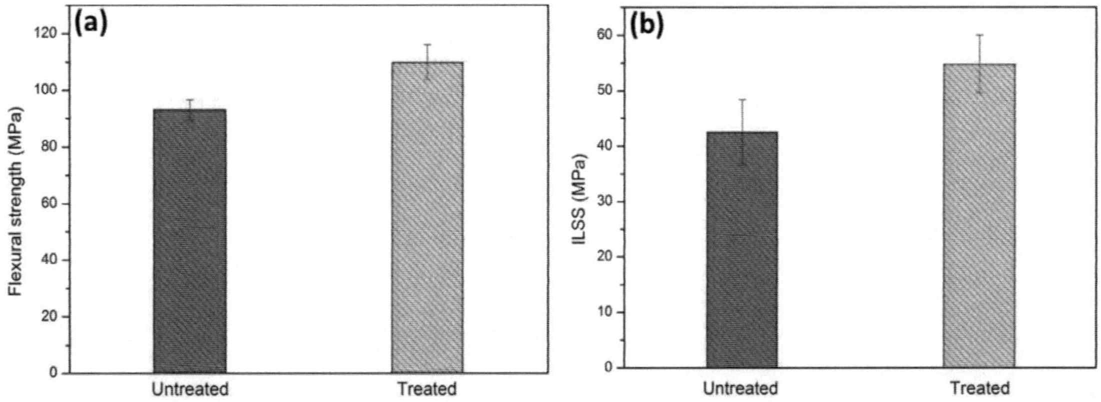

Figure 6. The mechanical properties of the aramid fibers reinforced vinyl epoxy composites.

3.6. Fracture morphology of the composites

To directly investigate the interfacial adhesion in aramid fibers/vinyl epoxy composites, the fractured surfaces of the composites was measured by SEM. The fracture surface morphologies of the aramid fibers/vinyl epoxy composites are shown in Figure 7. It can be seen from Figure 7a that the pull-out of the aramid fibers was relatively smooth with a small amount of resin. However, it is obvious that the entire fiber surface were wrapped with a thin layer of resin after $SCCO_2$ treatment (Figure 7b). These results showed that the interfacial adhesion of the aramid fibers treatment in $SCCO_2$ has been remarkably changed. The results agreed well with the above interfacial and mechanical properties.

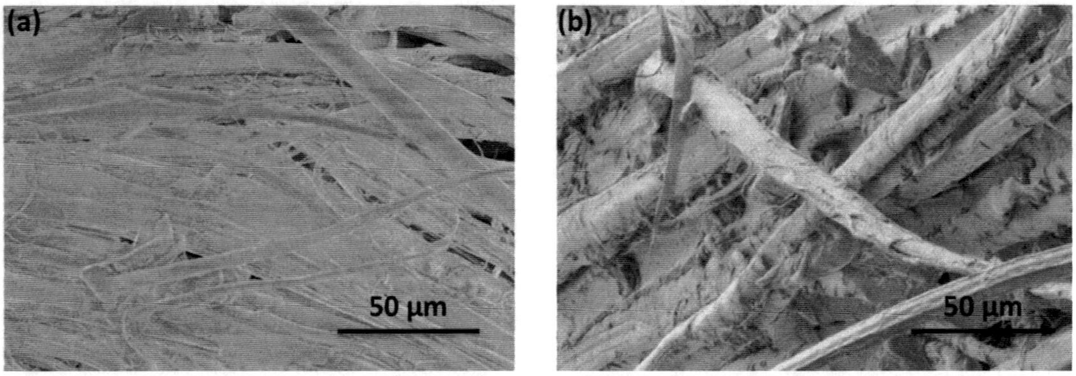

Figure 7. SEM images of the untreated (a) and $SCCO_2$ treated (b) fracture aramid fibers reinforced vinyl epoxy composites.

4. Conclusion

In summary, we developed an efficient method to improve the adhesion performances of the aramid fiber for composites. Isocyanate terminated liquid nitrile rubber was successfully introduced to the surface of the aramid fibers in $SCCO_2$. The adhesion properties of the aramid fibers were improved by 44.4% after $SCCO_2$ treatment with isocyanate terminated liquid nitrile rubber. Furthermore, SEM was also measured the fractured surfaces of composites, the results showed the interfacial adhesion of the aramid fibers after $SCCO_2$ treatment with isocyanate terminated liquid nitrile rubber for vinyl epoxy resin composites was greatly changed. We believe the processes provide a new approach to prepare composites with enhanced adhesion properties, making it possible to be used in preparation of other composites.

Acknowledgments
This work was supported by Nation Basic Research Program of China (973 Program) (2011CB606101), National Natural Science Foundation of China (Grant No. 51603120) and State Key Laboratory for Modification of Chemical Fibers and Polymer Materials, Donghua University (LK1602).

References

[1] Dibenedtto A T 2001 *Matter. Sci. Eng. A: Struct.* **302** 74
[2] Sa R, Yan Y, Wang L, Li Y, Zhang L Q, Ning N Y, Wang W C and Tian M 2015 *RSC Adv.* **5** 94351
[3] De Lange P J, Mäder E, Mai K, Young R J and Ahmad I 2001*Composites Part A* **32** 331
[4] Guo F, Zhang Z Z, Liu W M, Su F H and Zhang H 2009 *J. Tribol. Int.* **42** 243
[5] Li S, Han K Q, Rong H P, Li X Z and Yu M H 2014 *J. Appl. Polym. Sci.* **42050** 1
[6] Leal A A, Deitzel J M, Mcknight S H and Gillespie J W 2009 *Polymer* **50** 1228
[7] Kong H J, Yang P, Teng C Q and Yu M H 2015 *RSC Adv.* **5** 58916
[8] Du S M, Zhang J, Guan Y and Wan X H 2014 *Aust. J. Chem.* **67** 39
[9] Wang C X, Du M, Lv J C, Zhou Q Q, Ren Y, Liu G L, Gao D W and Jin L M 2015 *Appl. Surf. Sci.* **349** 333
[10] Lin J S 2002 *Eur. Polym. J.* **38** 79
[11] Imi K and Guil L 2004 *Compos. Sci. Technol.* **64** 2271
[12] Liu T M and Zheng Y S 2011 *J. Polym. Bull.* **66** 259
[13] Gu H 2009 *Mater. Des.* **30** 867
[14] Wang B, Duan Y G and Zhang J J 2016 *Mater. Des.* **103** 330
[15] Jia C X, Chen P, Liu W, Li B and Wang Q 2011 *Appl. Surf. Sci* **257** 4165
[16] Sun Y Y, Liang Q, Chi H J, Zhang Y J, Shi Y, Fang D N and Li F X 2014 *Fibers Polym.* **15** 1
[17] Xi M, Li Y L, Shang S Y, Li D H, Yin Y X and Dai X Y 2008 *Surf. Coat. Technol.* **202** 6029
[18] Liu L, Huang Y D, Zhang Z Q, Jiang Z X and Wu L N 2008 *Appl. Surf. Sci.* **254** 2594
[19] Sa R, Yan Y, Wei Z H, Zhang L Q, Wang W C and Tian M 2013 *ACS Appl. Mater. Interfaces* **5** 2062
[20] Du S M, Wang W B, Yan Y, Zhang J, Tian M, Zhang L Q and Wan X H 2014 *Chem. Commun.* **50** 9929
[21] Wang W B, Xiang Q, Guan Y, Zhang F, Zhang J, Yan C, Zhu Y D and Wan X H 2016 *J. Polym. Sci. Part A: Polym. Chem.* **54** 2050
[22] Wang L, Shi Y X, Sa R, Ning N Y, Wang W C, Tian M and Zhang L Q 2016 *Ind. Eng. Chem. Res.* **55** 12547
[23] Kikic I and Vecchione F 2000 *Curr. Opin. Solid State Mater. Sci.* **7** 399
[24] Zani F, Veneziani C, Bazzoni E, Maggi L, Caponetti G and Bettini R 2013 *Int. Pharm.* **450** 218
[25] King J W 2014 *Food. Sci. Technol.* **5** 215
[26] Choi Y S, Nesic S and Young D 2010 *Environ. Sci. Technol.* **44** 9233
[27] Baseri S, Karimi M and Morshed M 2014 *Fiber Polym.* **15** 161
[28] Zhao X, Hirogaki K, Tabata I, Okubayashi S and Hori T 2006 *Surf. Coat. Technol.* **201** 628
[29] Zheng H and Zheng L J 2014 *Fiber Polym.* **15** 1627
[30] Zheng H, Zhang J, Du B, Wei Q F and Zheng L J 2015 *J. Appl. Polym. Sci.* **41756** 1
[31] Kong H J, Teng C Q, Liu X D, Zhou J J, Zhong H P, Zhang Y, Han K Q and Yu M H 2014 *RSC Adv.* **4** 20599

Effects of various power process parameters on deposition efficiency of plasma-sprayed Al_2O_3-40% wt.TiO_2 coatings

Y J Wang[1], J Y Xu[1], Q H Zhao[1], Y Wang[1] and B Gao[1,2]

[1]School of Mechanical and Electrical, Guilin University of Electronic Technology, GL 541004, China

E-mail: bobiobest@126.com

Abstract. To investigate effects of various power process parameters on deposition efficiency, Al_2O_3-40% wt. TiO_2 is selected as raw material to be coated on Q235 steel substrate by air plasma spraying. Different variables of spraying current and voltage are designed, whose spraying power is from 12.8 KW to 16.8 KW and increment step is 0.8 KW. Deposition thickness is proposed as a simple method to characterize the deposition efficiency of coatings. Analysis of variance is used to observe the difference between two adjacent groups. It is shown that deposition efficiency increases with the increase of spraying current; by and large, it firstly rises and then decreases with the increment of spraying voltage. However, the effects of increasing the latter are much stronger on deposition efficiency.

Keywords: Deposition efficiency; Spraying voltage; Spraying current; Plasma spraying; Al_2O_3-40% wt.TiO_2

1. Introduction

Due to high quality of its fabricated coatings, plasma spraying is widely used in industry everywhere, such as in wear and corrosion condition. At the same time, there are tremendous advantages in deposition rate and efficiency in manufacturing [1-2]. For industrial production, economy is a factor to be seriously taken into consideration, which can be evaluated by an index called deposition efficiency (DE). However, at present, rare literatures, especially papers are publicly published on coating DE. In order to improve coating DE, a horn-section nozzle is added at the exit of spray flame, reducing the velocity of spraying flame flow and prolonging acceleration time in spraying flame [3]. On the basis of the foregoing research, the Laval nozzle structure is modified to be longer. It shows that the DE is 54% by using general plasma gun when spraying Al_2O_3-40% wt.TiO_2, however, the DE is 78% by using the improved nozzle. In other words, the DE of the latter is 44% higher than that of the former. Consequently, the porosity of the latter is 12.8%, which is 7.3% higher than that of the former [4]. Powder in the plasma flame is not fully melted may be the crucial factor responsible to low DE.

2. Experimental details

The plasma spraying equipment applied in this project is SX-80, which is the product of Guangzhou Sanxin Metal Co., Ltd., China, and the material powder is purchased from Xiandao Plasma Powder Co., Ltd., China, whose size is +45-15 μm in range. Scanning electron microscope (SEM) is used to snap section pictures of coatings.

Prior to plasma spraying, necessary dressing, degreasing, blasting, roughing and cleaning are executed on Q235 steel substrate.

Deposition thickness is proposed as a simple method to characterize DE, which can be qualitative and quantitative. It is assumed that powder feeding is stable, and coating porosity is ignored.

Some process parameters are set the same, the cycles of the robot (Yaskawa), motor voltage, gas flow of powder feeding is 3 times, 3 V, 200 L/h, respectively. The hydrogen, argon gas pressure, argon gas flow, spray distance, transverse velocity of the robot is 0.70 MPa, 0.85 MPa, 1600 L/h, 100 mm, 100 mm/s, respectively.

3. Results and discussion

3.1. Effects of spraying current on DE

To investigate the effects of various current, voltage is fixed at 40 V. Differently set current is shown in Table 1.

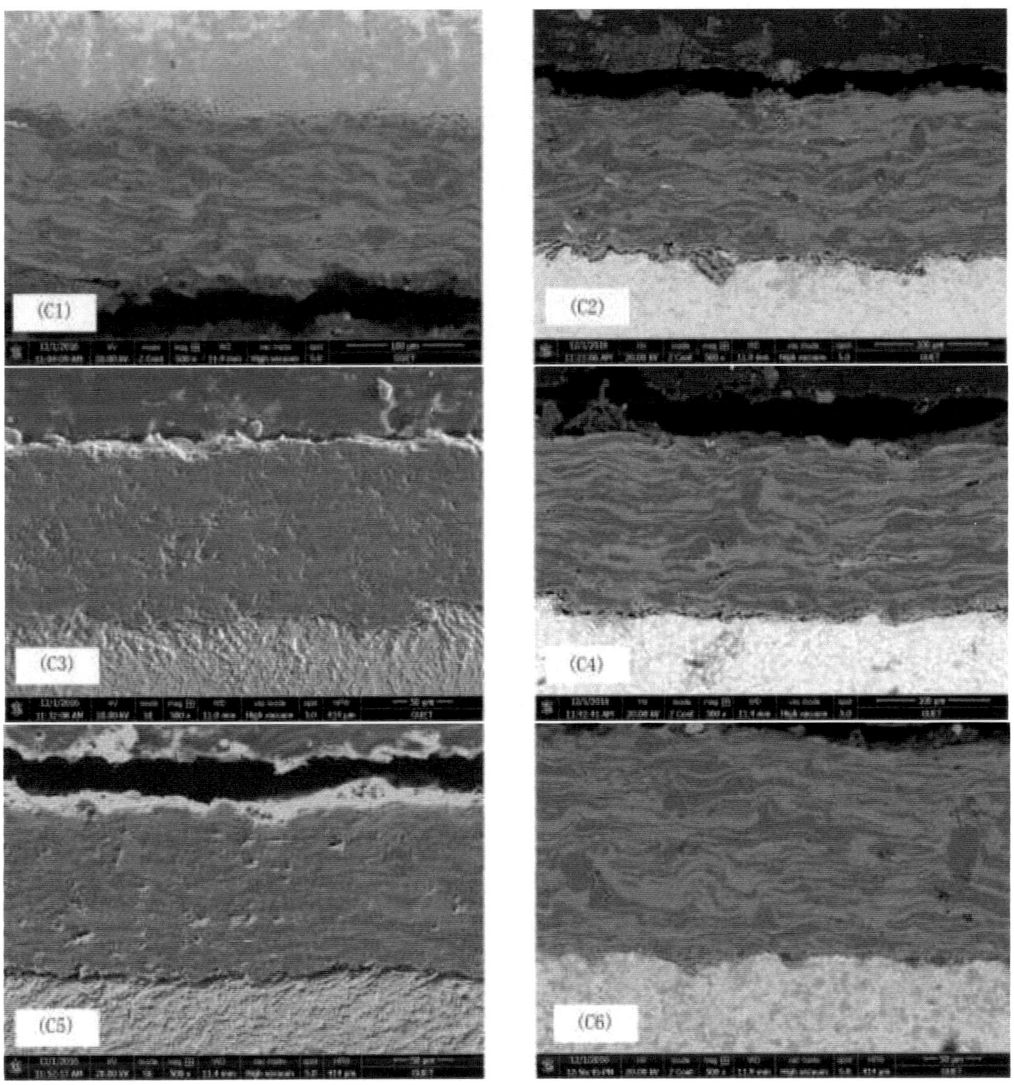

Figure 1. Coating section of different spraying voltage (500x).

IOP Conf. Series: Materials Science and Engineering **213** (2017) 012042 doi:10.1088/1757-899X/213/1/012042

Table 1. Spraying current variables and explanation.

No.	C1	C2	C3	C4	C5	C6
Value/ A	400	425	450	475	500	525

Figure 1 is sectional morphologies of C1-C6 coatings snapped by SEM in 500x scope. Thickness measurement is accomplished in digital image processing software ImageJ. To avoid measuring randomly, 15 auxiliary grid lines are added (Figure 2b), where Figure 2a is the section of C1, and measure lines are shown in Figure 2c. $\overline{x}$ is the mean value of 15 measure values, s is the standard error.

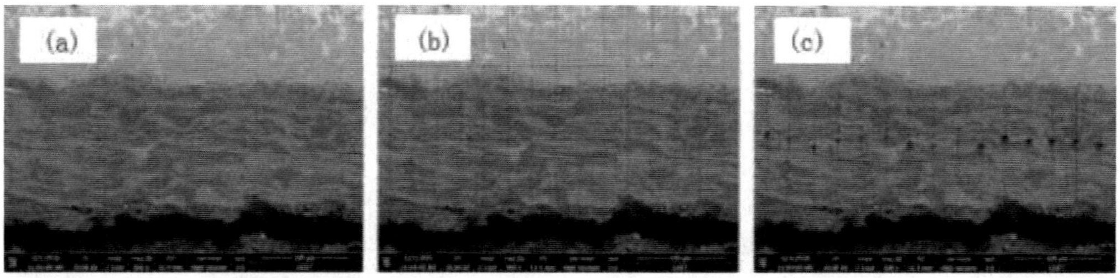

Figure 2. Procedures of measuring deposition thickness.

Table 2. Thickness of spraying current group (μm).

No.	C1	C2	C3	C4	C5	C6
$\overline{x}$	133.54	134.58	147.68	149.62	153.91	184.58
s	6.98	5.58	7.89	6.89	6.00	7.54

It can be seen from Table 2 and Figure 3 that coating thickness increases with the increase of spraying current, meaning deposition efficiency improves. The reason is that spraying power increases with the increase of current, and molten and semi-molten particles increase, while larger particles melt more fully, leading to higher coating deposition efficiency. At current C3-C5, it changes gradually in thickness. Molten particles compensate for the loss of vaporized particles. When it reaches C6, the mass of molten particles is much higher than that of vaporized particles.

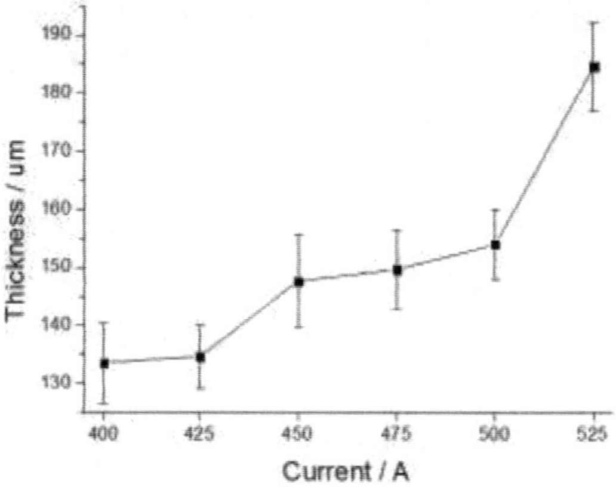

Figure 3. Effects of spraying current on deposition efficiency.

Analysis of variance (ANOVA) is used to observe difference in current variables. Firstly, normality tests in the variables are conducted, whose results are shown in Table 3. Secondly, Levene-test is used to verify whether the variance is homogeneous. The test result is 0.649, and the test of variance homogeneity passes.

Table 3. Shapiro-Wilk test.

group	C1	C2	C3	C4	C5	C6
statistic	0.968	0.959	0.908	0.969	0.96	0.948
P	0.823	0.668	0.125	0.842	0.701	0.491

Table 4. Lsd-t test of two adjacent groups.

i	j	i-j	standard error	statistic	Asp. P
C1	C2	-0.732	2.505	−0. 292	7.71e-1
C2	C3	-13.406	2.505	−5. 352	7.44e-7
C3	C4	-1.945	2.505	−0. 776	4.40e-1
C4	C5	-4.290	2.505	−1. 713	9.05e-2
C5	C6	-30.667	2.505	−12. 242	2.28e-20

Lsd-t test, one method of ANOVA, is used to compare difference between two adjacent groups, and results are shown in Table 4. When significance value (P, two-tailed) is less 0.05, significant difference is observed.

From the measure values, there is no significant difference between C1 and C2 group. The possible reason is that the current is small; the energy density of plasma flame is low, only small powder is molten or partially-molten, and large powder has little positive effect on DE [5-6]. Significant difference is observed between C2 and C3 group, semi-molten and molten particles increase with the increase of current, resulting in higher DE of C3. However, DE between C3 and C4 is not significant, the possible reason is that with the increase of spraying current, the size of molten or semi-molten powder particle tends to increase, but the small powder tends to vaporize at the time, and so are C4 and C5. However, the significance between C3 and C5 is 1.48e-2, there is significant difference between the two coatings. When the variable step is small, the difference is not significant. With the interval increases, significant difference appears. And then quantitative change leads to qualitative results. With the current increasing further, more and more large-size particles become molten or semi-molten, the vaporized small-size powder is finite, contributing to higher DE. That is why the coating thickness of C6 is thicker than that of C5.

3.2. Effects of spraying voltage on DE

To investigate the effects of various voltages, current is fixed at 400 A. Differently set voltage and the coating thickness are shown in Table 5. Coating section is is shown in Figure 4.

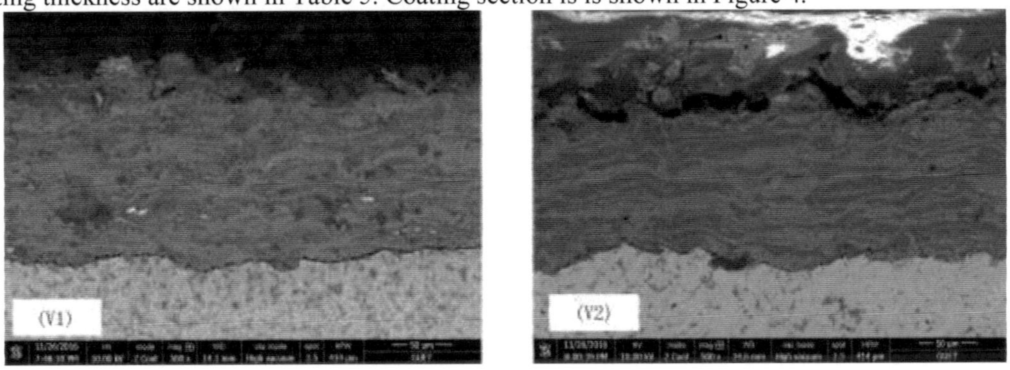

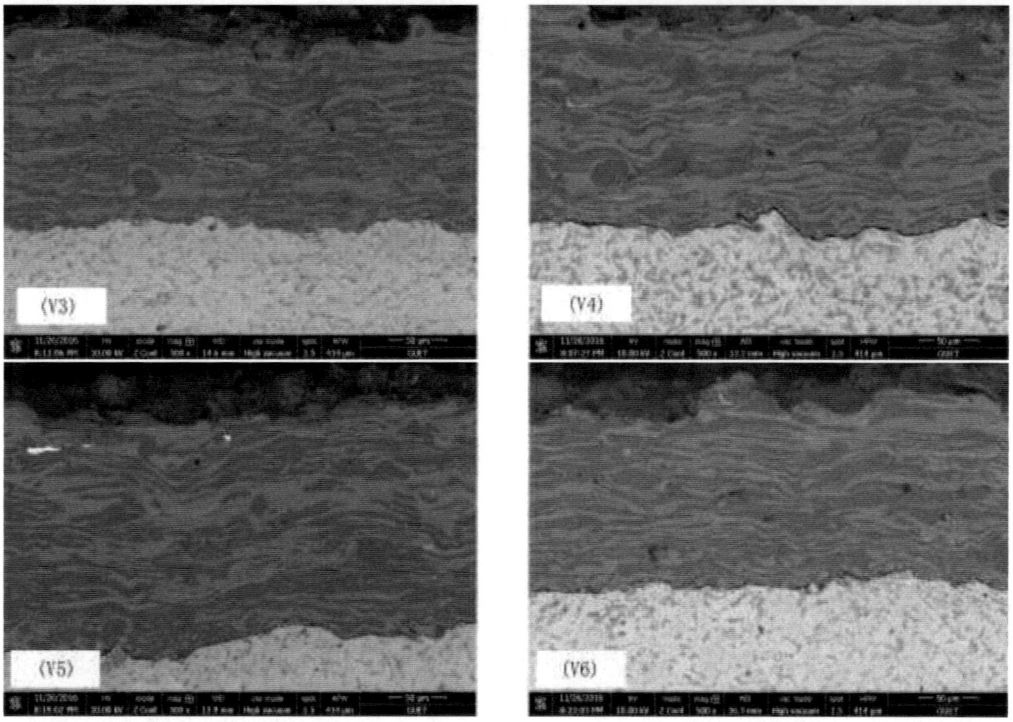

Figure 4. Coating section of different spraying voltage (500x).

As shown in Table 6, normality tests of V1-V6 pass. The significance value of Levene test is 0.442, which indicates the variance is homogeneous. Through variance analysis, the mean square between the groups is 8363.804, group mean square is 85.410, F is 97.926, and P is 1.54e-33, intimating that all the two groups are not statistically the same. The results of Lsd-t test in Table 7 indicate that the adjacent groups are significantly different.

Table 5. Explanation and thickness (μm) of spraying voltage variables.

	V1	V2	V3	V4	V5	V6
voltage / V	32	34	36	38	40	42
$\overline{x}$	*139.21*	*124.52*	*157.95*	*173.33*	*190.52*	*151.64*
s	6.40	11.07	8.76	6.67	10.42	10.92

Table 6. Shapiro-Wilk test.

group	V1	V2	V3	V4	V5	V6
P	0.823	0.668	0.125	0.842	0.701	0.491

Table 7. Lsd-t analysis of two adjacent groups.

i	j	(i-j)	standard error	statistic	Asp. P
V1	V2	14.690	3.375	4.353	3.76e-5
V2	V3	-33.429	3.375	-9.905	8.96e-16
V3	V4	-15.387	3.375	-4.559	1.73e-5
V4	V5	-17.189	3.375	-5.093	2.13e-6
V5	V6	38.882	3.375	11.521	5.63e-19

It can be seen from Table 5 and Figure 5 that in these measure values of 6 groups from V1-V6, a tendency is shown that with the increment of spraying voltage, it firstly rises and then decreases but V2 in coating's deposition thickness. Why V2 doesn't obey this law is that there may be random error in experiments. The reasons are similar to those of spraying current groups. When spraying voltage reaches 40V, the deposition efficiency comes to the maximum. When spraying voltage is lifted to 42 V, the result of deposition thickness is worse than that of 40 V. Larger spraying power leads to higher semi-molten and molten degree of big-size particles; however, it exacerbates the evaporation degree of small-size and middle-size particles. At the same time, overheated particles incline to splash when impacting substrate, resulting in further mass loss in deposition process[7]. When it comes to high spraying power, agglomeration phenomenon of overheated particles occurs, attributing to low DE[8].

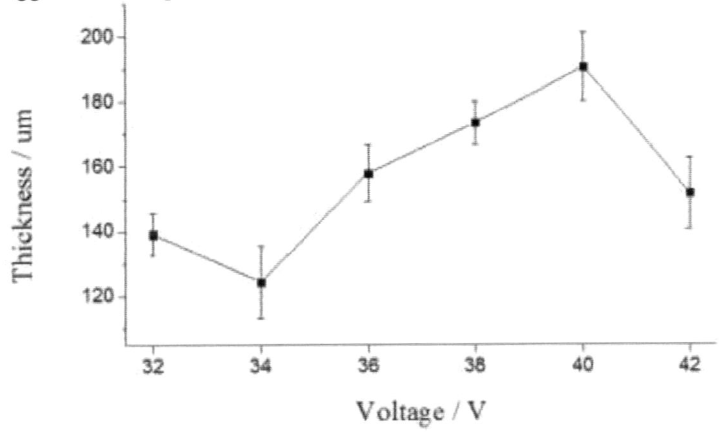

Figure 5. Effect of spraying voltage variable on coating thickness.

3.3. Difference between spraying current and voltage in affecting DE

The power increment of spraying current variable group is 0.8 KW (32V*25A=800 W), which is equal to that of spraying voltage variable group. Unlike current variable groups, the significance values of voltage variable groups are less than 0.05, so there is good reason to believe that the effect of increasing voltage is stronger than that of increasing current in the same step. Compared with spraying current groups, the peak value of DE appears earlier in spraying voltage groups. Hydrogen works as the auxiliary gas, through increasing whose gas flow, spraying voltage can be lifted. The mechanism is that hydrogen is molecule gas, whose heat transfer ability and thermal conductivity betters than that of monatomic gas. In addition, enthalpy of commonly-seen gas used in plasma spraying ranks $Ar<Ar+He<Ar+H_2<N_2<N_2+H_2$ in order [5].

4. Conclusions

Deposition thickness is proposed as a simple method to characterize the deposition efficiency. Conclusions can be drawn as follows.

(1) In the current variable groups, deposition efficiency increases with the increase of spraying current.

(2) In the voltage variable groups, but V2, there is a general tendency that deposition efficiency firstly increases and then decreases with the increase of spraying current.

(3) In the same step, the effect of increasing spraying voltage is stronger than that of increasing spraying current.

Acknowledgments

The authors acknowledge the financial support from Chinese National Nature Science Foundation (Grant No. 51361008) and encourage from Innovation Project of GUET Graduate Education (Project No. 2016YJCX105).

References

[1] Jia S K, Zou Y, Xu J Y, Wang J and Yu L 2015 *Trans. Nonferrous Met. Soc. China* **25** 175-83
[2] Wang Y J, Gao B and J Y 2017 *Hot Working Technology* **46** 42-7
[3] Schwenk A, Gruner H, Zimmermann S, Landes K and Nutsch G 2004 Improved nozzle design of de-Laval-type nozzles for the atmospheric plasma spraying *Proc. of ITSC (Osaka)* pp 600-5
[4] Deng C M, Zhou K S and Liu M 2010 *Chin. Surf. Eng.* **23** 19-23
[5] Davis and Associations 2004 *Handbook of Thermal Spray Technology* (USA: ASM International) ed J R Davis
[6] Song R G, Wang C and Lu G 2012 *Mater. Prot. Technol.* **45** 58-9
[7] Lech P 2011 *The Science and Engineering of Thermal Spray Coatings* (Beijing: China Machine Press)
[8] Wang H J, Guo Y M and Liu M 2008 *Nonferr. Metals(Extractive Metallurgy)* **2008** 53-6

Research on the corrosion inhibitors of zinc in hydrochloric acid

C X Sun[1,2], Y M Chen[1], H W Xu[1], C S Huang[1], M Zhang[1], J Y Wu[1], M Chen[1] and M Xue[1]

[1]Institute of Energy Research, Henan Academy of Science, No. 29 Garden Road, 450008 Zhengzhou R. P. China

E-mail: caixiasun@alum.imr.ac.cn

Abstract. Three organic compounds were tested as zinc corrosion inhibitors in hydrochloric acid: cetyltrimethyl ammonium bromide (CTAB), nicotini acid, bromohexadecyl pyridine. The static coupon test results indicate that CTAB and bromohexadecyl pyridine offer the best zinc corrosion protection, while nicotinic acid accelerates zinc corrosion. The polarization results indicate that CTAB, nicotinic acid and bromohexadecyl pyridine induce a positive shift in the E_0 of zinc in hydrochloric acid. A complex of CTAB and bromohexadecyl pyridine inhibits the corrosion of zinc in hydrochloric acid. SEM results indicate that the CTAB and bromohexadecyl pyridine formed a uniform and compact membrane on the surface of zinc that subsequently protects the zinc from effective corrosion.

1. Introduction

Zinc is commonly used in numerous industries and fields. It is often used as a sacrificial metal in coatings to protect other metals, particularly ferrous metals. The galvanized metals used in water-circulating systems often produce white rust, so the method of prevention of removal of corrosion is important[1]. Investigation into the corrosion tendencies of zince can lead to important applications in production and economics. The use of inhibitors is one of the most practical methods by which to protect zinc from corrosion, particularly in acidic mediums[2-3]. Organic inhibitors with electron-donating groups such as nitrogen, sulphur and oxygen often play an important role in zinc corrosion prevention[4-13].

The present work investigates the effects three organic compounds on the corrosion inhibition of zinc in 0.5M hydrochloric acid: cetyltrimethyl ammonium bromide (CTAB), nicotinic acid and bromohexadecyl pyridine. CTAB is a cationic surface active agent; nicotinic acid and bromohexadecyl pyridine are heterocyclic compounds. The effect of single inhibitor concentration on the anticorrosion efficient was investigated, as well as the anticorrosion efficiencies of CTAB, nicotinic acid and bromohexadecyl pyridine.

2. Experimental methods

2.1. Materials

CTAB(Tianjin Guangfu Fine Chemical Research insitute, AR); nicotinic acid(Tianjin kemi'ou Chemical Reagent Co., Ltd. AR); bromohexadecyl pyridine(Tianjin Guangfu Fine Chemical Research insitute, AR); hydrochloric acid(Tianjin Guangfu Fine Chemical Research insitute, AR).

2.2. Inhibitors prepared

The following inhibitors were prepared for experimental analysis:

1). 4.0g of CTAB were dissolved at 70°C. A volume of 100ml was set aside for experimental use.

2).1.5g of bromohexadecyl pyridine were disoolved at 50°C. A volume of 250ml was set aside for experimental use.

3).4.0g of acid were dissolved at 40°C. A volume of 250ml was set aside for later use.

Inhibitor solutions were then prepared according to the following mixtures:

Compound1: CTAB 200mg L^{-1} nicotinic acid 100mg/L and bromohexadecyl pyridine 60mg L^{-1}.

Compound 2: CTAB 20mg L^{-1} and bromohexadecyl pyridine 60mg L^{-1}.

2.3. Instruments

Potentiostat/galvanostat CS 2305, JSM-6510; scanning electron microscopy (SEM), BT224s Electronic Balance.

2.4. Experimental procedure

2.4.1. Weight loss test of zinc. The corrosion of zinc after exposure to inhibition solutions was evaluated according to the static hanging piece test, reference HG/T2387-2007. The corrosion inhibition test was conducted at 25°C for 96 hours in a testing solution of 0.5M hydrochloric acid. The experimental solution was comprised of 0.5M HCl and inhibitor solution. The corrosion rate was calculated according to the following formula:

$$K = \frac{W_1 - W_2}{S \cdot T} \quad (1)$$

where K is the corrosion rate, g $m^{-2} h^{-1}$; W_1 is the weight of the zinc plate before the test, g; W_2 is the weight of the zinc plate after the test, g; S is the initial area of the plate, m^2; and T is the time elapsed during the experiment, h.

2.4.2. Electrochemical test of zinc. Zinc electrodes were sealed in epoxy resin, except for an exposed crosssection at the working surface with an area of $1cm^2$. The counter electrode for the electrochemical test was made of platinum, while a saturated calomel electrode (SCE) was used as the reference electrode. The electrochemical studies were conducted on the Potentiostat/galvanostat CS2305. Linear polarization measurement was conducted with a potential perturbation of ±100mV around the open circuit potential (OCP). The electrolyte was prepared by adding various concentrations of different inhibitors to 0.5M hydrochloric acid solution.

2.4.3. SEM test. The zinc plates were observed on SEM JSM6510 after testing. The operation parameters were set to 20 kV acceleration voltages.

3. Results and discussion

3.1. Corrosion test results

Figure 1(a) displays the corrosion rates of zinc after exposure to various concentrations of CTAB inhibitor solutions. Results indicate that the introduction of CTAB (in the experimental range 0mg L^{-1}~600mg L^{-1}) can protect zinc immersed in 0.5M HCl from corrosion. The corrosion rate of zinc in 0.5M HCl solution is 17.5×10^{-4}g m^{-2} $h^{-1,}$ while the corrosion rate of zinc in 0.5M HCl in the presence of CTAB is reduced to below 6.6×10^{-4}g m^{-2} h^{-1}. Increases in CTAB concentration does not significantly alter the rate of zinc corrosion in HCl. Figure1(b) displays the corrosion rate of zinc in various concentrations of nicotinic acid inhibitor solutions. Results indicate that the nicotinic acid (in the experimental range 160mg L^{-1}~480mg L^{-1}) does not significantly alter the corrosion rate of zinc in HCl. When the concentration of nicotinic acid was 300 mg L^{-1} the corrosion rate (20.4 x10^{-4}g m^{-2} h^{-1}) is higher than that observed in 0.5m HCl (17.5x10^{-4}g m^{-2} h^{-1}). When the concentration of nicotinic acid is 480mg L^{-1} the corrosion rate is 14.01x10^{-4}g m^{-2} h^{-1}, which does not meet industrial requirements. Figure 1(c) displays the corrosion rate of zinc in various concentrations of bromohexadecyl pyridine inhibitor solution. Results indicate that the introduction of bromohexadecyl pyridine to 0.5M HCl solution can protect zinc from corrosion. The corrosion rate of zinc in 0.5M HCl solution decreases with increasing bromohexadecyl pyridine concentrations. When the bromohexadecyl pyridine concentration is 100mg/L the corrosion rate of zinc is 2.17x10^{-4}g m^{-2} h^{-1}.Figure 1(d) depicts the anti-corrosion efficiency of inhibitor mixtures in relation to zinc corrosion. Results indicate that compound 1, consisting of CTAB, nicotinic acid and bromohexadecyl pyridine, is more effective than any single inhibitor. The corrosion rate of zinc in HCl in the presence of compound 1 is 3.2x10^{-4}g m^{-2} h^{-1}. Additionally, a mixture of CTAB and bromohexadecyl pyridine without nicotinic acid is more effective than compound 1 and any single inhibitor.

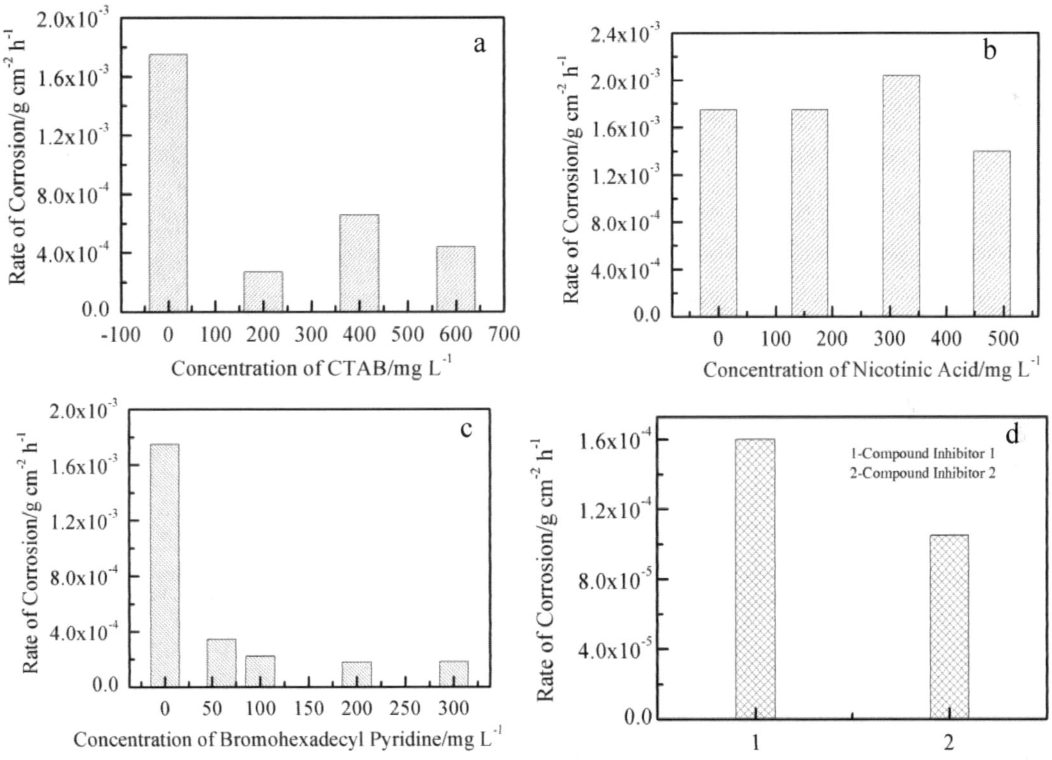

Figure 1. Corrosion rate of zinc in varying concentrations of inhibitor solutions (a) CTAB; (b) nicotinic acid; (c) bromohexadecyl pyridine; (d)compound.

3.2. Electrochemical test results

Table 1 displays the E_0 of zinc in the presence of various inhibitor solutions, obtained from Tafel fitting results. As shown in Table 1, the change of E_0 in the presence of CTAB and bromohexadecyl pyridine solutions is irregular with variations in inhibitor concentrations, while the change of E_0 in nicotic acid solution is regular. The addition of CTAB, nicotinic acid and bromohexadecyl pyridine all induce a positive shift in E_0 values. As shown in Table 2, the I_0 of zinc indicates better anticorrosion performance correlated to smaller values in different inhibitor solutions results indicate that the tendency of I_0 changes in the presence of CTAB is similar to changes in the presence of HCl alone.

Table 1. E_0 of zinc in various inhibitor solutions.

CTAB	E_0/V	Nicotic acid	E_0/V	Bromohexadecyl pyridine	E_0/V
0	-1.0384	0	-1.0384	0	-1.0384
200	-1.0027	160	-0.996	60	-0.96954
400	-1.0112	320	-0.99313	100	-0.98287
600	-1.0224	480	-0.97604	200	-0.98218
800	-1.0111	--	--	300	-0.97913
	Compound 1		-1.0008	Compound 2	-0.98972

Table 2. I_0 of zinc in various inhibitor solutions.

CTAB	I_0/A cm^{-2}	Nicotic acid	I_0/A cm^{-2}	Bromohexadecyl pyridine	I_0/A cm^{-2}
0	$6.8x10^{-4}$	0	$6.8x10^{-4}$	0	$6.8x10^{-4}$
200	$5.7x10^{-4}$	160	$6.9x10^{-4}$	60	$1.21x10^{-4}$
400	$2.4x10^{-3}$	320	$4.3x10^{-4}$	100	$1.7x10^{-4}$
600	$6.4x10^{-4}$	480	$5.28x10^{-4}$	200	$1.2x10^{-4}$
800	$2.8x10^{-3}$			300	$1.01x10^{-4}$
	Compound 1		$4.0x10^{-4}$	Compound 2	$3.9x10^{-5}$

The change in I_0 in the presence of nicotinic acid concentration is not significant, while the I_0 of zinc in inhibitor solutions is smaller than in the presence 0.5m HCl alone. Figure 2 shows the polarization curves of zinc in various inhibitor solutions. Results indicate that the introduction of CTAB, nicotinic acid and bromohexadecyl pyridine causes significant alteration in the cathode curve. Inhibitor compouns 1 and 2 cause obvious changes in anode and cathode reactions. The changes in open potential and cathode reaction may be caused by the film formed by the compound solution of CTAB, nicotinic acid, and bromohexadecyl pyridine.

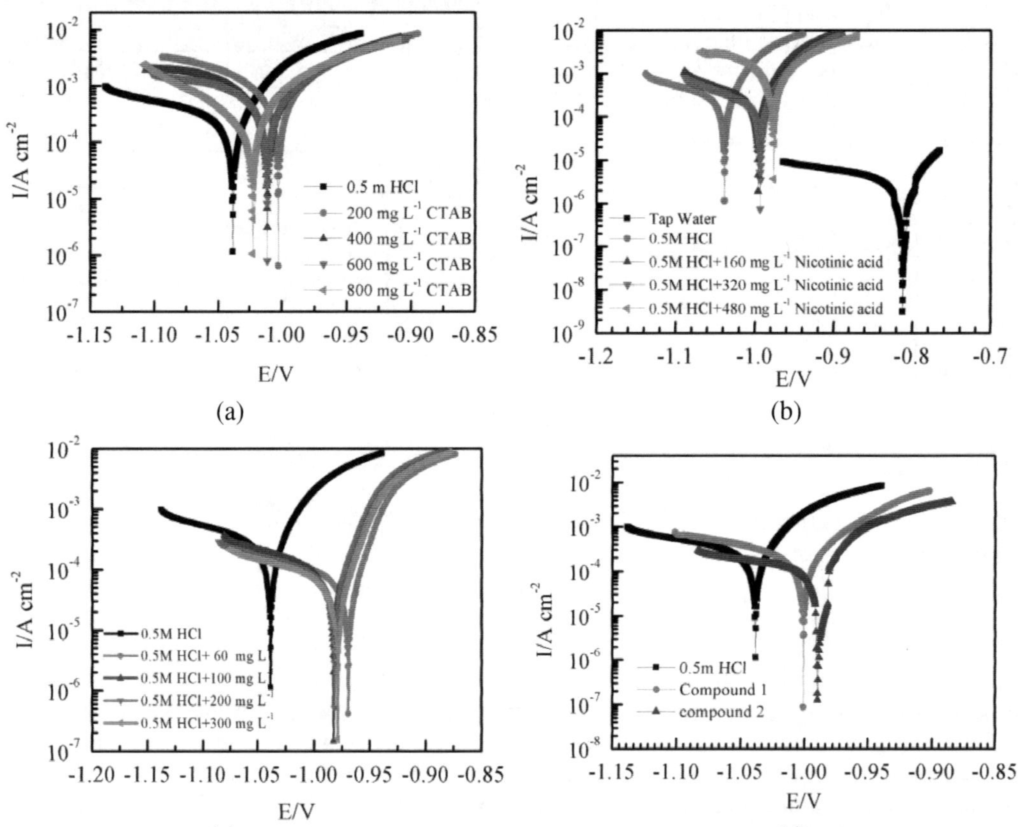

Figure 2. Polarization curves of zinc in (a) CTAB solution; (b) nicotic acid; (c) bromohexadecyl pyridine and (d) inhibitor compounds.

In summary, the corrosion and electrochemical test results indicated that compound 2 demonstrates better anti-corrosion performance than any other tested compound or single inhibitor solution.

3.3. Results of SEM test

Figure 3 depicts SEM photos of zinc after exposure to different inhibitor solutions in HCl for 96 hours at 25°C. Figure 3(a) demonstrates microcosmic scratches on the surface of the zinc. Figure 3(b) shows that when zinc is immersed in 0.5M HCl, a large number of corrosion pits formed on the surface of the zinc. The presence of CTAB in Figure 3(c), nicotinic acid in Figure 3(d) and bromohexadecyl pyridine in Figure 3(e) can form a uniform and compact film that prevents zinc from corrosion in the presence of HCl. Figure 4 depicts photos of zinc plate immersed in 0.5M HClalone and 0.5M HCl in the presence of compound 2 inhibitor solutions for 96 hours at 25°C. The results indicate the formation of a macroscopic film on the zinc plate.

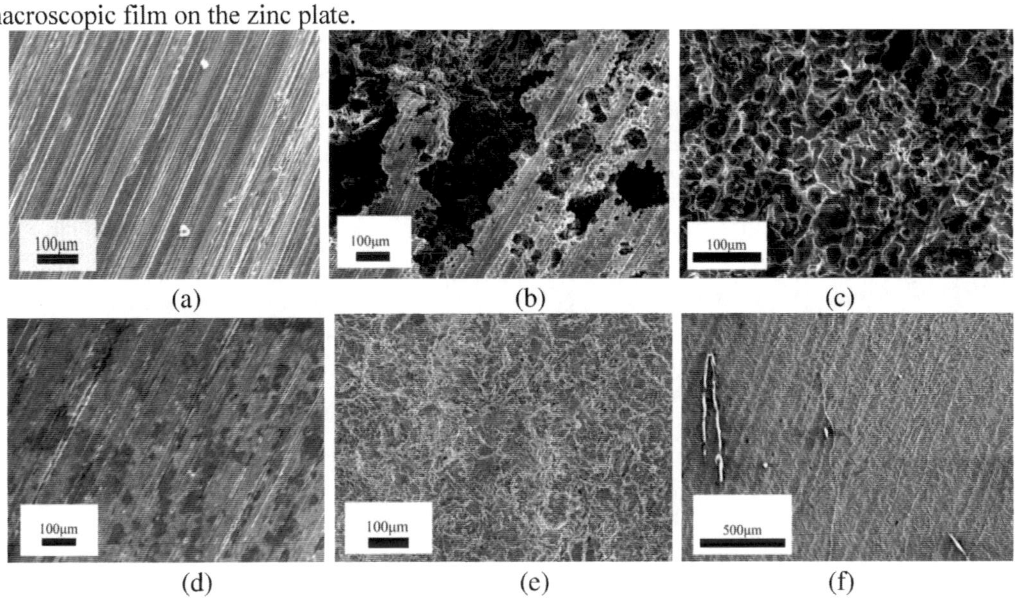

Figure 3. SEM photos of (a) pure zinc; (b) zinc in 0.5M HCl; (c) zinc in CTAB solution; (d) zinc in nicotinic acid solution; (e) zinc in bromohexadecyl pyridine solution; and (f) zinc in inhibitor compounds.

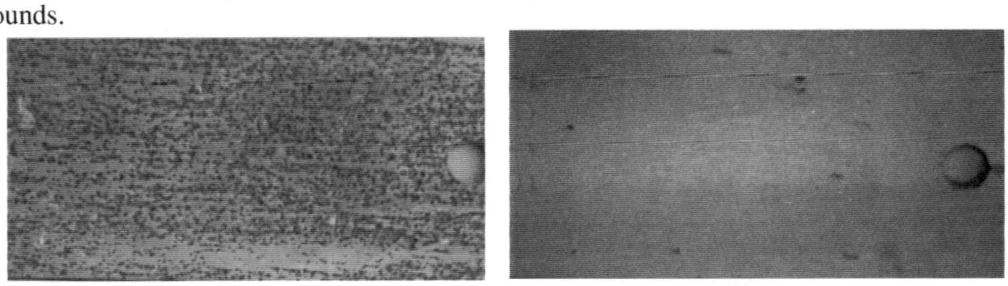

(a) (b)

Figure 4. Photos of zinc in (a) 0.5m HCl; and (b) inhibitor compound solutions.

4. Conclusion

1. Weight loss results indicate that CTAB and bromohexadecyl pyridine can protect zinc from corrosion in the presence of HCl. The weight loss does not alter markedly with increasing

concentrations of either inhibitor. The introduction of nicotinic acid displayed no effect on zinc corrosion.

2. Polarization and SEM results indicate that CTAB and bromohexadecyl pyridine form a compact and uniform membrane on the surface of zinc.

3.The weight loss tendency of zinc in the presence of various inhibitors result in similar I_0 values, which indicates the anti-corrosion performance of inhibitor solutions in reference to zinc.

Reference

[1] Fontana M W 2001 Method for treating water circulating systems *US patent* 6183649 B1
[2] Abdalah M 2003 *Corros. Sci.* **45** 2705-16
[3] Trabanelli G 1991 *Corros.***47** 410
[4] Gadallah A G and Moustafa H M 1992 *J. Appl. Electrochem.* **22** 644-8
[5] Morad M S 1999 *J. Appl. Electrochem.* **29** 619-26
[6] Maja M, Penazzi N, Farnia G and Sandona G 1993 *Electrochim. Acta* **38** 1453-9
[7] Juhel G, Beden B, Lamy C and Leger J M 1990 *Electrochim. Acta* **35** 479-81
[8] Zhang D S, Li L D, Cao L X, Yang N F and Huang C B 2001 *Corros. sci.* **43** 1627-36
[9] Manov S, Lamazouere A M and Aries L 2000 *Corros. Sci.* **42** 1235-48
[10] Aramaki K 2001 *Corros. sci.* **43** 1985-2000
[11] Agrawal Y K, Talati J D, Shah M D, Desai M N and Shah N K 2004 *Corros. Sci.* **46** 633-51
[12] Ein-eli Y, Auinat M and Starosvetsky D 2003 *J. Power sources* **114** 330-7
[13] Dobryszycki J and Biallozor S 2001 *Corros. Sci.***43** 1309-19

Molecular weight and distribution of ultra-high molecular weight poly (p-phenyleneterephalamide)

H J Kong[1,3], X M Ding[2], M M Qiao[2], Y Wu[2] and M H Yu[2]

[1]School of Materials Engineering, Shanghai University of Engineering Science, P. R. China

[2]State Key Laboratory for Modification of Chemical Fibers and Polymer Materials, College of Material Science and Engineering, Donghua University, P. R. China

E-mail: konghaijuan@sues.edu.cn

Abstract. The measurement of molecular weight (Mw) and distribution for ultra-high molecular weight poly (p-phenyleneterephalamide) (UHMWPPTA) is still a great challenge, because it is hardly dissolved in organic solvents normally to determine its Mw by gel permeation chromatography (GPC). In this paper, n-alkylated PPTAs with different molecular weight (including UHMWPPTA) were prepared by n-alkylation method. As the n-alkylated PPTAs with different length of alkyl chains, they can be dissolved in organic solvents such as tetrahydrofuran (THF). And the longer the alkyl side chains, the solubility is better. Molecular weight and of n-alkylated PPTAs were characterized by GPC with THF as eluent (MWGPC) and distibution was obtained by the weight molecular and number moleculr weight. The molecular weight of PPTA was measured in concentrated sulfuric acid by intrinsic viscosity measurement. The results showed a good linear relationship between them for PPTA with molecular weight from 10900 to 60800, which imply that molecular weight of UHMWPPTA could be measured by the GPC measurement of n-alkylation of PPTA.

1. Introduction

Poly (p-phenyleneterephalamide) PPTA fiber is a typical high performance fiber, with high intensity and modulus, high temperature-resistance and high chemical resistance. The molecular weight and its distribution is one of the key issues for high mechanical properties of aramid fiber. However, characterization of PPTA has been a difficult task, especially the ultra-high molecular weight of PPTA, because PPTA is hardly dissolved in the organic solvents and can only be dissolved in polar solvents, such as N,N-dimethylacetamide (often only in the presence of lithium salt), or strong acid, such as concentrated sulphuric acid or chloro-sulphonic acid[1]. Baird and Arpin et al investigated the intrinsic viscosity of dilute solutions of unfractionated PPTA in concentrated H_2SO_4 using light scattering and obtained the relationships between intrinsic viscosity and weight-average molecular weight of PPTA[2]. B. Chu et al characterized PPTA in concentrated sulphuric acid and determined the molecular weight distribution of PPTA by laser light scattering technique[3-6]. But the concentrated sulphuric acid would make the molecular weight of PPTA reduced and whose corrosive properties require special instruments. And it is hard to obtain the molecular-weight and its distribution from the gel permeation chromatography (GPC) characterization. Ogata et al n-alkylated the PPTA to determine their molecular weight distribution by using GPC with the THF as the solvent[7]. However, the measurement of molecular weight and its distribution of ultra-high molecular weight of PPTA is

IOP Conf. Series: Materials Science and Engineering **213** (2017) 012044 doi:10.1088/1757-899X/213/1/012044

still a great challenge. In this paper, a series of n-alkylated PPTA with different molecular weight (especially UMWPPTA) were prepared by n-alkylation method. The weight-average molecular weight and its distribution of n-alkylated PPTAs were characterized by the GPC measurement of n-alkylated PPTAs in THF (MWGPC) and intrinsic viscosity of PPTAs were measured in concentrated sulfuric acid (ηinh). The results showed a good linear relationship between MWGPC and ηinh for molecular weight from 10900 to 60800, which imply that molecular weight of SHMWPPTA could also be measured by the n-alkylation method.

2. Experimental

2.1. Materials

N-methy-2-pyrrolidone (NMP) and dimethyl sulfoxide(DMSO) were bought from Sinopharm Chemical Reagent Co. Ltd and reduced the water by reacting with calcium hydride under a nitrogen atmosphere and distilled under reduced pressure. LiCl was dried at 120 °C in vacuum. Potassium hydrogen dispersed in mineral oil from 25 to 35wt% purchased from of Acros, was used as received. Alkyl bromides were brought from Sino pharm Chemical Reagent Co. Ltd and were used as received. Ultra high molecular weight PPTA (SHMWPPTA) with intrinsic viscosity 9.0dl•g−1 was kindly produced in the production line of Hebei Silicon Valley Chemical Co, Ltd.

2.2. Preparation of PPTA

PPTA with different molecular weight was prepared by solution polycondensation of the terephthaloyl chloride with p-phenylenediamine in solvent of NMP containing 1.5wt% lithium chloride (LiCl) at -5°C for 20 min, and then at room temperature for 2 hours, then precipitated in water and exhaustively washed with water and dried in vacuum oven at 120°C for 6 hours[7].

2.3. N-alkylation of PPTA

N-alkylated PPTA was synthesized by the metallation reaction of PPTA with potassium in DMSO, followed by reacting with n-alkyl bromide, as shown in the following reaction scheme. Potassium hydrogen (25% suspension in mineral oil) was dissolved in DMSO under a nitrogen atmosphere and a clear solution was obtained. PPTA resin was added into the flask stirring to dissolve. The mixture was stirred for 4~5 h at 40 °C until all the PPTA had dissolved, giving a transparent deep red coloration, then alkyl bromides was added. After stirring for another time at 40 °C, n-alkylated PPTA products were precipitated in water. The products were filtered and washed with acetone, alcohol and water several times alternately. Finally, the product was dried in a vacuum oven at 100°C for 8 hours[8].

2.4. Measurements

Intrinsic viscosity of polymers in concentrated sulfuric (98 wt %) acid of concentration ($Cs=0.5g/dl^{-1}$) at 30 °C were measured by using a Ubbelhode viscometer. The structure of the products was characterized in KBr disks by using a FTIR spectrometer (Nicolet 8700). Element analysis instrument (ElementarVarioELIII) was used to measure the contents of elements C, H, N in different polymers to get the substitution degree of n-alkylation. The substitution degree was calculated as the equation (1) or (2)

$$DS_c = ((7C\%)/(6N\%) - 7)/n \tag{1}$$

$$DS_H = ((14H\%)/(N\%) - 4)/(2n + 1) \tag{2}$$

The molecular weight and distribution of the products were determined by Gel Permeation Chromatography (GPC) (Waters, MA). The test was carried out with a Waters e2695 pump, Waters 2414 refractive index detector and a Styragel HR5E column (5 μm). THFwas used as eluent with a flow rate of 1 ml/min. The sample was made with a concentration of 4 mg/mL, and the injection volume was 30 μl for each sample. Polystyrene (PS) standards were used to calibrate the curve[9].

IOP Conf. Series: Materials Science and Engineering **213** (2017) 012044 doi:10.1088/1757-899X/213/1/012044

3. Results and Discussion

3.1. Characterization of n-alkylated PPTA

As reported, the N-alkylation of PPTA can improve its solubility in organic solvent, such as THF. The molecular weight and distribution of PPTA can be measured by the results obtained from GPC. However, what alkyl group is more suitable? Therefore a series of n-alkylated PPTAs with different length of flexible alkyl chains were synthesized to study the effect of alkyl chains on the properties of n-alkylated PPTA. The FTIR spectra of PPTA and n-alkylated PPTA samples are shown in the Figure 1. A typical signal peak attribute to N-H in amide stretching vibration (3300 cm^{-1}) was almost absent in spectra of N-substituted polymers. Compared with curve of c, d, e, the absorption peak in the curve of b is weak, which may be caused by the degree of n-alkylation. The two strong absorption peaks at 2870 cm^{-1} and 2918 cm^{-1} were attributed to H-C- stretching vibrations bond, which demonstrated that distinctive of methylene ($-CH_2-$) groups in all N-substituted polymers. And the peaks at 1462 cm^{-1} and 725 cm^{-1} also indicated the presence of the alkyl group in N-alkylated polymers. The peaks at 1650 cm^{-1} (C=O stretching), 1540 cm^{-1} (CNH) and 1310 cm^{-1} (CNH) attributed to amide I, II, III bands in PPTA, respectively was reported[10] . In the cases of N-alkylated PPTAs, the amide of N-H at 3300cm^{-1} was almost absent.

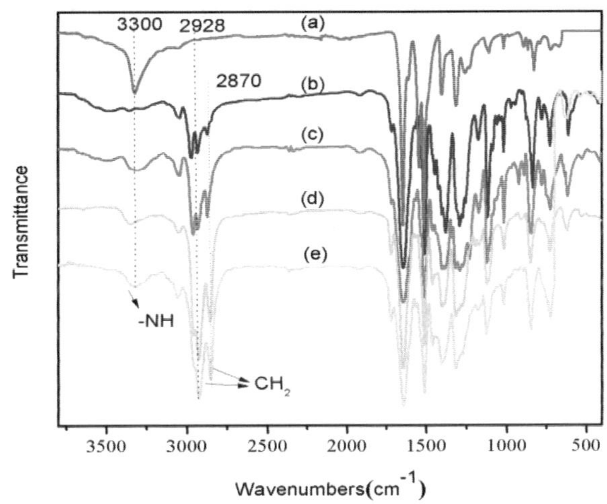

Figure 1. Spectras of PPTA and n-alkylated PPTA: (a) PPTA, (b) n-ethyl PPTA, (c) n-propyl PPTA, (d) n-butyl PPTA, (e) n-decylPPTA.

The solubility of PPTA and N-alkylated PPTAs in organic solvents, such as DMSO, THF and $CHCl_3$ were tested as shown in Table 1. N-alkylated PPTAs can be dissolved in DMSO, THF and $CHCl_3$, no matter what the alkyl groups, which implies that the solubility of n-alkylated PPTA in organic solvent is not affected by length of side chain alkyl group. However, the degree of substitution is reduced with the number of carbon atoms increasing (as shown in the Table 1), when the carbon atoms is 10, the substitution degree is about 85%. But there is no difference of the solubility between n-alkylated PPTA with shorter alkyl and high substitution degree and n-alkylated PPTA with longer alkyl and rather low substitution degree. This may be due to the small molecular chains is easier to react with PPTA in DMSO than long chains, and when a longer side substitution group was introduced in the PPTA backbone chains, hydrogen bondings between backbone molecular chains would be broken. The introduction of side chains decreased the chain-to-chain interactions such as hydrogen bonding between amide groups, which discrease the close packing and regularity of polymer chains, leading to an improvement in solubility and decrease in crystallinity.

Table 1. Degree of substitution and solubility of n-alkylated PPTAs.

Sample	Alkyl chain	Degree of substitution (mole%)	Solubility		
			THF	$CHCl_3$	DMSO
1			-[b]	-	-
2	n-ethyl	98.74	+[a]	+	+
3	n-propyl	97.89	+	+	+
4	n-butyl	95.21	+	+	+
5	N-decyl	85.69	+	+	+

[a] + soluble at room temperature;
[b] - insoluble

3.2. Molecular weight and distribution of n-alkylated PPTA

The metallization reaction of PPTA did not cause any degradation of the original PPTA was reported and there was no change in molecular weight of PPTA, the molecular-weight distribution of n-alkylated PPTA would be equal to the original PPTA even after the n-alkylation reaction[7]. As we found that the n-alkylated PPTA could be dissolved in the THF and $CHCl_3$, therefore GPC measurement was taken by using THF as the solvent. To learn the effect of the length of alkyl chains on the molecular weight and distribution of n-alkylated PPTA, we obtained the molecular weigth (Mn, Mw). The results in the Table 2 showed that the alkyl length has little effect on the results of molecular weight and weight distribution.

Table 2. Results of GPC measurements for n-alkylated PPTA PPTA (η_{inh}=4.6 dl·g^{-1}).

Alkyl chain	Mn (g·mol−1)	Mw (g·mol−1)	Mz (g·mol−1)	Mw/Mn	MwGPC (g·mol−1)
n-propyl	15800	27300	41100	1.73	20100
n-butyl	16300	28500	43900	1.74	19400
n-decyl	20900	35700	66300	1.71	19500

The molecular weight of a series of n-alkylated PPTA samples with different molecular weight were measured by using the GPC measurement in THF and intrinsic viscosity in concentrated sulfuric acid to determine the relationships between intrinsic viscosity and weight molecular weight. Molecular weight of alkylated PPTAs obtained by GPC analysis are plotted in Figure 2, the weight-averaged (Mw) and number-averaged(Mn) molecular weights of these polymers obtained by the GPC and intrinsic viscosity (η_{inh})were summarized in Table 3.

$$[\eta] = \mathbf{K}M_w{}^{\alpha} \tag{3}$$

$$[\eta] = 2764M_w{}^{1.386} \tag{4}$$

Table 3. Results of GPC measurements for n-alkylated PPTA.

	PPTA		N-alkylated PPTA				
No	η_{inh} (dl·g^{-1})	Mw_{GPC} (g·mol^{-1})	Alkyl chain	Mn (g·mol^{-1})	Mw (g·mol^{-1})	Mz (g·mol^{-1})	Mw/ Mn
1	2.6	10900	n-ethyl	8700	13500	19400	1.56
2	4.6	21100	n-propy l	16600	31000	49100	1.87

3	5.4	28500	n-butyl	21100	42000	67800	1.99
4	6.8	39100	n-hexyl	32000	64000	100700	2.00
5	9.0	60800	n-hexyl	50400	99600	158500	1.97

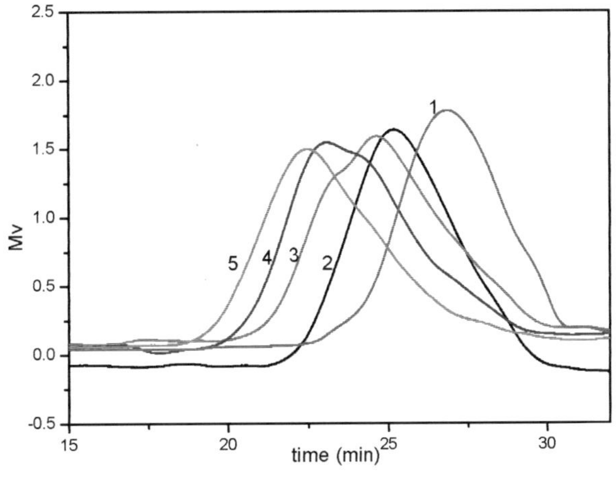

Figure 2. GPC results of n-alkylated PPTA samples with different molecular weight (η_{inh}dl·g^{-1}): 1--2.6, 2--4.6, 3--5.4, 4--6.8, and 5-9.0.

The intrinsic viscosity relates to the viscosity average molecular weight (M_w) according to the Mark-Houwink equation (3).The power exponent (α) and the K are estimated by using the method of least squares linear fit from the log-log plot of M_W and η_{inh}. According to the Figure 3 and we get the slope of the line, and the relationship of intrinsic viscosity of the PPTA and the molecular weight is to the power 1.386, and the constant K is 2764. We obtained the equation (4) which applies to the PPTA with the intrinsic viscosity from 2.6 dl·g^{-1} to 9.0 dl·g^{-1}. Figure 3 showed a good linear relationship between M_{WGPC} and η_{inh} for PPTA between $10900 \leq M_W \leq 60800$. Also ratios of Mw/Mn seemed to increase with increasing viscosities of the PPTA from 2.6 to 6.8 as shown in Table 3, and when the viscosity increase , the ratio is closer to 2.0 , which conforms to Flory's most probable distribution[11], which imply that molecular weight of UHMWPPTA can be measured by the GPC of n-alkylation PPTA.

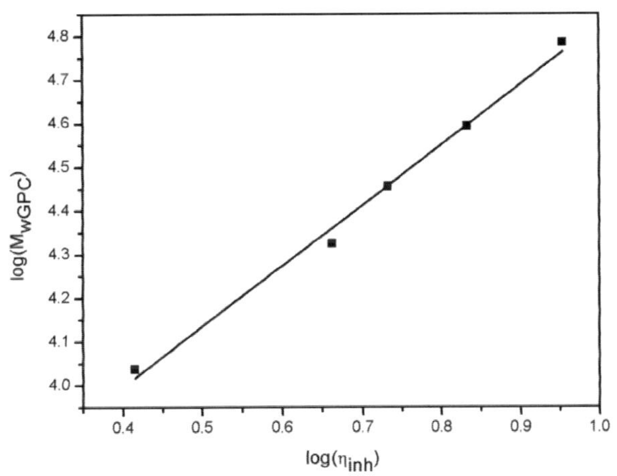

Figure 3. Molecular weight relationship of (M_{wGPC}) characterized by the GPC measurement with n-alkylated PPTAs in THF and (ηinh) intrinsic viscosity measurement of PPTA in concentrated sulfuric acid.

4. Conclusion

N-alkylated PPTA with various molecular weights and various alkyl groups, including ultra-high molecular weight PPTA were synthesized and characterized. The results from FTIR, The solubility characterization showed that the length of alkyl side chains has effect on molecular packing structure and substitution degree of the products. But as the substitution degree is different, there is no effect on their solubility in THF and $CHCl_3$. The weight-average molecular weight and its distribution of n-alkylated PPTA were characterized by GPC measurement in THF, and the intrinsic viscosity measurement of PPTA in concentrated sulfuric acid was measured. The results showed a good linear relationship between M_{WGPC} and ηinh of PPTA with molecular weight from 10900 to 60800, which imply that molecular weight of SHMWPPTA can be measured by the GPC measurement of n-alkylation PPTA.

Acknowledgement

This work was financially supported by the National Natural Science Foundation of China (Grant No.51603120, 11604204), State Key Laboratory for Modification of Chemical Fibers and Polymer Materials, Donghua University(LK1602) and the Open Project Program of High-Tech Organic Fibers Key Laboratory of Sichuan Province.

References

[1] Arpin M and Strazielle C 1977 *Polymer* **18** 591-8
[2] Baird D G and Smith J K. 1978 *J. Polym. Sci.: Polym. Chem. Edition* **16** 61-70
[3] Ying Q, Chu B, Qian R, Bao J, Zhang J and Xu C 1985 *Polymer* **26** 1401-7
[4] Chu B, Ying Q, Wu C, Ford J R and Dhadal H S 1985 *Polymer* **26** 1408-18
[5] Ying Q and Chu B 1986 *Macromolecules* **19** 1580-6
[6] Chu B, Ying Q, Wu C, Ford J R and Dhadwal H 1984 *Polymer Communications* (Guildford, England) **25** 211-3
[7] Ogata N, Sanui K and Kitayama S 1984 *J. Polym. Sci., Part A: Polym. Chem.* **22** 865-7
[8] Zhang R, Kong H, Zhong H, Liu J, Zhou J, Teng C, Ma Y and Yu M 2012 *Adv. Mater. Res.* 554-556 105-9
[9] Zuo J, Li S, Bouzidi L and Narine S S 2011 *Polymer* **52** 4503-16
[10] Motowo T T K 1981 *J. Polym. Sci.***19** 1133-45
[11] Naoya O K S and Shinichiro K 1984 *J. Polym. Sci. Polym. Chem.* **22** 865-7

IOP Conf. Series: Materials Science and Engineering **213** (2017) 012045 doi:10.1088/1757-899X/213/1/012045

Optimization of preparation process and characterization of carboxymethyl chitosan/sodium alginate hemostatic sponge

Z Hu[1,4], Q Q Ouyang[1], Y Cheng[1], P Z Hong[2], M N Liao[1], F J Chen[3] and S D Li[1,4]

[1]Faculty of Chemistry and Environment Science, Guangdong Ocean University, Zhanjiang 524088, China
[2]College of Food Science and Technology, Guangdong Ocean University, Zhanjiang 524088, China
[3]Faculty of Ocean and Meteorology, Guangdong Ocean University, Zhanjiang 524088, China

E-mail: huzhangqyx@126.com; sidongligdou@163.com

Abstract. Composite hemostatic sponge was prepared by vacuum freeze-drying using carboxymethyl chitosan and sodium alginate as the main materials and $CaCl_2$ as a crosslinking agent. On the basis of single factor experiments, an orthogonal experiment was carried out to optimize the preparation process of hemostatic sponge. The appearance, water absorption, porosity ratio, and in vitro hemostasis of the sponge were evaluated. The optimum conditions to prepare hemostatic sponge were obtained as follows: mass ratio of sodium alginate to carboxymethyl chitosan 4: 1, mass fraction of $CaCl_2$ 2%, and crosslinking temperature 30°C. The hemostatic sponge prepared under such conditions was off-white and porous. Its water absorption and porosity ratio were 3050% and 67.23%, respectively. Meanwhile, the hemostatic sponges had significant in vitro procoagulant activity. Therefore, the hemostatic sponge is expected to be developed as a novel medical material.

1. Introduction

With a special porous structure which facilitates the discharge of wound secretions, sponge is widely used in surgical hemostasis in many aspects, such as neurosurgery, dentistry, otorhinolaryngology and gynecology [1-4]. Many polymer materials can be used to prepare sponge. Particularly, the development of biomedical sponge with natural biopolymer has attracted much attention from researchers.

Chitosan, a linear polysaccharide consisting of 2-acetamido-2-deoxy-D-glucose linked by β-1, 4-glycosidic bonds, is the only natural alkaline polysaccharides up to day. Owing to its excellent biocompatibility, biodegradability, antibacterial and accelerating wound healing properties, chitosan has been widely used in delivery system for drugs, wound dressing and scaffolds for tissue engineering [5-7]. Carboxymethyl chitosan (CMCS) is a modified derivative of chitosan with high water solubility, and has great potential for application in the biomedical field [8-9].

Sodium alginate (SA), a seaweed polysaccharide extracted from brown algae, has good biocompatibility and hemostatic function and has been widely applied in tissue engineering [10-11]. Additionally, SA is particularly suitable for the treatment of burns and mechanical trauma due to its strong water absorptivity and gelation ability, which can relieve pain and prevent secondary bleeding from the wound by forming gels on the wound surface.

In this study, the composite sponge was prepared by freeze-drying using CMCS and SA as the matrix materials. The preparation conditions were optimized with water absorption as an evaluation index, and the related properties were examined to develop a new medical sponge material with excellent hemostasis.

2. Materials and methods

2.1. Materials

CMCS (degree of substitution, 63.5%) was purchased from Zhejiang Aoxing Biotechnology Co., Ltd (China). SA (chemically pure) was purchased from Zhengzhou Jianda Chemical Products Co., Ltd (China). The other reagents were of analytical grade.

2.2. Preparation of sponge

3% CMCS and 3% SA solutions were prepared, respectively, and then mixed at a certain mass ratio. $CaCl_2$ solution was slowly added at a 1/5 (v/w) ratio under stirring. Then, 10 g of the mixture was dispensed into a 5-cm glass culture dish and cross-linked at a constant temperature for 2 h. The dishes were then frozen for 24 h in a refrigerator at -20°C and dried in a vacuum freeze dryer for 18 h to obtain porous composite sponges.

2.3. Determination of water absorption

The sponge was placed in a desiccator containing anhydrous $CaCl_2$ and dried at room temperature to constant weight (W_0), and then immersed in distilled water to fully swell for 10 min. Afterwards the sponge was taken out from the water and weighed (W_1) after surface water was removed with filter paper. The water absorption was calculated as follows:

$$\text{Water absorption (\%)} = (W_1 - W_0)/W_0 \times 100\% \tag{1}$$

where W_0 and W_1 are the mass of the sponge before and after water absorption, respectively, in g.

2.4. Morphological observation

The appearance and pore structure of the sponge were observed by a DSC-TX10 digital camera (SONY Corporation, Tokyo, Japan) and a scanning electron microscope (SEM) JSM-6330F (JEOL Corporation, Tokyo, Japan).

2.5. Porosity ratio determination

A 50-mL volumetric flask was filled with ethanol to the calibration line, and weighed as W_1. A piece of sponge was weighed as W_0. The sponge was placed in the flask containing 2/3 volumetric ethanol and degassed to allow ethanol to penetrate into the pores of the sponge. And then ethanol was again added to the calibration line. The flask containing ethanol and sponge was weighed as W_2. Afterwards, the sponge soaked with ethanol was taken out and the flask containing the remaining ethanol was weighed as W_3. The porosity ratio of the sponge was calculated as equation (2):

$$\text{Porosity ratio (\%)} = (W_2 - W_3 - W_0) / (W_1 - W_3) \times 100\% \tag{2}$$

2.6. Procoagulant activity in vitro

The in vitro procoagulant activity of the sponge was determined according to the literature method [12]. The specific operations are as follows: 0, 20, 60, 100, 140, 180 mg of sponge fragments were placed in clean test tubes, respectively. The tubes were then placed in a water bath at a constant temperature of 37°C for 1 h. The anticoagulated rabbit blood (1 mL) was added to each of the above test tubes in turn and kept in the water bath. The tubes were tilted at 30° once every 30 seconds until the blood in the tubes did not flow. The coagulation times of blood in different tubes were recorded.

2.7. Statistical analysis

All the experiments were parallel performed in triplicate and the mean values were presented. All statistical analyses were performed using Statistica 7.0 (StatSoft Inc., USA).

3. Results and discussion

3.1. Single factor experiments

3.1.1. Effect of SA/CMCS mass ratio on water absorption of the sponge. The SA and CMCS were mixed at a mass ratio of 1: 1, 2: 1, 3: 1, 4: 1, and 5: 1, respectively. $CaCl_2$ solution with a mass fraction of 1.0% was slowly added at a 1/5 (v/w) ratio under stirring. Next, 10 g of the mixture was dispensed into a 5-cm glass culture dish and cross-linked at 25°C for 2 h. The dishes were then incubated for 24 h in a refrigerator at -20°C and dried in a vacuum freeze dryer for 18 h to obtain the sponge. The water absorption of the sponge was measured and the results were shown in Figure 1. The water absorption increased with the increase of the SA proportion, and peaked at a SA/CMCS mass ratio of 4: 1. Afterwards the water absorption decreased with further increase of the SA proportion. Therefore, three levels of SA/CMCS mass ratio of 3: 1, 4: 1, and 5: 1 were used for optimization of sponge preparation by orthogonal experiment.

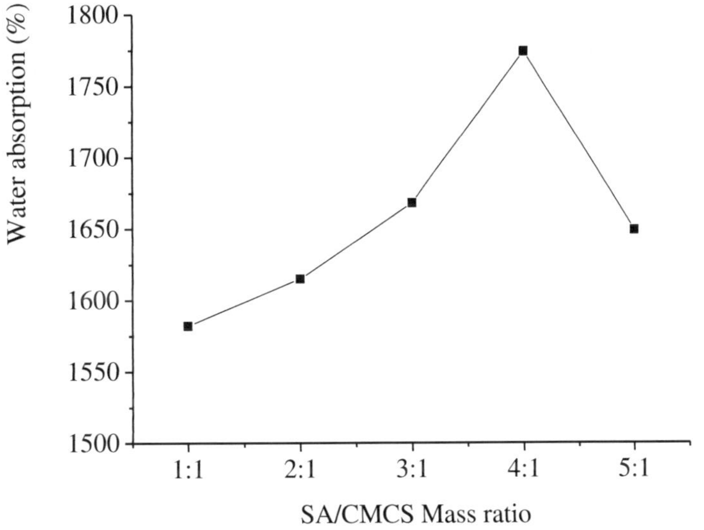

Figure 1. Effect of SA/CMCS mass ratio on water absorption of the hemostatic sponge.

3.1.2. Effect of mass fraction of $CaCl_2$ on water absorption of the sponge. The sponge was prepared at a SA/CMCS mass ratio of 3: 1 and 25°C with mass fraction of $CaCl_2$ 1.0%, 2.0%, 3.0%, 4.0% and 5.0%, respectively. The water absorption was measured and the results were shown in Figure 2. The water absorption of the sponge peaked at mass fraction of $CaCl_2$ 2.0%, and then decreased sharply with further increase of the mass fraction of $CaCl_2$. This may be because the too high calcium chloride concentration led to a high degree of cross-linking in the sponge, thus resulting in tightness of the sponge and consequently decrease in water absorption. Therefore, the hemostatic sponge preparation was optimized by orthogonal experiment with mass fraction of $CaCl_2$ of 1.0%, 2.0% and 3.0%.

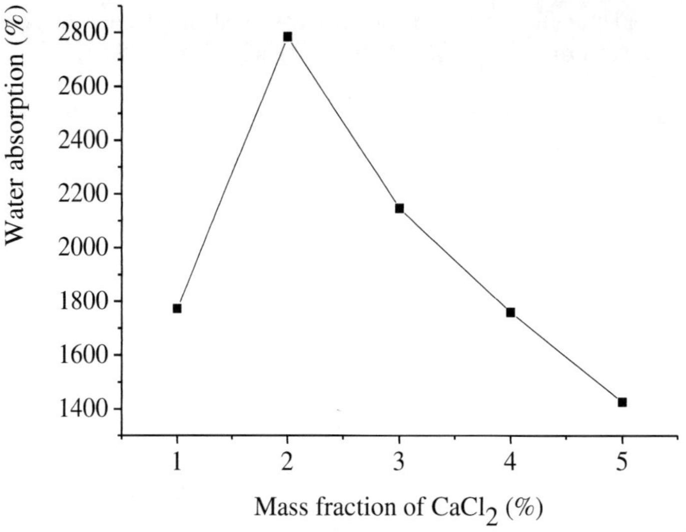

Figure 2. Effect of mass fraction of $CaCl_2$ on water absorption of the hemostatic sponge.

3.1.3. Effect of cross-linking temperature on water absorption of the sponge. The sponge was prepared by cross-linking for 2 h on standing at 25, 30, 35, 40, 45, and 50°C, respectively, with SA/CMCS mass ratio of 3: 1 and 1.0% $CaCl_2$ solution. The water absorption was measured and the results were shown in Figure 3. The water absorption of the sponge peaked at 35°C. It is because too high temperature will accelerate cross-linking and result in an excessive cross-linking degree, thus narrowing the pores in the sponge. Therefore, optimization of the hemostatic sponge preparation was carried out at 30, 35 and 40°C.

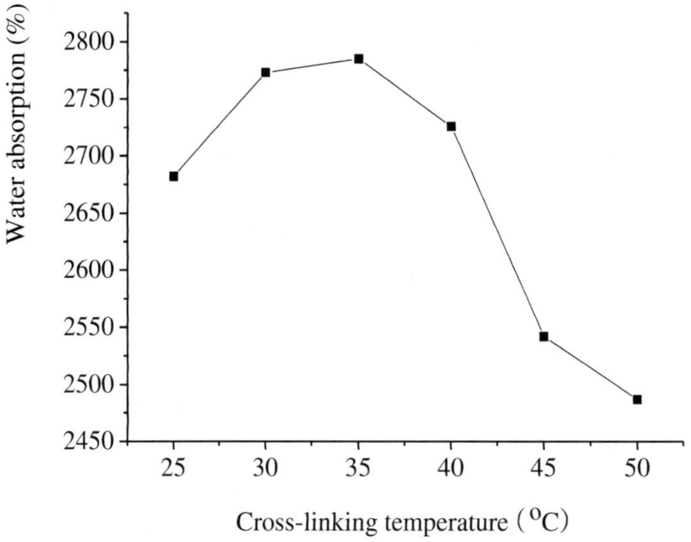

Figure 3. Effect of cross-linking temperature on water absorption of the hemostatic sponge.

3.2. Optimization of the hemostatic sponge preparation

Based on the results of the single factor experiments, an orthogonal experiment was designed with three factors and three levels by using water absorption as an evaluation index to optimize the hemostatic sponge preparation. The factors and levels are listed in Table 1 and the experiment design and results are shown in Table 2.

Table 1. Factors and levels for orthogonal experiment.

	1	2	3
A: SA/CMCS mass ratio	3:1	4:1	5:1
B: Mass fraction of $CaCl_2$ (%)	1.0	2.0	3.0
C: Cross-linking temperature (°C)	30	35	40

Table 2. Design, results and range analysis of L_9 (3^4) orthogonal experiment.

Run	Factors			Water absorption (%)
	A	B	C	
1	1	1	1	1907
2	1	2	2	2052
3	1	3	3	2019
4	2	1	2	2087
5	2	2	3	2991
6	2	3	1	2166
7	3	1	3	1531
8	3	2	1	2644
9	3	3	2	2020
k1	1993	1842	2239	
k2	2415	2562	2053	
k3	2065	2068	2180	
R	422	720	186	

Range analysis revealed that the ranking order of the factors influencing the water absorption of hemostatic sponge was B> A> C. In order words, mass fraction of $CaCl_2$ had the greatest influence, followed by SA/CMCS mass ratio and cross-linking temperature. According to the results of orthogonal experiment, the optimum preparation conditions were $A_2B_2C_1$, which was that SA/CMCS mass ratio 4: 1, mass fraction of $CaCl_2$ 2% and cross-linking temperature 30°C. As confirmed by three parallel experiments, the average water absorption of hemostatic sponge prepared under the optimum conditions was 3050%, representing strong water absorption capacity.

3.3. Morphology Studies

Macroscopic observation showed that the hemostatic sponge was off-white, tough and porous, meanwhile, it became swollen and soft after being soaked in water. SEM showed a clear reticular structure, pores with a uniform size and appropriate density (Figure 4).

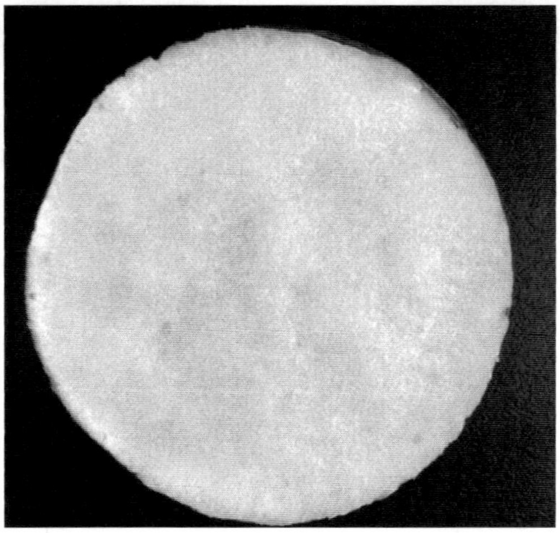

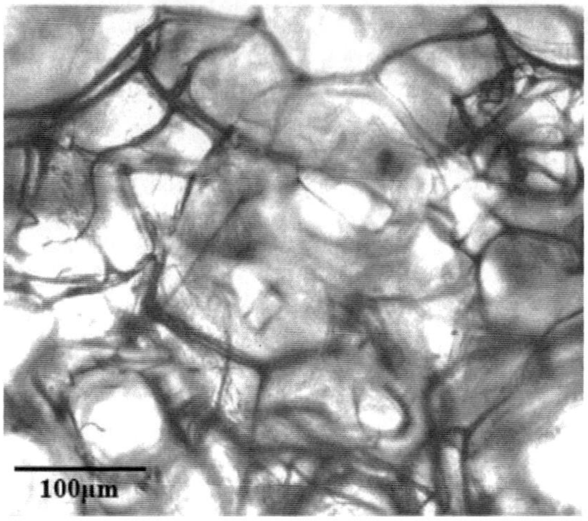

(a) (b)

Figure 4. The optical microscopy (a) and SEM micrograph (b) of the hemostatic sponge.

3.4. Sponge porosity ratio

The porosity ratio of the hemostatic sponge was investigated and the results were shown in Table 3. The average porosity ratio was 67.23% and the relative average deviation was 1.55%. The hemostatic sponge with a high porosity ratio provides enough pores and reticular space to absorb more blood. Moreover, it provides great contact surfaces between the sponge and the wound, which is conducive to the interaction between hemostatic materials and coagulation factors resulting in rapid hemostasis [13].

Table 3. Porosity ratio of the hemostatic sponge.

	1	2	3
W_0 (g)	0.013	0.018	0.016
W_1 (g)	56.52	56.52	56.52
W_2 (g)	56.43	56.42	56.43
W_3 (g)	56.22	56.15	56.19
Porosity ratio (%)	65.67	68.11	67.91
Average porosity ratio (%)		67.23	
Relative average deviation (%)		1.55	

3.5. In vitro procoagulant effect of the hemostatic sponge

Coagulation time is the time required for blood to clot outside of the body, which is an important indicator of procoagulant activity of the hemostatic materials. As shown in Table 4, the addition of sponge decreased the coagulation time of anticoagulated rabbit blood, indicating that the hemostatic sponge had significant procoagulant activity. Moreover, the coagulation time was significantly decreased with the increase of the sponge, demonstrating a dose-dependent relationship.

Table 4. Procoagulant effect of the composite sponge in vitro.

Mass of sponge (mg)	0	40	60	100	140	180

Time (mins)	>30 (not clotted)	7.0	4.5	3.0	1.5	1.0

4. Conclusions

The composite porous sponge was prepared by freeze-drying using CMCS and SA as the matrix. Based on the results of single factor experiments, an orthogonal experiment was performed to optimize the preparation conditions to develop the hemostatic sponge with excellent performance as a novel medical biomaterial.

Acknowledgments

The authors gratefully acknowledge the financial support by Guangdong Natural Science Foundation (2016A030308009 and 2016A030312004), Guangdong Scientific and Technological Planning Project (2015A020216019) and the Project of Enhancing School with Innovation of Guangdong Ocean University (2015KTSCX053, GDOU2016050255 and 2014KZDXM038).

References

[1] Deng C M, He L Z, Zhao M, Yang D and Liu Y 2007 Biological properties of the chitosan-gelatin sponge wound dressing *Carbohydr. Polym.* **69** 583-9
[2] Choi Y S, Hong S R, Lee Y M, Song K W, Park M H and Nam Y S 1999 Study on gelatin-containing artificial skin: I. Preparation and characteristics of novel gelatin-alginate sponge *Biomaterials* **20** 409-17
[3] Lai H L, Abu'Khalil A and Craig D Q 2003 The preparation and characterisation of drug-loaded alginate and chitosan sponges *Int. J. Pharm.* **251** 175-81
[4] Geiger M, Li R H and Friess W 2003 Collagen sponges for bone regeneration with rhBMP-2 *Adv. Drug Deliv. Rev.* **55** 1613-29
[5] Ravi K M N V, Muzzarelli R A A, Muzzarelli C, Sashiwa H and Domb A J 2004 Chitosan chemistry and pharmaceutical perspectives *Chem. Rev.* **104** 6071-84
[6] De Arruda I N Q, Pereira V A and Stefani R 2017 Application of chitosan matrix for delivery of rutin *J. Iran. Chem. Soc.* **14** 561-6
[7] Kamel N A, El-Messieh S L and Saleh N M 2017 Chitosan/banana peel powder nanocomposites for wound dressing application: Preparation and characterization *Mater. Sci. Eng. C* **72** 543-50
[8] Jiang Z, Han B, Liu W and Peng Y 2017 Evaluation on biological compatibility of carboxymethyl chitosan as biomaterials for antitumor drug delivery *J. Biomater. Appl.* **31** 985-94
[9] Jiang Z, Han B, Li H, Yang Y and Liu W 2017 Carboxymethyl chitosan represses tumor angiogenesis in vitro and in vivo *Carbohydr. Polym.* **129** 1-8
[10] Wrzeszcz K, Marycz K, Zimoch-Korzycka A and Jarmoluk A 2013 Crosslinking of sodium alginate to hydrogels for tissue engineering *Przem. Chem.* **92** 1018-22
[11] Ma G, Fang D, Liu Y, Zhu X and Nie J 2012 Electrospun sodium alginate/poly(ethylene oxide) core–shell nanofibers scaffolds potential for tissue engineering applications *Carbohydr. Polym.* **87** 737-43
[12] Chen Q 2011 *Research Methods in Pharmacology of Chinese Materia Medica* (Beijing: People's Health Publishing House) p 226
[13] Liu H, Zou Q and Wu D 2014 In vitro characteristics and hemostatic effect of *O*-carboxymethyl chitosan multi-hemostatic sponge *Herald Med.* **33** 850-52

Preparation and thermal properties of carbon Nanotubes/5-Amino-1H-Tetrazole energetic composites

C H Jia[1], S T Wang[1], Y C Li[1,2] and S P Pang[1]

[1]School of Materials Science and Engineering, Beijing Institute of Technology, Beijing 100081, China

E-mail: liyuchuan@bit.edu.cn

Abstract. Carbon nanotubes were prepared in large quantities by fluidized bed method and nitrogen-doped carbon nanotubes were prepared by chemical vapor deposition method. Carbon nanotubes/5-amino-1H-tetrazole composites were prepared by ultrasonic resonance method. The appearance characteristics and thermal properties of composites were analyzed by scanning electron microscope (SEM), transmission electron microscope (TEM), thermogravimetry (TG) and differential scanning calorimeter (DSC).

Keywords: Carbon nanotubes; Nitrogen-doped carbon nanotubes; 5-Amino-1H-tetrazole; Energetic composite

1. Introduction

In 1991, Iijima [1] discovered carbon nanotubes by using an electron microscope to observe the products of graphite-dc discharge. Carbon nanotubes are generally small in diameter, the outside diameter is generally several nanometers to several tens nanometers and inner diameter is just a few nanometers. Carbon nanotubes have excellent properties in physics and chemistry, especially the excellent mechanical and electrical properties [2].

The nanotube network in the nitrogen-doped carbon nanotubes has many structural defects due to the effect of the nitrogen doping. These structural defects include carbon vacancies and/or the inclusion of pentagons or heptagons within the carbon hexagonal network. The chemical environment of nitrogen atoms in the nanotubes were suggested that one nitrogen atom bonded to two carbon atoms, which were arranged as "pyridine-like" structures [3]; or one nitrogen atom bonded to three carbon atoms, which were arranged as "graphite-like" structures [4]. And the most common types of the nitrogen-dopants in the nitrogen-doped carbon nanotubes are substituted or accompanied by the incomplete-bonding defects [5-6]. Therefore, the existent of the large number of defects and reactive sites leads to a higher density and dispersion of functional groups on the surface of nitrogen-doped carbon nanotubes [7].

Composed with carbon nanotubes make energetic materials to achieve nano-level dispersion in the composites [8]. In addition, good thermal conductivity of carbon nanotubes can improve the heat transfer rate greatly in the reaction process of energetic materials [9].

In this paper, Nitrogen-doped carbon nanotubes, carbon nanotubes and 5-amino-1H-tetrazole composites were prepared by simple physical mixing and ultrasonic resonance methods [10]. Through

IOP Conf. Series: Materials Science and Engineering **213** (2017) 012046 doi:10.1088/1757-899X/213/1/012046

the control experiment, the effects of carbon nanotubes mass fraction on the thermal decomposition of the composites were studied.

2. Experimental

2.1. Materials and equipment

Materials: Carbon nanotubes were prepared according to literature [11-12] and nitrogen-doped carbon nanotubes were prepared according to literature [13-14]; 5-Amino-1-H-tetrazolium (5-ATz, **A**) and other chemical reagents and solvents were purchased and used as supplied without further purification. Analytical equipment: Scanning Electron Microscope (SU8010; Tokyo, Japan); Transmission Electron Microscope (JEM-2010; JEOL); Thermogravimetric analyzer (Netzsch STA 449 C; Selb, Germany); Differential Scanning Calorimeter (DSC Q2000; New Castle, USA).

2.2. The preparation of nanotubes/energetic composites

5-Amino-1H-tetrazole (**A**, 1.02 g, 12 mmol) was dissolved in 25 mL ethanol. Stirring for 20 min, the precipitate residue was removed by filtration and the saturated solution of **A** was obtained at room temperature. Then carbon nanotubes were added (mass fraction 10%), stirring for 10 min, ultrasonic resonance for 30 min, then solvent was evaporated to give the solid power (**B**). By the same method, we obtained the composites of **A** and nitrogen-doped carbon nanotubes with the mass fraction 5% (**E**). The composite materials of carbon tube mass fraction of 5% (**C**), 10% (**D**) were obtained by physical mixing method.

3. Results and discussion

3.1. SEM/TEM

The SEM image of nitrogen-doped carbon nanotubes (N-CNT) is shown in Figure 1; the TEM image of carbon nanotubes (CNT) is shown in Figure 2 and the TEM image of CNT/5-ATz is shown in Figure 3.

From Figure 2 and 3, the outer diameter of the non-loaded carbon nanotubes is about 3 nm, the TEM is clear and bright. The composite pipe diameter is thickened, about 6 nm, the tube is dark, and the local tube has granular solids.

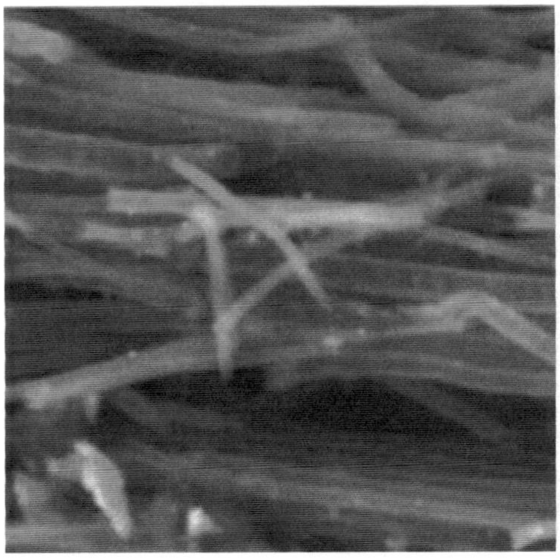

Figure 1. SEM image of N-CNT.

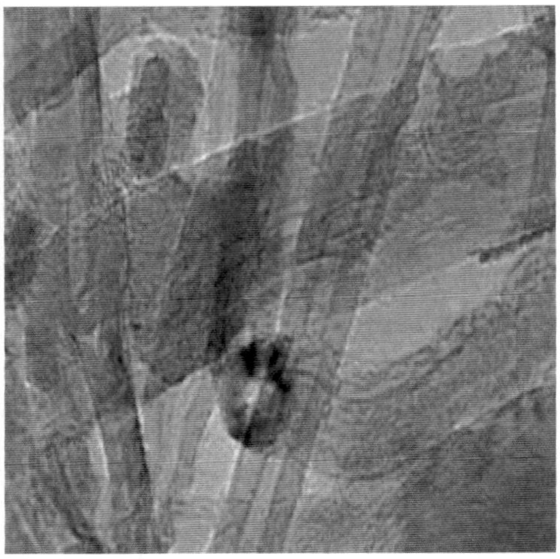

Figure 2. TEM image of CNT.

IOP Conf. Series: Materials Science and Engineering **213** (2017) 012046 doi:10.1088/1757-899X/213/1/012046

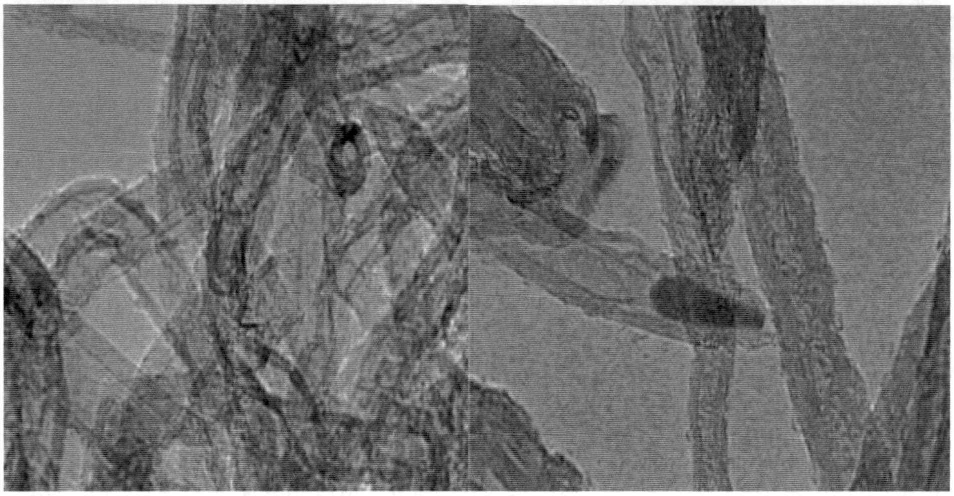

Figure 3. TEM image of CNT/5-ATz composites.

3.2. Thermogravimetric analysis

The TG and DSC curves of **A**, **B**, **C**, **D** and **E** under Ar atmosphere are shown in Figure 4. The DSC curves show that the thermal decompositions of all samples involved a single exothermic process. At the same time, the TG curves provided insight into the total weight losses from the 5-ATz (**A**) associated with this process [15]. Characteristic temperatures in the TG-DSC curves of the samples are shown in Table 1.

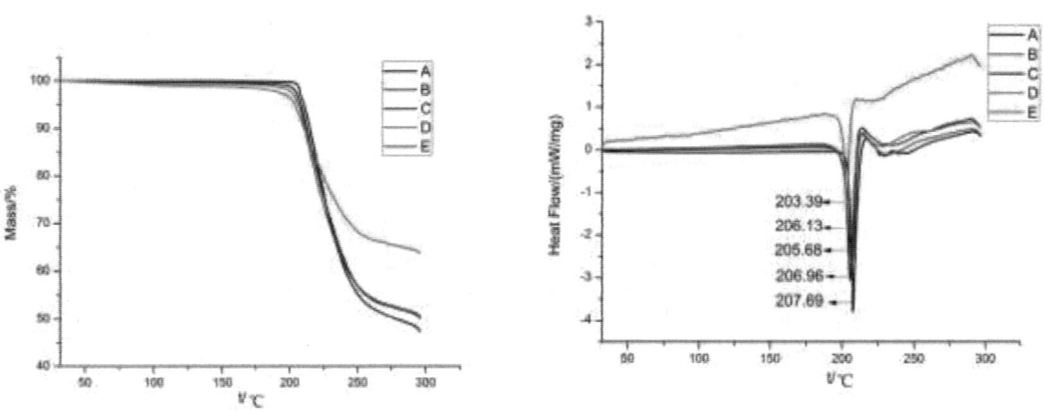

Figure 4. TG traces of the samples (L) and DSC traces of the samples (R).

Table 1. Characteristic temperatures in the TG-DSC curves of the samples.

	A	B	C	D	E
$^aT_0$/°C	203.82	181.03	185.34	189.35	174.32
^bT_P/°C	207.69	206.96	206.13	205.68	203.39

$^aT_0$ is the initial temperature of the TG curves; ^bT_P is the peak temperature of the DSC curves.

From TG data, it can be seen that the initial decomposition temperature of weightlessness process for **A** is 203.82 °C, for **B** is 181.03 °C, for **C** is 185.34 °C, for **D** is 189.35 °C, and for **E** is 174.32 °C, respectively. Comparing the data of **B** and **C**, compared with the sample, which were prepared by grinding the mixture of **A** and carbon nanotubes, the initial thermal decomposition temperature of the composites obtained by ultrasonic resonance method is about 4 °C lower. Comparing the data of **C** and **D**, the initial thermal decomposition temperature for the sample obtained by grinding the mixture of **A** and 5% carbon nanotubes is lower than the mixture of **A** and 10% carbon nanotubes about 4 °C. For **E**, the composite of **A** and nitrogen-doped carbon nanotubes, the thermal decomposition temperature was obviously lower than **A** 30 °C; compared with **B** it also decreased by 7 °C. Through the above analysis, the following conclusion were drawn: the excellent heat conductivity of carbon nanotubes reduces the thermal decomposition temperature of **A** in the composite materials; the presence of nitrogen in the nitrogen-doped carbon nanotubes makes it better for conducting thermal conductivity. Therefore, the reduction of the thermal decomposition temperature of **A** after forming the composite is more obvious. This result is consistent with other reports [16-18].

From DSC data, it can be seen that compared with **A**, the maximum peak temperatures of heat decomposition for **B**, **C** and **D** were not obviously changed; the exothermic peak temperature of thermal decomposition process for nitrogen-doped carbon nanotubes composites is about 4 °C in advance compared with the else. Compared the DSC exothermic data with the other groups of carbon nanotube composites, the nitrogen-doped carbon nanotubes have greatly increased on the release of heat because of containing nitrogen.

4. Conclusions

The thermal conductivity of carbon nanotubes reduces the thermal decomposition temperature of 5-amino-1H-tetrazole. Compared with the carbon nanotubes, the nitrogen-doped carbon nanotubes have much better conductance about thermal conductivity due to nitrogen element in the carbon nanotubes, make the thermal decomposition temperature of the 5-amino-1H-tetrazole, which in composites, decreases more obviously and the heat of the decomposition was improved greatly. In addition, compared with the composites formed by physical mixing, the composites formed by ultrasonic resonance make the thermal decomposition temperature of 5-amino-1H-tetrazole much lower.

Acknowledgements

The authors gratefully acknowledged the financial support from the National Natural Science Foundation of China (21576026 and U153062).

References

[1] Iijima S 1991 *Nature* **354**(6348) 56-8
[2] Wang J, Zhao Y T and Wang Q L 2005 *Journal of Panzhihua University* **22**(5) 113-5 (*In Chinese*)
[3] Czerw R *et al* 2001 *Nano Lett.* **1**(9) 457-60
[4] Liao L, Fang P and Pan C 2011 *Journal of Nanoscience and Nanotechnology* **11**(2) 1060-7
[5] Li Y C, Teng P Y, Yeh C H, Koshino M, Chiu P W and Suenaga K 2015 *Nano Lett.* **15** 7408
[6] Lee W J, Lim J and Kim S O 2017 *Small Methods* **1** 1600014
[7] Terrones M *et al* 2004 *Mater. Today* **7**(10) 30-45
[8] Jin M M and Luo Y J 2014 *Acta Armamentarii* **35**(6) 822 (*In Chinese*)
[9] Wang J N, Li X D and Yang R J 2006 *Chinese Journal of Explosives & Propellants* **29**(2) 44-7 (*In Chinese*)
[10] Li X X, Li X D and Yang R J 2004 *Journal of Energetic Materials* **12** 143 (*In Chinese*)
[11] Cui C J, Qian W Z and Wei F 2011 *Acta Physico-Chimica Sinica* **27**(10) 2462–8 (*In Chinese*)
[12] Wen Q, Qian W Z and Wei F 2008 *Chinese Journal of Catalysis* **29**(7) 617 (*In Chinese*)
[13] Chai Y, Yu L G, Wang M S, Zhang Q F and Wu J L 2006 *Acta Scientiarum Naturalium Universitatis Pekinensis* **42**(1) 89-92 (*In Chinese*)
[14] He W, Wang L M, Cai W, Tang C and Yao H 2016 *Journal of Materials Engineering* **44**(2) 49-55 (*In Chinese*)
[15] Kumar R, Sharma S, Pathak D, Dhiman N and Arora N 2017 *Solid State Ionics* **305** 57-62

IOP Conf. Series: Materials Science and Engineering **213** (2017) 012046 doi:10.1088/1757-899X/213/1/012046

[16] Zhang B Y, Jiao Q J, Ren H, Jia R J, Zhang K, Wu G N and Liu G Q 2011 *Chinese Journal Energetic Materials* **19**(4) 401-4 (*In Chinese*)
[17] Gu K Z, Li X D and Yang R J 2006 *Chinese Journal of Explosives & Propellants* **29**(1) 48-51 (*In Chinese*)
[18] Zeng G Y, Lin C M, Zhou J H and Pan L P 2012 *Chinese Journal of Explosives & Propellants* **35**(6) 55-7 (*In Chinese*)

IOP Conf. Series: Materials Science and Engineering **213** (2017) 012047 doi:10.1088/1757-899X/213/1/012047

Effect of the rotational speed of on the surface quality of 6061 Al-alloy welded joint using friction stir welding

T T Feng[1], X H Zhang[1], G J Fan[1,2] and L F Xu[1]

[1]Shandong Provincial Key Laboratory of Horticultural Machineries and Equipments, Shandong Agriculture University, Tai'an 271018, China

E-mail：fttao79@163.com

Abstract. The rotational speed of the stir-welding head is an important technological parameter in friction stir welding (FSW) process. For investigating the effect of the rotational speed of the stir-welding head on the surface quality of the welded joint, in this study, the weld tests were conducted under different rotational speeds (in which the welding speed was fixed), and then the effects were analyzed using the heat-fluid analysis model established. The test results revealed that cracks or grooves could be observed on the welded joint at small rotational speeds; with the increase of rotational speed, the weld surface became bright and clean; as the rotational speed further increased, the surface of the welded joint may be over burnt. Through analysis, it can be observed that appropriate increasing the rotational speed of the stir-welding joint increased the heat input in welding; meanwhile, fewer materials participated in the formation of weld, the material's flowability was improved, and the resistance that impeded the advance of the stir-welding needle was reduced, thereby improving the quality of the welded joint.

1. Introduction

Friction stir welding (FSW) is a novel solid-state bonding technology and shows many advantages compared with the traditional welding methods. FSW can weld many metals that were hardly welded in the past [1-6]. However, when using FSW, an unreasonable setting of some technological parameters would lead to poor-quality welded joint and further affect its practical use. A high-speed rotation of the stir-welding head is the guarantee for the success implementation of FSW since it is tightly correlated with the heat input during welding. It is still unclear whether the quality of the welded joint can be improved by enhancing the rotational speed of the stir-welding joint. This paper conducted FSW weld tests at different rotational speeds and established the heat-fluid model for exploring the reason why the variation of rotational speed of the stir-welding head can affect the quality of welded joint.

2. Technological test of FSW

6061 aluminum (Al) alloy plates with a size of 150 mm×50 mm×6.3 mm were used in the present test. The weld tests were conducted under different conditions, i.e., the rotational speed of the stir-welding head was set at 450 r/min, 650 r/min and 850 r/min, respectively, during which the welding speed was fixed as 60 mm/min. Figure 1 shows the appearance of the welds. As shown in Figure 1(a), at a rotational speed of 450 r/min, the poor-quality weld was formed, accompanied by irregular surface and many grooves; at a rotational speed of 650 r/min, favorable weld with a smooth surface can be

IOP Conf. Series: Materials Science and Engineering **213** (2017) 012047 doi:10.1088/1757-899X/213/1/012047

observed in Figure 1(b); at a rotational speed of 850 r/min, as shown in Figure 1(c), rough selvedge was produced, and over burning-induced blurring can also be found on the weld surface.

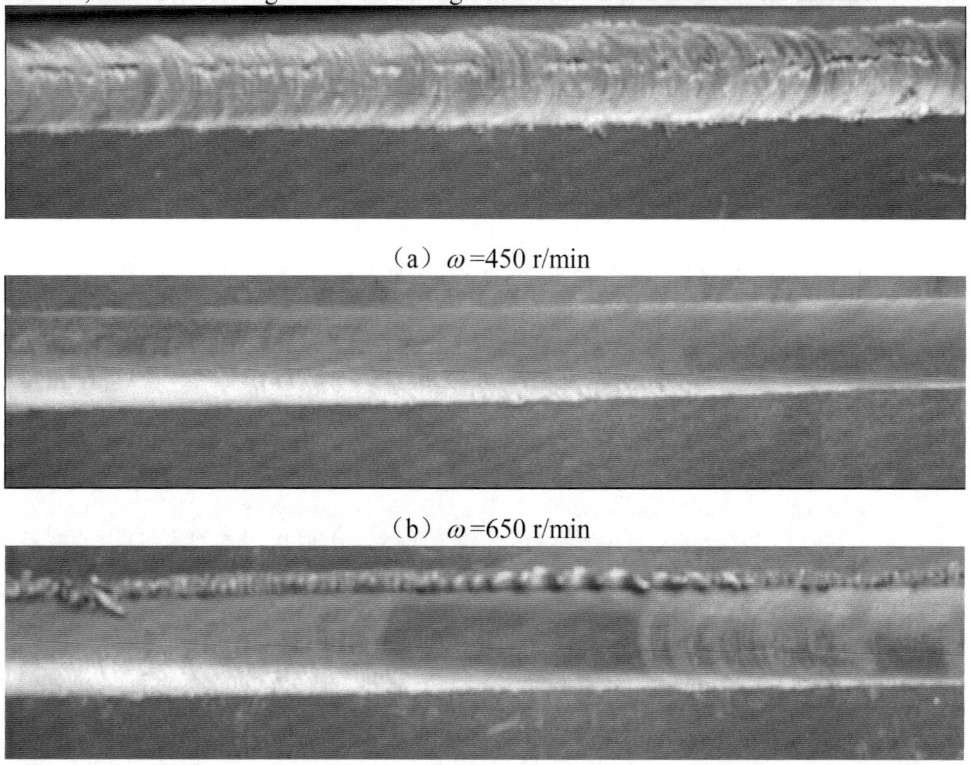

(a) ω=450 r/min

(b) ω=650 r/min

(c) ω=850 r/min

Figure 1. Effect of different rotating speed on joint quality.

3. Establishment of heat-flux analysis model

During traditional welding process, heat input is related to the applied welding current and voltage. By contrast, FSW is a kind of machining methods and shows no direction connection with the welding specifications. To analyze the rotational speed of the stir-welding head on the quality of the welded joint, this study established the heat-flux analysis model based on fluid dynamics theory. According to this model, the material was regarded as the laminar, viscous and non-Newtonian fluid that flowed around the cylinder [7], and heat-fluid coupling condition at quasi-stable state was then analyzed by solving the fluid's continuity equation, momentum equation and energy equation.

The heat input in welding is sourced from the friction heat in the contact region between the stir-welding head and the workpiece as well as the plastic-deformation-induced heat. At the quasi-stable state, the friction shearing stress at any a point in the contact region equals to the shearing stress during the occurrence of plastic deformation [7]. In this study, the deformation heat produced in the processing was also taken into account.

The heat produced by the friction between the shaft shoulder and the workpiece in the contact region as well as the plastic deformation can be written as:

$$q_{zj} = C_f(\frac{\pi n R}{30} - U\sin\theta)\tau_{yield} \quad (1)$$

The heat produced by the friction between the side of the stir-welding needle and the workpiece in the contact region as well as the plastic deformation can be written as:

$$q_{zc} = C_f(\frac{\pi n r}{30} - U\sin\theta)\tau_{yield} \tag{2}$$

The heat produced by the friction between the bottom of the stir-welding needle and the workpiece in the contact region as well as the plastic deformation can be written as same as Equation (1).

In Equation (1) and Equation (2), Cf denotes the ratio of the heat produced in the contact region between the stir-welding head and the workpiece to the heat transferred to the workpiece and was set as 0.85 in this study (i.e., = 0.85); n denotes the stir-welding head's rotational speed; U denotes the workpiece's movement speed (which was equal to the welding speed but in an opposite direction); R denotes the distance between any a point in the contact region between the shaft shoulder and the workpiece and the rotational axis of the stir-welding head; r denotes the radius of the stir-welding needle; θ denotes the angle between the line connecting any a point in the contact region between the shaft shoulder and the workpiece and the rotational axis and the workpiece's movement direction; τ_{yield} denotes the shearing stress when the material is yielded. The relation between yield strength and shearing stress can be expressed as:

$$\tau_{yield} = \sigma_s / \sqrt{3} \tag{3}$$

On the other hand, the material's yield strength decreases with rising temperature [5]. When the temperature increases to some value, the yield strength is reduced to 0 and no heat is produced in the contact region between the stir-welding head and the workpiece; then, the temperature stops increasing.

As stated above, this study adopted 6061 Al alloy for analysis. Al alloy has a melting point of 855 K, whose density shows slight variation with the temperature and approximately equals to 2700 kg/m3. Through fitting, the variations of heat conductivity coefficient, specific heat Cp and yield strength with temperature can be expressed as:

$$\lambda = 25.22 + 0.3978\mathrm{T} \tag{4}$$

$$C_p = 929.3 - 0.627T \tag{5}$$

$$\sigma_s = \begin{cases} 182.16 + 0.71544T - 0.00134T^2 (T \le 477K) \\ 3671.57 - 14.290T + 0.01870T^2 (T \le 855K) \end{cases} \tag{6}$$

In the calculation model, the stir-welding head rotated counterclockwise, and the workpiece moved along the positive direction of x-axis from left to right at a constant welding speed of 50 mm/min. To reduce the calculation time, only part of the workpiece was calculated. As shown in Figure 2, the cylindrical region, with a dimension of 52 mm× 6.3 mm× 50 mm, was taken into calculation. The x-axis length was twice of the diameter of shaft shoulder.

The shaft shoulder, the side and the bottom of the stir-welding needle were regarded as the heat-input boundaries, as the heat-flux formula described in Equation (1) and Equation (2). The upper and lower surfaces, as well as the advancing and backward sides of the stir-welding needle, were set as the moving walls, with a speed equivalent to the workpiece's movement speed. The fluid's inlet was set as the speed's inlet boundary condition, and the velocity equaled to the movement speed of the workpiece; its outlet boundary was set as the pressure's outlet boundary condition. The upper surface was exposed to the air and regarded as a heat-convection boundary, with a heat convection coefficient of 50 W/(m2•K); the lower surface was in contact with the base plate, with a heat conduction coefficient of 500 W/(m2•K); both the advancing and backward sides were partly in contact with the air and partly in contact with the fixture, and the heat conduction-convection coefficient was set as 200 W/(m2•K) while the temperature was set as 300 K.

The calculation region was then modeled using Gambit. Next, the model was meshed and then imported into Fluent (fluid dynamics software); the calculation was conducted using implicit, linear and separate resolver and the mesh was discredited. During the solving process, the pressure was calculated according to the standard discrete equation, the momentum equation was calculated using the second-order upwind equation, and the energy equation was calculated using the first-order upwind equation.

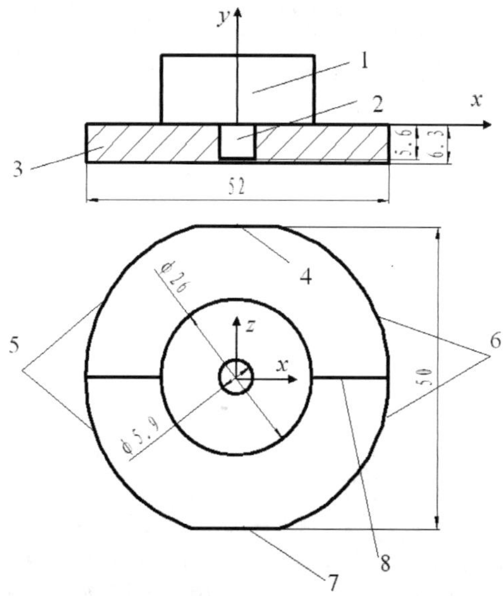

1 shaft shoulder 2 stir-welding needle 3 workpiece 4 forwarding side 5 inlets 6 outlets 7 retreating side

Figure 2. Schematic of computational zone.

4. Results and discussion

4.1. Effect of the rotational speed on peak temperature

Figure 3 showed the peak temperatures when the rotational speed of the stir-welding head was set as different values, during which the welding speed was set at 50 mm/min. It can be observed that, at a small rotational speed, the peak temperature increased significantly with the increase of rotational speed; as the rotational speed increased to 600r/min, the peak temperature almost reached a stable value. This is because that, when the welding speed and the size of the stir-welding head are fixed, the heat input in welding process is related to the rotational speed of the stir-welding head and the maximum shearing stress of the material. With the increase of the rotational speed, the material's temperature rose, while the material's maximum shearing stress remained unchanged due to temperature increment; therefore, peak temperature remained constant. Regarding energy consumption, the rotational speed of the stir-welding head should not exceed 600r/min. During FSW, the peak temperature is 80%~90% of the melting point [7]; accordingly, if only the heat quantity is taken into account, a rotational speed of 200r/min still can satisfy the welding requirements.

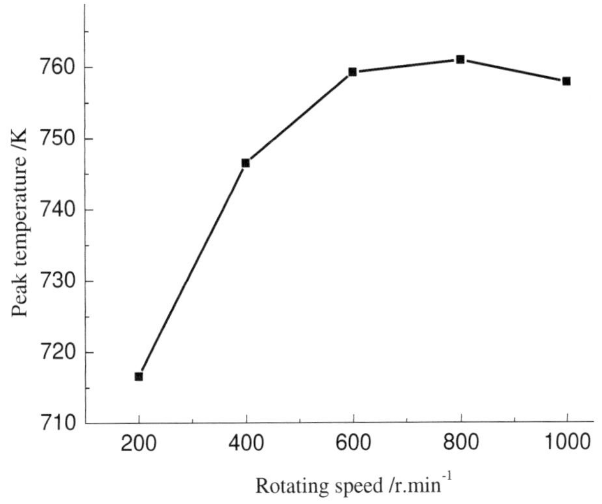

Figure 3. Effect of different rotating speed on peak temperature.

At a small rotational speed, less heat was input, and the material was poor in flowability; however, more materials should be welded from the front of the stir-welding needle, and the back of the stir-welding head and some were unable to migrate, thereby producing cracks or grooves. With the increase of the rotational speed of the stir-welding head, the heat-input increased, the material's flowability was significantly improved and fewer materials needed to be migrated, thereby enhancing the quality of the welded joint. As the rotational speed of the stir-welding head further increased, although heat input remained unchanged, fewer materials were migrated; accordingly, the heat was focused on few materials and resulted in over burning.

4.2. Effect of the rotational speed on longitudinal force

Figure 4 showed the longitudinal forces when the rotational speed of the stir-welding head was set at different values, during which the welding speed was also set at 50 mm/min. With the increase of the rotational speed, the longitudinal force further decreased, and the curve began to flatten. As the rotational speed increased, the heat input in welding increased continuously, the material's temperature and strain rate increased but the flow stress decreased, and the resistance that impeded the advance of the stir-welding head decreased; therefore, materials were more easily migrated from the front to the back of the stir-welding needle. However, when the rotational speed of the stir-welding head was increased to some value, the temperature was no longer changing, while the material's strain rate continued increasing, thereby slowing down the material's flow stress. From the perspective of the tool's service life, the rotational speed should be increased, but it may lead to poor-quality welded joints and increase energy consumption.

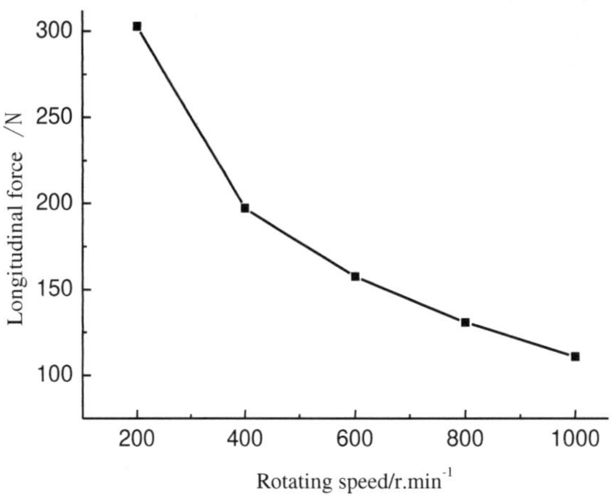

Figure 4. Effect of different welding speed on longitudinal force.

5. Conclusions

Appropriately increasing the rotational speed of the stir-welding head improved the surface quality of the welded joint; however, at a higher rotational speed, the poor quality welded joint was produced. The increasing of the rotational speed of the stir-welding head would lead to produce more heat input; therefore, fewer materials participated in the formation of the weld, the material's flowability was improved, and the forward resistance decreased. The setting of the rotational speed of the stir-welding head should take overall consideration. Both the quality of the welded joint and the tool's service life should be balanced.

Acknowledgement

The authors would like to thank the Shandong Province Key Projects of Research and Development (No. 2016GNC112006) and the Shandong Agriculture University Innovation Team for the financial support.

References

[1] Song M and Kovacevic R 2003 Thermal modeling of friction stir welding in a moving coordinate system and its validation *Int. J. Mach.Tools Manu.* **43** 605-15
[2] Schmidt H and Hattel J 2005 Modeling heat flow around tool probe in friction stir welding *Sci. Technol. Weld. Joi.* **10**(2) 176-86
[3] Colegrove P A, Shercliff H R and Zettler R 2007 Model for predicting heat generation and temperature in friction stir welding from the material properties *Sci. Technol. Weld. Joi.* **12**(4) 284-97
[4] Schmidt H B and Hattel J H 2008 Thermal modeling of friction stir welding *Scripta Mater.* **58** 332–7
[5] Chen C M and Kovacevic R 2003 Finite element modeling of friction stir welding—thermal and thermo-mechanical analysis *Int. J. Mach.Tools Manu.* **43** 1319-26
[6] Deng X M and Xu S W 2004 Two-Dimensional Finite Element Simulation of Material Flow in the Friction Stir Welding Process *J. Manu. Processes* **6**(2) 125-33
[7] Feng T T and Zhang X H 2013 Model for three-dimensional heat transfer and plastic flow of friction stir welding processing *Transactions of the China Welding Institution* **34**(7) 105-8

Rubber mixing process and its relationship with bound rubber and crosslink density

A Hasan[1,4], Rochmadi[2], H Sulistyo[2] and S Honggokusumo[3]

[1]Chemical Engineering Dept, State Polytechnic of Sriwijaya, Palembang, Indonesia
[2]Chemical Engineering Dept, Gadjah Mada University, Yogyakarta, Indonesia
[3]Gapkindo (The Indonesian Rubber Association), Jakarta, Indonesia

Email: abu_hasan@polsri.ac.id

Abstract. This research studied the relationship between bound rubber and crosslink density based on rubber mixing process. Bound rubber was obtained after natural rubber was masticated and mixed with rubber chemicals and filler while crosslink density was collected after rubber compound was vulcanized. Four methods are used and each method refers to four ways of incorporating carbon black during mixing. The first method, after rubber was masticated for 5 minutes, the addition of rubber chemicals and filler was done simultaneously. Rubber was masticated for 1 minute and continued mixing of rubber chemicals and filler where mixing was different from first method. This was the second method. The third method was the same as the second method but the filler used N 660 while in the second method N 330. The last method is not the same as the first and second, the rubber is only masticated for 3 minutes and then mixed with filler and followed by rubber chemicals sequentially. The results showed that bound rubber and crosslink density were influenced by mixing and mastication process. Bound rubber dropped and crosslink density was relatively stable in the first three mixing methods for increasing carbon black at the beginning of the mixing process. Bound rubber and crosslink density stated opposite results in the fourth mixing method. The higher the bound rubber the lower the crosslink density. Without regard to mixing methods, there is a non-linear relationship between bound rubber formation and crosslink density determination

1. Introduction

Rubber mixing process (RMP) is one of the determinations of physical and chemical properties of vulcanized rubber. RMP starts from rubber mastication and then followed by rubber chemicals and/or filler mixing into rubber. Rubber compound as a RMP product forms bound rubber and its vulcanizate contains crosslink density. Bound rubber is reinforcement level in rubber vulcanized which is formed from physical bonding of filler and rubber molecules. Bound rubber is predisposed by filler content [1], the type of filler [2], storage time [3] and the mixing time [4]. Crosslink density is the level of vulcanization of rubber backbone and sulfur as vulcanizing agent or reactant [5-8]. Crosslink density is worked on structure and carbon black's surface chemistry [9-10] while the type and content of filler affect cure characteristics [9-12]. Bound rubber and crosslink density greatly affect physical and chemical behaviors of vulcanized rubber. To produce rubber with the specific physical and chemical properties, requires a certain RMP anyway. The order of mixing rubber chemicals and fillers can affect the physical properties such as abrasion resistance of vulcanized rubber. RMP was studied by carbon black addition into rubber in the first footpace and then oil was put into rubber after the carbon black fused, abrasion was significantly bigger compared with that of the compound prepared by adding oil together with carbon black in rubber [13] . The change in abrasion resistance is more due to the increase of bound rubber than that of the crosslink density. The extent to which the relationship between the bound rubber and crosslink density is not widely studied. Wang (1998) states that if the active center of carbon black is littered with rubber chemicals, especially the type of antioxidants such as amine it is very difficult to be replaced by rubber molecules [4]. This means that the bound rubber being dropped. This fact can be considered analogous to that of affairs between the bound rubber and crosslink density. When bound rubber up on the other way the crosslink density to be down. Choi (2007) studied bound rubber based on carbon black reinforced rubber compounds and crosslink density came from rubber vulcanizates, but in this study the relation of bound rubber and crosslink density did not discuss [14]. It is therefore in this study not only studied the effect of RMP upon bound rubber formation and crosslink density analysis but also its relation both.

2. Experimental

2.1. Rubber formula

Natural rubber RSS-1 as Indonesian's product and carbon black type N 330 and 660 from Cobot Indonesia were main raw materials in the investigation as well as sulfur, TBBS, TMQ, ZnO, $C_{18}H_{36}O_2$, and softener as rubber chemicals. Together with bound rubber level was determined by toluene. All ingredients were commercial commodities and non-treated materials.

Table 1. Natural Rubber Formula.

No	Ingredients	phr
1	RSS-1	100
2	ZnO	5
3	$C_{18}H_{36}O_2$	2
4	Carbon Black type N 330/660	50
5	Softener	5
6	TMQ	2
7	TBBS	0.5
8	Sulfur	2.5

Table 2. Mixing process consecution of A.

No	Ingredients	Time, minutes
1	RSS-1	5
2	ZnO	2
3	$C_{18}H_{36}O_2$	
4	TMQ	1
5	CB_1	2
6	CB_2 + oil	4
7	TBBS	1
8	Sulfur	1

2.2. Equipment

Equipment used in this research were open two roll mill, hot press apparatus and glassware for determination of bound rubber and crosslink density

2.3. Mastication and milling process

Rubber and ingredients listed in Table 1 were masticated and milled on mixing temperature of 60°C. This milled rubber formula followed the mixing procedure stated on Table 2. These rubber samples were called A. The addition sequence of carbon black followed consecution on Table 5 code 1, so this sample called A1. Rubber was masticated about 5 minutes, continued by puting into practice of ZnO and $C_{18}H_{36}O_2$ and milled for 2 minutes. Then TMQ was added into the mix and milled for 1 minute. Carbon black at amount of 10 phr was farthered into the mix and milled for 2 minutes. This process is called CB_1. The rest of carbon black about 40 phr was combined into the mix together with oil and milled about 4 minutes. It is regarded as CB_2+oil. Then TBBS and sulfur addition needed 1 minute for milling time each.

The formula based on Table 1, a mixing process was also conducted in relation to the consecution and the time each consecution needed, as shown in Table 3. The sample produced was called B. Rubber was masticated for 1 minute, followed by $C_{18}H_{36}O_2$ addition and milled for 1 minute. Then sulfur and ZnO were added into the mix and milled for 2 minutes. The amount of carbon black 10 phr was added into the mix and milled for 3 minutes. This process is called CB_1. The rest of carbon black about 40 phr was added into the mix together with oil and milled for 7 minutes. It is regarded as CB_2+oil. Then TBBS addition into the mix was milled for 6 minutes and TMQ was needed 1 minute for milling time. Carbon black addition process into the rubber was applied using the method performed to the consecution of Table 5. The samples were designated as B1, B2, B3 and B4.

Table 3. Mixing process consecution of B.

No	Ingredients	Time, minutes
1	RSS-1	1
2	$C_{18}H_{36}O_2$	1
3	Sulfur	2
4	ZnO	
5	CB_1	3
6	CB_2 + oil	7
7	TBBS	6
8	TMQ	

Table 4. Mixing process consecution of D.

No	Ingredients	Time, minutes
1	RSS-1	3
2	CB_1	2
3	CB_2 + oil	5
4	ZnO	2
5	$C_{18}H_{36}O_2$	
6	TMQ	1
7	TBBS	1
8	Sulfur	1

Formula on Table 1 was used to mixing process too following mixing process such listed on Table 4. Rubber was masticated for 3 minutes. Carbon black was added after that and milled for 7 minutes. Carbon black was affixed into the in the amount of 10 phr out of the total amount of 50 phr. Such process was called CB_1 and needed 2 minutes for milling. The rest of 40 phr was affixed into the rubber afterwards together with oil. It was regarded as CB_2+oil and needed 5 minutes for milling. ZnO and $C_{18}H_{36}O_2$ was added after that and milled for 2 minutes, followed by TMQ addition, sulfur and TBBS milled for 1 minute each. This sample was called D1. After mastication and mixing process, bound rubber determined and rubber compound then were vulcanized on temperature of 150°C and followed analysis of crosslink density.

Bound rubber testing using rubber compound 1 gram was rendered in 50 ml toluene solvent for 4 days. Dissolution performed at room temperature. After dilution, the rubber compound was then dried at a temperature of 110°C to constant weight under vacuum. Initial sample and sample after dissolution were then weighed. Bound rubber is calculated in a way that used by Choi [2], Choi et al. [3], Choi [4], Qian et al. [15], Wang et al. [16], and Wolft [17] as follows:

$$\text{Boundrubber} = \frac{W_{fg} - W[\frac{m_f}{(m_f+m_p)}]}{W[\frac{m_p}{(m_f+m_p)}]} \text{ x } 100\ \% \tag{1}$$

Where W is initial sample weight, W_{fg} carbon black load and gel, m_f filler mass in the compound, dan m_p the polymer mass in the compound.

Testing of crosslink density used swelling technique. Vulcanized rubber about 1 gram was soaked in the toluene solvent for 24 hours at room temperature, then weighed weight of vulcanized rubber in an expanding after a surface residual solvent removed using tissue paper. Vulcanized rubber was dried for 24 hours at a temperature of about 110°C under vacuum conditions. After that, vulcanized rubber was left at room temperature for 24 hours, and then weighed weight in the dry state. Crosslink density is calculated using the equation Parks [18] the following:

$$Q = \frac{(\text{swelling weight} - \text{dried weight})}{\left(\frac{\text{original weight x 100}}{\text{formula weight}}\right)} \tag{2a}$$

$$\text{Crosslink density} = \frac{1}{Q} \tag{2b}$$

IOP Conf. Series: Materials Science and Engineering **213** (2017) 012048 doi:10.1088/1757-899X/213/1/012048

Table 5. The Carbon Black Enlargement into Mixing Process.

Code	CB_1	CB_2+oil
1	10 phr	40 phr+oil
2	20 phr	30 phr+oil
3	30 phr	20 phr+oil
4	40 phr	10 phr+oil

3. Results and discussion

The bound rubber samples A, B, C, D, and D4 as well as crosslink density are shown in Figure 1-6. Figure 1 shows that the bound rubber down significantly as well as crosslink density is almost stable from A1 to A4. The trend is almost the same as the curve in Figure 1 occurs also in the curves in Figure 2. RMP of A and B are not significant impact on crosslink density and even tend to be stable, otherwise bound rubber to be down sharply.

In the RMP for samples A and B illustrated in Figure 1 and 2, ingredients mixing were simultaneously performed, this can cause the surface of carbon black tainted by rubber chemicals so it does not provide the increase of bound rubber. Here, carbon black also can not properly serve as a catalyst, so that the reaction of sulfur with rubber backbone is not so influenced by the RMP. Thus crosslink density showed no significant change despite initial mixing of more carbon black in the RMP. Bound rubber B is greater than that of A. It is caused by mastication time B for 1 min less than that of A for 3 min. Low rubber mastication time results high shearing force of rubber. High shearing force can solve aggregate into fine particles of carbon black that is better than low rubber shearing force. Bound rubber B is greater than that of C (see Figure 3). This is caused by the difference in size of the carbon black particles used. Compound B employs carbon black with a small particle size and C uses carbon black particle greater than that of B. The smaller the particle size the easier carbon black dispersed in rubber, so that the bound rubber bigger.

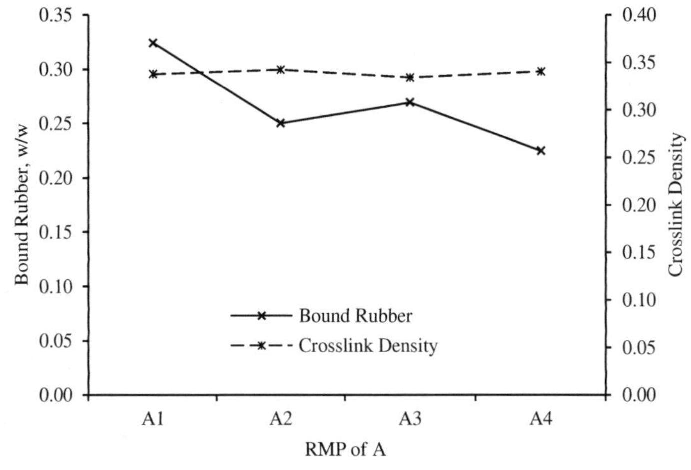

Figure 1. Effect of RMP upon formed bound rubber and determined crosslink density of A. Milling temperature of 60°C.

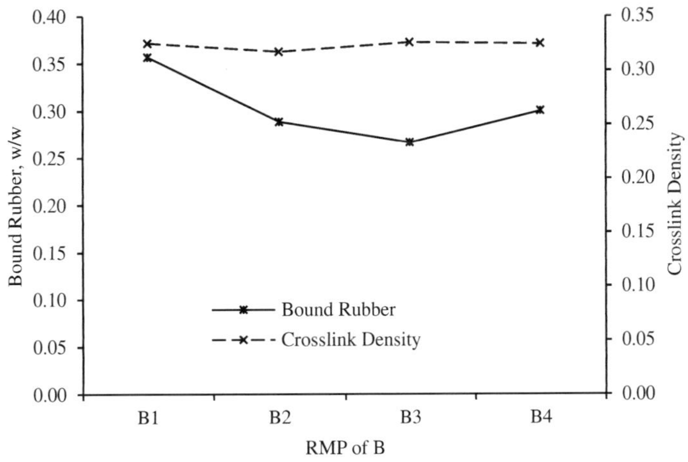

Figure 2. Effect of RMP upon formed bound rubber and determined crosslink density of B. Milling temperature of 60°C.

RMP influences of bound rubber formation and crosslink density detemination on sample D shown in Figure 4. Bound rubber has a tendency down and instead crosslink density rises from D1 until D4. RMP on sample D mixing process of carbon black conducted in advance than that of rubber ckemicals so that carbon black surface is not stained by rubber chemicals yet. Rubber molecules adsorption on carbon black surface get better until the bound rubber formation rises. Therefore the bound rubber D somewhat larger than those of A and B. Increase bound rubber from D1 until D4 caused disturbed vulcanization reaction so that crosslink density goes down from D1 until D4. The increase of rubber bound was due to the more and more carbon black is mixed in early RMP. Here, high shearing force causes carbon black aggregates break down into finer particle size until particles are easily dispersed into the rubber. In consequence, more rubber backbone was adsorbed on carbon black surface as well as bound rubber rised from D1 until D4.

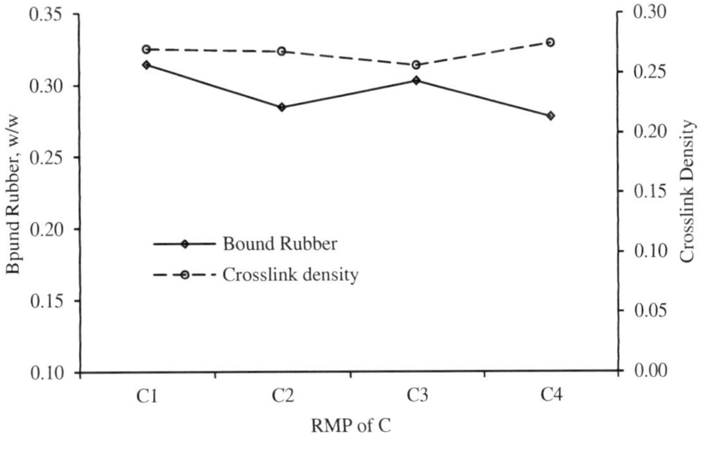

Figure 3. Effect of RMP upon formed bound rubber and determined crosslink density of C. Milling temperature of 60°C.

Bound rubber formation and crosslink density analysis were contradictive as based on Figure 4, the experiment was then carried out to study how phr carbon black rubber compound and phr sulfur in vulcanized rubber influenced bound rubber and crosslink density formation (Figure 5-6). From Figure 5, the contradiction of bound rubber and crosslink density (Figure 4) could be described. Figure 5 gives information that innumerable of carbon black (phr) into the rubber causes high confortable bound rubber and it shows crosslink density decreased. High phr of carbon black into rubber compound causes high surface area and it results in high bound rubber. High bound rubber can decrease active center of carbon black and causes the reaction of rubber backbone and sulfur is stained . In this matter, the crosslink density becomes low.

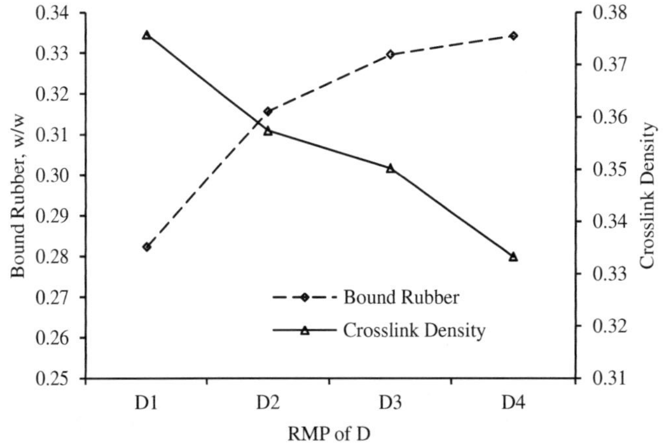

Figure 4. Effect of RMP upon formed bound rubber and determined crosslink density of D. Milling temperature of 60°C.

Figure 6 gives an information how phr of sulfur into rubber compound affects crosslink density and bound rubber formation contradictively. The main reason why the contradiction exists is the same as like described on Figure 5. High phr of sulfur into rubber compound causes high reaction rate and decreases the active center of carbon black. This fact results the decrease of bound rubber prior to the increase of crosslink density. The facts stated on Figure 4-6 give the relation crosslink density and bound rubber. Bound rubber becomes low in contras crosslink density is gained.

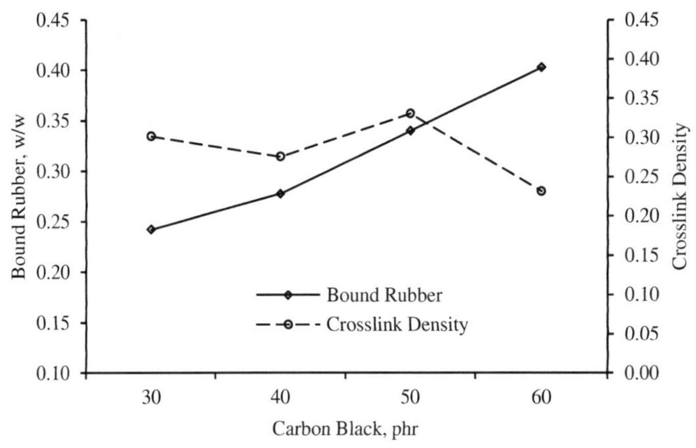

Figure 5. Effect of carbon black content against bound rubber and crosslink density of D4. Milling temperature of 60°C.

Regardless how to process the rubber compound A, B, C or D, and refer to the facts illustrated on Figure 4-6, the curve that relate crosslink density/bound rubber and bound rubber were existed such illustrated us on Figure 7.

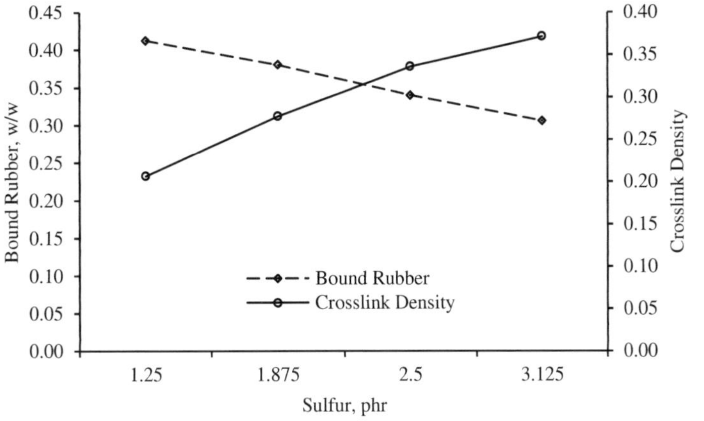

Figure 6. Effect of sulfur content against bound rubber and crosslink density of D4. Milling temperature of 60°C.

Figure 7 gives the information that there is a non linear relation between crosslink density with bound rubber. Plot result shows the relation between crosslink densities with bound rubber that follows following equation:

$$y = -1{,}128\ln(x) + 0.1856 \tag{3}$$

y is crosslink density/bound rubber and x bound rubber.

It can be concluded that the crosslink density is more formed when bound rubber formation is low, in the other words at a certain rubber bound, crosslink density becomes constant. According to Lorenz dan Parks [19], the filler may affect the properties of vulcanizate either through the interaction between the filler and the vulcanizing agent as well as reciprocal action between filler and rubber. When the reciprocal action between filler and curing agent occurs, a number of curing agents or accelerators cannot react with rubber. Kraus [20] stated that softener and oils can reduce the amount of crosslinking or reduce the efficiency of sulfur reaction. This also shows that there is an interaction between softener and oil with vulcanizing agent. From both of these facts it is stated that crosslink density is decreasing and otherwise the bound rubber rises.

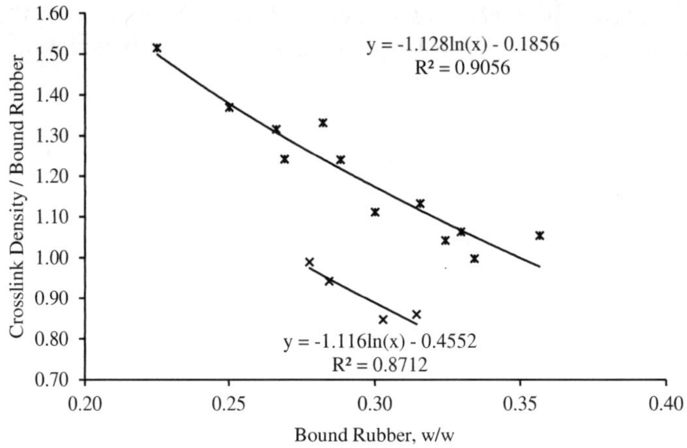

Figure 7. Relationship between bound rubber and crosslink density.

According to Lorenz and Parks [19], the filler can affect the properties of vulcanized rubber through interactions between the filler and a curing agent or interaction between filler with rubber. When interaction between filler and curing agent case, a part of curing agent or accelerator can not cause a reaction with the rubber. Kraus [20] stated that the softener and oil can reduce the amount of crosslinking or decrease the efficiency of the use of sulfur. This matter shows that there is an interaction between softener and oil with curing agent. From these facts indicate that the crosslink density rubber down and instead bound up.

4. Conclusion

RMP affected bound rubber formation. More carbon black is added at the initiation of RMP resulted in the decrease of bound rubber on sample of A and B. These were caused by sequence of rubber mixing and rubber chemicals as well as carbon black simultaneously. If carbon black was mixed at the beginning of RMP followed by rubber chemicals mixing into rubber, such as on sample of D, so the formation of bound rubber increased as well as the increase of large amount of carbon black mixed at the beginning of RMP. RMP also influenced the crosslink density on sample of D but it had small effect upon sample of A and B. Without regard to mixing methods, there is a non-linear relationship between bound rubber and crosslink density.

Acknowledgment

I would like to express my gratitude to Directorate General of Higher Education Republic of Indonesia (in the scheme of Doctor Dissertation Grant 2010) and Overseas Seminar Program 2017, Directorate General of Strengthening Research and Development, Ministry of Research, Technology, and Higher Education, Indonesia

References

[1] Choi S S 2004 *J. Appl. Polym. Sci.* **93** 1001-6
[2] Choi S S, Hwang K J and Kim B T 2005 *J. Appl. Polym. Sci.* **98** 2282-9
[3] Choi S S 2002 *Polym. Test.* **21** 201
[4] Wang M J 1998 *Rubber Chem. Technol.* **71** 520-89
[5] Bateman L 1963 *The Chemistry and Physics of Rubberlike Substances: Study of the Natural Rubber Producers' Research Association* (London: MacLaren&Sons ltd.) Chapter 15
[6] Morton M 1997 *Rubber Technology* 3rd ed (New York: Van Nostrand Reinhold) Chapter 4
[7] Billmeyer F W 1984 *Textbook of Polymer Science* 3rd ed (New York: Jhon Weley & Sons) Chapter 19
[8] Indian Rubber Institute 2000 *Rubber Engineering* (New York: McGraw-Hill) Chapter 6

[9] Leblanc J L 2002 *Prog. Polym Sci.* 27(4) 627-687
[10] Chen C H, Koenig J L, Shelton I L and Collins E A 1981 *Rubber Chem. Technol.* **54**(4) 734-750
[11] Ansarifar A, Nijhawan R, Nanapoolsin T and Song M 2003 *Rubber Chem. Technol.* **76**(5) 1290-310
[12] Fei Z, Long C, Qingyan P and Shugao Z 2012 Influence of carbon black on crosslink density of natural rubber *J. Macro. Sci. Part B: Physics* **51**(6) 1208-17
[13] Wang M J 2006 *Int. Rubber Conf. (Lyon France)* pp 16-8
[14] Choi S S, Kim J C, Ko J E, Cho Y S and Shin W G 2007 *J. Ind. Eng. Chem.* **13(6)** 1017-22
[15] Qian S, Huang J, Guo W and Wu C 2007 *J. Macro Sci. Part B: Physics* **46** 453-66
[16] Wang T, Wang M J, Shell J and Tokita N 2000 *Kautsch Gummi Kunstst.* **53** 497
[17] Wolff S, Wang M J and Tan E H 1993 *Rubber Chem. Technol.* **66**(2) 163-77
[18] Parks C R 1982 *Rubber Chem. Technol.* **55** 1170
[19] Lorenz O and Parks C R 1961 *J. Polym. Sci.* **50(154)** 299-312
[20] Kraus G 1957 *Rubber Chem. Technol.* **30(3)** 928-51

Flux pinning properties of GdBCO bulk through the infiltration and growth process

Y F Zhang[1,3], J J Wang[1], X J Zhang[1], C Y Pan[1], W L Zhou[1], Y Xu[1], Y S Liu[1] and M Izumi[2]

[1]Department of Mathematics and Physics, Shanghai University of Electric Power, 28 Xuehai Road, New Pudong District, Shanghai 201300, P.R.China
[2]Tokyo University of Marine Science and Technology, Lab of Applied Physics, Japan 135-8533

Email: 2009000018@shiep.edu.cn

Abstract: $REBa_2Cu_3O_{7-\delta}$(RE123 or REBCO, RE=rare earth elements, Gd, Y, Nd, etc.) bulk high temperature superconductors (HTS) have been used in lots of aspects, such as in magnetic levitation, *et al.*, owing to the performance of high magnetic flux trapping. GdBCO superconductor bulk with 25 mm diameter has been successfully fabricated by top-seeded infiltration and growth (TSIG) method. We chose $YBa_2Cu_3O_{7-\delta}$ (Y123) particles as the liquid source, which provide enough liquid sources during the growth and encourage the growth along a-b plane of GdBCO bulk. Then the existence of Y123 liquid source partly decreases the effect of the sub-grain boundaries in a-growth sectors and improves the properties of GdBCO bulk. The shape of the trapped field is close to circle. The critical current density of C2 and B2 (J_C) enhances. The superconducting transition temperature (T_C) is around 94.5K in the different position and keeps the superconducting properties. It is the important experimental data for the engineering applications of the superconductor bulk.

1. Introduction

The trapped magnetic field is 17T in $YBa_2Cu_3O_{7-\delta}$ (Y123 or YBCO) superconducting bulks with a diameter of 26mm, which is higher than that in the traditional magnets [1]. The application of $REBa_2Cu_3O_{7-\delta}$ (RE123 or REBCO, RE=rare earth elements, Gd, Y, Nd.etc.) bulk high temperature superconductors (HTS) in power system, such as in magnetic separation systems and magnetic levitation [2-4], is attracting the researcher owing to its high critical density (J_C) and magnetic flux trapping performance. The top seeded melt growth (TSMG) is a conventional method to fabricate single-domain RE123 bulks. While the top seeded infiltration and growth (TSIG) process is popular recently [5-8], the TSIG method resolve the followed problems, such as shape distortion, shrinkage of the final sample and leakage of the $Ba_3Cu_5O_8$ (035) liquid phase etc.. Zhou *et al.* successfully processed GdBCO bulk of 32 mm in diameter by $YBa_2Cu_3O_y$ (Y123) liquid source through the modified TSMG method. The enough growth along the a-b plane has been allowed due to the Y123 liquid source, which depressed the accumulation of Gd_2BaCuO_5 (Gd211) particles, so as to dramatically improve the properties of the GdBCO bulk.

In this article, we successfully fabricated the single domain GdBCO bulk of 25mm in diameter with the Y123 liquid source of 5mm thickness by TSIG method. The superconductivity properties, trapped field and critical current density, of the GdBCO bulk have been studied.

2. Experimental

An initial composition of Gd123 + 40 mol% Gd211 + 10 wt% Ag_2O + 0.5 wt% Pt for commercial pure powders of Gd123, Gd211, Ag_2O and Pt were mixed thoroughly and then made a cylinder pellet with the diameter of 25 mm and the thickness of 12 mm. Commercially pure Y123 powders were chose to press into a pellet with the diameter of 25 mm and the thickness of 5 mm as the liquid source. The pellet of Y123 liquid source was put under the GdBCO precursor. Then we place the entire arrangement onto the Al_2O_3 sheet into the box furnace. The Y_2O_3 pellet of 3 mm in thickness was put onto the sheet, avoiding the reaction of the Al_2O_3 and the Y123. The temperature profile is as followed. The maximum value of temperature (T_{max}) was set to 1079 °C, about 70 °C above the peritectic temperature (T_p) . And the end of the reaction temperature was set to 987 °C, about 20°C below the T_p. With a cooling rate of 0.3°C/h, a cold Nd123/MgO thin film was employed as a seed crystal to growth the bulk [8]. After growth, the annealing process was carried out in flowing oxygen , the sample was first heated to 450°C in 5 h and hold for 40 h, then the temperature was slowly decreased to 350°C in 140 h, 300°C in 30 h. Eventually it was cooled down to room temperature. The process has been reported elsewhere [9-10].

In order to measure the trapped magnetic flux density, we chose a 1.0 T magnetic field and cooled down the bulk to liquid nitrogen temperature. Thirty minutes later, the applied field was removed and the distribution of the trapped magnetic flux density was obtained by the Hall probe sensor. DC magnetization measurements were carried out using Physical Property Measurement System (PPMS).The small rectangular samples in a 2 mm×2mm×1 mm size were cut from the bulk to study the DC magnetization. There are two specimens under seed position, labeled as C1, C2, and the other specimens are under the boundary, labeled as B1, B2. J_c–B curves were deduced under the extended Bean's critical state model [11].

3. Results and discussion

As shown in Figure 1, single domain GdBCO bulk with Y123 liquid source of 5mm thickness is successfully processed by TSIG. The Nd123 seed remains stable and there is no obvious melting phenomenon. It is clear that the diameter of the liquid source Y123 is smaller than that of GdBCO bulk. Y123 power decomposes at the peritectic temperature so as to provide the liquid source of GdBCO crystal growth [12]. Then the size of Y123 pellet shrinks as we can see. It may affect the superconducting properties of GdBCO bulk.

Figure 1. Top view of GdBCO bulk processed with Y123 liquid source in the thickness of 5mm.

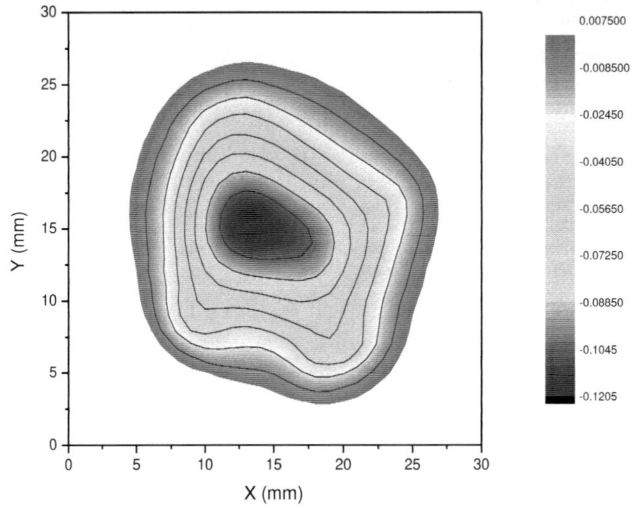

Figure 2. Trapped flux density of GdBCO bulk processed with Y123 liquid source in the thickness of 5mm.

The distribution of the trapped magnetic flux density of the GdBCO bulk with the thickness of 5mm Y123 liquid source was shown in Figure 2. A circle trapped field distribution can be seen clearly, which shows the high performance of the superconducting bulk [7]. As one of the most important parameters for the practical application, the trapped field was measured and the maximum flux density value is 0.12T, which is higher than that without Y123 liquid source [9-10].

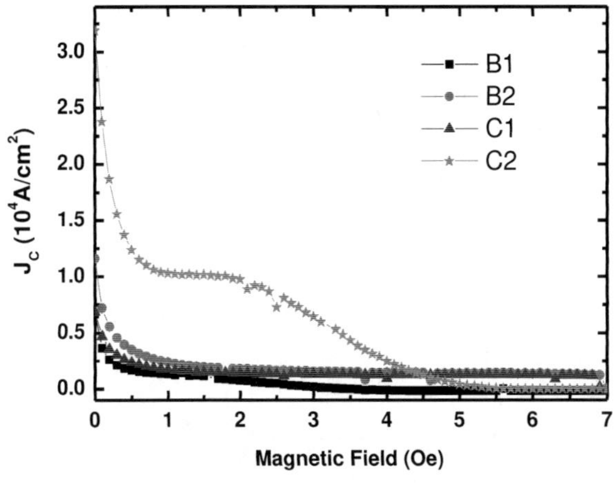

Figure 3. J_C-B curves of the samples in the different positions of the GdBCO bulk processed with Y123 liquid source in the thickness of 5mm at 77 K.

Figure 3 shows the J_C–B curves of the specimens which located at different positions B1, B2, C1 and C2. We can see that the maximum critical current density value (J_{max}) is 3.24×10^4A/cm^2 for the specimen taken from C2. As is well known, the growth of c direction is fast than that of a-b plane in the melt-textured superconducting bulk. The Y123 liquid source inhibits the growth of c direction and keeps the rich Gd211 particles in the specimen C2, which improves the superconducting properties of C2 specimen. The J_C of specimen B2 is higher than that of specimen B1. We speculate that the existence of Y123 liquid source avoids the deficient of liquid source during the growth and decreases the growth of Gd211 particles at the boundary sector. The sub-grain effect has been inhibited. The Gd211 particles in the suitable size and the number increase the J_C of the specimen B2.

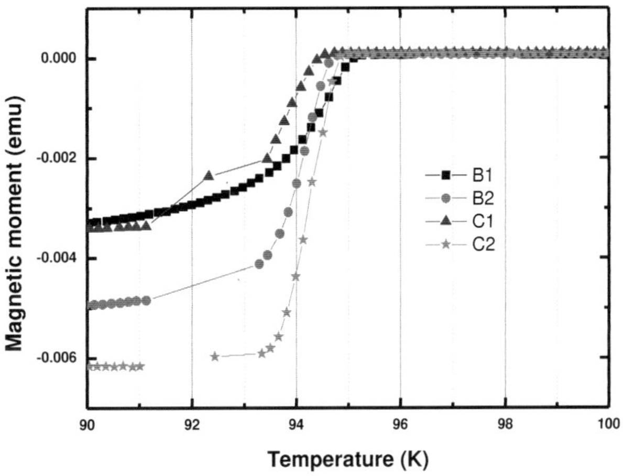

Figure 4. Superconducting transition temperature (T_C) curves of the samples in the different positions of the GdBCO bulk processed with Y123 liquid source in the thickness of 5mm.

Figure 4 shows the superconducting transition temperature, T_C, curves under the external magnetic field. The T_C is around 94.5 K, which exhibits the high quality of the superconducting bulk. The narrow transition width appears in the specimen C2 and B2, which shows the better superconducting properties. The result is corresponded with that of J_C. It can be obtained that the Y123 liquid source can enable the sufficient growth of the sample along *a-b* plane. Employing the Y123 liquid source can be effective to process the well-textured single domain GdBCO bulk in large size without the leakage of the 035 liquid phase [5, 13].

4. Conclusion

The single domain GdBCO bulk superconductor with the diameter of 25 mm has been successfully growth by the TSIG technique. The Y123 pellet with 5mm thickness solves the leakage problem of liquid source and encourages the growth along a-b plane of GdBCO bulk. So that it increases the accumulation of Gd211 particles in C2 and obviously decreases the effect of sub-grain boundaries. Thus, the properties of superconductor bulk have been improved. After detailed investigations in trapped flux, critical current density and critical transition temperature of the samples, it can be concluded that the method can be effective to fabricate single domain GdBCO superconductor bulks with large scale in the future.

Acknowledgments

Project supported by the National Natural Science Foundation of China (Grant No. 11004129), the Scientific Research Foundation for the Returned Overseas Chinese Scholars, Education Ministry of China (SRF for ROCS, SEM), the Innovation Program of Shanghai Municipal Education Commission, China (Grant No. 11YZ197, 12ZZ174), the "Chen Guang" Project of Shanghai Educational Development Foundation, china (Grant No. 12CG63), and the Shanghai University Scientific and Cultivation for Outstanding Young Teachers in Special Fund, China (Grant No. sdl10005).

IOP Conf. Series: Materials Science and Engineering **213** (2017) 012049 doi:10.1088/1757-899X/213/1/012049

References

[1] Tomita M and Murakami M 2003 *Nature* **421** 517
[2] Matsuzaki H, Kimura Y, Ohtani I, Izumi M, Ida T, Akita Y, Sugimoto H, Miki M and Kitano M 2005 *IEEE Trans. Appl. Supercond.* **15** 2222-5
[3] Chen Y L, Chan H M, Harmer M P, Todt V R, Sengupta S and Shi D 1994 *Physica C* **234** 232-6
[4] Reddy E S and Rajasekharan T 1998 *Supercond. Sci. Technol.* **11**(5) 523
[5] Yang W M, Zhi X, Chen S L, Wang M, Li J W, Ma J and Chao X X 2014 *Physica C* **496** 1-4
[6] Li J W, Yang W M, Wang M, Guo Y X and Zhong L F 2015 *J. Supercond. Nov. Magn.* **28** 1725
[7] Zhou D F, Shogo H, Li B Z, Noudem J and Izumi M 2014 *Supercond. Sci. Technol.* **27** 044015
[8] Zhou D F, Izumi M, Miki M, Felder B, Ida T and Kitano M 2012 *Supercond. Sci. Technol.* **25** 103001
[9] Zhang Y F, Izumi M, Xu Y and Kimura Y 2010 *J. Phys: Conference Series* **234** 012052
[10] Zhang Y F, Izumi M, Kimura Y and Xu Y 2009 *Physica C* **469** 1169-71
[11] Chen D X and Goldfarb P R 1989 *J. Appl. Phys.* **66** 2489
[12] Zhou D F, Shogo H, Li B Z, Xu K, Noudem J and Izumi M 2013 *Supercond. Sci. Technol.* **26** 015003
[13] Cardwell D A, Shi Y H, Babu N H, Pathak S K, Dennis A R and Iida K 2010 *Supercond. Sci. Technol.* **23** 034008

Polyazulene based materials for heavy metal ions detection

A Oprisanu[1], E M Ungureanu[1,4], R Isopescu[1], L Birzan[2], M Mihai[1] and C Vasiliu[3]

[1]University "Politehnica" of Bucharest, Gheorghe Polizu 1-7, 011061, Sector1, Bucharest, Romania

[2]Institute of Organic Chemistry "C. D. Nenitzescu" of Romanian Academy, Spl. Independentei 202B, PO Box 15-258, 71141-Bucharest, Romania

[3]National Institute of Research and Development for Optoelectronics, INOE-2000, 409 Atomistilor Str., 77125, P.O. Box MG-5, Bucharest-Magurele, Romania

E-mail: em_ungureanu2000@yahoo.com

Abstract. Azulene is a special monomer used to functionalize electrodes, due to its spontaneous electron drift from the seven-membered ring to the five-membered ring. The seven-membered ring of the molecule may act as electron acceptor, while the five-membered ring – as electron donor. This leads to very attractive properties for the synthesis of functional advanced materials like: materials with nonlinear optical and photorefractive properties, cathode materials for lithium batteries, or light emitting diodes based on organic materials. Azulene derivatives have been used rarely to the metal ions electroanalysis. Our study concerns the synthesis and electrochemical characterization of a new azulene based monomer 4-(azulen-1-yl)-2,6-bis((E)-2-(thiophen-3-yl)vinyl)pyridine (**L**). **L** has been used to obtain modified electrodes by electrochemical polymerization. Poly**L** films modified electrodes have been characterized by cyclic voltammetry in ferrocene solutions. The complexing properties of poly**L** based functional materials have been investigated towards heavy metals (Pb, Cd Hg, Cu) by preconcentration – anodic stripping technique in order to analyze the content of these cations from water samples.

Keywords: Azulene derivatives; Chemically modified electrodes; Heavy metal ions detection; Electrochemical characterization

1. Introduction

Functionalized conducting polymer films have received considerable attention due to their great potential as analytical electrode materials [1-2]. Stable polymer film modified electrodes containing covalently attached dithiocarbamate [3] or EDTA [4] moieties, able to sense copper, cadmium or lead, have been produced by chemical derivatization of films previously coated by electropolymerization onto electrodes. Incorporation of complexing fragments into polymeric materials can alternatively be done by doping or ion exchange procedures [5]. However, the most efficient approach towards chelating-modified electrodes lies in the direct electropolymerization of complexing monomers, producing in one step stable functionalized polymers films with controlled sizes and sites concentrations. This straightforward approach has, however, seldom been applied to the determination of trace metals and only few examples can be found in literature [6-7].

Azulenes have a five-member (electron-rich) cyclic moiety connected with a seven-member (electron-poor) cyclic moiety. Azulene derivatives present an irreversible electrooxidation, and an

IOP Conf. Series: Materials Science and Engineering **213** (2017) 012050 doi:10.1088/1757-899X/213/1/012050

irreversible [8] or quasi-reversible reduction [9]. By electrooxidation polymeric films can be obtained in order to build electrochemical sensors, as shown in our recent works [10]. This research studies aimed to obtain modified electrodes for heavy metal ions detection.

There are different methods for determination of heavy metals from water samples based on iron oxide/graphene composite [10], bismuth nanoparticle-porous carbon paste [11], and zinc oxide-graphene composite [8]. The approach in our group is focused on complexing modified electrodesbased on azulene derivatives [11]. In order to develop new sensors for heavy metals 4-(azulen-1-yl)-2,6-bis((E)-2-(thiophen-3-yl)vinyl)pyridine (**L**),with the formula shown in Figure 1, has been synthesized and electropolymerized to get poly**L** modified electrodes.

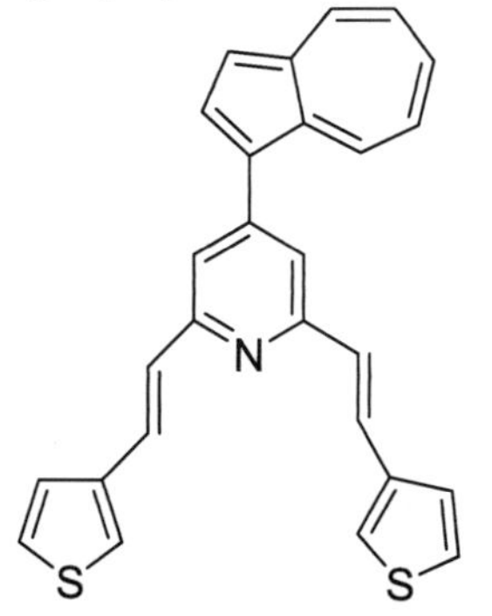

Figure 1. Structure of **L.**

The investigated compound has been characterized by electrochemical techniques, and the preparation of the modified electrodes has been studied. The heavy metals recognition based on poly**L** modified electrodes has been tested.

2. Experimental

Acetonitrile (CH_3CN) and tetrabutylammonium perchlorate (TBAP) from Fluka were used as received for solvent and supporting electrolyte, respectively. The azulene derivative (L) was synthesized according to previously described methods [7]. The metal salts used for recognition were commercial products: mercury (II) acetate, cadmium nitrate tetrahydrate, and lead (II) nitrate from Sigma Aldrich, copper (II) acetate monohydrate from Fluka. The electrochemical characterization of L was performed in 0.1M TBAP, CH_3CN.

The electrochemical experiments were carried out using a potentiostat PGSTAT 12 AUTOLAB connected to a three-electrode cell. The working electrode was a glassy carbon electrode from Metrohm (diameter of 3 mm). Its active surface was polished before each determination with diamond paste (1 μm) and cleaned with the solvent. A platinum wire was used as auxiliary electrode, and the reference electrode was dependent on the solvent: Ag/Ag^+ in acetonitrile (Ag/ 10mM $AgNO_3$, 0.1M TBAP, CH_3CN) or Ag/AgCl in water. The applied potential was finally referred to the potential of the ferrocene/ferricinium redox couple (Fc/Fc^+) which in our experimental conditions was 0.07 V.

Cyclic voltammetry (CV) curves were recorded at different scan rates (between 0.1 V/s and 1.0 V/s). Differential pulse voltammetry (DPV) curves were recorded at 0.01V/s with a pulse height of 0.025 V and a step time of 0.2 s. Rotating disk electrode (RDE) curves were recorded at 0.01 V/s. The anodic and cathodic CV, DPV and RDE scans have been recorded starting from the stationary potential.

The heavy metals ions detection was carried out in 0.1 M acetate buffer at pH = 5.5 which has been used as supporting electrolyte. The acetate buffer was prepared from 0.2 M acetic acid and 0.2 M sodium acetate solutions. The detection was studied using a glassy carbon disk (3 mm diameter) modified with polyL as working electrode, and Ag/AgCl, 3 M KCl from Metrohm as reference electrode. The heavy metals detection has been performed from synthetic heavy metals solutions with concentrations between 10^{-4} and 10^{-8} mol/ L, prepared by successive dilutions from a stock solution containing all cations at the same concentration (10^{-2} M).

IOP Conf. Series: Materials Science and Engineering **213** (2017) 012050 doi:10.1088/1757-899X/213/1/012050

All electrochemical experiments have been conducted at 25°C under argon atmosphere.

3. Results and discussion

3.1. Electrochemical characterization of L

The electrochemical experiments were performed by differential pulse voltammetry (DPV), cyclic voltammetry (CV), and rotating disk electrode (RDE) in 0.1 M TBAP, CH_3CN in millimolar solutions of **L**.

DPV and CV curves at different concentrations (0 – 3.75mM) of **L** are presented in Figure 2. Three anodic and four cathodic main peaks are distinguished in DPV. They were denoted in the order of their apparition in anodic (a1 – a3) or cathodic (c1 – c4) scans. DPV peaks potentials are given in Table 1. In CV some peaks are not well distinguished, they appearing as shoulders (c2 and c4). The peak currents increase with concentrations in CV. In DPV curves this increase is valid for the first anodic and cathodic peaks, but for several peaks it is not valid (a2 and c4). Figure 3 shows the linear dependences of the peak currents on **L** concentration. Table 2 presents the equations and correlation coefficients for these CV and DPV dependences.

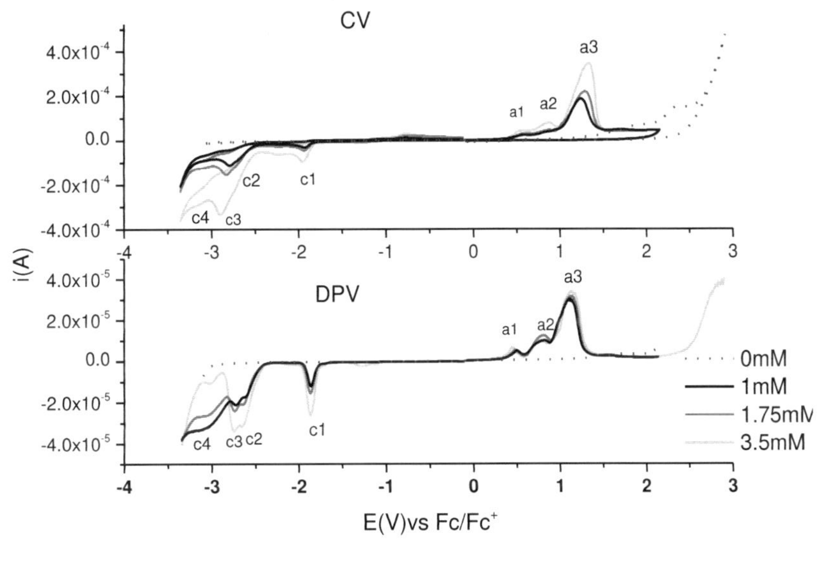

Figure 2. CV (up) and DPV (down) curves on glassy carbon (3 mm in diameter) for **L** at different concentrations in 0.1M TBAP, CH_3CN.

Table 1. Peak potentials (V) vs Fc/Fc^+ from DPV and CV curves for L (1 mM).

Peak	Method	
	CV	DPV
a1	0.569 (i)*	0.501
a2	0.861 (i)*	0.800
a3	1.245 (i)*	1.118
c1	-1.930 (i)*	-1.865
c2	-	-2.624
c3	-2.789	-2.724
c4	-3.077	-3.09

*r - reversible process; q - quasi-reversible process; i - irreversible process

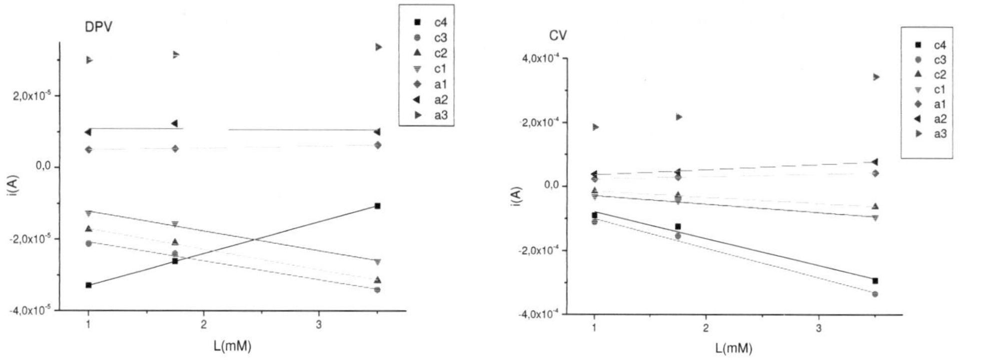

Figure 3. Dependences of the peak currents on **L** concentration in CV and DPV.

Table 2. Equations and correlation coefficients for CV and DPV peak currents.

Method	Equation*	Correlation coefficient
CV	$i_{peak\ a1}$ (A) = 1.51E-5 + 7.91E-6· [**L**]	0,999
	$i_{peak\ a2}$ (A) = 1.96 E-5+ 1.64E-5· [**L**]	0,968
	$i_{peak\ a3}$ (A) = 1.14E-4 + 6.47E-5· [**L**]	0,979
	$i_{peak\ c1}$ (A) = -2.64E-6 - 2.6E-5· [**L**]	0,986
	$i_{peak\ c2}$ (A) = 5.14E-6 – 1.92E-5· [**L**]	0.997
	$i_{peak\ c3}$ (A) = -8.57E-6 – 9.19E-5· [**L**]	0.976
	$i_{peak\ c4}$ (A) = 4.88E-6 – 8.36E-5 · [**L**]	0.963
DPV	$i_{peak\ a1}$ (A) = 4,35E-6 + 5.95E-7· [**L**]	0.985
	$i_{peak\ a2}$ (A) = 1.115-5 -1,63E-7· [**L**]	-0,954
	$i_{peak\ a3}$ (A) = 2,87E-5 + 1,46E-6· [**L**]	0.972
	$i_{peak\ c1}$ (A) = -6.75E-6 – 5.46E-6· [**L**]	0.984
	$i_{peak\ c2}$ (A) = -1,14E-5 – 5.7E-6· [**L**]	0.996
	$i_{peak\ c3}$ (A) = -1,55E-5 – 5,23E-6 · [**L**]	0.985
	$i_{peak\ c4}$ (A) = -4,18E-5 – 8,91E-6 · [**L**]	0.999

*$i_{peak}$is expressed in A, while [**L**] is expresed in mmol/L (mM)

RDE curves have been were recorded for different concentrations of **L**. In Figure 4 are shown the curves obtained at **L** concentration of 1,75mM at different rotation rates (500 – 1500 rpm). In the anodic scans the curves present 3 anodic processes a1, a2 and a3, denoted according to the corresponding peaks in DPV (shown below). RDE a1 peak currents increase with increasing the rotation rate, but this increase is stopped at potentials higher than a2, when the current drops suddenly to values closer to zero, and they are kept constant over a large scan range. This drop is characteristic for the electrode coverage with insulating films. Consequently, the aspect of RDE curves has confirmed the coverage of the electrode with an insulating layer, fact that has been noticed in DPV curves. The process a3 appears rather higher in RDE curves, but shows no variation with the rotation rate of **L**, being attributed to the accumulation of oxidation products from the supporting electrolyte.

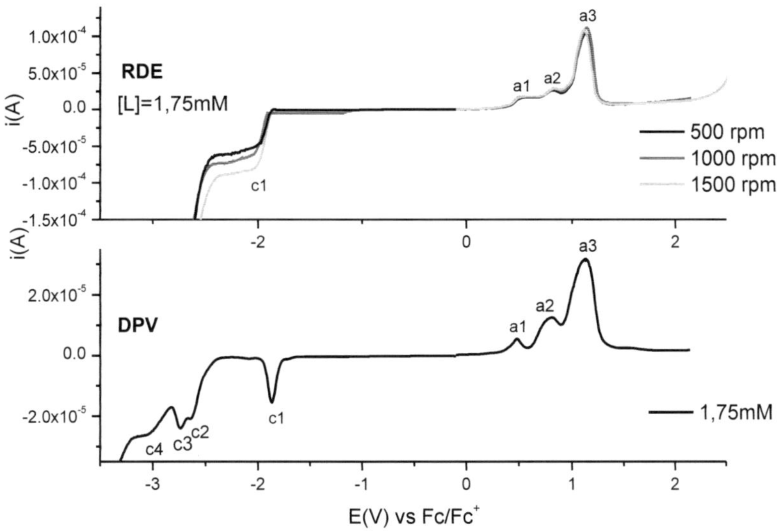

Figure 4. RDE and the corresponding DPV curveson glassy carbon (3mm in diameter) at different rotation rate (rpm) for **L** (1,75mM) in 0.1M TBAP, CH_3CN.

Figure 5 shows the CV curves for different scan rates (0.1 V/s – 1.0 V/s) in the domains of the first anodic and cathodic peaks, as well as on different scan domains (0.1 V/s), in comparison with the corresponding DPV curve recorded for the same concentration of **L** (1.75mM). They enable establishing the irreversible (i) or quasi-reversible (q) character of each process. Table 1 also shows their peak potentials from CV and DPV curves.

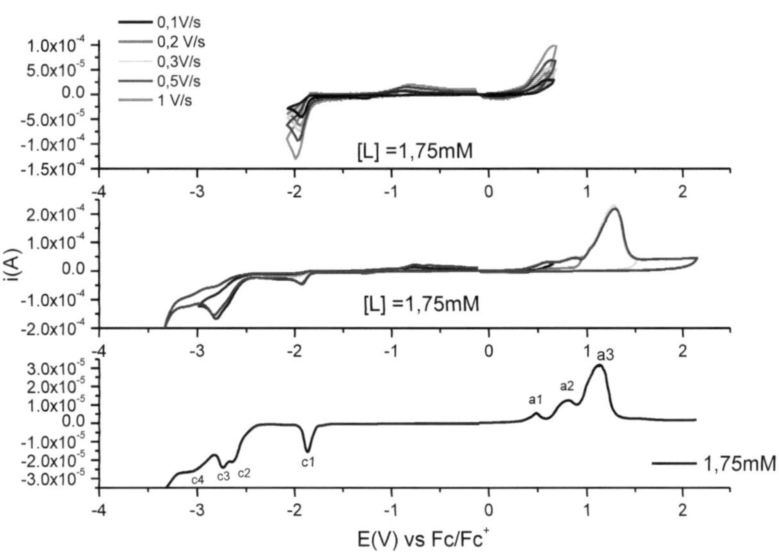

Figure 5. CV curves at different scan rates and scan domains and DPV curve at the concentration of 1.75mM for **L** in 0.1M TBAP,CH_3CN.

IOP Conf. Series: Materials Science and Engineering **213** (2017) 012050 doi:10.1088/1757-899X/213/1/012050

*3.2. Modified electrodes based on poly**L***

Poly**L** film were obtained by successive scanning or by controlled potential electrolysis (CPE) in **L** solutions (1.75mM) in the domain of the anodic processes identified to lead to insulating films. Evidence on the films formation on the electrode surface was obtained by transferring the modified electrodes in a solution of ferrocene (0.5 mM) in 0.1M TBAP, CH_3CN. Figure 6 shows the CV curves obtained after the transfer of the modified electrode obtained by successive potential scanning in the domain of the first anodic peak. It can be seen that the signal for ferrocene couple on the modified electrode is not very different from that recorded on the bare electrode, indicating the formation of a thin layer film.

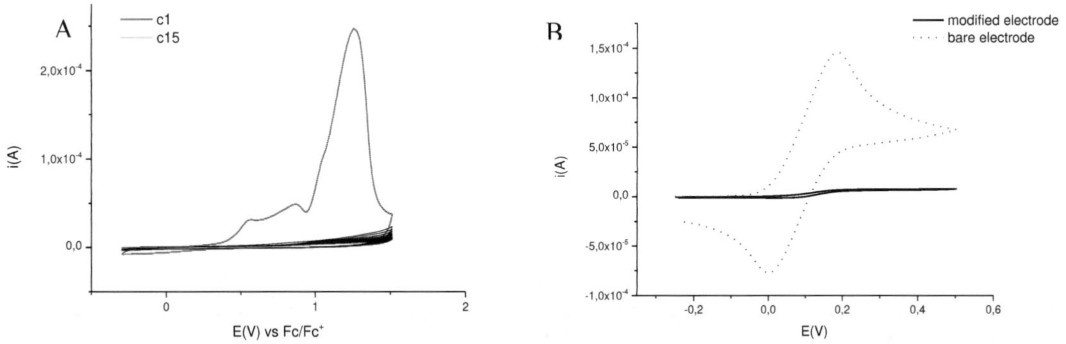

Figure 6. A. CV curves (0.1 V/s) for the preparation of the modified electrode by successive scans in 1,75mM solution of **L** in 0.1M TBAP, CH_3CN; **B**. and the corresponding CV curves (0.1V/s) obtained after the transfer of the modified electrode in 1mM ferrocene solution.

In Figure 7 are given the CV curves of ferrocene recorded on different modified electrodes obtained by CPE. The electrolyses were performed at 1.61 V vs Fc/Fc^+ using different electropolymerization charges (0 – 1.5mC). All signals indicate coverage of the electrode with films, because the signal is modified in comparison with that on bare electrode. The modified electrode obtained by CPE at the potential of 1.61V when the charges are growing from 0.5mC to 1.5mC the signal increases (when thicker films are expected to be formed).

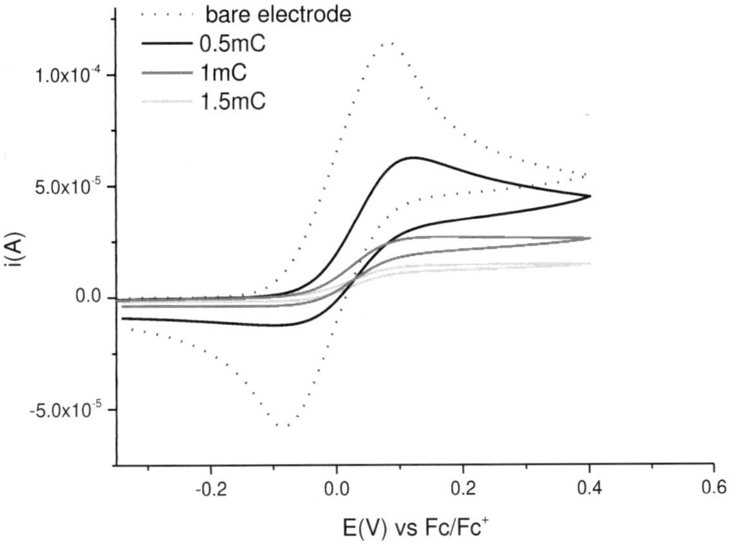

Figure 7. CV curves (0.1 V/s) of a modified electrode in 1,75mM ferrocene transfer solutions in 0.1M TBAP, CH_3CN; the modified electrodes were prepared by CPE at 1.61 V using different electropolymerization charges.

3.3. Heavy metals sensing

In order to sense the heavy metals polyL glassy carbon modified electrodes were prepared by controlled potential electrolysis at 1.61 V using 1mC from L solutions (1 mM).

Modified electrodes by CPE were introduced into the transfer cell containing 0.1M acetate buffer pH 5.5. The electrode equilibration was done during 15 CV cycles in the domain between -0.9 V and +0.6 V, and the overoxidation during 15 CV cycles between -0.2 V and +1.2 V. Then the modified electrodes were introduced in the assay solutions containing heavy metals ions, and they were maintained for 15 minutes under magnetic stirring. A potential of -1.2 V was applied during 120 s for their reduction, then the electrode was polarized in anodic sense at a scan rate of 0.01 V/s. The stripping currents for the metals dissolution were recorded. The stripping peaks at 10^{-6} M appeared at -0.56V, -0.164 V, and -0.8 V for Pb, Cu, and Cd respectively.

From the heavy metals tested (Cd, Pb, Cu, Hg) the recognition was suitable for lead, copper and cadmium, in the order given by the intensity of their analytical signals. In Figure 8 are shown the results obtained for the modified electrodes introduced in solutions containing heavy metal cations in different concentrations. Based on the identification of different heavy metal potential values, it can be concluded that the obtained signals can be attributed to Cu (II) and Pb (II). The other ions are not seen in the stripping. The response of the modified electrode is controlled by the concentration of these species in accumulation solutions. The detection limits for Pb (II) and Cu (II) were lower than 10^{-6} M. Studies about optimizing the performances of the polyL modified electrodes in view of better sensing are underway.

IOP Conf. Series: Materials Science and Engineering **213** (2017) 012050 doi:10.1088/1757-899X/213/1/012050

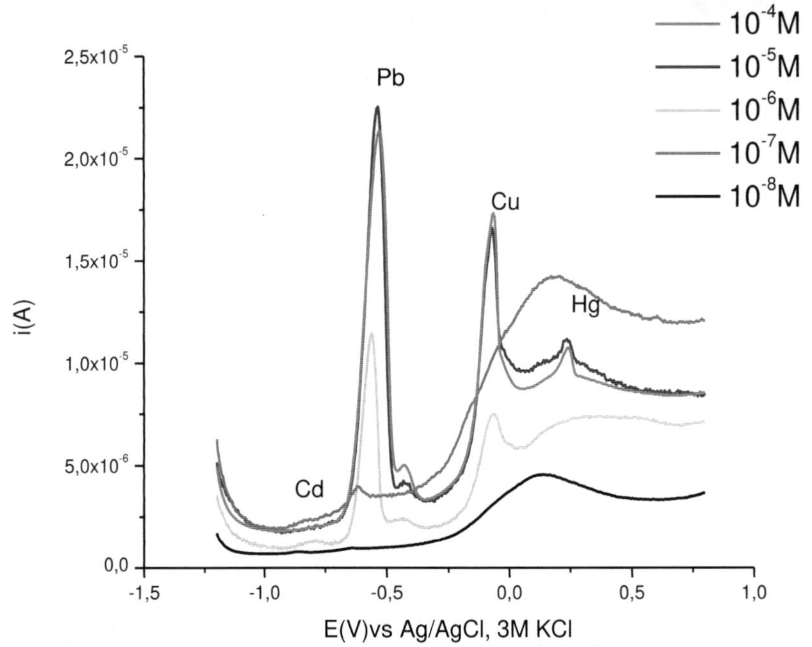

Figure 8. Anodic stripping curves recorded on poly**L** modified electrodes after immersion in acetate buffer at pH 5.5 (accumulation time 15 min) containing different concentrations of in Cd(II), Pb(II), Cu(II) and Hg(II) ions; the modified electrodes were obtained by CPE (1.61 V, 1mC).

Several experiments using poly**L** modified electrodes prepared in the same conditions were performed for Pb^{2+} detection in order to evaluate the characteristic parameters for its detection by this method.

4. Conclusions

The electrochemical characterization of 4-(azulen-1-yl)-2,6-bis((E)-2-(thiophen-3-yl)vinyl)pyridine (**L**) was performed by cyclic voltammetry, differential pulse voltammetry and rotating disk electrode methods. The anodic and cathodic processes have been established, and the conditions for obtaining electrodes modified with **L** have been identified. The films formation on the surface of the electrode was confirmed by the transfer of the modified electrode in ferrocene solutions, when the CV signal was found attenuated.

Chemically modified electrodes have been prepared by electropolymerization of **L** at positive potentials. The modified electrodes were tested for the recognition of heavy metal cations: Cd(II), Pb(II), Cu(II) and Hg(II). The best results were obtained for lead and for copper. The detection conditions have to be optimized. Using the present data, the new modified electrode can be used for analysis of lead and copper in water with a content of these heavy metals higher than 10^{-6} M.

Acknowledgments

The authors are grateful for the financial support from: UEFISCDI Romanian projects ID PN-II-RU-TE-2014-4-0594 contract no. 10/2014, PN-II-PT-PCCA-2013-4-2151 contract no. 236/2014, and China-Romania bilateral project Contract no. 68BM / 2016.

References

[1] Imisides M D, John R, Riley P J and Wallace G G 1991 *Electroanalysis* **3** 879

[2] Deronzier A, Moutet J C and Coord 1996 *Chem. Rev.* **147** 339
[3] O'Riordan D M T and Wallace G G 1986 *Anal. Chem.* **58** 128
[4] Rahman M A, Won M S and Shim Y B 2003 *Anal. Chem.* **75** 1123
[5] Wier L M, Guadalupe A R and Abruiia H D 1985 *Anal. Chem.* **57** 2009
[6] Buica G O, Bucher C, Moutet J C, Royal G, Saint-Aman E and Ungureanu E M 2009 *Electroanalysis* **21** 77
[7] Ugo P, Sperni L and Moretto L M 1997 *Electroanalysis* **9** 1153
[8] Essaidi Z *et al* 2011 *Optical Materials* **33** 1387
[9] Osaka T *et al* 1987 *J. Electrochem. Soc.* **134** 2645
[10] Szlachcic P *et al* 2015 *Dyes and Pigments* **114** 184
[11] Buica G *et al* 2013 *J. Electroanal. Chem.* **693** 67